20 002 520 52

AF606476

PACKAGING OF ELECTRONIC SYSTEMS

A Mechanical Engineering Approach

McGraw-Hill Series in Mechanical Engineering

Jack P. Holman, *Southern Methodist University*
John R. Lloyd, *Michigan State University*
Consulting Editors

Anderson: *Modern Compressible Flow: With Historical Perspective*
Arora: *Introduction to Optimum Design*
Bray and Stanley: *Nondestructive Evaluation: A Tool for Design, Manufacturing and Service*
Dally: *Packaging of Electronic Systems: A Mechanical Engineering Approach*
Dieter: *Engineering Design: A Materials and Processing Approach*
Eckert and Drake: *Analysis of Heat and Mass Transfer*
Heywood: *Internal Combustion Engine Fundamentals*
Hinze: *Turbulence*
Hutton: *Applied Mechanical Vibrations*
Juvinall: *Engineering Considerations of Stress, Strain, and Strength*
Kays and Crawford: *Convective Heat and Mass Transfer*
Kane and Levinson: *Dynamics: Theory and Applications*
Martin: *Kinematics and Dynamics of Machines*
Phelan: *Fundamentals of Mechanical Design*
Raven: *Automatic Control Engineering*
Rosenberg and Karnopp: *Introduction to Physics*
Schlichting: *Boundary-Layer Theory*
Shames: *Mechanics of Fluids*
Sherman: *Viscous Flow*
Shigley: *Kinematic Analysis of Mechanisms*
Shigley and Mischke: *Mechanical Engineering Design*
Shigley and Uicker: *Theory of Machines and Mechanisms*
Stoecker and Jones: *Refrigeration and Air Conditioning*
Vanderplaats: *Numerical Optimization: Techniques for Engineering Design, with Applications*

PACKAGING OF ELECTRONIC SYSTEMS
A Mechanical Engineering Approach

James W. Dally

Department of Mechanical Engineering
University of Maryland

McGraw-Hill Publishing Company

New York St. Louis San Francisco Auckland Bogotá Caracas
Hamburg Lisbon London Madrid Mexico Milan Montreal
New Delhi Oklahoma City Paris San Juan São Paulo
Singapore Sydney Tokyo Toronto

This book was set in Times Roman by Science Typographers, Inc.
The editors were John Corrigan and John M. Morriss;
the production supervisor was Louise Karam.
The cover was designed by Tony Paccione.
Project supervision was done by Science Typographers, Inc.
Arcata Graphics/Halliday was printer and binder.

PACKAGING OF ELECTRONIC SYSTEMS
A Mechanical Engineering Approach

1 2 3 4 5 6 7 8 9 0 HAL HAL 9 5 4 3 2 1 0

ISBN 0-07-015214-4

Library of Congress Cataloging-in-Publication Data

Dally, James W.
Packaging of electronic systems: a mechanical engineering approach / James W. Dally.
p. cm.—(McGraw-Hill series in mechanical engineering)
Includes bibliographies and index.
ISBN 0-07-015214-4
1. Electronic packaging. I. Title. II. Series
TKL7870.D25 1990
621.381′046—dc20 89-12399

ABOUT THE AUTHOR

James W. Dally obtained a bachelor of science degree and a master of science degree, both in mechanical engineering, from the Carnegie Institute of Technology. He obtained a doctoral degree in mechanics from the Illinois Institute of Technology. He has taught at Cornell University, the Illinois Institute of Technology and the University of Rhode Island. From 1984 to the present he has been a professor in the Department of Mechanical Engineering at the University of Maryland.

Professor Dally has also worked at the Mesta Machine Company, Armour Research Foundation and IBM, Federal Systems Division. He is a fellow of the American Society of Mechanical Engineers, Society for Experimental Mechanics and American Academy of Mechanics. He is a member of the American Society for Engineering Education. He was elected to the National Academy of Engineering in 1984.

Professor Dally has coauthored five books; *Experimental Stress Analysis*, *Photoelastic Coatings*, *Instrumentation for Engineering Measurement*, *Static and Dynamic Photoelasticity and Caustics*, and *Packaging of Electronic Systems*. He has written over 100 scientific papers and holds 5 patents.

CONTENTS

PREFACE

This book has been written for engineers to serve as a first text on the packaging of electronic systems. The material has been written for an engineering student or for a practicing professional working as a mechanical or electrical engineer with a firm producing electronic instruments or systems. The engineering student should have completed fundamental courses in the engineering sciences, thermal sciences and materials as prerequisites. The practicing professional will probably be at the early stages of his or her career and be more concerned with the technical details of the design rather than the business strategy of a product line.

This book is an introduction to packaging electronic systems and it covers a very broad range of topics from the physics of semiconductors to the design of advanced high performance heat exchangers. To accommodate this breadth, we have divided the text into three independent parts. The first, Part I, which includes three chapters, deals with foundations for design. Chapter 1 is descriptive of the entire field, covering in some detail the mechanical design issues that arise in developing an electronic system. Also covered are busines aspects of this industry, particularly as they are related to a rapidly changing technology, which leads to early obsolescence of an existing product line, and exacting requirements for investing in new product development. Chapter 2 describes electronic components with emphasis on semiconductor devices. Importance is placed on silicon technology and the new developments in logic and memory circuits with higher levels of integration. We try to show in this chapter the dynamics of the technology and the emerging design problems associated with the introduction of VLSI. The most important problems are handling high I/O counts and dissipating very large heat flux with exceedingly small temperature differences. Chapter 3 covers circuit analysis. The conventional methods of analysis of ordinary ac and dc circuits are briefly reviewed. However, the emphasis is placed on transmission lines where inductance, resistance and capacitance is distributed along the length of the line. This development is new to most mechanical engineers and it is critical to their understanding of propagation delay, line charging and pulse reflections that must be taken into account in the design of even moderate performance electronic systems.

Part II deals with the three basic levels of packaging electronic components. The chip carrier, the first level package, is treated in detail in Chapter 4. Both pin in-hole and surface-mounted chip carriers are covered. The printed circuit board, the second level package, is described in Chapter 5. The treatment is often descriptive and provides the opportunity to introduce the terminology used in the industry to depict circuit board features. Some of the very difficult problems associated with component placement and trace routing are introduced in this chapter with simple examples. The manufacturing process associated with production of circuit boards is described in Chapter 6. This process must be well understood if one is to design circuit boards suitable for a quality product. The final chapter in Part II covers third level packaging. The third level is a loose term that describes many components encountered when one encloses a number of printed circuit boards in instrument cases or cabinets. We treat connectors, back panels, cabling and both commercial and military enclosures in some detail.

Analysis methods commonly employed to predict performance of systems are covered in Part III. Chapter 8 deals with heat transfer by conduction. Basic conduction equations are reviewed and applications to problems arising in conducting the heat generated on a chip to the heat sink are given. Heat transfer by radiation and convection are treated in Chapter 9. We have not covered heat transfer by boiling since its use in the industry to date has been extremely limited. The final chapter covers vibration of components and circuit boards. Here the emphasis is on simple analytical procedures, which give insight on methods to reduce transmissibility coefficients and, if possible, to avoid resonance conditions. We recognize that failure by fatigue is a major problem in systems exposed to vibration, and we treat methods of fatigue analysis and introduce fragility charts.

The need for a book of this type became apparent to me in 1983 when I became responsible for mechanical design of electronic systems at one of the laboratories of a large well respected computer firm. My experience at that time was on the machine design aspect of mechanical engineering with particular emphasis in mechanics. I attempted to resolve my experience deficiency by going to the library, intending to read a dozen or so books by as many authors. My hope was to cover the field using the combined experiences of the authors with their different viewpoints. I was surprised (shocked is a better word) to find nearly a complete void. Except for a book or two on heat transfer and one on vibrations pertaining to electronics, the library shelves were empty. Not to be daunted, I turned to the professional societies and found that the American Society of Mechanical Engineers (ASME) did not yet recognize packaging as a technical area. (ASME has recently organized a technical division on packaging and is now publishing a technical journal covering the field.) The next step involved the trade journals. I found several with many interesting articles, however, the articles were mostly descriptive and nearly devoid of analytical content. I came to the conclusion that practically no one believed in analytical descriptions where equations were used to illustrate either good design or bad design concepts of electronic assemblies.

This fact amazed me at the time because the electronics business was very large (over $200 billion) and growing at the rate of about 15% per year. I did not understand then and still do not understand now, how a business can become so large and so important and be virtually ignored by the technical societies and the academic community.

When I joined the faculty at the University of Maryland sometime later, one of my highest priorities was to initiate an introductory level course on electronic packaging into the curriculum. I have taught this course entitled Mechanical Design of Electronic Systems to seniors and first year graduate students seven times over the past four years. Over the same period of time I also taught another course, a standard offering on machine design. As the course developed with each teaching experience, I attempted to present the material in the same way we cover material in lecturing a class in machine design. Because the two topics differ considerably in detail, the parallel was not perfect. However, we can introduce the fundamentals in part I, cover the design aspects of the components in part II and show analysis procedures in considerable detail in part III.

The book does differ to some degree from the conventional design textbook commonly found in mechanical engineering. I have attempted to introduce some business aspects of design. Of particular importance here is the fact that the market price of a product drives the design. I have also tried to cover some of the thought process associated with design by describing both design and manufacturing aspects (good and bad) in the narrative parts of the text. The exercises that are given at the end of each chapter are markedly different from those found in the usual engineering textbook. They require the student to do much more than plug and chug. We require the student to sketch, to prepare graphs describing solution space, to conceive new designs better than existing designs, to write engineering briefs, to interpret solutions and draw conclusions related to design merits and to make judgments based on business aspects. These exercises will stretch the experience base of most students and, indeed, they may stretch the experience base of some instructors. Care should be taken in assigning the exercises because they range in difficulty from trivial to impossible. Hopefully, the more difficult problems can be used to stimulate classroom discussion and the easier ones will not lull the student into a false sense of security.

I want to thank Norm Roos at IBM for his introduction to this field and for his patience as I slowly came up to speed. Thanks are in order to Bill Fourney and Bill Walston at the University of Maryland for their help in scheduling this course as a technical elective. I appreciate the many suggestions made by two graduate students, Greg Braunberg and John Berger, who read the entire text and made many helpful suggestions. Harry Charles, Johns Hopkins University; Alan Kraus, Naval Postgraduate School; and Bob Rowlands, University of Wisconsin–Madison, made valuable comments in their reviews of this manuscript. But most of all I want to thank the 200+ students who were courageous enough to take an elective course without an assigned textbook on a topic with an impossible title. The feedback they provided was essential in developing the sequence and scope of the material selected for inclusion in the text.

James W. Dally

LIST OF ACRONYMS

CAD	Computer-aided design
CAE	Computer-aided engineering
CC	Chip carrier
CCCC	Controlled collapsible chip connection
CEEE	Common electronic equipment enclosure
CMOS	Complimentary metal oxide semiconductor
CPU	Central processing unit
CRT	Cathode ray tube
ECL	Emitter coupled logic
FET	Field effect transistor
FRU	Field replaceable unit
IC	Integrated circuit
I/O	Input-output
LCC	Leadless chip carrier
LSI	Large scale integration
MECL	Motorola emitter coupled logic
MIL-STD	Military standard
MLB	Multilayered board
MOS	Metal oxide semiconductor
MOSFET	Metal oxide semiconductor field effect transistor
MSI	Medium scale integration
MTBF	Mean time between failures
NMOS	*N*-type metal oxide semiconductor
PCB	Printed circuit board
PMOS	*P*-type metal oxide semiconductor
PPM	Parts per million
PRN	Priority ranking number
PROM	Programmable read only memory
PTF	Precision thin film

PTH	Plated through hole
PVC	Polyvinyl chloride
PWB	Printed wiring board
RAM	Random access memory
ROM	Read only memory
SIP	Single in-line package
SMT	Surface mount technology
SSI	Small scale integration
TAB	Tape automated bonding
TCM	Thermal conduction module
TTL	Transistor transistor logic
ULSI	Ultra-large-scale integration
UV	Ultraviolet
VLSI	Very-large-scale integration
WSI	Wafer scale integration

PACKAGING OF ELECTRONIC SYSTEMS

A Mechanical Engineering Approach

PART I

FOUNDATIONS OF MECHANICAL DESIGN OF ELECTRONIC SYSTEMS

CHAPTER 1

INTRODUCTION

1.1 GOALS AND OBJECTIVES

This book has been prepared to serve as a text for students and entry level persons beginning to design electronic systems. The coverage begins at the interface between electronic engineering and mechanical engineering and pertains to the mechanical and manufacturing issues that arise in developing a new electronic system. The treatment is quite broad, starting with the integrated circuits (the chip) and proceeding through the many levels of packaging involved in developing a complete electronic system. The material is often highly descriptive, particularly when compared to the mathematical treatments presented in more mature subjects in mechanics or mechanical design. However, the descriptive material is important to introduce the essential vocabulary, which is full of acronyms, and to present the wide array of electronic components that the mechanical or electrical engineer must deal with in the design process.

To organize the material and to facilitate understanding by the reader, this book is divided into three parts. Part 1 covers background material including semiconductor physics, analog circuit theory, digital circuit theory and transmission line theory. The objective of this background information is to give the engineers involved in packaging the basic understanding as to why and how electronic circuits operate. Experience has shown that this background information greatly enhances communication between the mechanical and electrical engineering functions in a development project and enhances the opportunity for parallel product development instead of sequential development.

Part 2 involves packaging beginning at the first level where the chip is housed in its carrier and extending through the higher levels to the design of the cabinets and instrument panels. The word "packaging" is poorly understood by

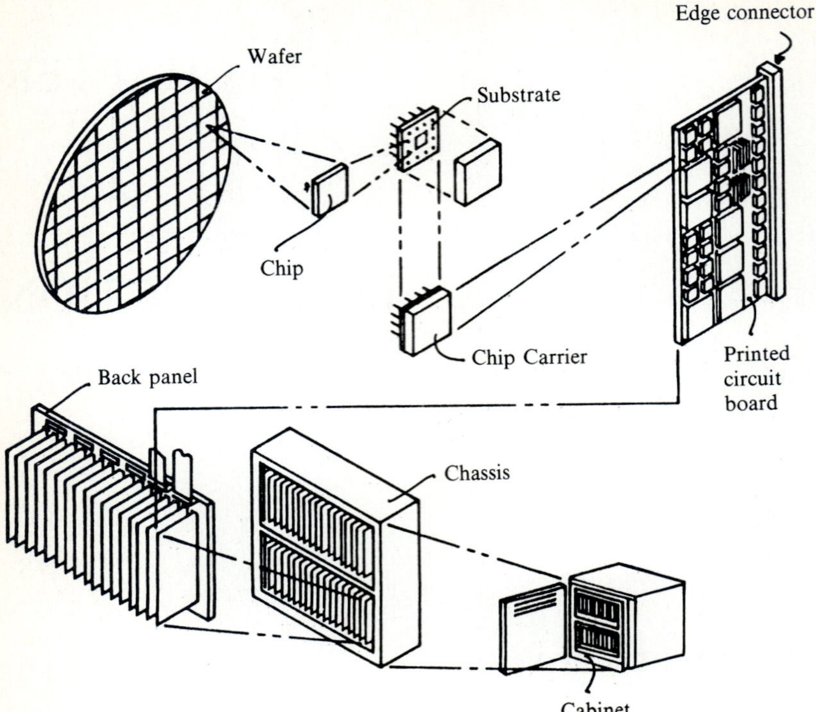

FIGURE 1.1
A schematic illustration of several levels of packaging involved in an electronic system.

the engineering community where it is often confused with the design of a container to prevent damage during shipment of a product. Packaging of electronic systems refers to the placement and connection of many electronic and electromechanical components (sometimes thousands) in an enclosure that protects the system from the environment and provides easy access for routine maintenance. An example, which describes some of the important features of packaging, is illustrated in Fig. 1.1. Note that the packaging process starts with the chip, which has been fabricate on a wafer of silicon. After testing, the chip is housed in a chip carrier and small wires are used to electrically connect the chip to the carrier (the first level of connections). Next, the chip carriers are placed on a circuit board and connected together (second level) with wiring traces that have been formed by photoetching the circuit board. Connectors on the circuit boards are then inserted into contacts on a back panel, which carries the higher level connections that permit communication from one circuit board to the next. Cables are shown that connect the power supply to the back panel and that bring the input and output (I/O) signal into the unit. Finally, the entire array of circuit boards, back panels, power supplies and cables are housed in a cabinet. The packaging aspect of the design of an electronic system is extremely important because about one-half of the cost of a system is involved in packaging and the other half of the cost is for the individual electronic devices.

Part 3 deals with the environmental aspect of packaging. In addition to carefully packing thousands of electronic components into a stylish and functional cabinet, packaging involves the protection of these components from the environment. Heat management is perhaps the most important of the environmental considerations because the operating temperature of the chips markedly affects the reliability of the circuit board and the availability of a system. For electronic systems destined for use in the field by either the military or by industry, shock and vibration represent harsh environments that must be accommodated in the design of the hardware. Finally, noise is an important consideration, particularly in air-cooled equipment where one or more fans are used to move significant volumes of air.

In this presentation, simple examples that emphasize basic theoretical approaches and which illustrate sound design concepts will be selected. In actual practice, the real problems will be much more complex but the applicable theory and the design concepts are the same. In many design offices special codes or software exists for handling some of the complexity associated with very large electronic systems. We will recognize some of these codes in this book but we will not describe them in any detail because they change rapidly with time.

1.2 MECHANICAL DEVELOPMENT IN DESIGN OF ELECTRONIC SYSTEMS

A mechanical development department is usually responsible for packaging electronic systems in most industrial firms involved in producing an extensive line of electronic products. Because an electronic system is usually quite complex, involving several different disciplines, multidepartment organizations are usually established to handle the logical flow of paper, CAD tapes and other information generated in a typical development. An example of one organization is shown in Fig. 1.2. The development process is initiated (in large firms at least) by corporate planners who identify customer needs and predict market trends and technological advances. With this information as a basis they prepare the specifications for

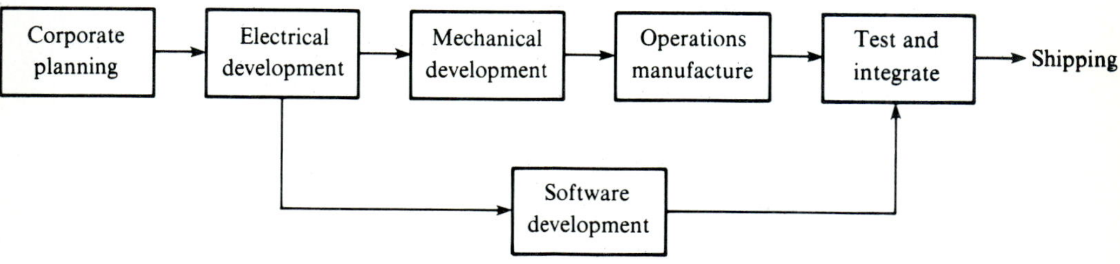

FIGURE 1.2
Typical development organization showing information flow and discipline interfaces.

a new product indicating performance requirements, market estimates on volume and price and other higher level design objectives. Development budgets and schedules are also part of the output from the corporate planners.

The project moves from the planners to electronic development where the system architecture is established and then the detailed analog and digital circuits are designed. The output from electronic development is passed on to mechanical development in the form of circuit diagrams, logic diagrams, component lists, wiring lists and ground rules for component layout and shielding requirements for the cables. The output from the electronic development department also goes to the computer science department where the code necessary to operate the system is written.

The mechanical development department packages the system by sizing and designing circuit boards, placing components on each board and connecting the appropriate pins on each component. Back panels and power distribution busses are designed in conjunction with gates, chassis drawers and cabinets. Connectors and cables are designed to ensure connectivity as well as access for routine maintenance. A cooling system is designed to carry away the heat that is dissipated by the many different electronic devices employed in constructing high performance systems. Operating panels that are appreciated by the operator and the customer are developed to be both attractive and functional. The system is designed to accommodate the environment as described by the controlling specifications. The output from the mechanical development department is in the form of drawings, CAD tapes, artwork or glass masters and reports describing analyses predicting response of the system to the environment. The output from mechanical development is transmitted to operations.

Operations designs and builds the tooling necessary to manufacture the components, subassemblies, circuit boards, cables, etc. The circuit boards, back panels, subassemblies, cabinets, etc. are fabricated and then assembled into the final product. This manufacturing process is quite challenging, particularly during the assembly and test of the first few units. Design errors are discovered during this period and close cooperation between mechanical development and operations is essential to develop design modifications that correct the errors. Design for enhanced manufacturing is a critical objective in developing a product known for its quality as well as for its performance.

The prototype moves from operations to unit test, where the performance of the circuits and the software are evaluated. This phase, often called debugging, requires close cooperation between the electrical engineers and the software personnel. The mechanical engineer is involved in recording the electrical changes (soft wires) so that the changes in connections can be made on revised circuit board artwork and cable drawings. Unit evaluation may involve shock, vibration, temperature cycling or other qualification-type testing. If failure occurs with the hardware, the mechanical engineer is responsible for the diagnostics and for the design modification required to improve the package to adequately protect the system from the environment imposed during qualification testing.

1.3 MECHANICAL DESIGN ASPECTS OF PACKAGING

There are seven main issues involved in the design of an electronic system as indicated:

1. Connections
2. Thermal management
3. Manufacturing
4. Maintenance
5. Shock and vibration
6. Ergonomics
7. Environment

Each of these issues markedly affects the design, and care and attention is necessary in every area to insure that the design specifications are met or exceeded.

1.3.1 Connections

In advanced electronic systems it is easy to define at least six levels of connections. We start with the bonding pads on the chip and connect these pads to the input or output (I/O) leads on the chip carrier. Second, we connect the leads from the chip carrier to the printed circuit board (PCB). This connection involves the design of the solder joint and the trace routing on the PCB. Next, connections are made between the different PCBs that exist in the system. There are several different approaches to connecting PCBs depending primarily on the complexity and size of the system. For small simple systems containing only a few cards, the PCBs are usually connected with edge card connectors that are cabled together. However, for large systems that may contain several hundred cards, the PCBs are connected together with edge card connectors that are inserted into a back panel. Multilayers of circuit traces, which are photoetched on each layer of the back panel, serve as a very compact form of cabling. Fourth, the back panels, which usually are housed in a drawer or a gate, must be connected together so that one subsystem can communicate with another subsystem. Because the number of connections, at this fourth level, is reduced to primarily subsystem I/O and power, the signal connections are made with cables and the power connections made with bus bars capable of handling high currents. The fifth level of connection involves wiring the various gates or drawers to the I/O connectors on the cabinet. Cable harnesses for signal I/O and bus bars for power distribution are commonly employed. The sixth and final level of connection occurs in very large systems where cabinets and or work stations are connected together. The methods employed for cabinet to cabinet connections depend largely on the

FIGURE 1.3
The Cray 1 super computer contains 278,000 logic chips, 1056 circuit boards, 24 back panels and 12 power supplies. (*Courtesy of Cray Research Inc.*)

distance between cabinets and the speed required in transmitting signals. For short distances these connections are usually made with coaxial cable or twisted wire pairs. For longer distances, only coaxial cable is sufficient. For very long distances a dedicated set of cables becomes very expensive and transmission is accomplished with common carriers such as phone lines or microwave transmission. Recent advances in transmission with fiber optics has, in some instances, permitted the development of dedicated transmission busses using these very pure glass fibers.

Very large digital systems contain millions of gates and literally thousands of chips that must be connected together. For example the Cray I super computer, designed in the early 1970s and shown in Fig. 1.3, contains 278,000 logic chips each with 16 leads. These logic chips were placed on 1056 PCBs that were inserted horizontally into 24 back panels. The back panels were supported in a vertical position by the columns shown in Fig. 1.3. The Cray I weighs in at 10,500 lb and originally sold for about $8 million.

Clearly the relatively simple task of connecting a lead or a wire from point A to point B becomes extremely difficult when we must consider many thousands or even millions of wiring points. In addition, and even more important, is the reliability of the connections. Each of the Cray circuit boards just described contains about 20,000 joints where leads are either soldered or welded together. If even one of these joints fails over the life of the product, the entire board will fail and the system may malfunction. With the number of wiring joints exceeding a million in even very simple systems, meticulous attention must be given to every aspect of the design and manufacture of the connection system.

1.3.2 Thermal Management

Heat is generated at several locations in electronic systems. The power supplies, where the ac line voltage is converted to the various dc supply voltages required, are a significant source of the heat generated. Also, the I^2R losses that occur at each chip and along the wiring result in additional generated heat. The heat load to be dissipated depends to a significant degree on the type of product. For high performance computers and signal processors, the heat load is large and elaborate cooling methods are employed. In small relatively simple systems the heat generated is small, and natural or free convection is often sufficient to transfer the heat from the system to the environment. Regardless of the complexity, the heat management system employed is extremely important since this system controls the temperatures of the microcircuits on the chips. These chip temperatures in turn control the reliability of the electronic system. The equation predicting the failure rate for an individual component follows the Arrhenius model given by

$$\lambda(T_j) = A + Be^{-C[1/(T_j+273)-1/298]} \tag{1.1}$$

where A, B and C are constants that depend on the chip and chip carrier technologies and T_j is the junction temperature on the chip in degrees Celsius. With the very large number of components used in a typical electronic system, it is essential that the junction temperatures of the most critical of the components be minimized to enhance reliability of the system and to improve the availability. Availability is the percent of the total time that the system is up and operational.

The methods of cooling used in design depend on the product price and performance. For simple low cost systems with small heat loads (for example a typical dot matrix printer) the heat load is dissipated with natural convection. No added costs are involved and the heat flux is low enough to be accommodated with natural convection without incurring large junction temperatures. For slightly more complex systems with higher packing densities and higher heat loads, small fans are employed to give the improved heat transfer coefficients associated with forced convection. As we move up the price and performance scale, conduction cooling is employed where chilled water or freon is used in cold plates to transfer the heat to the environment. The highest performance IBM mainframe computers, the Japanese super computers and two of the Cray super computers use conduction to maintain the very low junction temperatures required for extremely high reliability.

Some applications require deployment of electronic systems for extended periods of time in space; communication and surveillance satellites are examples. In the vacuum of space, heat transfer to the environment by convection or conduction is not possible and radiation methods must be employed for the final step in dissipating the heat.

All of these heat management systems must be designed in conjunction with an electrical system that often requires that electrical insulation be placed in the path of the heat flow. The conflict in the electrical requirements for insulation and the mechanical requirements for a good conductor of heat makes the design of an efficient heat transfer system very challenging.

1.3.3 Manufacturing

We noted in Fig. 1.2 that mechanical development and operations that include manufacturing were usually two separate organizations. However, the design of a product crosses all of the organizational interfaces. It is particularly important that the design of each subassembly in a product be accomplished so that it is compatible with each step in the manufacturing process. This is much more difficult to achieve than might be imagined, because many designers have little experience in manufacturing and many manufacturing engineers have never designed a product.

An obvious example of design for manufacturing is the use of common first level packages on a circuit board. First level packages (chip carriers) are available with several different types of lead structures such as pin in-hole, gull wing surface mount, *J* lead surface mount, leadless, etc. If different types of chip carriers are used on the same circuit board the manufacturing process becomes much more difficult. Because each type of package requires different assembly equipment and different soldering equipment and/or procedures, the circuit board may have to pass through several different lines to complete the assembly. This added processing is certain to increase cost, decrease yield and degrade the quality of the board.

Layout details represent another example of design for manufacturing. Are the first level packages placed on the PCB with the same orientation and with sufficient space between the packages? The orientation is important because the assembly machines may not rotate a package prior to insertion. If the orientation is not consistent, it is necessary to rotate the board and this requires a second pass through the assembly machine. If sufficient space is not provided between the chip carriers, the gripper on the assembly machine can produce interference and assembly is impossible. In this case it is necessary to redesign the circuit board.

Dimensioning is another area that affects manufacturing in a significant way. Drawings should be dimensioned to accommodate the user. For example the drawing of a heat frame that supports a circuit card should be prepared and dimensioned for a toolmaker who is responsible for the dies used to fabricate the frame. A second drawing should be prepared and dimensioned for the inspector who will certify the dimensional accuracy of the parts after they are produced. The dimensioning on these two drawings will be different so as to facilitate the task of the toolmaker and the task of the inspector.

Examples of design for manufacturing are numerous and they all illustrate that close cooperation between operations and mechanical development is essential if a high quality and cost effective electronic system is to be produced on schedule.

1.3.4 Maintenance

Mechanical design markedly affects the ability to properly service a product in the field. This is particularly important for a surprisingly large number of different types of electronic systems. Have you ever been in line at the local bank when its computer went down? How long did you wait for it to become operational? Failures will occur at random when the reliability of the system is not adequate. To minimize delay and inconvenience, prompt service is essential. Prompt service is no accident and minimizing downtime requires careful design, extensive training of service personnel and an adequate inventory of spare parts.

The general design approach to quality maintenance is to design the system with a number of different field replaceable units (FRUs). These may be PCBs or power supplies or other subassemblies. One does not attempt to repair the single component that failed because it takes too long to find that component and the number of spare parts required for field service is too large to be practical. Time to access the FRU is important and that time is often controlled by the mechanical design. Drawers that open to expose banks of PCBs are often employed to reduce access time to a minimum. Extraction and insertion forces for the PCB are also important. The average service person can exert about 35 lb in engaging the edge card connector on the PCB into its contacts on the back panel. At about 6 oz/pin for edge connectors, one is limited to about 100 pins before it becomes necessary to incorporate some form of assist (lever and/or jack screws) to aid the service personnel in extracting and inserting the cards.

Cable wiring is another area where long delays are frequently encountered in servicing. Cable harnesses frequently contain 100 or more wires that lead from one difficult to access location to another difficult to access location. After finding the failed cable, the service person does not want to take the time to cut the faulty cable from the harness, replace it and then retie the harness. Instead, the design should incorporate 5–10% spare wires that can be used as substitutes. Repair consists of disconnecting the ends of the faulty wire and reconnecting the ends of one of the available spare wires.

Thermal warning systems are another area of design for maintenance. When forced convection or conduction methods are used to cool an electronic system, failure of a fan or a pump is possible. This failure can result in overheating of major portions of the system within a short period of time. To prevent the resulting damage, a thermal warning system, which alerts the operator to initiate a controlled shutdown is incorporated into the system. A second system, which is automatic, is also incorporated to exercise an uncontrolled shutdown in the event of operator error. These safety systems should be installed on expensive products.

Again other examples could be cited to illustrate the importance of designing the mechanical aspects of the system to permit rapid and complete servicing. Close cooperation between mechanical development and the field service organization is essential during the design process to insure a product capable of quality servicing.

1.3.5 Shock and Vibration

Shock and vibration are important at three different times in the life of an electronic system. First, during the manufacturing and assembly process when the product or its subassemblies are being moved from one manufacturing station to another. Abusive handling or dropping the product can produce very high accelerations which will deform or fracture devices and cause failure of the system to function. Second, in transporting the system through the distribution chain until it is finally installed at the customers location. Because the distribution chain may be long, with the product handled at several locations, the normal approach is to enclose the product in a shock proof container. This shock proof container deforms under impact when the product is dropped and limits the *G* loads transmitted through the enclosure to the product. Third, both shock and vibration environments may be encountered in service. Of course, in the typical office surroundings a computer system is not subject to adverse environments, but an advanced signal processor on a military platform may indeed be exposed to severe vibrations and/or shock on repeated occasions and for extended periods of time. It is the design for hostile shock and/or vibrations associated with military applications that will be covered in this treatment.

In specifying a product for military applications the procuring agency will define the vibratory environment through the use of a standard military specification. For example the shipboard electronic systems are often designed to satisfy the vibration specification defined in MIL-Standard 167-1, which describes the *G* levels and frequencies as indicated in Fig. 1.4. The frequency range for submarine equipment is quite low, 4–34 Hz, because the excitation is from the propeller that rotates at a relatively low angular velocity. The design approach in this application is to construct stiff structures with relatively high natural frequencies to avoid the possibility of a resonance frequency occurring in the cabinet, any circuit board or major subassembly within the frequency range specified.

The specifications on the vibratory environment for aircraft electronics are entirely different. The range of frequencies is much wider—from 15–2000 Hz and with the weight limitations imposed on aircraft electronics, it is impossible to design with sufficient rigidity to avoid resonances. For this reason the design

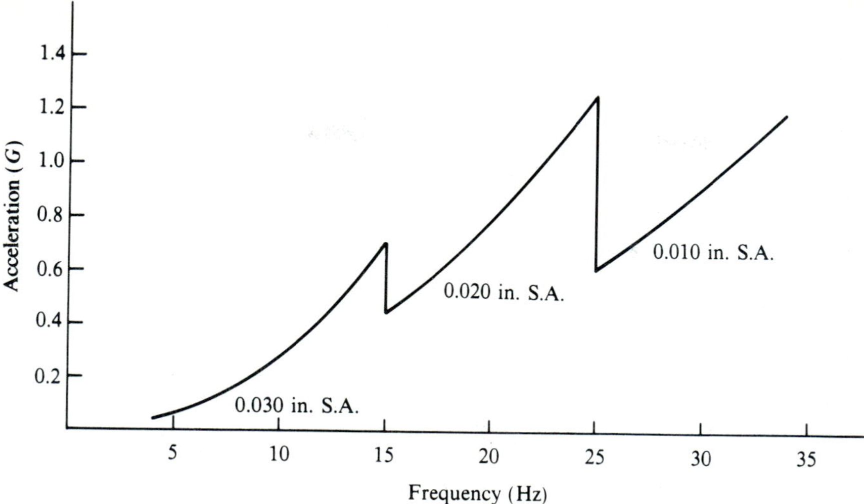

FIGURE 1.4
Vibration specification from MIL-STD-167-1 for electronic systems installed on submarines.

approach is to accept the resonances that will occur at several different frequencies and limit the force and displacement transmission factors. With limited forces it is then possible to design the circuit boards and subassemblies with sufficient fatigue endurance to avoid failure during the period of exposure.

Upon completion of the prototype, it is subjected to a closely controlled qualification test where the product is exposed to the shock or vibration environment as specified. Damage that often occurs includes fatigue of structural components that support the electronic subassemblies, lead and solder joint failures due to fatigue, pin breakage in connectors, cable chafing and shear of bolts and pins due to shock imposed loads. Clearly the design of military hardware with demanding shock and/or vibration specifications is a challenging problem that requires an extensive knowledge of this subject. We will show the basics of the theory covering shock, vibration and fatigue in treating this area of design.

1.3.6 Ergonomics

Ergonomics involves the relationship between people and the machine. The basic idea is to design the electronic system to reduce the physical effort required by the operator in its usage. The examples of ergonomics in design of electronic systems are numerous. The design of a keyboard for a computer terminal should clearly take into account the need for concave key shape with tactile feedback. Spacing between the keys, the angle of the keyboard, the ability to change this angle and the height of the keyboard all affect the productivity of the operator and the quality of the output.

There are three main areas of ergonomics that markedly affect mechanical design: the operator panel layout, the operator work station and the environment —noise and lighting. Layout involves placing switches, meters, lights and control knobs in logical positions where the operator's task is made as easy as possible. For example, the meter that displays the state of a variable should be located adjacent to the control knob for that variable. Then the operator can monitor the meter and the control knob together as an adjustment is made. Controls should follow common convention—clockwise to increase voltage or current. Switches toggle upward in the on position in the United States, but remember that they toggle downward for the on position if the product is designed for a customer in Europe.

Work station design involves operator comfort and is much more important than generally considered, particularly if the operator is expected to be on station for extended periods of time. If possible the operator should be comfortably seated as it increases his or her attention span and alertness. The chair should be designed with sufficient adjustments to adapt to the user and not vice versa. Whereas the mechanical designers may not have the responsibility for designing chairs, they should be sufficiently knowledgeable in anthropometrics to select an appropriate chair and then to design the operator–machine interface to accommodate the range of operators that will be positioned in that chair.

The office environment involves noise and lighting. Noise can be very distracting and prevent complete concentration on the task at hand. If the noise is sufficiently loud and the operator is exposed for extended periods of time, hearing losses and other physical disorders result. Lighting involves the proper intensity and distribution of light at the work station. Of particular importance is glare from instrument panels and displays such as CRTs. Glare must be avoided by the use of antireflection coatings or by the positioning of the light sources to avoid the reflections producing glare. When glare is permitted, displays cannot be monitored without eye strain and/or error.

1.4 RANGE OF PRODUCTS

The products produced and marketed as electronic systems cover an extremely wide range. Some products are advanced, complex and high in price. Others are very simple, produced in millions of units and very competitive in cost. These differences in the products require, at least to some degree, a change in the design approach and in some cases major changes in the methods employed.

At the risk of oversimplification, we will divide the entire market of electronic products into three general classes, namely the high end, intermediate and low end products. As the name implies the high end products are high performance, high cost, long life and usually relatively low volume. The primary design objective with the high end product is performance and while cost is also important, it is secondary to performance. Large main frame computers, super computers and advanced signal processors for military applications are products in this category. Extremely high reliability, often achieved with redundant subsystems, is a characteristic of these products. Prices in the range from $1–10 million per unit with annual sales of about 10–100 units are typical. Select laboratory instrumentation usually associated with health care facilities or major national laboratories also fall into this category.

The intermediate product line is much broader with more lower cost products included in this category. Mini- or midline computers, less critical military systems, most laboratory instruments and most special purpose data processing systems are the typical products. Performance is quite important but so is cost. Any gain in performance must be carefully balanced by the extra expenditure required to achieve the incremental gain. Product volume is larger and hence the design is coordinated even more closely with manufacturing. The product life is in the intermediate range so the design tends toward a flexibility that permits product upgrades by making periodic changes in the model. Reliability remains important but usually not at the cost of major redundant subsystems. Prices usually range from $100,000–$1,000,000 per unit in the intermediate range product.

The low end product is the largest segment of the market, the most competitive, the least profitable, the lowest cost and the most demanding from the design point of view. Products such as work stations, microcomputers, consumer electronics, automotive electronics, home appliances, office equipment and etc. are typical examples. Annual product volume is very high and manufacturing and price drive design. Performance is always important but in this product it must be achieved with the absolute minimum increase in cost. Quality is achieved by the close coupling of design and manufacturing. Reliability and maintainability are important, but often these design goals are supplemented with a policy that dictates a complete replacement of the unit for several days during servicing.

While the variety of product complicates the task of writing a text to address the entire field, it stimulates the designer to address design in a market-oriented manner and to consider the most important of the essentials—quality, price, performance and reliability.

1.5 BUSINESS ASPECTS

The electronics system business is and has been the fastest growing segment of the manufacturing and service industries in the United States during the past 20 years. Annual sales, depending on the definitions imposed, are in the neighborhood of $400 billion per year and growing at an annual rate of 10–15%. It is a very competitive business since it requires relatively little capital to introduce a new product and to start a new company. Also, funds may be provided by venture capitalists seeking to share the equity in the firm. New ideas for new products are essential. The successful ideas and products developed by existing and/or new companies drive the growth in the industry.

Paramount to the growth of the business has been the technological advances that seem to occur on schedule in the development of new and

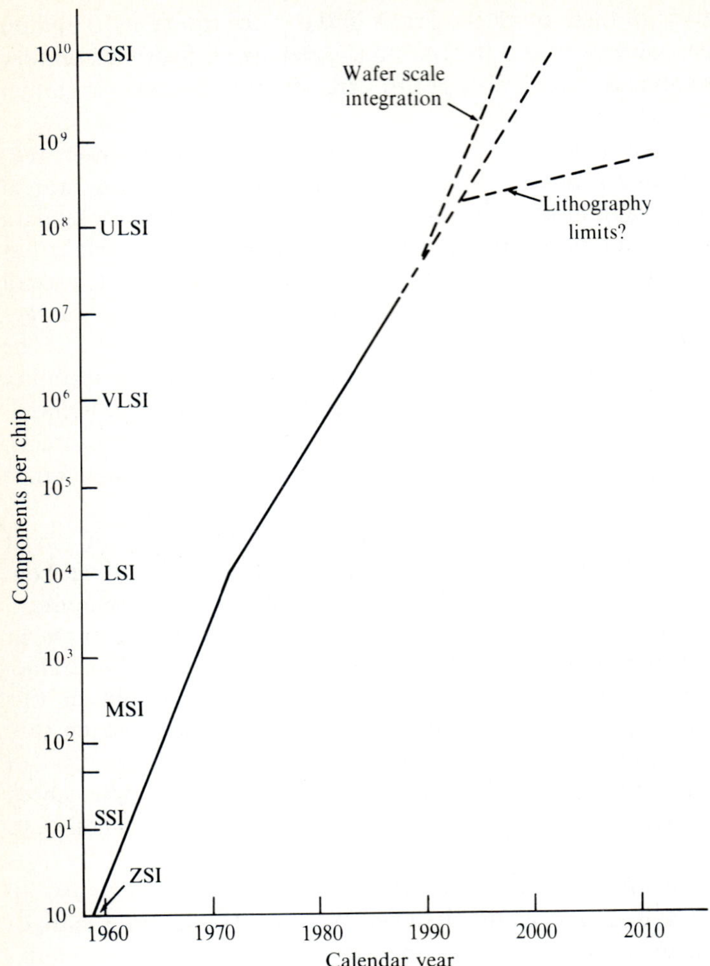

FIGURE 1.5
Past progress made in increasing component count on a chip and projections for future progress.

advanced integrated circuits (ICs). This progress, illustrated in Fig. 1.5, shows the steady increase in the number of components that can be placed on the small area of a silicon chip. As more and more components are placed on a single chip, several advantages result that permit the development of entirely new products and the marked improvement of existing products. Most important is the reduced cost per memory bit or logic gate. A new advanced chip contains either many more memory cells or logic gates than the previous generation of that type of chip. Also, as the chip becomes more dense the speed of the switching is improved and the processing rate of the product increases.

We are experiencing a rapidly changing business that is driven by new developments in IC technology, by new ideas for product development and by new requirements for information processing. These developments will lead to improvements in the way we process and store information and will provide the basis for continuous enhancement in the standard of living for society as a whole.

REFERENCES

The following periodicals provide current articles that will aid the student in understanding the acronyms and in identifying the components involved in electronic system design:

1. *IBM Journal of Research and Development*, International Business Machine Corporation, Armonk, N.Y.
2. *Electronic Design*, Hayden Publishing Company, Hasbrouck Heights, N.J.
3. *Printed Circuit Fabrication*, PMS Industries, Alpharetta, Ga.
4. *Printed Circuit Design*, PMS Industries, Alpharetta, Ga.
5. *Electronic Packaging and Production*, Cahners Magazine Division of Reed Publishing, Newton, Mass.
6. *Circuit World*, Wela Publications Ltd., Scotland, U.K.

Other interesting references include:

7. *Erogonomics Handbook*, International Business Machines Corporation, Armonk, N.Y., SV04-0224-00.
8. Cunniff, P. F.: *Environmental Noise Pollution*, Wiley, N.Y., 1979.

EXERCISES

1.1. Inspect the computer system that you personally use most of the time. Describe the keyboard features that help you type more accurately and those features that help you from becoming tired as you type. What do you think of the keyboard layout? Is it optimized in any sense? Clearly it is not, but do you know the story behind why the keyboard layout is so poor.

1.2. Are the switches to your PC placed on the rear panel instead of on the front panel where they would be more available? Why are the switches placed in a location that is not easily accessible?

1.3. Why are the vents for the cooling air placed on the side panels of your PC? The top of the PC provides a larger area and would permit including more vents for enhanced air flow with a lower head loss.

1.4. The sound pressure level issued by a product is measured in decibels. What is the typical sound pressure level produced by a rock and roll band? This is clearly too high for any product, but what are acceptable sound pressure levels? What common product usually exceeds the acceptable limits in an office environment?

1.5. List the different types of connectors with which you are familiar.

1.6. Take the cover off of your PC and identify as many of the subassemblies as possible.

1.7. Without removing a circuit card, identify as many of the components on that card as possible. If you don't follow these instructions and you do remove the card for a better view make sure you wear a grounding strap on your wrist. The strap will prevent electrostatic damage to one or more of the components.

1.8. Estimate the number of connections on this PC circuit card.

1.9. Describe the thermal management system in five different electronic products that you encounter on a day to day basis.

1.10. Prepare a graph of λ, the failure rate, as a function of junction temperature for T_j ranging from 20–150°C. Let $A = B = C = 1$ for this initial determination. Discuss the effect of the higher temperatures on the failure rate.

1.11. Give an example of good design for manufacturing of an electronic product. If you can't think of one for an electronic product, give an example for an automotive product.

1.12. Give an example of poor design for manufacturing for electronic products.

1.13. If you drop a box on the floor from say a height of 3 ft, determine the deceleration during impact. What assumptions did you have to make in your analysis?

1.14. What is your opinion of the ergonomics design of the chair in your office or your classroom? Why do you think management selected the chair you use?

1.15. Give an example of a high end product and estimate its cost.

1.16. Give an example of an intermediate product and estimate its cost.

1.17. Give an example of a low end product and estimate its cost.

1.18. What is the least expensive electronic product that you can identify? Is it produced in a large volume? How many electronic components does it contain? Is it designed for ease of manufacturing? Is it designed for ease of maintenance?

1.19. What is a venture capitalist? Do they serve an important function in the development of small business in the United States?

1.20. The 1M bit memory chip was introduced into new products by select companies in 1988. When do you estimate that the 4M memory chip will be introduced?

1.21. If your PC memory card contains 640 kbytes, how many memory chips are installed on the board if the chips are:

(*a*) 64 kbit
(*b*) 256 kbit
(*c*) 1 Mbit
(*d*) 4 Mbit

Show also the relative size of the card for the four cases.

CHAPTER 2

ELECTRONIC COMPONENTS AND SEMICONDUCTOR DEVICES

2.1 INTRODUCTION

In packaging electronic systems it is often necessary to design cabinets, drawers or gates to house the components and to permit ready access for maintenance. Printed circuit boards (PCBs) are designed to support hundreds of small components with thousands of interconnections. A cooling system is incorporated into the cabinets to dissipate the heat generated by the electronic devices. Interconnections and heat removal with small thermal penalties are the two most important features in the design of a highly reliable product. There is a wide variety of hardware both mechanical and electrical that must be included in the design of even a relatively simple electronic system. A partial listing of some of the most common components used in design is:

Discrete transistors	Power supplies
Logic-type ICs	Transformers
Memory-type ICs	Plasma panels
Resistors	Cathode ray tubes
Capacitors	Vacuum tubes
Inductors	Disk drives
Potentiometers	Tape drives
Relays	Lamps
Switches	Fans
Circuit breakers	Cold plates
Connectors	Cable harnesses

The emphasis in this chapter is placed on the packaging of microelectronic circuits that are used extensively in the design of high performance digital systems, laboratory instrumentation and automated manufacturing systems. To package microelectronic circuits it is necessary to understand in at least a qualitative manner the functional behavior of microelectronic devices. For this reason the basic principles of semiconductor theory are reviewed and the operation of semiconductor diodes and transistors is described. The coverage is then extended to introduce logic gates and the two basic gate technologies (bipolar and MOSFET) that are commonly employed today. The last part of the chapter treats the scale of integration with projections into the future, which will affect packaging strategies into the next decade. Of particular importance is the identification of significant changes in packaging design that will occur because of the introduction of new electronic devices in the coming decade. These changes in packaging design will be driven by higher I/O count and higher heat dissipation that will be typical of the newer high performance components that will be released in the coming decade. Higher gate count and heat dissipation is mandatory in order to fully utilize the logic devices produced with VLSI and emerging ULSI technologies.

2.2 CONDUCTORS, INSULATORS AND SEMICONDUCTORS

Conductivity in materials depends on the structure of the atoms of elements that are combined to give an alloy and the resulting atomic bonding between these atoms. Consider for example the metal aluminum with the atomic structure illustrated in Fig. 2.1. The aluminum atom has a full inner shell (the K shell) and a full L shell with two and eight electrons filling these shells. However, the outer M shell contains only three electrons and they are loosely bound to the nucleus. These M shell electrons are nearly free and they act as negative charge carriers in conducting current in a wire made of aluminum. The resistance R of a conductor in the shape of a wire or a rectangular conductor (trace) formed on the surface of

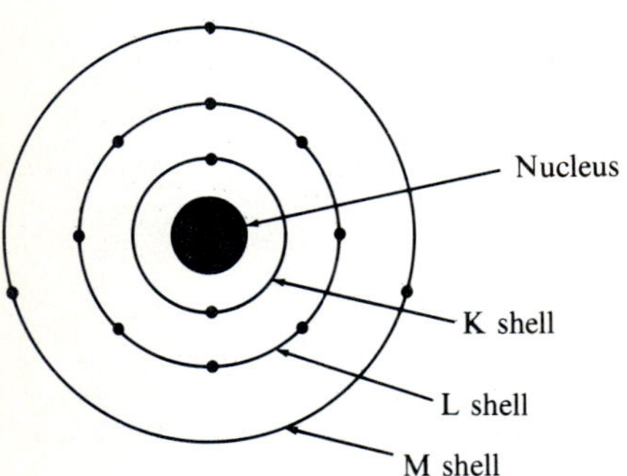

FIGURE 2.1
Atomic structure for aluminum.

TABLE 2.1
Resistivity ρ of select materials

Material	Classification	Resistivity (Ω-cm)
Silver	Conductor	1.63×10^{-6}
Copper	Conductor	1.72×10^{-6}
Aluminum	Conductor	2.83×10^{-6}
Nickel	Conductor	6.9×10^{-6}
Platinum	Conductor	9.8×10^{-6}
Silicon	Semiconductor	1.56×10^{5}
Aluminum oxide	Insulator	1×10^{15}
Silicon oxide	Insulator	1×10^{14}
Epoxy	Insulator	1×10^{15}
Polyethelene	Insulator	1×10^{18}

a PCB is given by:

$$R = \rho L/A \tag{2.1}$$

where L is the length of the conductor (cm)
A is the cross section area (cm^2)
ρ is the resistivity (Ω-cm)

The resistivity ρ depends on the atomic structure of the element and if there are a large number of loosely bound electrons in the outer shell, the resistivity is low as indicated in Table 2.1. It is evident from the results shown here that the resistivity of different types of materials can vary over a very wide range. For metal conductors ρ is of order 10^{-6} Ω-cm but for insulators ρ is of order 10^{12} Ω-cm or higher.

The atomic structure of insulating materials clearly shows the reason for the large values of ρ. Consider the atomic structure of SiO_2 shown in Fig. 2.2. Silicon has a full K shell, a full L shell and four electrons in the outer M shell. Oxygen has a full K shell and six of eight electrons necessary to fill the L shell. The

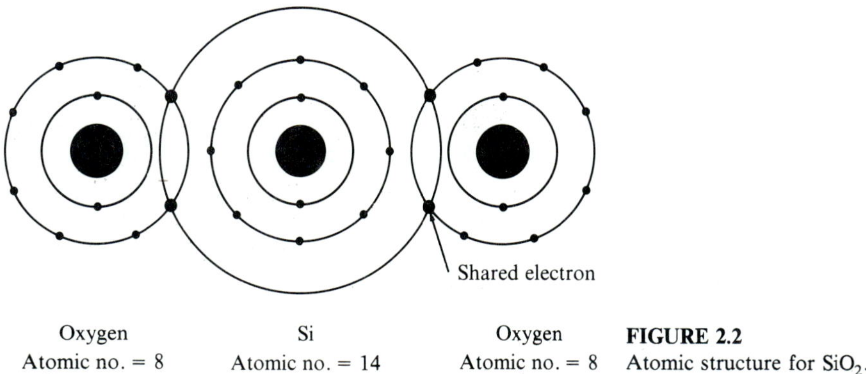

FIGURE 2.2
Atomic structure for SiO_2.

FIGURE 2.3
Covalent bonding in silicon.

silicon atom combines with two oxygen atoms to provide the correct number of electrons to fill the outer shells of the oxygen atoms when the electrons are shared. The shared electrons are covalently bonded and no free electrons are available to act as charge carriers. Materials like SiO_2, Al_2O_3, MgO, BN, Si_3N_4 and BeO, which are classified as ceramics, are all covalently bonded with complete outer shells. They all act as insulators with very high resistivity.

Elements such as carbon, silicon and germanium each have four electrons in their outer shells. The atoms in these elements form covalent bonds with four neighboring atoms to form stable structures similar to that shown for silicon in Fig. 2.3. The electrons are not loosely bound due to the covalent structure and the electrons are not available to act as charge carriers. However, the atomic structure is not perfect and some of the electrons have sufficient energy to jump from the valence state to the conduction state. For this reason materials such as silicon, germanium and gallium arsenide are classified as semiconductors and exhibit resistivity much higher than the metal conductors but much lower than the ceramic insulators. The resistivity of silicon is 1.56×10^5 Ω-cm and is due to the presence of one electron out of 2×10^{13} that has sufficient energy (1.1 eV) to move from the valence band to the conduction band. The ability of these electrons to carry charge is termed intrinsic conduction.

2.3 EXTRINSIC SEMICONDUCTORS

The intrinsic semiconducting capabilities of silicon were described in the previous section. The semiconducting properties of silicon, germanium or gallium arsenic can be modified by changing the atomic structure of single crystals of these materials. To show the modification of the crystal lattice, consider the structure of silicon with its covalent bonding as shown in Fig. 2.3. If one or more of the silicon atoms in the lattice is replaced with an impurity atom, the conducting properties of the modified structure are changed. Impurity atoms called dopants are of two basic types: those elements with five valence electrons in their outer shell such as phosphorus, arsenic and antimony and those elements with three valence electrons in their outer shells including boron, aluminum and gallium.

If phosphorus with its five valence electrons is introduced into the silicon structure, the fifth electron does not become involved in the covalent bonding. This extra electron, an extrinsic charge, is relatively free to carry current. The resistivity of the doped silicon is lower than that of pure silicon. The silicon with dopant elements from the V column of the periodic table are classified as type-*N* semiconductors since they contain extra electrons, which are negative charge carriers.

If silicon atoms in the crystal lattice are replaced with a dopant atom with three valence electrons in its outer shell such as boron, then covalent bonding occurs between silicon and boron but the outer shell is not filled and a single vacancy or hole exists. This hole is an accepter of electrons and in this mode it acts as a positive charge carrier. The semiconductors with dopant elements selected from column III in the periodic table are classified as type-*P* because of the positive charge that is carried by the holes that exist and move through the atomic lattice.

Conduction occurs in a semiconductor when either the electrons or the holes move due to an applied electric field. The resistivity of the semiconductor is given by

$$\rho = 1/(eN\eta) \tag{2.2}$$

where N is the number of charge carriers
e is the charge on the carrier
η is the mobility of the carrier

The velocity v of the charge carrier is dependent on

$$v = \eta E \tag{2.3}$$

where E is the electric field. It is interesting to note that the velocity of the electrons in N-type silicon is about three times larger than the velocity of the holes in P-type silicon under the same applied field. The reduced mobility of the holes is due to the mechanics of the motion of the holes. The manner in which the holes move through the lattice is presented in Fig. 2.4. The movement of a hole is shown to be due to the movement of an electron from an adjacent atom to fill the hole while at the same time creating a new hole displaced by one lattice spacing. This cumbersome motion of the holes accounts for their lower velocity through the lattice structure.

The resistivity of both N- and P-type silicon can be controlled over a wide range 10^2–10^{-3} Ω-cm by adjusting the amount of dopant (its concentration) that is added to the silicon as indicated in Fig. 2.5. The ability to control the resistivity is of critical importance in fabricating diodes, transistors or resistors in silicon. Methods have been developed for producing single crystal silicon with dopant impurities of different concentrations introduced at local sites on the surface of a wafer cut from a single crystal.

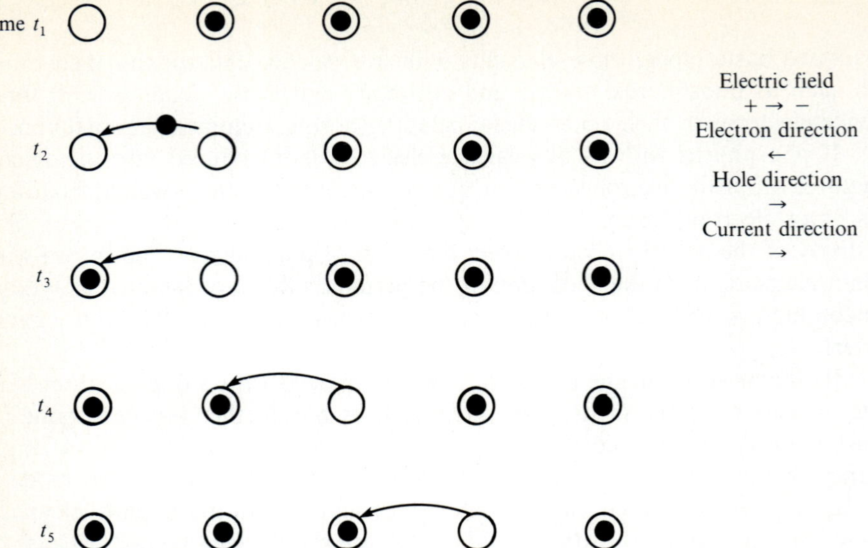

FIGURE 2.4
Movement of a hole to the right by a sequence of electron movements to the left.

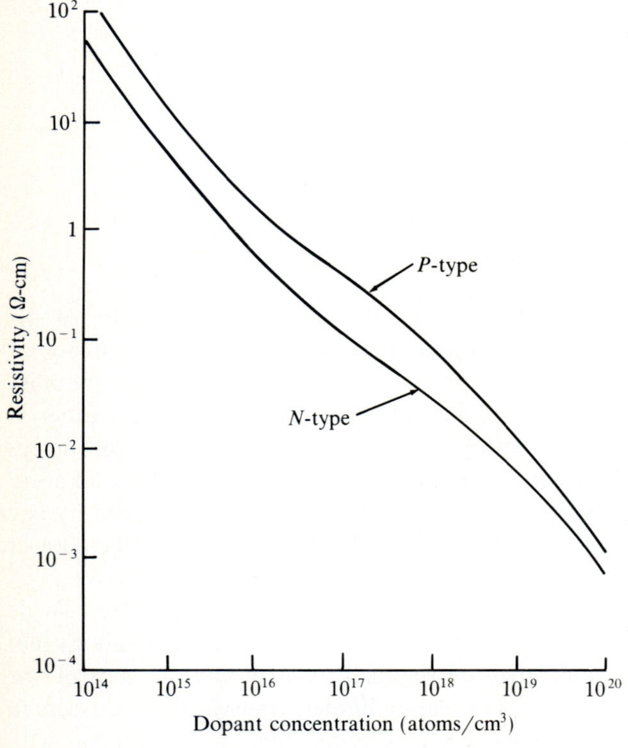

FIGURE 2.5
Resistivity of *P*- and *N*-type silicon as a function of dopant concentration.

2.4 THE *P-N* JUNCTION

An interface between *P*- and *N*-type silicon can be formed by taking *N*-type material and diffusing boron into one of the surfaces of the material to create a local region of *P*-type material. This procedure produces a *P-N* junction, which is illustrated schematically in Fig. 2.6*a*. Note, that the concentration of the dopant on both sides of the junction can be adjusted as indicated in Fig. 2.6*b*. The *P*-type material contains holes that act as acceptors of electrons and the *N*-type material contains electrons that serve as donors to fill these holes.

At the interface of the *P* and *N*, the holes and electrons combine and eliminate each other forming a thin region free of charge carriers. The layer where the holes have accepted the donor electrons is termed the depletion region, indicating that the available charge carriers have been depleted. The remaining holes in the *P* material create a negative charge on the electrode opposite the

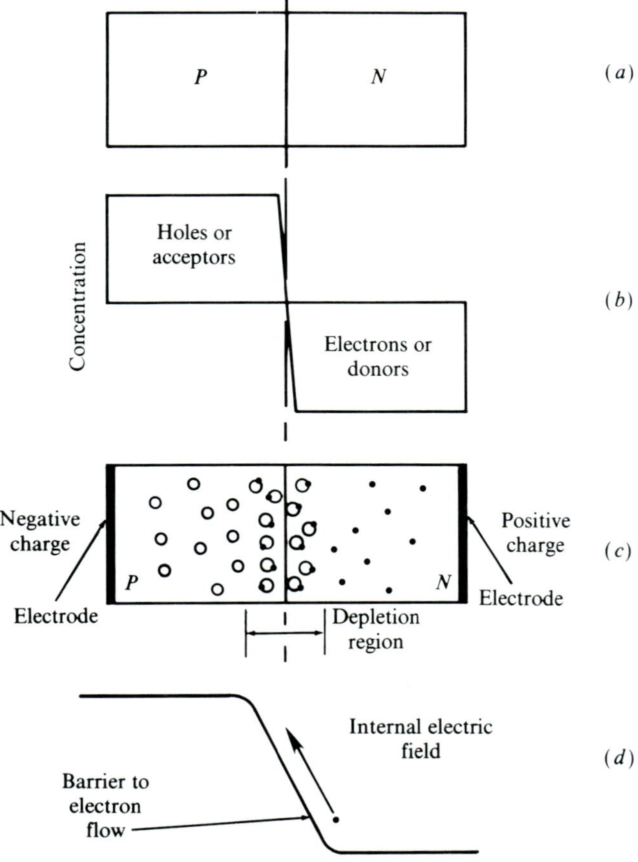

FIGURE 2.6
Characteristics of a *P-N* junction.

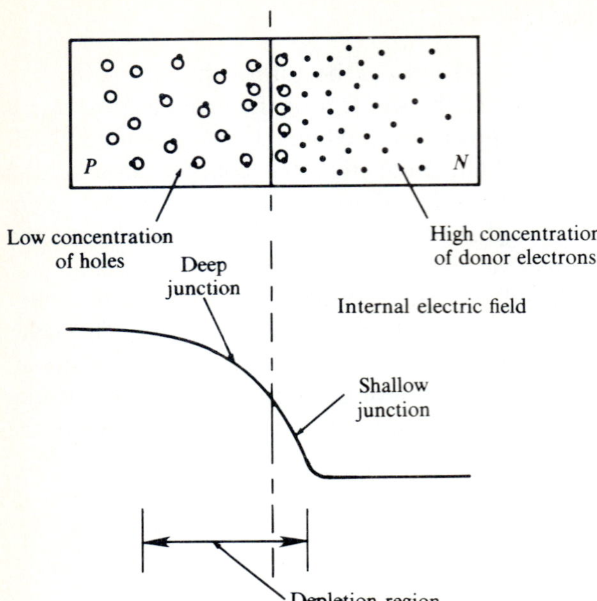

FIGURE 2.7
Influence of dopant concentration on the thickness of the *P-N* junction.

junction and the remaining electrons in the *N* material create a positive charge on its electrode. In this state the junction is electrically neutral as indicated in Fig. 2.6*c*.

The presence of positive charge carriers (holes) on one side of the depletion layer and negative charge carriers on the other side of the layer develops an internal electric field across the junction. Essentially the depletion layer acts like pure silicon capable of only intrinsic conduction and it limits the flow of electrons across the junction.

The thickness of the depletion layer depends on the concentration of the impurity dopants. High concentrations lead to thin layers and lower concentrations yield thick layers. This influence of concentration on the thickness of the depletion layer is illustrated in Fig. 2.7. The ability to control the thickness of the depletion layer is useful in the development of diodes and transistors.

2.5 SEMICONDUCTOR DIODES

The simplest semiconductor device is the diode and it is based on the properties of a *P-N* junction. A diode is a device that permits the flow of current in one direction and blocks the flow in the other direction. To show the essential features of a semiconductor diode, consider the *P-N* junction as described in Section 2.4 and place a positive voltage V^+ on the *P* electrode and a negative voltage V^- on the *N* electrode. This arrangement, which is shown in Fig. 2.8*a*, represents the diode under a forward bias voltage. The positive voltage on the *P* electrode repels the positively charged holes and drives them toward the junction.

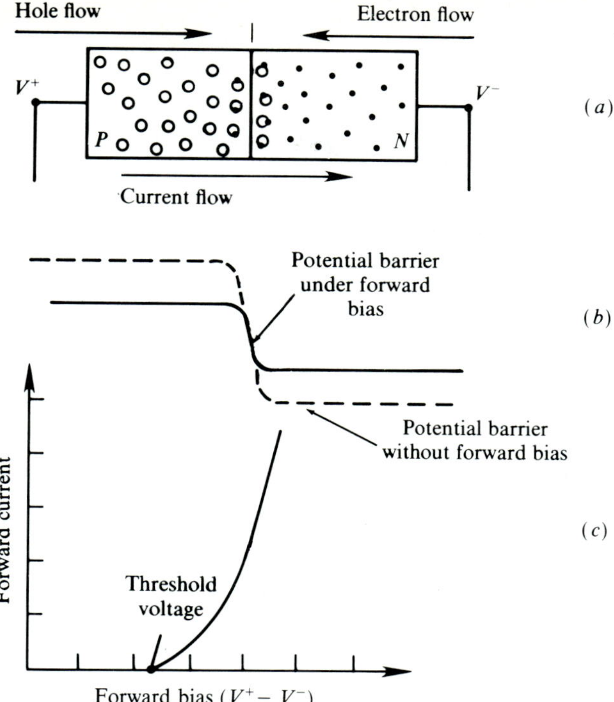

FIGURE 2.8
Characteristics of a semiconductor diode under forward bias voltage.

The negative voltage on the *N* electrode repels the negatively charged electrons driving them toward the junction. Additional recombination of holes and electrons takes place, the depletion layer becomes thinner and the internal electric field, which serves as a barrier to conduction, is reduced as illustrated in Fig. 2.8*b*. Further increases in the forward bias ($V^+ - V^-$) overcome the potential barrier, permitting the flow of electrons from the *N* material across the junction to the *P* electrode and the flow of holes from the *P* material across the junction to the *N* electrode. The forward current is shown as a function of the forward bias in Fig. 2.8*c*. It should be noted that no current flows until the potential barrier is overcome by the forward bias voltage.

Next, consider the same semiconductor diode but with the polarity reversed so that the V^+ is applied to the *N* electrode and the V^- is applied to the *P* electrode as indicated in Fig. 2.9*a*. In this case the electrons in the *N* material are attracted to its electrode and the holes in the *P* material are attracted to its electrode. The depletion layer is widened and the potential barrier due to the internal field increases as is shown in Fig. 2.9*b*. Current flow due either to the motion of electrons or holes across the junction cannot occur because of this increase in the potential barrier due to the reverse bias. Some leakage current

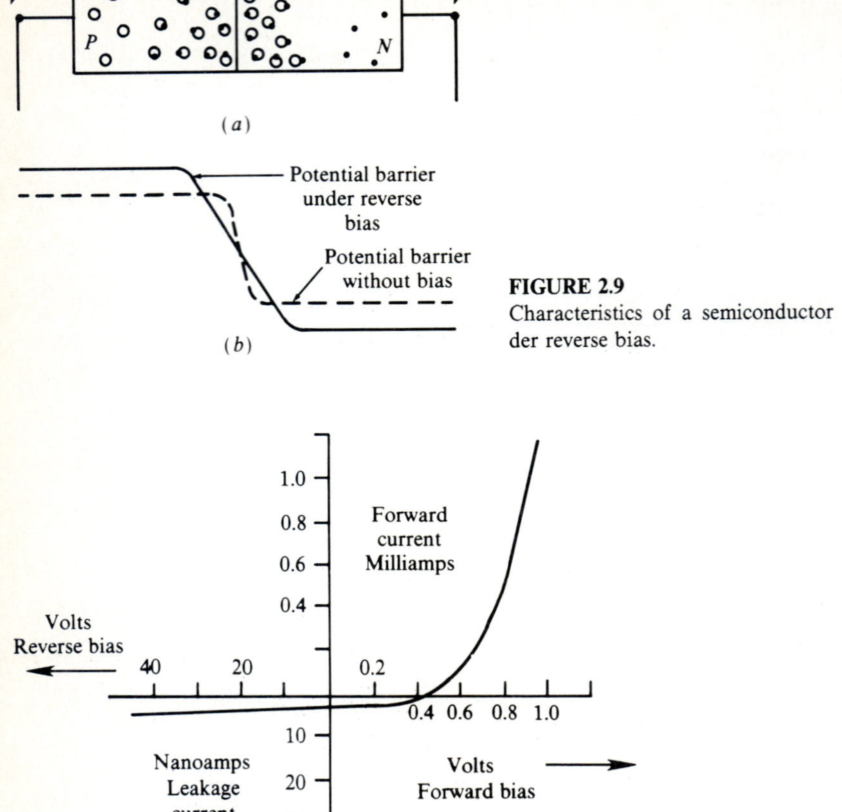

FIGURE 2.9
Characteristics of a semiconductor diode under reverse bias.

FIGURE 2.10
Forward and leakage current as a function of voltage bias on a *P-N* junction diode.

does flow but this is due to intrinsic conduction resulting from lattice imperfections. Typical leakage currents are measured in nanoamperes whereas the forward currents are measured in milliamperes. Forward and reverse currents are shown as a function of voltage bias for a typical semiconductor diode in Fig. 2.10.

2.6 TRANSISTORS

Transistors are solid state switches that can be open to effectively block current flow or closed to permit current flow as illustrated in Fig. 2.11. Solid state

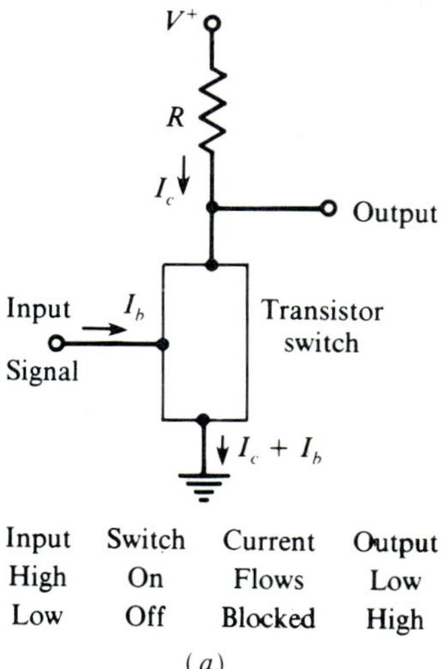

Input	Switch	Current	Output
High	On	Flows	Low
Low	Off	Blocked	High

(a)

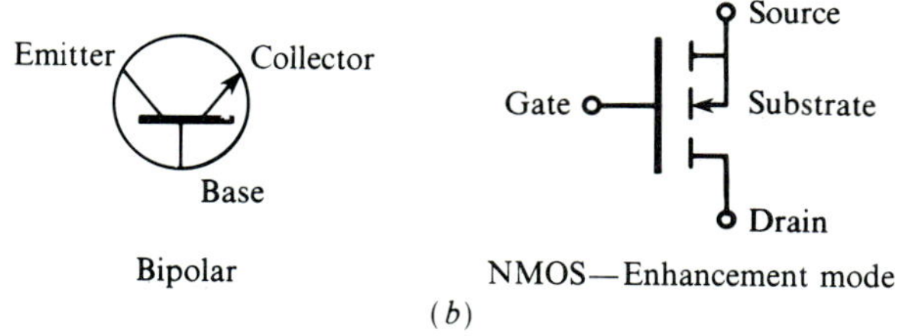

(b)

FIGURE 2.11
(a) A transis'or acting as a solid state switch. (b) Symbols for bipolar and FET transistors.

switches differ in two important aspects from mechanical switches. First, the transistor has no moving parts and is activated by a control signal. As a consequence of this feature, the switch can be activated very quickly with typical switching times of the order of 1–10 ns. Second, the transistor acts as a current amplifier since the current passing through the switch $I_c + I_b$ is 10–100 times larger than the control current I_b required to activate the transistor. These two characteristics of the transistor permit it to be used extensively in digital logic circuits where logic design requires an extremely large number of switches to execute even simple digital functions. Transistors are also used extensively in analog circuits where the amplification of currents is extremely important.

There are two basic types of transistors that are in common use today, namely the bipolar type and the field effect transistor (FET). Key features of both of these basic transistors will be described in the next section.

2.6.1 Bipolar Transistors

Consider first a *N-P-N* bipolar transistor as defined in Fig. 2.12*a*. The three elements have solid state interfaces and electrodes are placed on each element to provide for electrical connections to the emitter, base and collector. Note, that

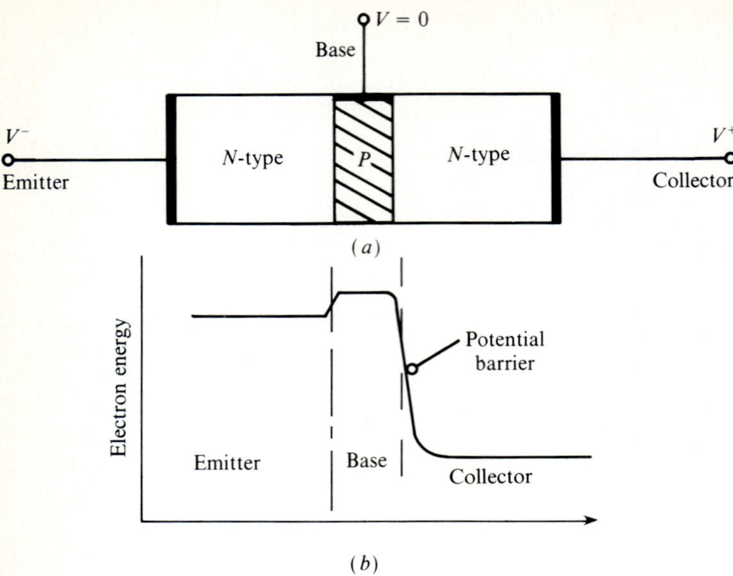

FIGURE 2.12
Bipolar transistor acting as an open switch.

the *P*-type Si in the base is relatively thin compared to the *N* material used in the collector and emitter. This arrangement corresponds to that of two back-to-back diodes that share a base as a common element.

The base is lightly doped and the depletion regions at the *P-N* junctions on both sides of the base are wide and effectively cover the entire thickness of the base when V^+ is applied to the collector and V^- is applied to the emitter. The depleted *P* material in the base acts as a potential barrier to conduction between the emitter and the collector as shown in Fig. 2.12*b*. The transistor acts as an open switch since the base–collector diode is operating in reverse bias and because the forward bias on the base–emitter diode is less than the threshold voltage required for forward current to be conducted.

If a positive voltage is applied to the base electrode, the forward bias on the emitter–base junction increases and the potential barrier at this interface is overcome. Electrons from the emitter are injected into the base. These electrons find the base depleted and diffuse through it, and then they are attracted to the positively charged collector electrode. Thus, it is clear that a small positive voltage on the base turns the transistor switch on, electrons flow from the emitter to the collector and current flows from the collector to the emitter. A simple circuit presented in Fig. 2.13 shows the action of the transistor as a switch. The potentiometer at the left permits the base to emitter voltage V_{BE} to be varied. If V_{BE} is less than the threshold voltage of the transistor, the switch will be open $V_{CE} = V_s$ and the forward current is blocked. However, when V_{BE} is greater than

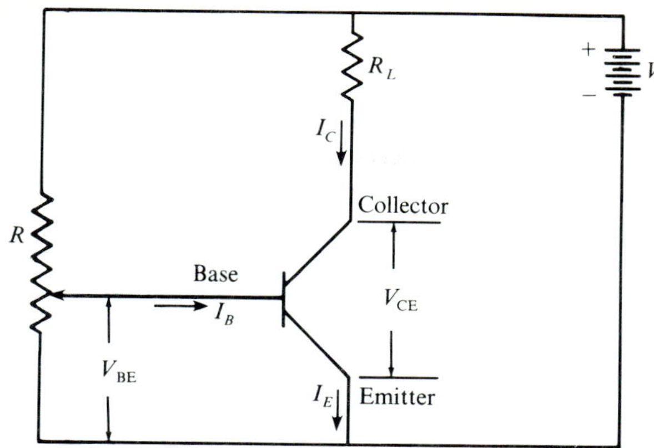

FIGURE 2.13
Circuit showing the adjustment of the base voltage to the transistor until it acts as a closed switch and conducts.

the threshold voltage, the switch will be closed, V_{CE} approaches zero and the forward current I_c flows.

An analysis of the circuit shown in Fig. 2.13 with Kirchhoff's law indicates that the current I_E may be written as

$$I_E = I_B + I_C \tag{2.4}$$

The current gain G for the transistor is defined as

$$G = I_E/I_B \tag{2.5}$$

The current gain is dependent primarily on the thickness and area of the base and the degree of doping on both sides of the *P-N* junctions. Typical current gains of 10–100 are commonly achieved in bipolar transistors.

The illustration of the *N-P-N* bipolar transistor presented in Fig. 2.12 is a schematic used to show the concept of back-to-back diodes with forward and reverse bias. The actual construction details of a transistor differ significantly from this simple diagram. Production techniques for integrated circuits are based on placing many thousands of transistors as well as other components such as resistors, capacitors and diodes on a wafer cut from a very large single crystal of Si. The wafers used in production facilities today are usually from 4–6 in. (100–150 mm) in diameter and 20–30 mils (0.5–0.75 mm) thick. The transistors are arranged in a planar array on one surface of the wafer. The *N-P-N* structure, described previously, is formed through the surface of the wafer using diffusion or ion implantation techniques to vary the local concentration of the *N*- and *P*-type impurities. A cross section of a single *N-P-N* bipolar transistor more representative of current technology is represented in Fig. 2.14.

The structure in Fig. 2.14 is developed beginning with a wafer cut from a *P*-doped silicon crystal. Islands of N^+ (high concentration of N impurities) and more localized islands of N and P are formed by a series of production steps that utilize advanced lithographic processes to define feature sizes that are measured in micrometers or fractions of micrometers. A layer of SiO_2, which

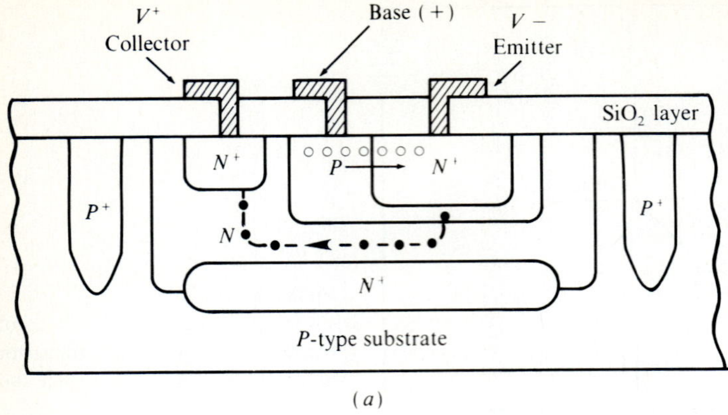

(a)

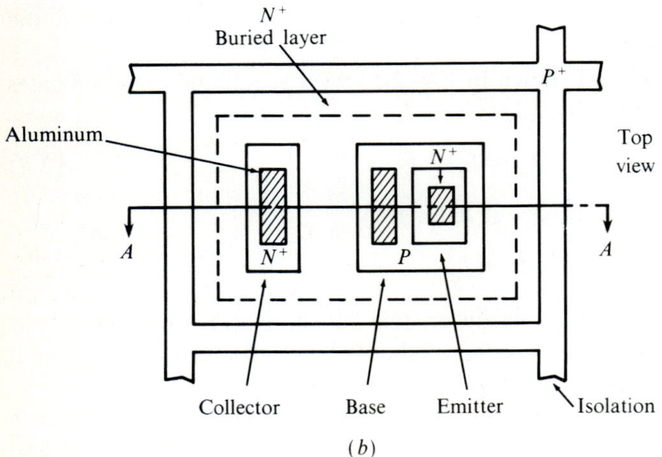

(b)

FIGURE 2.14
Details of an N-P-N bipolar transistor fabricated on a wafer of silicon.

serves as an insulator to prevent surface currents, is formed by oxidizing the top surface of the wafer. Channels to the N^+ and P islands are etched through the SiO_2 layer and filled with a vapor-deposited aluminum, which provides the electrodes for the collector, base and emitter. Deep islands of P^+ are placed around the structure to provide transistor-to-transistor isolation. The electron flow from the N^+ emitter region, through the thin P base to the N and N^+ collector is shown in Fig. 2.14a.

A top view of the structure is illustrated in Fig. 2.14b where the planar dimensions of the islands of doped silicon are indicated. As manufacturing processes are improved, the dimensions of the features are reduced and the

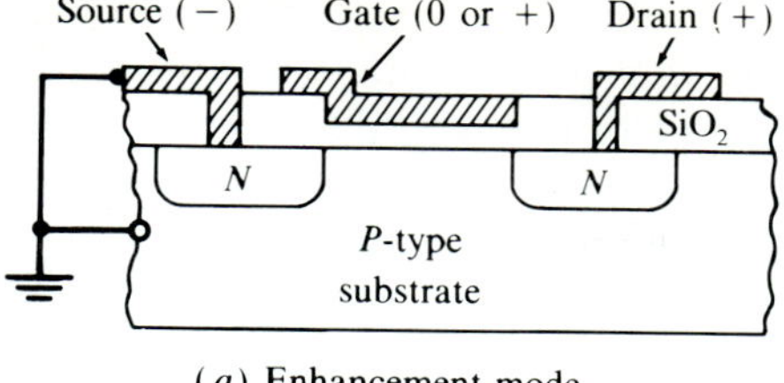

(a) Enhancement mode

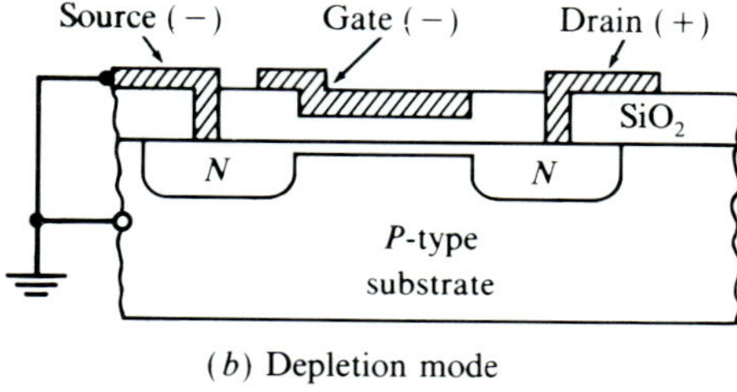

(b) Depletion mode

FIGURE 2.15
Cross section of a NMOS transistor showing key features.

overall planar area required for a transistor, capacitor, resistor or a gate is decreased. With feature sizes on junctions and channels of about 1 μm, the area of Si necessary for a bipolar transistor with its isolating boundary is of the order of 2×10^{-10} m^2, which permits the placement of about 120,000 transistors on a 5 × 5 mm size chip.

2.6.2 Metal Oxide Semiconductor Field Effect Transistor

The metal oxide semiconductor field effect transistors (MOSFET) devices were developed after the bipolar transistor. As the name implies, these transistors utilize field effects to control the flow of electrons or holes to perform the switching action. A typical structure of a NMOS transistor, which is fabricated by implanting *N*-type islands in a *P*-type substrate, is presented in Fig. 2.15*a*. After the two islands are implanted in the substrate, the entire surface is covered with an oxide layer. The *P* substrate between the two islands forms the channel region. An electrode is placed over the SiO_2 layer to form the gate, which is essentially a small capacitor coupled to the channel region. The electrode to one *N* island (the source) is connected to the *P* substrate and then it is grounded. The electrode to the other *N* island (the drain) is connected to a positive supply voltage. The voltage on the gate controls the switching action of the transistor. When the gate voltage is zero, the positive voltage on the drain attracts the electrons in the *N* material at the source, but these electrons cannot flow through the *P* channel because it contains many holes that combine with the electrons to produce a depletion layer that blocks the further flow of electrons. In this condition of zero or negative gate voltage, the NMOS transistor acts as an open switch. However, if a positive voltage is applied to the gate, the holes in the *P* channel are repelled and the electrons from the source can travel across the *P* channel to the drain and the current flows in the opposite sense. With a positive gate voltage the transistor acts as a closed switch. This type of NMOS transistor action is called the enhancement mode because the positive gate voltage enhanced the flow of electrons from the source to the drain.

NMOS devices are fabricated in a different way to operate in the depletion mode. The depletion mode NMOS transistor is illustrated in Fig. 2.15*b*. The primary difference between fabrication detail is the presence of a thin layer of *N* material under the gate that connects the source and the drain. In this configuration a zero gate voltage permits the normal passage of electrons through the *N* channel from the source to the drain. However, the application of a negative voltage to the gate repels the electrons from the thin *N* channel converting it to an insulating channel blocking the flow of electrons from the source to the drain and effectively opening the transistor switch.

PMOS transistors may also be fabricated to operate in a manner similar to the NMOS transistors as is illustrated in Fig. 2.16. In this case two islands of *P* material are placed in a *N* substrate. With PMOS the polarities of the source, gate and drain are reversed and the charge carriers are holes rather than electrons. The PMOS transistors are not commonly employed because their switching time is 2–3 times longer than the NMOS transistors. The longer switching time is due to the lower mobility (velocity) of the holes in switching

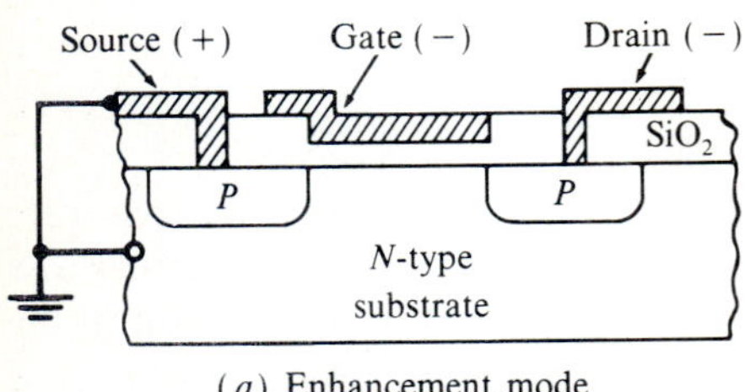

(*a*) Enhancement mode

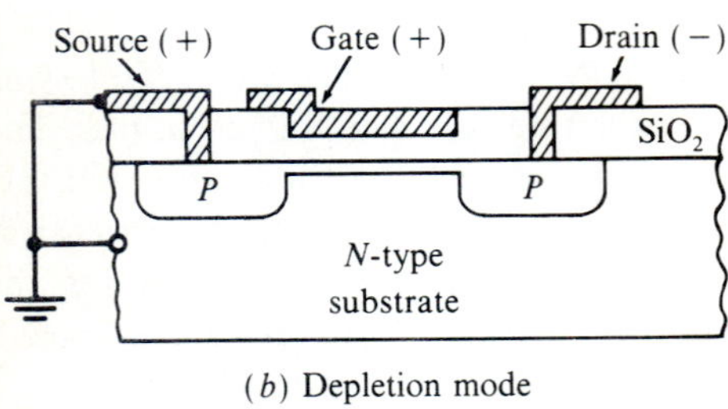

(*b*) Depletion mode

FIGURE 2.16
Cross sections of a PMOS transistor showing key features.

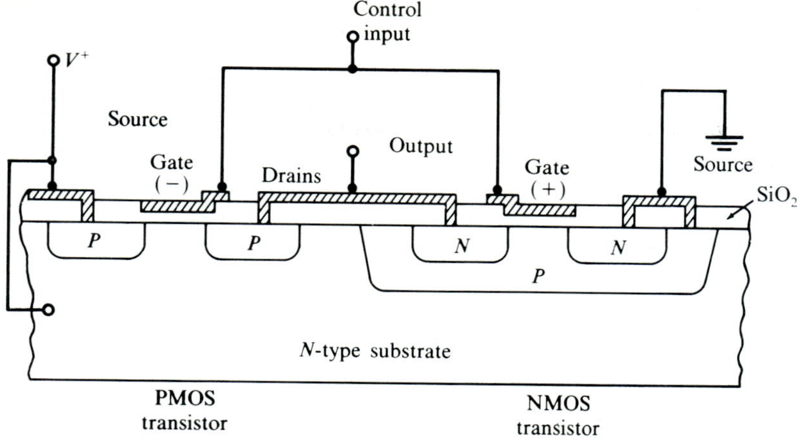

FIGURE 2.17
A CMOS transistor consists of a series-connected pair of NMOS and PMOS transistors.

PMOS in comparison to the velocity of the electrons that are utilized in NMOS switching.

A third type of MOS transistor is the complementary MOS device known as CMOS. This device incorporates both a NMOS and a FMOS transistor connected in series as shown in Fig. 2.17. The two gates are connected together and are activated by a single control voltage. The two drains are also connected together to provide an output signal that can be either high or low. The two sources are wired separately with the source on the NMOS transistor grounded and the source on the PMOS transistor connected to the V^+ supply. This series connection of the two transistors results in a very small power dissipation in the steady state. The two transistors require control signals of opposite polarity to conduct and as a consequence one or the other of transistor switches is always open. Since one or the other of the transistors is open in the steady state, the V^+ and the V^- supplies are not connected together except for the period during switching from one state to the other. When the control signal is negative, the PMOS transistor conducts and the output goes high; however, the current flow is very small because of the high impedance and the low voltage difference between the source and the output. When the control signal is positive, the NMOS transistor conducts and the output goes low. Again the current flow is minimized due to the high impedance and low potential difference between the gate and the source. Large current flow and power losses occur only during the switching of the CMOS transistor. During switching a conducting path exists between the V^+ supply and the ground or V^- as one transistor is turned on and the other is turned off. The power dissipated in a CMOS transistor depends on the frequency of operation of the transistor. As the frequency increases, the power dissipation increases.

2.7 COMPARISON OF TRANSISTOR TYPES

Bipolar transistors are low impedance devices that require relatively high currents to operate and dissipate large amounts of heat. Bipolar transistors switch rapidly, and for equal feature size and power density the bipolar devices can be operated at a higher frequency than the MOS devices. A comparison of the switching delay time for bipolar, NMOS and CMOS, presented in Fig. 2.18, shows the superiority of bipolar devices in switching speeds for gate powers exceeding 0.1 mW. Bipolar transistors must be isolated to prevent interaction with adjacent devices on the chip. This isolation is achieved by using isolation barriers as demonstrated in Fig. 2.14; however, the placement of these barriers utilizes a sizable area on the chip. This placement reduces the number of bipolar transistors that can be placed on a chip and the density (number of components per area) is lower with bipolar than with MOS technology. The diffusion depths in bipolar are relatively deep and this reduces the current density and improves the reliability and enhances the power capacity of the off chip drivers.

The MOSFET devices exhibit a high impedance and consequently they operate with lower currents. The isolation of the MOSFET transistors is inherent and no isolation barriers are required. This feature saves chip area and component density is higher than that which can be achieved with bipolar technology.

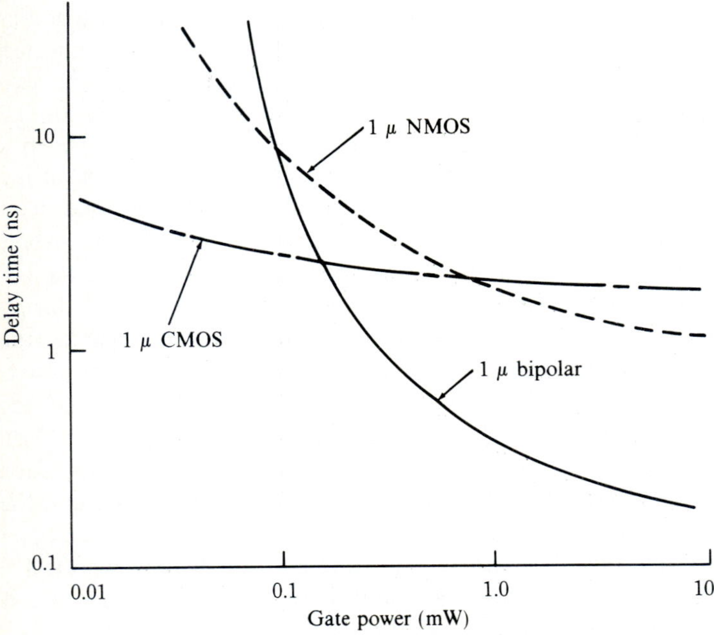

FIGURE 2.18
Power delay relationships for bipolar, NMOS and CMOS showing the superior speed for bipolar at high power levels.

Processing of NMOS or PMOS is generally easier than with bipolar and the design scales readily to smaller feature sizes. The channels are shallow and parallel to the surface and their length is dependent on feature size. This lateral deployment adds capacitance and limits the switching speed. The very shallow diffusion layers yield small cross-sectional areas for current flow near the surface at relatively high current density.

PMOS transistors are rarely used because their switching speed is low relative to say NMOS transistors. The mobility of holes as charge carriers is much less than the mobility of electrons and since PMOS depends largely on the movement of holes, it is inherently slow. NMOS operates with electrons as charge carriers and is much faster than PMOS. A comparison of the speed of NMOS and bipolar transistors shown in Fig. 2.18 indicates that bipolar transistors are much faster than NMOS for power levels exceeding 0.1 mW. The primary advantages of NMOS over bipolar is density, ease in scaling to smaller feature sizes and less complexity in manufacturing.

CMOS transistors consist of two series-connected transistors, one PMOS and the other NMOS. In this arrangement the forward current from the V^+ supply to ground occurs only during the switching operation and not during steady state in either the high or the low mode. This feature greatly reduces the power dissipation of the CMOS transistors and relatively high switching speed can be achieved at low power levels. Indeed, for average powers of less than about 0.2 mW, CMOS compares favorably with bipolar technology in switching speed. Since feature size is critical in determining MOS switching speeds, one may anticipate significant improvements and wider usage of CMOS as feature size decreases below 1 μm. Also, this trend toward CMOS will be driven by the problems of accommodating the very large power dissipated by bipolar transistors while maintaining junction temperatures at a sufficiently low value to insure high reliability and system availability.

2.8 LOGIC GATES

In most digital instrumentation or computers logic gates are employed in large numbers to perform complex operations at extremely high rates. The circuits

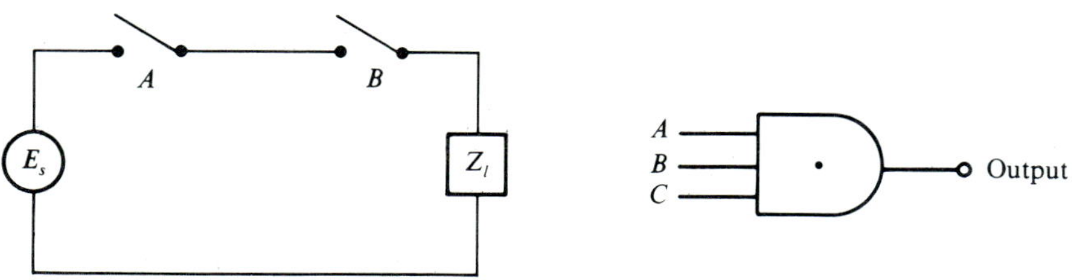

FIGURE 2.19
Circuit representing the AND gate and its symbol.

TABLE 2.2
Truth table for the AND gate $A \cdot B = T$

Switches or inputs		Output
A	B	T
0	0	0
0	1	0
1	0	0
1	1	1

involved are large; they often contain 10^5–10^7 logic gates. However, even if these circuits are large they are simple because they contain only three different types of basic gates. The basic gates include the AND, OR and NOT. Other more complex logic elements are often used but they consist of combinations of these three basic gates.

The AND gate may be represented by the circuit shown in Fig. 2.19 where two switches A and B are placed in the line from the source to the load. The voltage E_s is applied to the load only if switch A *and* switch B are both closed. The possibilities for the AND gate are listed in a truth table shown in Table 2.2. Note, that (0) is used to represent a false statement and (1) to represent a true

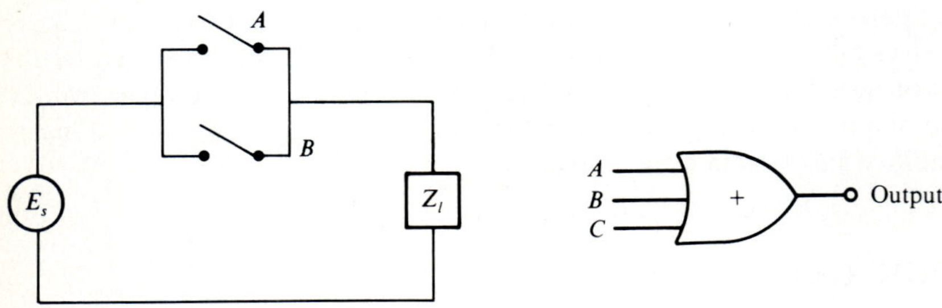

FIGURE 2.20
Circuit representing the OR gate and its symbol.

TABLE 2.3
Truth table for the OR gate $A + B = T$

Switches or inputs		Output
A	B	T
0	0	0
0	1	1
1	0	1
1	1	1

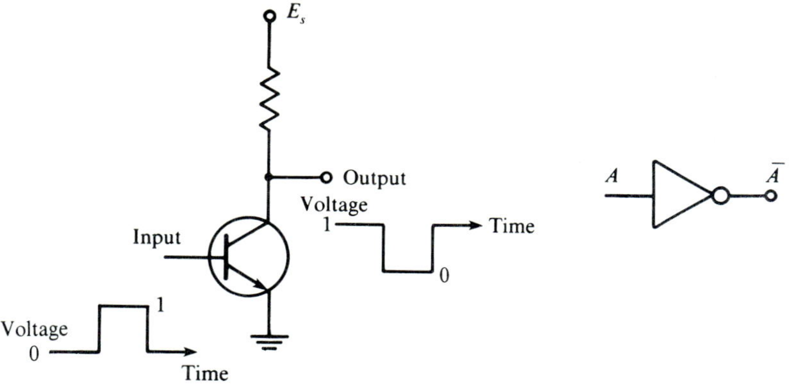

FIGURE 2.21
Circuit representing the NOT gate and its symbol.

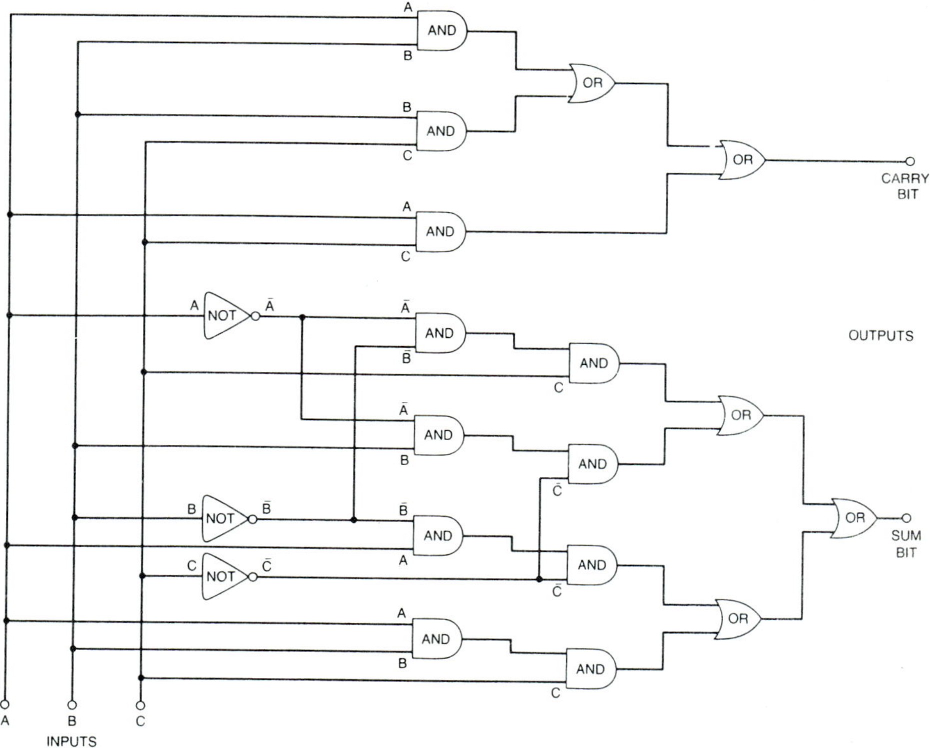

FIGURE 2.22
This binary adder is a functional element of a digital system and is constructed from many simple AND, OR and NOT gates.

statement. With regard to the voltage applied to the load, (1) indicates that it is true that E_s is applied to the load.

The OR gate is represented by the circuit given in Fig. 2.20 where two switches A and B are placed in parallel in the line between the voltage source and the load. When A is closed OR when B is closed, the voltage is applied to the load ($T = 1$). The truth table for a two switch OR gate is presented in Table 2.3.

The NOT gate, which is illustrated in Fig. 2.21, is an inverter. In this case the mechanical switch has been replaced by a transistor that is turned on (closed) by a positive input voltage. If the input signal to the transistor is say (0), the transistor acts as an open switch, no current flows and the output voltage is E_s or (1). When the input signal goes to (1), the transistor conducts, acting like a closed switch, and the output is grounded, giving the low state or (0). It is clear from this description that when the input is high (A), the output is low ($\bar{A}$) and changing the input to low ($\bar{A}$) results in an output that is high (A).

These basic gates are arranged in circuits to perform digital functions. For example, the arrangement of AND, OR and NOT gates shown in Fig. 2.22 represents a binary adder. A digital system is composed of many of these digital functions and may contain a million or more of the simple basic gates. The number of chips use to build the logic circuits depends on the scale of integration used to fabricate the chip. With VLSI (very large scale integration) it is possible to place on the order of 10^4 gates on a single chip, thus permitting the development of extremely large digital systems with only 100–1000 chips.

2.9 GATE TECHNOLOGIES

Transistors are used to replace mechanical switches shown in the logic gates represented in Figs. 2.19 and 2.20. These transistors and other electrical components such as resistors, capacitors and diodes are implanted on the surface of a silicon wafer to produce an integrated circuit (IC). Since the advent of the transistor in the early 1950s, there has been an evolution in the design of logic gates to improve their performance. The objectives of the design changes have been to increase the switching speed of the gate, to improve density by reducing the number of components required for the gate or by reducing the chip area needed for the components, to reduce the noise generated in switching, to improve the drive capability and to reduce power requirements.

2.9.1 Bipolar Gates

The various families of designs for logic gates based on bipolar technologies are listed in Table 2.4. Of these many families, transistor–transistor logic (TTL) and emitter-coupled logic (ECL) are the most commonly used today. A typical NAND gate for the TTL family, illustrated in Fig. 2.23, shows the characteristic features of the design. Note that three emitters share a common base and a collector, and that all of the emitters must be high for the transistor $T1$ to

TABLE 2.4
Families of logic gate design using bipolar technology

TRL	Transistor–resistor logic	Mid 1950s	Discrete components
DTL	Diode–transistor logic	Late 1950s	Diodes replace resistors
RTL	Resistor–transistor logic	Mid 1960s	Integrated circuits
TTL	Transistor–transistor logic	1970s	Wide transistor use
STL	Schottky transistor logic	1970s	Added Schottky diode
ECL	Emitter-coupled logic	1970s	Nonsaturating
IIL	Integrated injection logic	1970s	No resistors

conduct. This sharing of the base and collector reduces the number of components, improves performance and reduces the power required to operate.

ECL technology is the fastest of the bipolar logic gates because it utilizes nonsaturating current steering transistors and reduces the time necessary to charge a saturating-type transistor. This delay time in a saturating-type transistor commonly used in TTL logic is illustrated in Fig. 2.24. A schematic circuit diagram showing an ECL gate, presented in Fig. 2.25, indicates the key features employed in the design. The collector is grounded and the voltages placed on the base and emitter are negative. This arrangement insures that the collector–base junction is in reverse bias and that base saturation is avoided.

A detailed description of this circuit is beyond the scope of this text; however, it is possible to note that the input side of the ECL gate is similar to a differential amplifier where the input voltages A and B are compared to the reference voltage on T_3. Depending on the results of this comparison, the current I_E will flow through either T_2 or T_3; thus, these two transistors are known as current steering transistors. The resulting voltage drops: either $I_E R_2$ controls T_4 or $I_E R_3$ controls T_5. This dual feature permits two types of gates, the OR and NOR, to be incorporated into the same circuit.

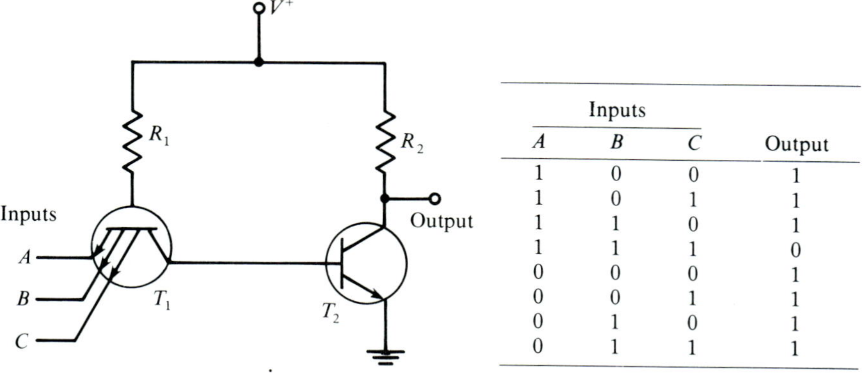

Inputs			
A	*B*	*C*	Output
1	0	0	1
1	0	1	1
1	1	0	1
1	1	1	0
0	0	0	1
0	0	1	1
0	1	0	1
0	1	1	1

FIGURE 2.23
A TTL NAND gate showing three emitters sharing a common base and a collector.

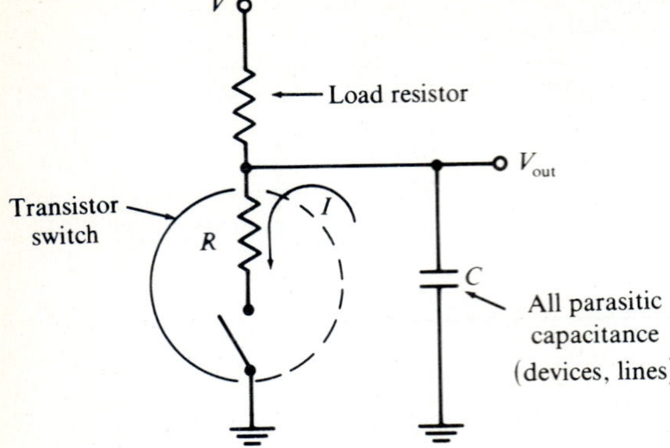

FIGURE 2.24
Time delay in operating a saturated transistor is dependent on the RC constant of the circuit loop.

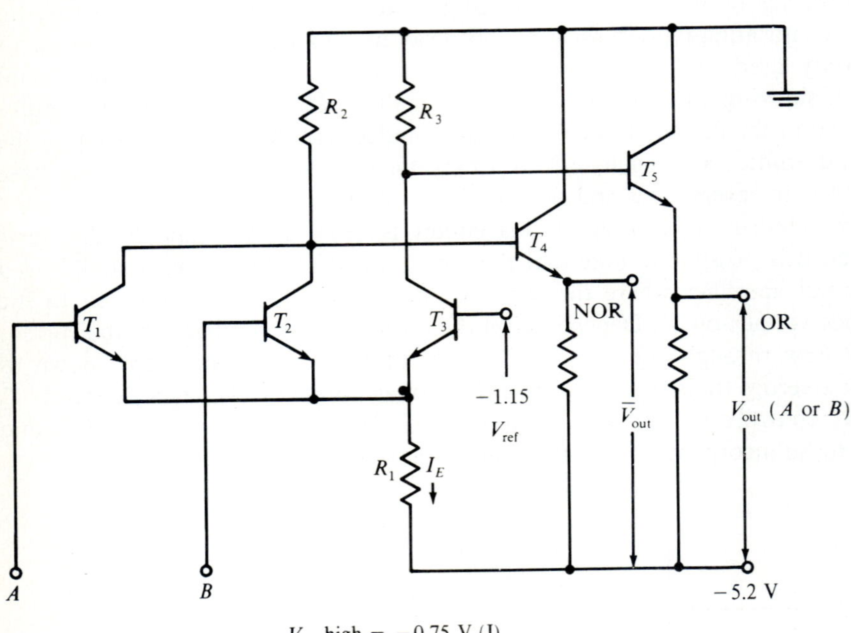

FIGURE 2.25
Schematic circuit showing the nonsaturated operation of T_2 and T_3, the current steering transistors in ECL logic.

ECL, in addition to the advantage of good switching speed, has good noise immunity and good drive capability. The most significant disadvantages are its large power dissipation and its incompatibility with TTL-type gates.

2.9.2 MOSFET Gates

MOSFET circuits were introduced after bipolar gates, in the late 1960s, with a FET gate incorporating a resistive load as shown in Fig. 2.26*a*. The resistor of this circuit was replaced with a second MOSFET transistor as shown in Fig. 2.26*b* to lower the resistance of the gate and to improve performance. Next, the enhancement-type load transistor was replaced with a depletion-type transistor as shown in Fig. 2.26*c* to accomplish switching at lower gate voltages. Finally, two complementary transistors, one PMOS and the other NMOS, were connected in series as illustrated in Fig. 2.26*d* to minimize power dissipation. This configura-

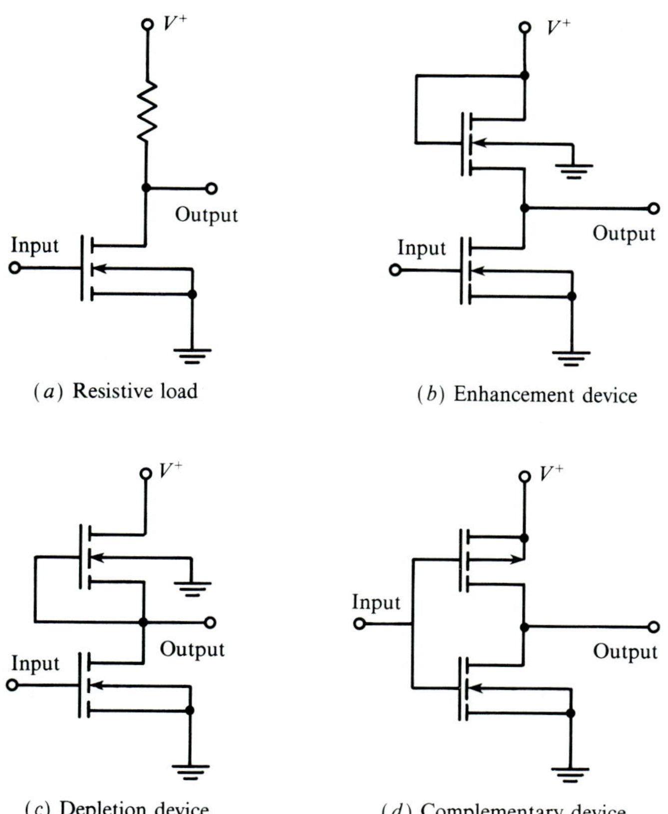

FIGURE 2.26
Progressive development of MOSFET logic devices.

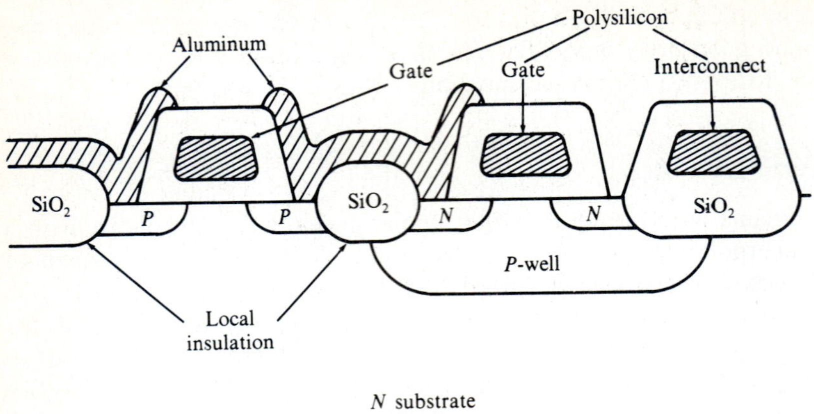

FIGURE 2.27
Fabrication details for CMOS transistors.

tion is the most popular of the MOSFET devices in use today and it is known as CMOS.

When the two complementary transistors are connected in series they operate from a single input and very little power is dissipated because the two transistors require input signals of opposite polarity for conduction to occur. For this reason they are never both on at the same time and the supply voltage V^+ is never connected to ground through the transistors. The power dissipated depends on the frequency of switching with 1 mW at 1 MHz typical in CMOS gates with a 2 μm feature size.

The fabrication details of a CMOS gate, which are presented in Fig. 2.27, show the use of polysilicon (a highly doped amorphous silicon) as the gate material and the use of relatively shallow *N* and *P* channels. Since isolation channels between transistors are not required, the density (gates/unit area) that can be achieved in MOS technology is higher than that which can be obtained with bipolar technology.

2.10 POWER DELAY PRODUCT

The power delay product is a figure of merit, which can be used to judge the performance of a transistor. The power refers to the heat dissipated by the transistor and the delay pertains to the switching time. Because the design goal is to minimize both the power and the delay, high performance is associated with a very low power delay product. Typical delay time–power relationships for NMOS, CMOS and bipolar transistors with 1 μm feature size are shown in Fig. 2.18. These curves show that switching time can be decreased by increasing the power dissipation particularly for bipolar transistors. The reduced switching time

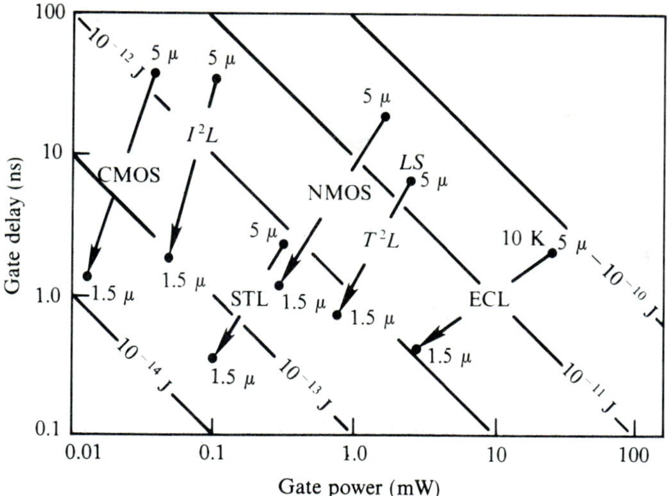

FIGURE 2.28
Reducing feature size improves the power delay product for all logic technologies.

is important because it permits circuits to be driven at high frequencies; however, the higher powers complicate the problem of cooling the chips to maintain the required reliability.

Another method of reducing both delay and power is to reduce the size of the channels when the chip is manufactured. The effect of reducing feature size from 5 to 1.5 μm on the power delay product for several different types of logic gates is illustrated in Fig. 2.28. It should be noted that the switching speeds associated with feature sizes of 1–1.5 μm are in the subnanosecond range.

2.11 SCALE OF INTEGRATION

Since the introduction of integrated circuits in 1959, there has been a rapid increase in the number of components (transistors, resistors, capacitors and diodes) that can be placed on a single chip. The dramatic increase in the number of components, which is presented in Fig. 2.29, is due to a decrease in feature size and an increase in the size of the area used for the chip. Today, an advanced chip may support as many as a million components with feature sizes of about 1 μm. Continued progress in reducing feature size to about 0.25–0.5 μm is anticipated in the coming decade until limits on the lithographic process slow further development.

The increase in components per chip reflects the scale of integration of the components and circuits placed on a single die of silicon. The scale of integration began in 1959 and continues to develop as indicated in Table 2.5. The reduction in the feature size permits a marked increase in the components per chip and affects the electrical characteristics of the transistors. To illustrate the effect of

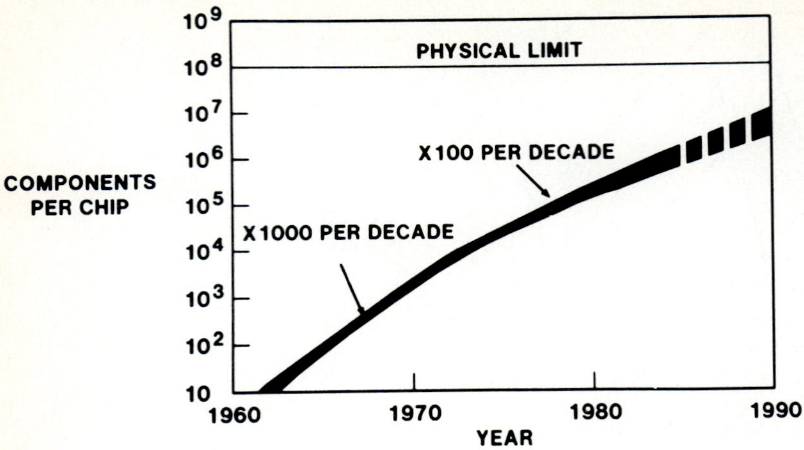

FIGURE 2.29
Components placed on a single chip as a function of time. (*Copyright, 1984, AT & T Bell Laboratories. Reprinted by permission*)

feature size and its reduction, consider making the smallest feature on the chip twice as small. The change in the electrical parameters of the transistor due to this reduction in feature size is given in Table 2.6.

The very favorable changes due to scaling down the feature size includes reduced device power, lower power-delay product, lower currents and improved reliability due to the increased number of on chip connections. According to these scaling laws the chip power does not change with feature size; however, in actuality the chip power has increased. The chip areas have increased and the scaling laws are not followed precisely in fabrication or in specifying the supply voltages. A more realistic scaling rule for chip power for bipolar transistors would be $K^{1.5}$ rather than K^0, where K is the scaling factor. The improved performance of dense logic chips with 10^3–10^4 gates is striking and every effort is

TABLE 2.5
Definition of the scale of integration

Scale	Notation	Components / chip	Year
Zero scale	ZSI	1	1959
Small scale	SSI	2–50	1960–1965
Medium scale	MSI	50–1000	1965–1969
Large scale	LSI	10^3–10^5	1969–1980
Very large scale	VLSI	10^5–10^7	1980–1990
Ultra large scale	ULSI	10^7–10^9	Future
Giga scale	GSI	$\geq 10^9$	Future

TABLE 2.6
Changes in electrical parameters of a transistor due to reducing feature size by a factor of $K = 2$

Parameter	Scale factor	Change
Supply voltage	K^{-1}	$\frac{1}{2}$
Current	K^{-1}	$\frac{1}{2}$
Fields	K^{0}	1
Devices/chip	K^{2}	4
Device power	K^{-2}	$\frac{1}{4}$
Chip power	K^{0}	1
Device delay	K^{-1}	$\frac{1}{2}$
Delay power product	K^{-3}	$\frac{1}{8}$
Line resistance	K^{1}	2

being made by the international semiconductor industry to develop logic chips with a very high gate count and memory chips with a very high bit count.

Two problems arise due to the increased gate count on a logic chip. First, the number of input and output (I/O) connections that must be made to the chip increases dramatically, making the design of the chip carrier much more difficult. Second, the chip power, particularly for the bipolar transistors, increases to the point where very innovative chip cooling techniques become mandatory. To illustrate this point consider three different bipolar logic chips with 3000, 10,000 and 40,000 gates that could be fabricated with feature size of about 1.5, 1.0 and 0.5 μm, respectively. With 3–5 mW per gate plus additional power for the off chip drivers, power dissipation of 35, 45 and 150 W/chip may be anticipated as the gate count increases from 3000 to 40,000 gates. These very high powers will require markedly different methods for transferring the heat generated to the environment.

2.12 I/O COUNT AND RENT'S RULE

Bonding pads are provided around the perimeter of the chip to permit input and output signals to enter or leave the chip. Bonding pads are also provided for the supply voltages and the ground lines that supply the power to the chip. These bonding pads, shown in Fig. 2.30, are terminals for the interconnections required to wire a group of chips together to provide a digital logic function or, if sufficient chips are involved, a complete electronic system.

The number of I/O increases with the number of gates and it is important to estimate the number required on the chip, on the circuit board or the subassembly, which consists of several circuit boards, and a back panel. A common method for estimating the I/O count is to utilize Rent's rule, an empirical equation developed in the early 1950s. Rent's rule relates gate count

FIGURE 2.30
Top surface of a chip showing the bonding pads that serve as the chip I/O terminals. (*Courtesy of International Micro Industries, Inc.*)

and I/O count for random logic connections as

$$N_{\mathrm{I/O}} = aN_G^b \tag{2.6}$$

where $N_{\mathrm{I/O}}$ is the number of I/O
N_G is the number of gates involved
a is a constant between 0.5 and 1.5
b is a constant between 0.4 and 0.8

The value of the exponent b in Rent's rule is very important because it has a marked effect on the $N_{\mathrm{I/O}}$ particularly when N_G is large as shown in Table 2.7. The value of the exponent b required for a given system depends strongly on performance and the digital function. For high performance systems with large digital functions and with a large amount of random logic, a value of b between 0.6 and 0.7 can be anticipated. For lower performance systems or if the digital function can be partitioned onto a single custom designed chip with a large degree of structured logic, then b between 0.5 and 0.6 is usually adequate. The IBM thermal conducting module in the 3081 computer, which contains 100 logic chips with about 700 gates/chip for a total $N = 70{,}000$ gates, is designed with an 1800 pin grid array. This configuration corresponds to a $b = 0.67$. The Motorola 68,000 microprocessor with about 5000 gates is mounted in a dual in-line package

TABLE 2.7
$N_{I/O}$ as a function of N_G for $a = 1$ and different values of b

	$N_{I/O}$			
N_G	$b = 0.5$	0.6	0.7	0.8
10	3	4	5	6
50	7	10	15	23
100	10	16	25	40
500	22	42	77	144
10^3	32	63	126	251
5×10^3	71	166	388	910
10^4	100	251	631	1585
5×10^4	234	660	1947	5743
10^5	316	1000	3162	10000

(DIP) with 64 leads corresponding to a $b = 0.5$. This chip represents a lower performance level; however, the performance is improved to a considerable degree by using structured logic, where major digital functions are fabricated on a single chip. Structured logic markedly lowers I/O requirements and improves performance of the system since lead length is also reduced.

It is clear that the new LSI and VLSI chips will have a much higher I/O count than the SSI and MSI chips of the past and that design for performance will require chip carriers and circuit boards to accommodate $N_{I/O}$ between 200 and 500 or higher.

2.13 MEMORY DEVICES

Digital logic circuits process extremely large amounts of information and it is necessary to store large amounts of data during and after processing to fully utilize the capabilities of a digital system. A large number of different storage devices are employed in the process including:

- Memory chips
  - RAM—random access memory
  - ROM—read only memory
  - PROM—programmable read only memory
- Magnetic disks
  - Floppy
  - Hard
- Optical disks
- Magnetic tapes

Storage of data in memory requires the capability to write or to transfer information to the memory devices and to read or retrieve this information from memory at some later time. All of the memory devices in the preceding list have

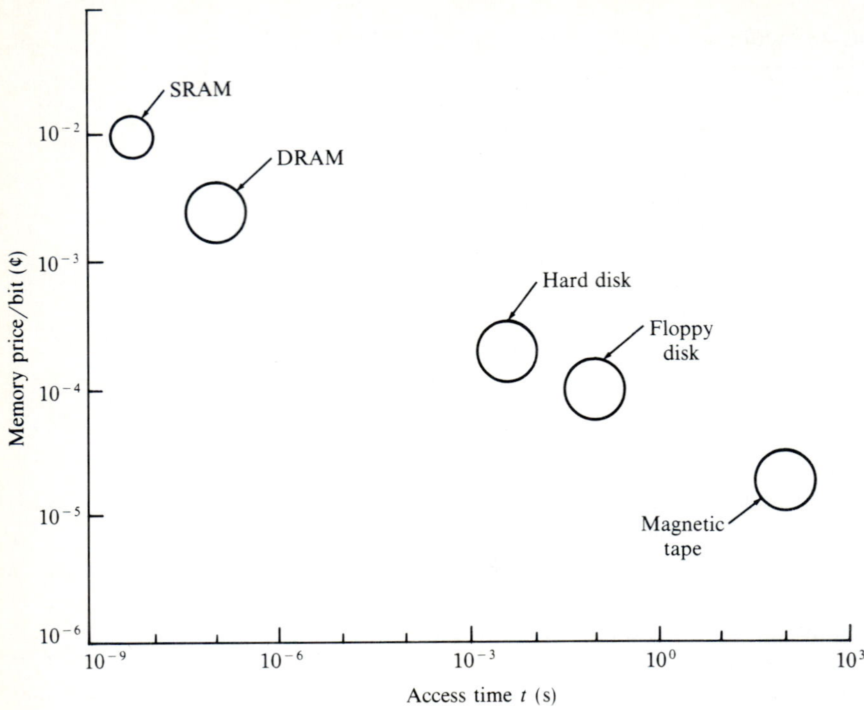

FIGURE 2.31
Access time–price trade-off for memory devices.

the capability of reading and writing. However, the ROM and PROM are basically used only for reading because the information stored in these devices is essentially permanent and is written to memory in initializing the digital system.

The important characteristics of memory devices include storage capacity, access time, reliability and cost. Access time and cost of the various devices trade off as is indicated in Fig. 2.31. The lowest cost device is a tape where information is stored by magnetizing a thin layer of a magnetic material that is coated on the surface of a tape. The access time is very long, ranging from about 10–100 s, because of the time required to position the tape relative to the read or write heads. The disk drives provide significantly better access time 10–100 ms while retaining very high capacity and very low cost. The improved performance is achieved by using flying read–write heads that are positioned rapidly over the entire surface of the rotating disks. Disk memories use a hard disk for very high capacities 10–100M bytes and short access times and floppy disks for lower storage capacity 200K–1M bytes and longer access time.

For higher speed applications it is not possible to use electromechanical means (moving magnetic surfaces with read–write heads) to store information since the access time is so long that the performance is degraded. For these applications the information is stored electronically with access times that range from 10–100 ns. The dynamic RAM is the most common way to read and write

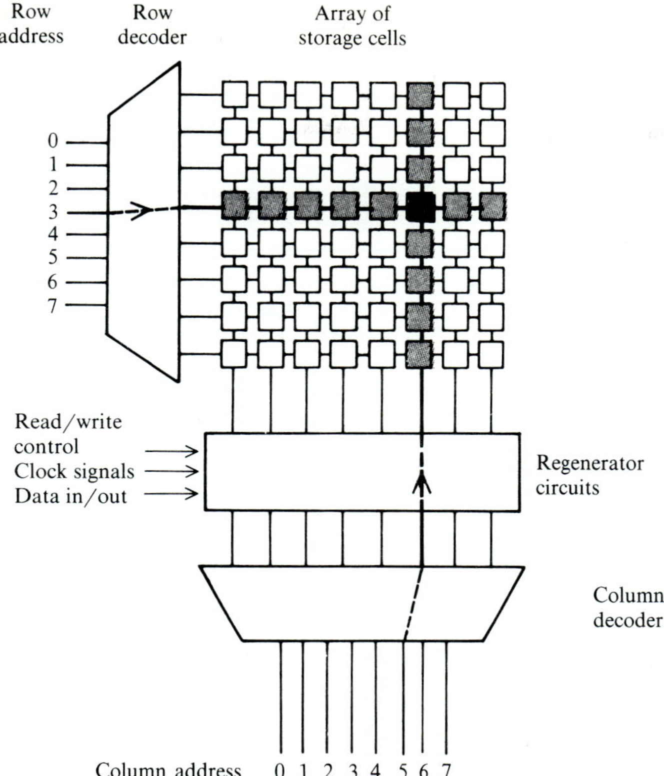

FIGURE 2.32
Array of storage cells for a dynamic RAM.

data with relatively low cost. The dynamic RAM consists of an array of storage cells as illustrated in Fig. 2.32. Each cell provides a memory location where either 0 or 1 may be stored. A particular cell is located by a row and a column address and is activated only when both the row and column lines are high. The switching of a unit cell is illustrated in Fig. 2.33 where a MOS transistor is shown connected to row and column selection lines.

The information is stored in the unit cell by charging a small capacitor (10–15 fF). If the capacitor is charged the unit cell is storing 1, but if there is no charge the unit cell is storing 0. Because the unit cell loses charge due either to leakage or voltage loss that occurs during reading, it must be recharged periodically (every 1–2 ms the entire array of the cell is replenished). The RAM is controlled by a regeneration circuit that periodically samples each cell and replenishes the capacitors as required. After replenishment the regeneration circuit permits reading and writing to the cell as identified by the row and column selection lines.

The capacity of dynamic RAMs has increased at a rapid rate while the cost per bit has dropped dramatically over the past 20 yr. In 1988 the cost of a 256

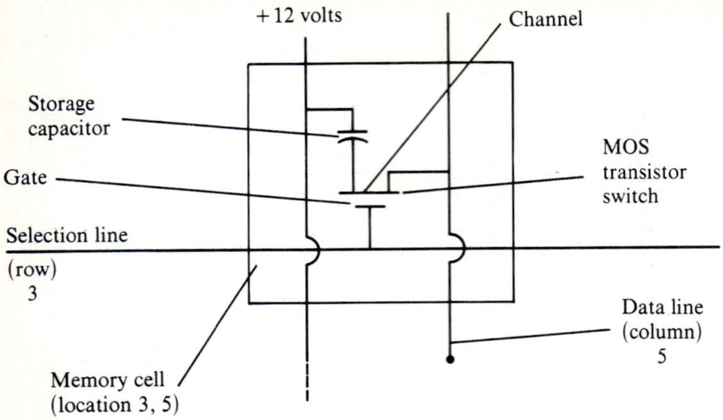

FIGURE 2.33
A unit storage cell of a dynamic RAM.

Kbit (262,144 × 1 bits) RAM chip was available at about $2.50 or at a cost of 1¢/Kbit. The chip carrier for the RAM is almost always a DIP because the number of I/O required for even large capacity RAM is low. Heat generation is also small because only one cell is active (replenish, read or write) at any instant. The primary requirement in packaging memory chips is to achieve high density in a location in close proximity to the logic circuits that address the RAM.

REFERENCES

1. Select chapters in *Microelectronics*, W. H. Freeman, San Francisco, Calif., 1977, which include: R. N. Noyce, "Microelectronics," p. 2–11; J. D. Meindl, "Microelectronic Circuit Elements," p. 12–25; W. C. Holton, "The Large Scale Integration of Microelectronic Circuits," p. 26–39; D. A. Hodges, "Microelectronic Memories," p. 54–65.
2. Malmstadt, H. V. and C. G. Enke: *Digital Electronics for Scientists*, Benjamin, Menlo Park, Calif., 1969.
3. F. Keller, M. J. Skove and W. E. Gettys, *Physics*, McGraw-Hill, N.Y., 1989.
4. W. F. Smith, *Principles of Materials Science and Engineering*, McGraw-Hill, N.Y., 1989.
5. J. B. Russell, *General Chemistry*, McGraw-Hill, N.Y., 1980.

EXERCISES

2.1. Determine the resistance of a No. 22 gage copper wire of length (*a*) 3 m, (*b*) 6 in., (*c*) 10 ft, (*d*) 100 yd.

2.2. A conducting trace 10 mil wide and 3 in. long is produced by photoetching a laminated circuit board with a copper cladding having a thickness of 2 oz/ft^2. Determine the resistance of the trace.

2.3. Prepare a drawing of the atomic structure of insulating materials such as (*a*) Al_2O_3, (*b*) MgO and (*c*) Si_3N_4. Draw all of the shells in each atom and identify the shared electrons involved in the covalent bonding.

2.4. Prepare a drawing of the atomic structure of germanium showing the shells in adjacent atoms and identifying the electrons involved in the covalent bonding evident in this structure.

2.5. Prepare a drawing of the atomic structure of gallium arsenide (GaAs) that is also used as a semiconductor. Identify the shells and describe the electrons that serve to covalently bond the two atoms.

2.6. Describe the difference between intrinsic and extrinsic conduction.

2.7. Reference a periodic table and list the elements that are considered as (*a*) type III elements and (*b*) type V elements.

2.8. Prepare a drawing of the lattice structure of silicon (Si) with a dopant atom boron (B) included in the lattice. Describe the conducting characteristics of this new material. Will the resistivity of the new material depend on the number of boron atoms included in the silicon lattice.

2.9. Repeat Exercise 2.8 but use arsenic (As) as the dopant.

2.10. A batch of *P*-type silicon is to be formulated with a resistivity of 10^{-1} Ω-cm. If the total batch has a volume 10 liter, determine the volume of boron that is mixed with the silicon prior to melting.

2.11. Repeat Exercise 2.10 but change the resistivity to 5 Ω-cm.

2.12. For a *P-N* junction describe the depletion layer. Indicate how the thickness of the depletion layer is adjusted in a semiconductor device.

2.13. What is the mechanical equivalent to a diode.

2.14. Reference Fig. 2.8 and note the bias voltages applied to the electrodes. Explain the direction of hole, electron and current flow shown in this illustration.

2.15. For a *P-N* junction forming a diode, define forward bias and reverse bias. Why is a leakage current observed in a diode subjected to reverse bias. How large are the leakage currents in comparison to the forward currents.

2.16. Consider the transistor switch shown in Fig. 2.11 and use Ohm's law to explain why the output voltage is either high or low when the switch is off or on.

2.17. A wafer 6 in. in diameter is used to fabricate chips with a die size of 5 × 7 mm. Determine the number of dies that can be processed on each wafer. If the yield from the process is 42%, estimate the number of good dies that can be obtained from each wafer. If 20 wafers are processed in a boat of wafers, find the number of good dies expected.

2.18. Prepare a table showing the voltage (high or low) on the source, gate and drain and the resulting switch state for a NMOS transistor of (*a*) enhancement mode and (*b*) depletion mode.

2.19. Repeat Exercise 2.18 for a PMOS transistor.

2.20. In Fig. 2.17 a section view of a CMOS transistor that utilizes series-connected NMOS and PMOS is shown. Both of these transistors operate in the enhancement mode. Prepare an equivalent drawing for a CMOS transistor fabricated from NMOS and PMOS transistors operating in the depletion mode.

2.21. Perform a trade-off analysis citing the relative advantages and disadvantages of bipolar, NMOS and CMOS transistors.

2.22. The EXCLUSIVE-OR gate is shown in Fig. E2.22. Construct a truth table for this logic element.

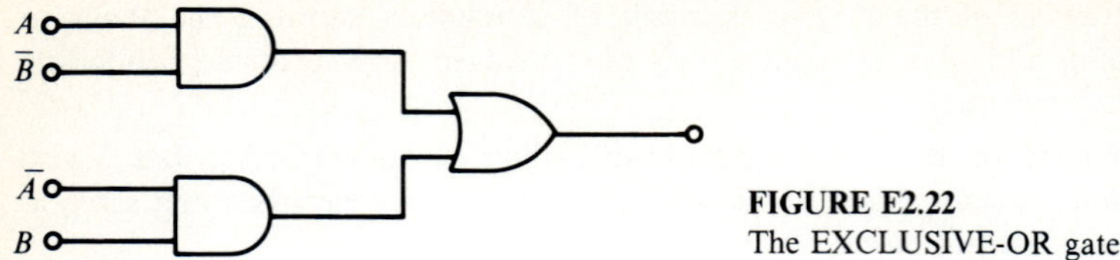

FIGURE E2.22
The EXCLUSIVE-OR gate.

2.23. A logic gate half-adder is shown in Fig. E2.23. Trace inputs AB of 00, 01, 10 and 11 and give the corresponding outputs on the sum and carry terminals.

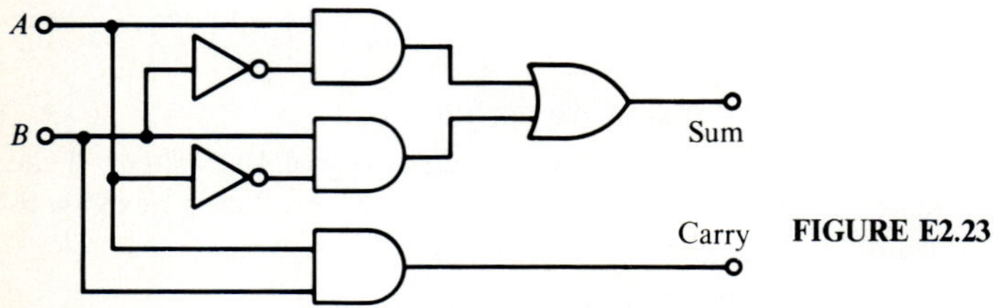

FIGURE E2.23

2.24. A binary adder shown in Fig. 2.22 has three inputs A, B and C. For inputs of 000, 001, 010, 011, 100, 101, 110 and 111 trace through the circuit and indicate the output on the sum and carry bits.

2.25. Discuss the relative merits of TTL and ECL gates and then describe the disadvantages for each.

2.26. Describe the advantages and disadvantages of CMOS used in the construction of logic circuits.

2.27. What is the unit used for the power delay product.

2.28. Estimate the power-delay product for a CMOS gate with 1.5 μm feature size. Also indicate the delay time and the gate power for this gate.

2.29. Repeat Exercise 2.28 for an ECL gate with the same feature size.

2.30. Estimate when the components placed on a single chip will reach 10^8.

2.31. Following Rent's rule estimate the number of I/O required on a circuit card with 10,000 logic gates. Take $a = 1$ and $b = 0.5$.

2.32. For the same circuit card as described in Exercise 2.31, plot the I/O required if b is varied from 0.3–0.9. Keep $a = 1$ in this exercise.

CHAPTER 3

CIRCUIT ANALYSIS

3.1 INTRODUCTION

Logic circuits, used widely in electronic systems, involve gates that are switched from say a low state to a high state to produce a voltage pulse with a rise time t_r as illustrated in Fig. 3.1. The frequency components of the pulse depend on the rise time and the predominate frequencies can be determined from a Fourier series expansion representing the pulse. The Fourier series can be written as

$$V(t) = \sum_{n=0}^{\infty} (A_n \cos n\omega t + B_n \sin n\omega t) = \sum_{n=0}^{\infty} C_n e^{j\omega t} \tag{3.1}$$

where A_n, B_n and C_n are coefficients determined to fit the pulse. Exercises given at the end of the chapter illustrate the changes in frequency content of the pulse with variations in the rise times. If the gate switches with a very short rise time t_r, about 1 ns or less, then very high frequencies must be transmitted without significant distortion if the fidelity of the pulse is to be maintained. On the other hand, if the gate switches slowly with $t_r = 1$ μs or more, then the predominate frequencies are much lower. The pulse shape is easier to maintain and ringing (oscillation) with transient voltages which over- and/or undershoot the steady state voltage is minimized.

Two very different types of circuit analyses are employed in circuit design depending on the frequencies and the distance of propagation that are involved. For frequencies of less than about 100 MHz with relatively short propagation distances, a traditional analog circuit analysis is usually sufficient to adequately describe the circuit behavior. For frequencies above 100 MHz with longer propagation distances, the circuit wiring becomes much more critical. Distributed

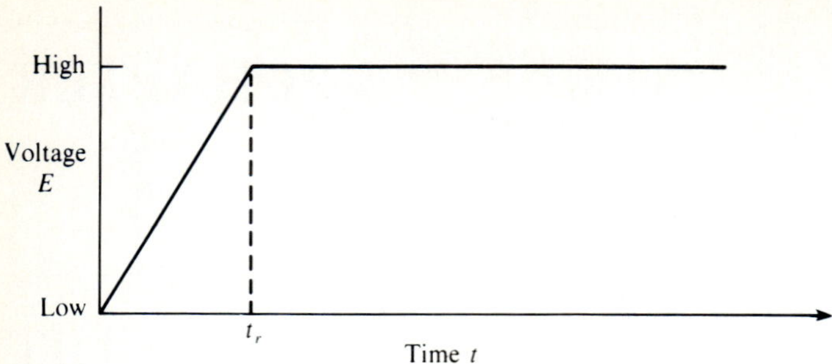

FIGURE 3.1
Voltage pulse produced by switching a logic gate from the low state to the high state.

inductance and capacitance present in all circuit lines (wires) become significant at high frequencies and affect the behavior of the circuit. Transmission line theory is introduced to account for the effects of distributed capacitance and inductance on the observed behavior of pulses propagating over extended circuits.

3.2 THE ANALOG METHOD OF ANALYSIS

The analog method of analysis is usually employed to treat sinusoidal signals where the frequency is fixed. This approach can also be used to advantage in treating pulse-like signals associated with digital systems, providing the predominant frequencies involved in the Fourier representation of the pulse are less than about 100 MHz. The upper limit of the analog method of analysis depends on the length of the lines, the magnitude of the distributed inductance and capacitance and the criticality of the arrival times of the pulses involved.

As an example of the analog method of analysis, consider the circuit shown in Fig. 3.2. This circuit is driven by a power supply that provides a sinusoidal input voltage E_i and current I_i, which can be written as

$$E_i = E_a e^{j\omega t} \quad \text{and} \quad I_i = I_a e^{j\omega t} \tag{3.2}$$

where E_a and I_a are amplitudes of voltage and current, respectively, and ω is the circular frequency of the signal.

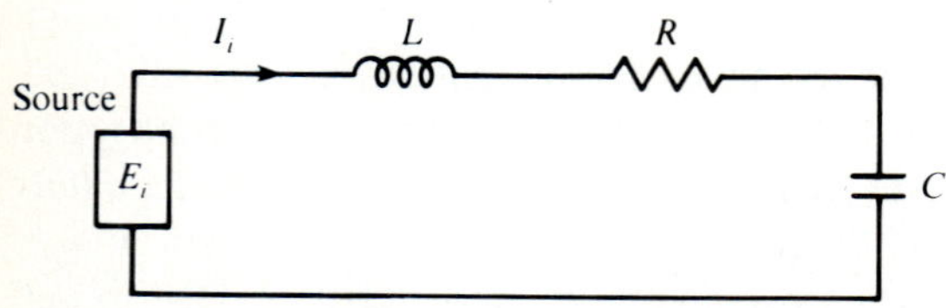

FIGURE 3.2
An analog circuit with discrete components including a resistor R, inductor L and capacitor C.

To analyze this circuit determine the voltage drop across each discrete component and neglect the effect of the wiring by assuming that the wiring impedance is small compared to the impedance of the discrete components R, C and L. The voltage drop across the resistor is then given by

$$E_R = I_i R = I_a R e^{j\omega t} \tag{3.3}$$

The voltage drop across the inductor is given by

$$E = L\,dI/dt \tag{3.4}$$

Differentiating Eq. (3.2) and substituting the result into Eq. (3.4) gives

$$E_L = j\omega I_a L e^{j\omega t} \tag{3.5}$$

The voltage drop across the capacitor is

$$E_C = q/C \tag{3.6}$$

where q is the charge. Since

$$I = dq/dt \tag{3.7}$$

it is clear from Eqs. (3.2), (3.6) and (3.7) that

$$E_C = \frac{1}{C}\int I\,dt = \left(\frac{I_a}{j\omega C}\right)e^{j\omega t} = -j\left(\frac{I_a}{\omega C}\right)e^{j\omega t} \tag{3.8}$$

If the voltage drops across the components are summed and equated to the input voltage, we obtain the following relation between the amplitudes of the voltage and current:

$$E_a = [R + j(\omega L - 1/\omega C)]\,I_a \tag{3.9}$$

Examination of Eq. (3.9) shows that a complex function with real and imaginary parts is involved in the relation between E_a and I_a. If this complex function is divided into its real and imaginary parts as illustrated in Fig. 3.3, a

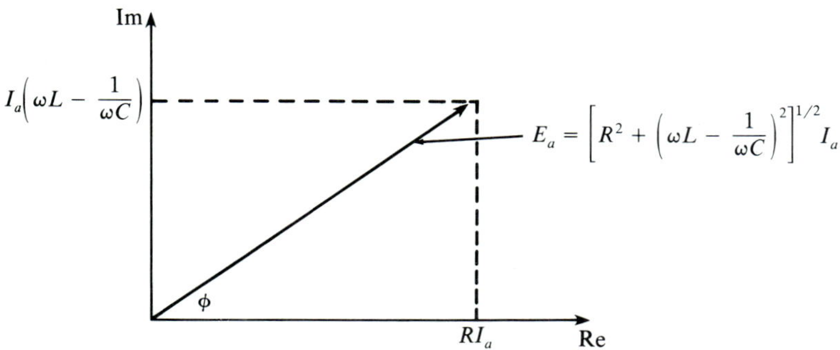

FIGURE 3.3
Representation of the impedance components on the complex plane.

more functional expression for E_a in terms of I_a is obtained:

$$E_a = ZI_a \tag{3.10}$$

where

$$Z = \left[R^2 + (\omega L - 1/\omega C)^2\right]^{1/2} \tag{3.11}$$

is the total impedance of the circuit. While Eqs. (3.10) and (3.11) define the amplitudes of the voltage and the current, the phase of the voltage relative to the current remains to be determined. Reference to Fig. 3.3 shows the phase angle ϕ, which is given by

$$\phi = \tan^{-1}\left(\frac{\omega L - 1/\omega C}{R}\right) \tag{3.12}$$

If $\phi > 0$ as depicted in Fig. 3.3, the voltage leads the current and the complete expression for $E(t)$ is

$$E = \left[R^2 + (\omega L - 1/\omega C)^2\right]^{1/2} I_a e^{j(\omega t + \phi)} \tag{3.13}$$

Equation (3.13) shows the amplitude of the voltage and the relative phase of the voltage with respect to the current. Since the frequency ω is limited, the effects of distributed inductance and capacitance due to the circuit wiring may be neglected. The results obtained are independent of the wiring parameters such as its type, length and gage. Probes attached to the circuit to measure voltage and current from the source with signals displayed on an oscilloscope would display I_a, E_a and the phase angle ϕ. These values are controlled by the values of L, R and C associated with the discrete components if the distributed inductance, resistance and capacitance associated with the wiring is small. For relatively low frequencies and short lines the discrete values of R, C and L associated with individual components in the circuit are dominant and the traditional analog approach is adequate to describe circuit behavior.

3.3 TRANSMISSION LINE THEORY

At very high frequencies or for very long lines, it is necessary to consider the wiring between components because the distributed inductance and capacitance of the wiring affects the behavior of the circuit. To show the effect of distributed electrical parameters, consider an element dx of the wiring, illustrated in Fig. 3.4, that contains a voltage source on one end and an open (nothing) on the other end. In this circuit L' is the distributed inductance, R' is the distributed resistance and C' is the distributed capacitance. These quantities are all specified per unit length of the line. The term G' is the inverse of the distributed resistance

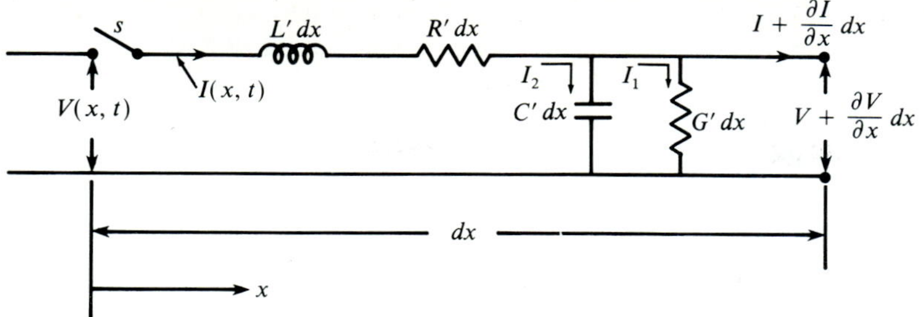

FIGURE 3.4
Equivalent circuit for a transmission line of length dx.

to ground per unit length of line:

$$G' = 1/R'_g \tag{3.14}$$

where R'_g is the distributed resistance to ground of the insulation separating the two conductors.

To determine the voltage and the current flow in this simple line segment, close the switch at $t = 0$ and write the voltage drop over the length dx of the line as

$$V - (V + \partial V/\partial x)\, dx = (R'I + L'\, \partial I/\partial t)\, dx$$

which reduces to

$$\partial V/\partial x = -(R'I + L'\, \partial I/\partial t) \tag{3.15}$$

Consider next, the current flow and note that

$$I - [I + (\partial I/\partial x)\, dx] = I_1 + I_2 = (G'V + C'\, \partial V/\partial t)\, dx$$

which can be simplified to give

$$\partial I/\partial x = -(G'V + C'\, \partial V/\partial t) \tag{3.16}$$

Differentiate Eq. (3.15) with respect to x and Eq. (3.16) with respect to t to obtain

$$\partial^2 V/\partial x^2 = -(R'\, \partial I/\partial x + L'\, \partial^2 I/\partial x\, \partial t) \tag{3.17}$$

$$\partial^2 I/\partial x\, \partial t = -(G'\, \partial V/\partial t + C'\, \partial^2 V/\partial t^2) \tag{3.18}$$

Finally, substitute Eqs. (3.16) and (3.18) into Eq. (3.17) to find

$$\partial^2 V/\partial x^2 = R'G'V + (R'C' + L'G')\, \partial V/\partial t + L'C'\, \partial^2 V/\partial t^2 \tag{3.19}$$

This partial differential equation describes the voltage current relationship in a transmission line in terms of its distributed parameters R', G', L' and C'. It is known as the telegrapher's equation and is the basic relation governing transmission line theory. Transmission line theory is used to characterize the behavior of pulses propagating over circuit lines from one digital device or subsystem to another when the frequency components are high or if the distances involved are large.

3.4 SINUSOIDAL SIGNAL PROPAGATION ON A TRANSMISSION LINE

To explore the behavior of a transmission line with an open end, consider the case of a sinusoidal power supply that provides an input voltage and an input current, which may be written as

$$V(x,t) = V(x)e^{j\omega t} \quad \text{and} \quad I(x,t) = I(x)e^{j\omega t} \tag{3.20}$$

Substituting Eq. (3.20) into Eqs. (3.15) and (3.16) gives

$$dV/dx = -(R' + j\omega L')I(x) \tag{3.21a}$$

$$dI/dx = -(G' + j\omega C')V(x) \tag{3.21b}$$

Differentiate Eq. (3.21a) with respect to x and substitute Eq. (3.21b) into the resulting equation to yield

$$d^2V/dx^2 - (R' + j\omega L')(G' + j\omega C')V = 0 \tag{3.22}$$

This is an ordinary second-order differential equation, which has a solution of the form

$$V(x) = Ae^{-\lambda x} + Be^{\lambda x} \tag{3.23}$$

The expression for λ is given by the roots of the characteristic equation as

$$\lambda = [(R' + j\omega L')(G' + j\omega C')]^{1/2} \tag{3.24}$$

Substituting Eq. (3.23) into Eq. (3.20) gives

$$V(x,t) = (Ae^{-\lambda x} + Be^{\lambda x})e^{j\omega t}$$

which can be rewritten as

$$V(x,t) = Ae^{(j\omega t - \lambda x)} + Be^{(j\omega t + \lambda x)} \tag{3.25}$$

At this point, it is important to interpret the solutions given by Eqs. (3.24) and (3.25) because they describe the two most important features of transmission lines. The results of Eq. (3.25) indicate that the sinusoidal voltage divides into two parts $Ae^{(j\omega t - \lambda x)}$ and $Be^{(j\omega t + \lambda x)}$. The part associated with $Ae^{(j\omega t - \lambda x)}$ is the forward wave, which propagates as a damped sinusoid along the positive x axis. The part associated with the $Be^{(j\omega t + \lambda x)}$ is the backward wave that is also a

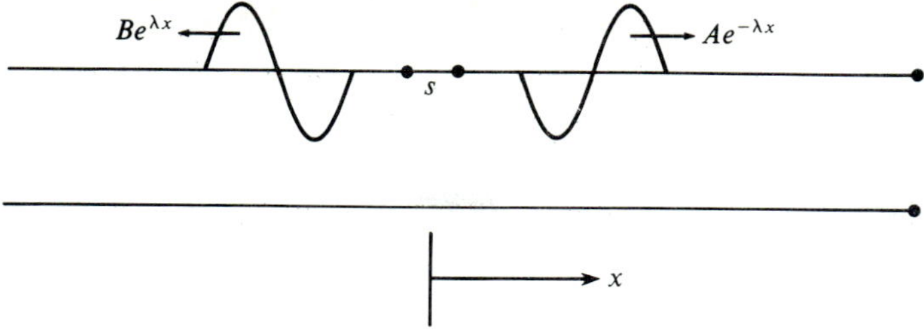

FIGURE 3.5
Propagation of sinusoidal voltage signals in the positive and negative directions along a transmission line.

damped sinusoid propagating along the negative x axis. The waves are illustrated in Fig. 3.5. The amplitude of each sinusoid is given by either $Ae^{-\lambda x}$ or $Be^{\lambda x}$.

The term λ given in Eq. (3.24) shows that the amplitude of the sinusoid decreases as the signal propagates along the line. This decrease in amplitude with propagation in either the positive or negative x direction is due to two different effects, namely, attenuation and dispersion. The term λ is often expressed as

$$\lambda = \alpha + j\beta \tag{3.26}$$

to describe both of these effects. In Eq. (3.26) α is the attenuation constant and β is the phase constant. It is possible to determine α and β in terms of L', R', C' and G' by equating Eqs. (3.24) and (3.26) to obtain

$$\beta^2 = (1/2)\left\{-(R'G' - \omega^2 L'C') \pm \left[(R'G' - \omega^2 L'C')^2 + \omega^2(R'C' + L'G')^2\right]^{1/2}\right\}$$

$$\alpha^2 = (1/2)\left\{(R'G' - \omega^2 L'C') \pm \left[(R'G' - \omega^2 L'C')^2 + \omega^2(R'C' + L'G')^2\right]^{1/2}\right\} \tag{3.27}$$

For a transmission line that is classified as a low loss line where $\omega L' \gg R'$ and $\omega C' \gg G'$, the relations for α and β reduce to

$$\alpha = 0 \quad \text{and} \quad \beta = \omega\sqrt{L'C'} \tag{3.28}$$

From Eq. (3.28) it is evident that the decay of the sinusoid propagating along a low loss line is due to the phase constant β as the attenuation constant α has vanished.

The current $I(x)$ is determined by substituting Eq. (3.23) into Eq. (3.21a) and solving to obtain

$$I(x) = (Ae^{-\lambda x} - Be^{\lambda x})/Z_0 \tag{3.29}$$

where

$$Z_0 = [(R' + j\omega L')/(G' + j\omega C')]^{1/2} \tag{3.30}$$

The term Z_0 is the characteristic impedance of the transmission line. Examination of Eq. (3.29) indicates that two waves carry the current along the line, with the forward wave I_f propagating in the positive x direction with an amplitude $Ae^{-\lambda x}/Z_0$ and the backward wave I_b propagating in the negative x direction with an amplitude of $-Be^{\lambda x}/Z_0$. The negative sign on the backward current term indicates that the sign of I_b is opposite the sign of the backward term V_b for the voltage.

For a low loss line, the characteristic impedance becomes

$$Z_0 = \sqrt{L'/C'} \tag{3.31}$$

The characteristic impedance is purely resistive (i.e., Z_0 is a real number) and the attenuation constant α is zero.

3.5 TERMINATION OF TRANSMISSION LINES

In the previous section the behavior of sinusoidal signals propagating along an infinitely long transmission line was described. In practical applications transmission lines are of finite length and they are terminated in some manner at both ends. The method of termination markedly affects the behavior of the signals and it is important to examine the effects of several different types of line termination.

3.5.1 Open Circuit Termination

Consider a transmission line of length $\mathscr{L}$ as illustrated in Fig. 3.6. The analysis of this line is simplified by placing the voltage source at $x = -\mathscr{L}$ and by placing the open end at $x = 0$. The open circuit condition at $x = 0$ implies that

$$I(0) = 0 \tag{a}$$

Substituting this result into Eq. (3.29) with $x = 0$ gives

$$I(0) = (A - B) = 0 \tag{b}$$

or

$$A = B \tag{c}$$

FIGURE 3.6
Open circuit termination of a transmission line of length $\mathscr{L}$.

Now, substitute Eq. (c) into Eqs. (3.23) and (3.29) to obtain

$$V(x) = A(e^{-\lambda x} + e^{\lambda x}) \tag{d}$$

$$I(x) = A(e^{-\lambda x} - e^{\lambda x})/Z_0 \tag{e}$$

The impedance at the source Z_{so} due to the open line is

$$Z_{so} = V(-\mathscr{L})/I(-\mathscr{L}) \tag{f}$$

Substituting Eqs. (e) and (d) into Eq. (f) gives

$$Z_{so} = Z_0(e^{\lambda \mathscr{L}} + e^{-\lambda \mathscr{L}})/(e^{\lambda \mathscr{L}} - e^{-\lambda \mathscr{L}}) = Z_0 \coth \lambda \mathscr{L} \tag{3.32}$$

For short transmission lines of the type usually found in electronic systems $\lambda \mathscr{L} \ll 1$ and Eq. (3.32) may be approximated by

$$Z_{so} = Z_0/\lambda \mathscr{L} \tag{g}$$

By substituting Eqs. (3.24) and (3.30) into Eq. (g), we obtain

$$Z_{so} = 1/\mathscr{L}(G' + j\omega C') \tag{h}$$

For a low loss line with $G' \ll \omega C'$, it is clear that

$$Z_{so} = -j/wC'\mathscr{L} \tag{i}$$

The results of Eqs. (h) and (i) indicate that the transmission line will not exhibit an infinite impedance to the source even though it is open ended. Current will flow in the line through the distributed capacitance between the line and the ground and through the distributed resistance R'_g to ground.

3.5.2 Short Circuit Termination

Consider the transmission line shown in Fig. 3.7, which is terminated by a short across its end located at $x = 0$ and a sinusoidal voltage source at $x = -\mathscr{L}$. At $x = 0$ the presence of the short indicates that

$$V(0) = 0 \tag{a}$$

Substituting Eq. (a) into Eq. (3.23) gives

$$V(0) = A + B = 0 \tag{b}$$

and

$$A = -B \tag{c}$$

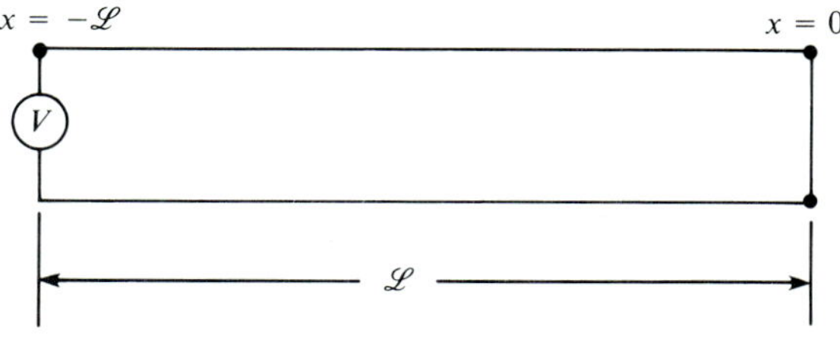

FIGURE 3.7
Short circuit termination of a transmission line of length $\mathscr{L}$.

Substituting Eq. (c) into Eq. (3.29) yields

$$V(x) = A(e^{-\lambda x} - e^{\lambda x}) \tag{d}$$

$$I(x) = A(e^{-\lambda x} + e^{\lambda x})/Z_0 \tag{e}$$

The impedance at the source Z_{ss} due to the short circuit at the end of the line is

$$Z_{ss} = V(-\mathscr{L})/I(-\mathscr{L}) = Z_0(e^{\lambda\mathscr{L}} - e^{-\lambda\mathscr{L}})/(e^{\lambda\mathscr{L}} + e^{-\lambda\mathscr{L}}) = Z_0 \tanh \lambda\mathscr{L} \tag{3.33}$$

Again for short lines $\lambda\mathscr{L} \ll 1$ and Eq. (3.33) can be approximated by

$$Z_{ss} = Z_0\lambda\mathscr{L} \tag{f}$$

Substituting Eqs. (3.24) and (3.30) into Eq. (f) gives

$$Z_{ss} = \mathscr{L}(R' + j\omega L') \tag{g}$$

In this instance, the impedance of the short circuited line depends on the distributed resistance, the distributed inductance, the frequency of the signal and the length of the line.

3.5.3 Transmission Lines with Arbitrary Termination Impedance

In this case consider the transmission line with an arbitrary termination impedance Z_L placed at $x = 0$ and a sinusoidal voltage source at $x = -\mathscr{L}$ as shown in Fig. 3.8. At the termination location,

$$V(0)/I(0) = Z_L \tag{a}$$

From Eqs. (3.23), (3.29) and (a), it is evident that

$$Z_L = Z_0(A + B)/(A - B) \tag{b}$$

Solving Eq. (b) for the ratio B/A gives

$$B/A = (Z_L - Z_0)/(Z_0 + Z_L) = p \tag{3.34}$$

The impedance at the source Z_s due to the combined effect of the line and the

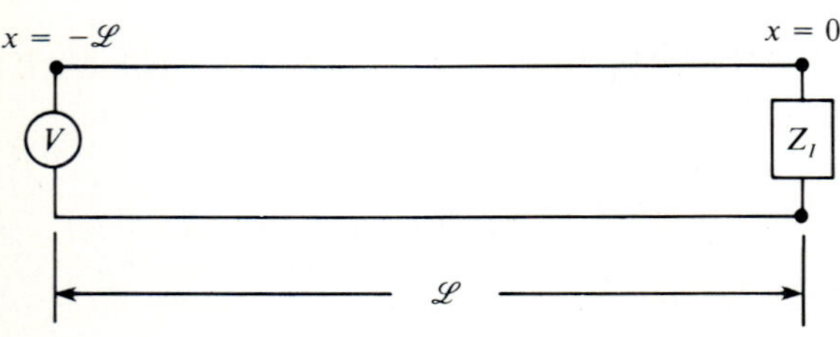

FIGURE 3.8
Transmission line of length $\mathscr{L}$ with an arbitrary termination impedance Z_L.

termination load is

or

$$Z_s = V(-\mathscr{L})/I(-\mathscr{L})$$

$$Z_s = Z_0(Ae^{\lambda\mathscr{L}} + Be^{-\lambda\mathscr{L}})/(Ae^{\lambda\mathscr{L}} - Be^{-\lambda\mathscr{L}}) \tag{c}$$

Substituting Eq. (3.34) into Eq. (c) yields

$$Z_s = Z_0(e^{\lambda\mathscr{L}} + pe^{-\lambda\mathscr{L}})/(e^{\lambda\mathscr{L}} - pe^{-\lambda\mathscr{L}}) \tag{3.35}$$

which can also be written as:

$$Z_s = Z_0(Z_L + Z_0 \tanh \lambda\mathscr{L})/(Z_L \tanh \lambda\mathscr{L} + Z_0) \tag{3.36}$$

The behavior of this line clearly depends on the value of p given by Eq. (3.34). For $Z_L \to \infty$, the open circuit case, $p = 1$ and Eq. (3.35) reduces to Eq. (3.32). For $Z_L = 0$, the short circuit case, $p = -1$ and Eq. (3.35) reduces to Eq. (3.34). For $Z_L = Z_0$, the matched termination case, where the impedance of the load is equal to the characteristic impedance of the line, $p = 0$ and Eq. (3.36) becomes

$$Z_s = Z_0 = Z_L \tag{3.37}$$

In this case the impedance at the source is exactly the same as the load impedance.

3.6 PULSE PROPAGATION ALONG A LOW LOSS TRANSMISSION LINE

For a low loss line where R' and G' may be neglected, the telegrapher's equation (3.9) reduces to the wave equation

$$\partial^2 V/\partial t^2 = c^2\, \partial^2 V/\partial x^2 \tag{3.38}$$

where $\quad c = 1/\sqrt{L'C'}\quad$ is the velocity of propagation (3.39)

A solution for the wave equation, which is well suited for describing pulse propagation, is expressed as a sum of two functions f and g:

$$V(x, t) = f(x - ct) + g(x + ct) \tag{3.40}$$

The first function in this equation, $f(x - ct)$, represents a wave that propagates at a constant amplitude in the positive x direction as t increases so that the argument $(x - ct)$ remains constant. The second function, $g(x + ct)$, represents a wave propagating in the negative x direction with a constant amplitude. The velocity of propagation of both of the waves is c, which is defined in Eq. (3.39).

The current I is determined by substituting Eq. (3.40) into Eq. (3.15) and utilizing the low loss conditions to obtain

$$\partial f/\partial x + \partial g/\partial x = -L'\, \partial I/\partial t \tag{a}$$

Integrating Eq. (a) to obtain $I(x, t)$ gives

$$I = (1/cL')[f(x - ct) - g(x + ct)] \tag{b}$$

Substituting Eqs. (3.31) and (3.39) into Eq. (b) leads to

$$I = (1/Z_0)[f(x - ct) - g(x + ct)] \tag{3.41}$$

which can be expressed as

$$I = (I_f - I_b)/Z_0 \tag{3.41a}$$

From this result it is evident that the propagation of current pulses can be described with forward and backward waves. The current I_f associated with the forward wave is

$$I_f = f(x - ct)/Z_0 = V_f/Z_0 \tag{3.42}$$

where V_f is the voltage associated with the forward wave. The current in the backward wave, I_b is

$$I_b = -g(x + ct)/Z_0 = -V_b/Z_0 \tag{3.43}$$

where V_b is the voltage of the backward wave. It should be noted in Eq. (3.43) that the sign of the current is opposite that of the voltage for the pulses propagating in the negative x direction.

3.7 EFFECT OF TERMINATION ON PULSE PROPAGATION

The behavior of a pulse propagating along a transmission line depends on the termination at both ends of the line. To simplify the initial analysis, consider that the line is terminated at the input end with a voltage source that exhibits an output impedance $Z_s = Z_0$ as shown in Fig. 3.9. Matching the source impedance to the characteristic impedance of the line eliminates the reflection of the pulses from the source end and reduces the complexity of the analysis. Attention will be focused on the pulse propagating down the line to the load end and the behavior of the pulse when it reaches the position $x = \mathscr{L}$. Three cases will be considered including the open end, the short circuited end and the end with an arbitrary resistive load.

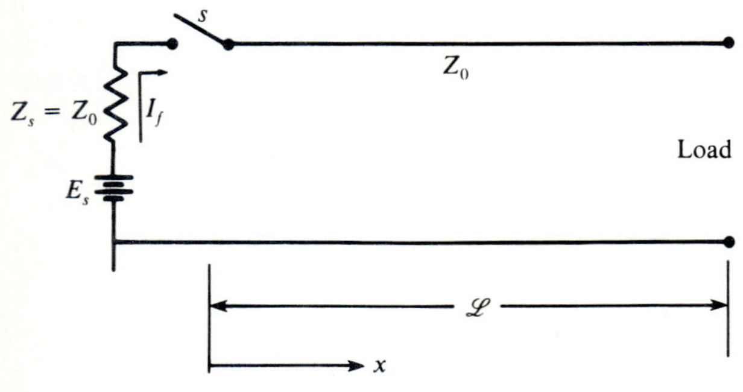

FIGURE 3.9
Transmission line with a dc voltage source with an impedance $Z_s = Z_0$.

3.7.1 Open-Ended Line

Examine the transmission line of length $\mathscr{L}$ shown in Fig. 3.9 with a source voltage E_s and an open end at $x = \mathscr{L}$. When the switch S is closed at $t = 0$, the transmission line at the position $x = 0$ is subjected to a voltage $V = E_s$. Substituting this initial condition into Eq. (3.40) gives:

$$f(x - ct) + g(x + ct) = E_s$$

or

$$V_f + V_b = E_s \tag{a}$$

From Fig. 3.9 and Eq. (3.42) it is evident that

$$V_f = I_f Z_0 \tag{b}$$

Kirchhoff's law gives

$$E_s = I_f Z_s + I_f Z_0 = I_f(Z_s + Z_0) \tag{c}$$

From Eqs. (b) and (c) it is clear that

$$I_f = E_s/(Z_s + Z_0) \tag{d}$$

$$V_f = [Z_0/(Z_s + Z_0)]\,E_s \tag{e}$$

with $Z_s = Z_0$. Then

$$V_f = E_s/2 \qquad \text{at } t = 0 \tag{f}$$

$$I_f = V_f/Z_0 = E_s/2Z_0 \qquad \text{at } t = 0 \tag{g}$$

For the time $0 \leq t \leq \mathscr{L}/c$ the voltage and the current pulses propagate down the line at a constant velocity c as indicated in Fig. 3.10*b*. When the pulses reach the end of the line at $t = \mathscr{L}/c$, a reflection occurs that satisfies the boundary condition for the open end, which is $I = 0$. Substituting $I = 0$ into Eq. (3.41a) gives

$$I = (I_f + I_b)/Z_0 = 0 \tag{h}$$

or

$$I_b = -I_f = -E_s/2Z_0 \quad \text{at } t = \mathscr{L}/c \tag{i}$$

From Eq. (3.43)

$$V_b = -Z_0 I_b = E_s/2 \quad \text{at } t = \mathscr{L}/c \tag{j}$$

For the time $\mathscr{L}/c \leq t \leq 2\mathscr{L}/c$ both the forward and the backward waves are propagating as indicated in Fig. 3.10*d*. The voltages are superimposed to give

$$V = V_f + V_b = E_s/2 + E_s/2 = E_s \tag{k}$$

$$I = I_f + I_b = E_s/2Z_0 - E_s/2Z_0 = 0 \tag{l}$$

At $t = 2\mathscr{L}/c$ the pulse front returns to the source end and no additional reflections occur and a steady state is reached. A voltage $V = E_s$ exists along the entire line indicating that the line has been charged and the supply voltage has

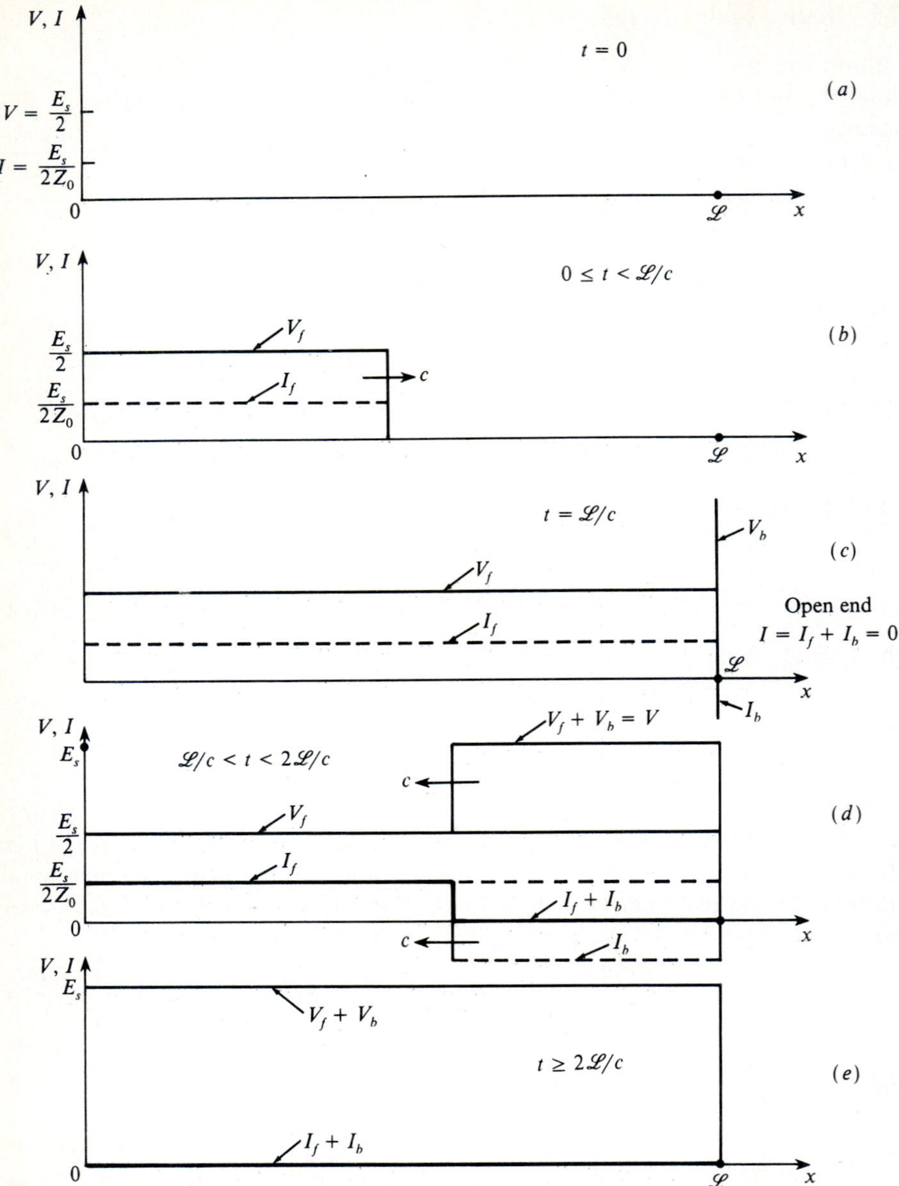

FIGURE 3.10
Distribution of voltage and current along the open-ended line at various times during the charging process.

been established. The energy $\mathscr{E}$ that has been supplied to charge the line is

$$\mathscr{E} = C'\mathscr{L}E_s^2/2 \tag{3.44}$$

Also, in the steady state for $t \geq 2\mathscr{L}/c$ no current is flowing in any portion of the line. This fact is due to the open end and the infinite impedance due to the distributed capacitance when $\omega = 0$.

3.7.2 Short Circuit at the Load End

If the transmission line is terminated with a suitable conductor between the high and low lines at $x = \mathscr{L}$, the behavior of the propagating current and voltage pulses is changed during reflection. For the initial period of propagation $0 \leq t < \mathscr{L}/c$, while the waves are propagating down the line, the solution is identical with the open circuit case as illustrated in Fig. 3.10a and 3.10b. The V_f and I_f waves have not reached the end of the line and are not yet affected by the termination.

When the waves reach the short circuit at $x = \mathscr{L}$ and $t = \mathscr{L}/c$, the voltage and current pulses reflect to satisfy the boundary condition $V(\mathscr{L}) = 0$. Then by Eq. (3.40)

$$V = V_f + V_b = 0 \tag{a}$$

which gives

$$V_b = -V_f = -E_s/2 \quad \text{for } t \geq \mathscr{L}/c \tag{b}$$

and by substituting Eq. (b) into Eq. (3.43) we obtain

$$I_b = E_s/2Z_0 \quad \text{for } t \geq \mathscr{L}/c \tag{c}$$

The distribution of the voltage and the current along the transmission line after reflection from the short circuited end is presented in Fig. 3.11a. At $t = 2\mathscr{L}/c$, the reflected pulses reach the source and since $Z_s = Z_0$, no reflection occurs and steady state is achieved. In the steady state $V(x) = 0$ and $I(x) = E_s/Z_s$ for time $t > 2\mathscr{L}/c$ as illustrated in Fig. 3.11b.

3.7.3 Arbitrary Resistive Load at the Line End

If the transmission line is terminated at $x = \mathscr{L}$ with a resistive load where $Z_L = R_L$, the magnitude and sign of the reflected pulses depend on the magnitude of Z_L in comparison with Z_0. Again, during the period of initial propagation $0 \leq t \leq \mathscr{L}/c$ the current and voltage propagate down the line with $V_f = E_s/2$ and $I_f = E_s/2Z_0$. At $t = \mathscr{L}/c$ the pulses reach the end of the line and encounter the terminating load $Z_L = R_L$. The boundary condition that governs the reflection process is

$$I = V/Z_L = (V_f + V_b)/Z_L \tag{a}$$

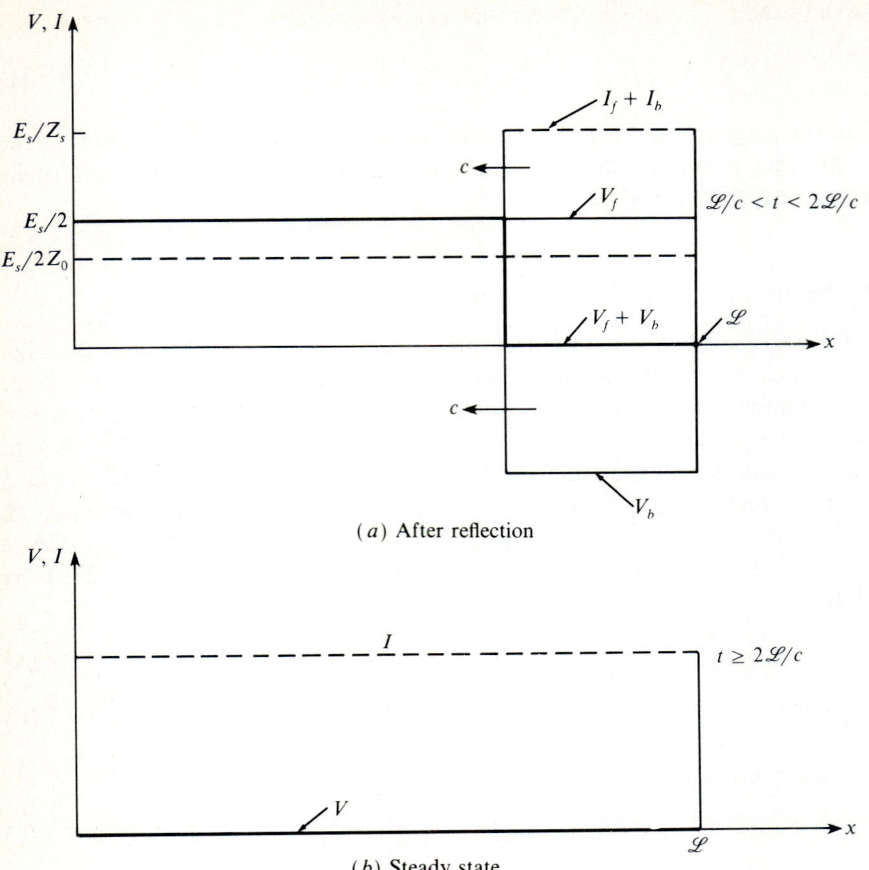

FIGURE 3.11
Distribution of voltage and current along the short circuited line.

Note, from Eqs. (3.42), (3.43) and (a) that

$$V_f + V_b = Z_L(I_f + I_b) \tag{b}$$

and
$$(E_s/2) - Z_0 I_b = Z_L[(E_s/2Z_0) + I_b] \tag{c}$$

Solving Eq. (c) for I_b yields

$$I_b = (E_s/2Z_0)[(Z_0 - Z_L)/(Z_L + Z_0)] \tag{d}$$

and
$$V_b = -Z_0 I_b = -(E_s/2)[(Z_0 - Z_L)/(Z_L + Z_0)] \tag{e}$$

It is clear that the signs of the reflected voltage and current pulses I_b and V_b depend on the magnitude of Z_L with respect to Z_0. Three cases should be considered, which include (a) $Z_0 = Z_L$, (b) $Z_0 < Z_L$ and (c) $Z_0 > Z_L$. For case

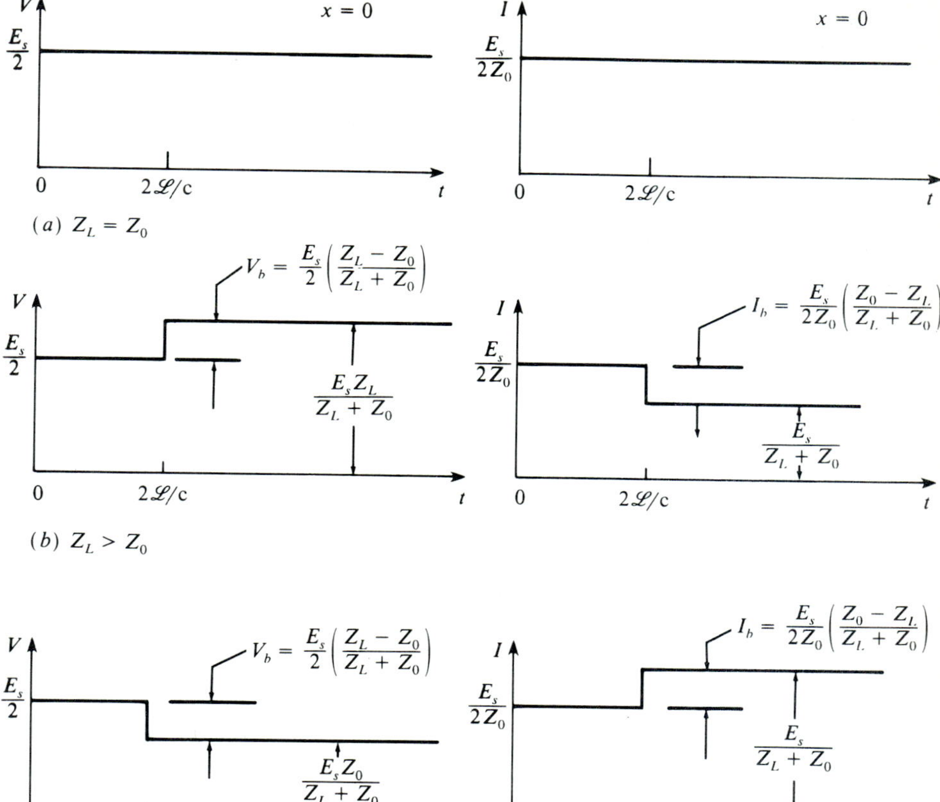

FIGURE 3.12
Voltage and current as a function of time with a resistive load $Z_L = R_L$ at the end of the line where $x = \mathscr{L}$.

(a) where the load impedance matches the characteristic line impedance there is no reflection and $V_b = I_b = 0$. In this case the voltage and current for $t > \mathscr{L}/c$ are given by

$$V(x) = E_s/2 \quad \text{and} \quad I(x) = E_s/2Z_0 \tag{3.45}$$

The voltage and current distributions along the line for this case are presented in Fig. 3.12*a*.

For case (b) where $Z_0 < Z_L$, reference to Eqs. (d) and (e) indicates that the reflected current will be negative and the reflected voltage will be positive. The combined forward and backward components for $\mathscr{L}/c \le t \le 2\mathscr{L}/c$ are

$$V = V_f + V_b = E_s Z_L/(Z_L + Z_0) \tag{3.46}$$

$$I = I_f + I_b = E_s/(Z_L + Z_0) \tag{3.47}$$

For case (c) where $Z_0 > Z_L$, Eqs. (d) and (e) show that I_b will be positive and V_b will be negative. The combined forward and backward pulses give the voltage and current as indicated in Eqs. (3.46) and (3.47). Graphs showing the voltage and current distributions for the cases where $Z_0 = Z_L$ are presented in Fig. 3.12*b* and 3.12*c*.

3.8 INFLUENCE OF A FINITE RISE TIME ON PULSE SHAPE

In the treatment of pulse propagation presented in Section 3.7, the switch S was closed, instantaneously producing a pulse with a rise time $t_r = 0$. In real circuits switches do not close instantaneously and the pulses formed by transistors switching from one state to another exhibit a finite rise time. A pulse formed by switching a transistor can be idealized with a terminated ramp function as illustrated in Fig. 3.13.

The behavior of ramp-fronted pulses propagating along a transmission line depends on the termination at both ends of the line and the ratio of the rise time t_r to the propagation time t_p. The value of t_p is given by

$$t_p = \mathscr{L}/c \tag{3.48}$$

Consider as an example a transmission line with a supply providing a voltage E_s through an impedance $Z_s = Z_0$ and a resistive load $R_L < Z_0$. If the line is sufficiently long so that $t_p > t_r$, the ramp-fronted pulse will propagate down the line in the manner shown in Fig. 3.13*a*. When the pulse reaches the load at $x = \mathscr{L}$, it will reflect and a second ramp-fronted pulse is generated which propagates in the negative x direction. The reflected voltage pulse V_b is given by

$$\begin{aligned} V_b &= -(E_s/2)[(t - t_p)/t_r][(Z_0 - Z_L)/(Z_L + Z_0)] \quad &\text{for } t_p \le t \le t_p + t_r \\ V_b &= (E_s/2)[(Z_0 - Z_L)/(Z_L + Z_0)] \quad &\text{for } t > t_p + t_r \end{aligned} \tag{3.49}$$

The distribution of the voltage along the line at four different times after reflection is presented in Fig. 3.13. It is evident from these results that the voltage at the source end of the line will vary as a function of time as shown in Fig. 3.14*a*.

If the rise time is long with say $t_r = 2t_p$, then the reflection at $x = \mathscr{L}$ occurs before E_s is fully applied. The voltage at the input end of the line increases from 0 to $E_s/2$, achieving its peak value at $t = 2t_p$. At this time the reflected wave arrives at the source and the voltage decays linearly with time during the interval $2t_p \le t \le 4t_p$ until it achieves its steady state condition as given by Eq. (3.46).

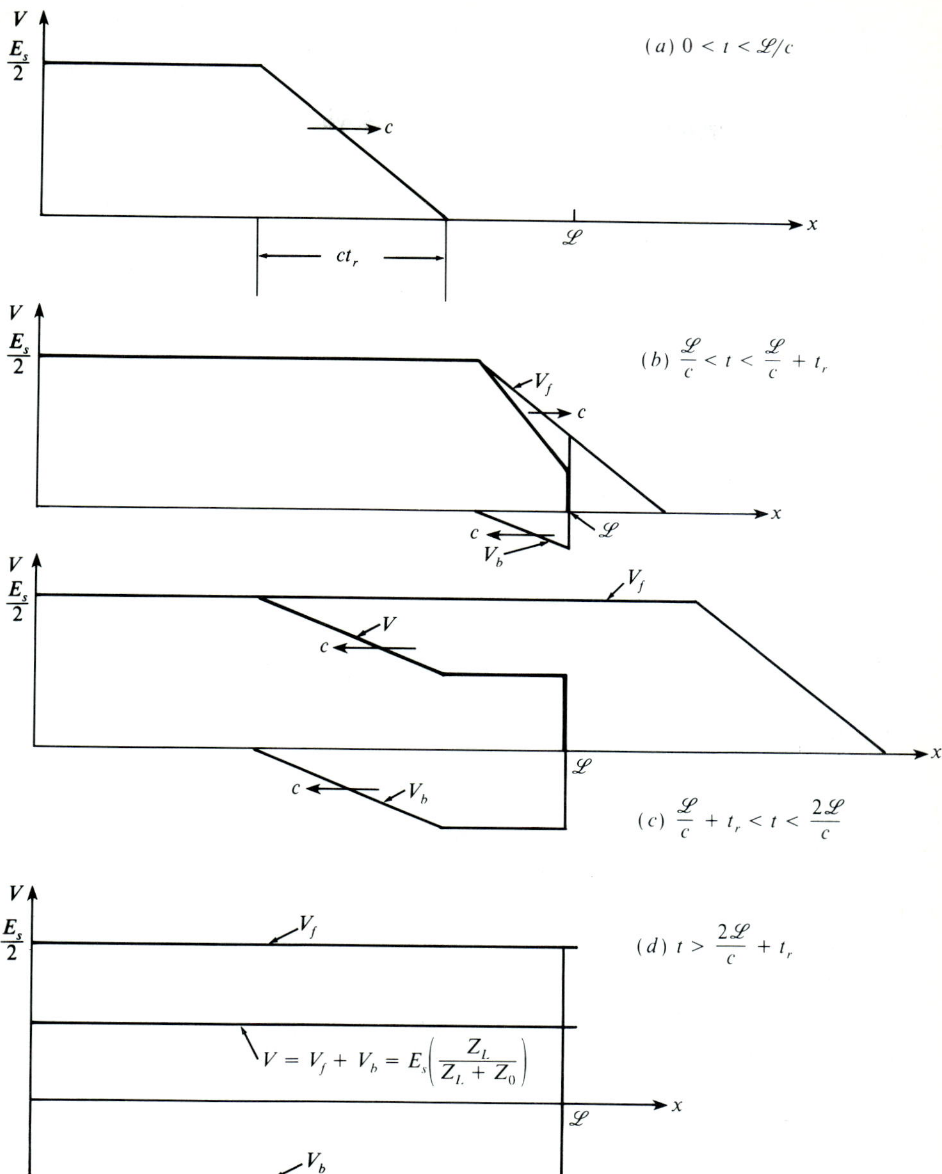

FIGURE 3.13
Voltage distribution along a line with a ramp-fronted pulse at four different times during the reflection process.

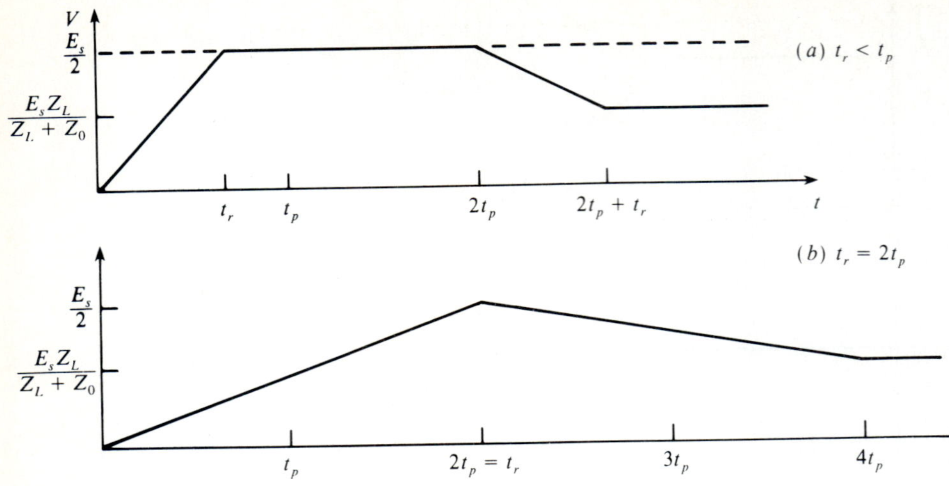

FIGURE 3.14
Voltage as a function of time for $Z_L < Z_0$ and different rise times t_r.

It is important to note that finite rise times, associated with switching, produce a delay in achieving steady state conditions on the line at both the source and the load. These line delays must be minimized in order to achieve high rates of signal processing in high performance computers.

3.9 ARBITRARY RESISTIVE LOADS AT BOTH ENDS OF THE LINE

In the previous sections dealing with pulse propagation the source impedance was matched to the characteristic impedance of the line and reflections did not occur at $x = 0$. Consider here the more general case where $Z_s \neq Z_0$ and $Z_L \neq Z_0$ with the line not properly terminated at either end as indicated in Fig. 3.15. When the switch S is closed at $t = 0$, the forward voltage V_f and current I_f are given from Eqs. (d) and (e) from Section 3.7.1 as

$$I_f = E_s/(Z_s + Z_0) \tag{3.50}$$

$$V_f = [Z_0/(Z_0 + Z_s)]\, E_s \tag{3.51}$$

This pulse propagates up and down the line and reflects from both the load and the supply ends many times before steady state conditions are achieved. To show the voltage at either end of the line during this transition period it is convenient to define reflection coefficients C_V for the voltage and C_I for the current as

$$C_V = V_r/V_i \quad \text{and} \quad C_I = I_r/I_i \tag{3.52}$$

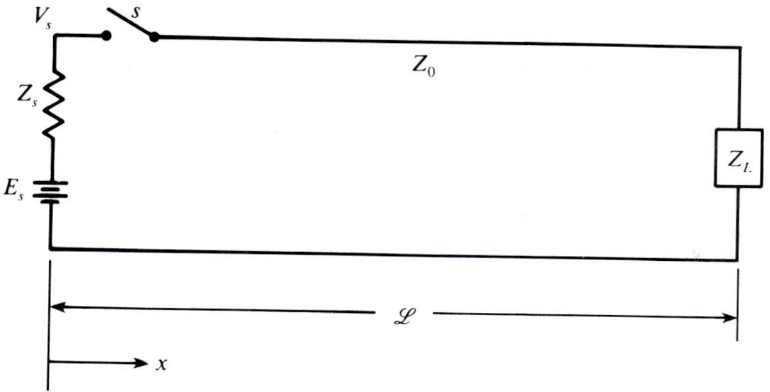

FIGURE 3.15
Transmission line with a dc source with an impedance $Z_s \neq Z_0$ and a load impedance $Z_L \neq Z_0$.

where the subscripts r and i refer to the reflected and incident pulses, respectively. Following the procedure described in Section 3.7.3, it is evident that Eqs. (3.52) may be written as

$$C_{VL} = (Z_L - Z_0)/(Z_L + Z_0) \quad \text{for load end} \tag{3.53}$$

$$C_{Vs} = (Z_s - Z_0)/(Z_s + Z_0) \quad \text{for source end} \tag{3.54}$$

and

$$C_I = -C_V \tag{3.55}$$

Note, that the reflection coefficient depends on the termination at the end of the line (source or load) where the reflection occurs. The reflection coefficients can be either positive or negative depending on the magnitudes of Z_s, Z_L and Z_0.

Consider the first reflection that occurs at the load end of the line at $t = \mathscr{L}/c$. The reflected voltage is given by Eq. (3.52) as

$$V_{r1} = C_{V\mathscr{L}} V_i = C_{V\mathscr{L}} V_f \tag{a}$$

and the total voltage at $x = \mathscr{L}$ is

$$V_1 = V_{r1} + V_f = V_f(1 + C_{VL}) \tag{b}$$

The second reflection takes place at the source where $V_{r1} = V_i$ and Eq. (3.52) gives

$$V_{r2} = C_{Vs} V_i = C_{Vs} C_{VL} V_f \tag{c}$$

and

$$V_2 = V_{r2} + V_1 = V_f(1 + C_{VL} + C_{Vs} C_{VL}) \tag{d}$$

This process continues as the pulse propagates up and down the line reflecting from the ends. The reflected pulse amplitudes diminish in size and a steady state voltage V_{ss} is developed along the entire line after a large number of reflections:

$$V_{ss} = [Z_L/(Z_L + Z_s)]\, E_s \tag{3.56}$$

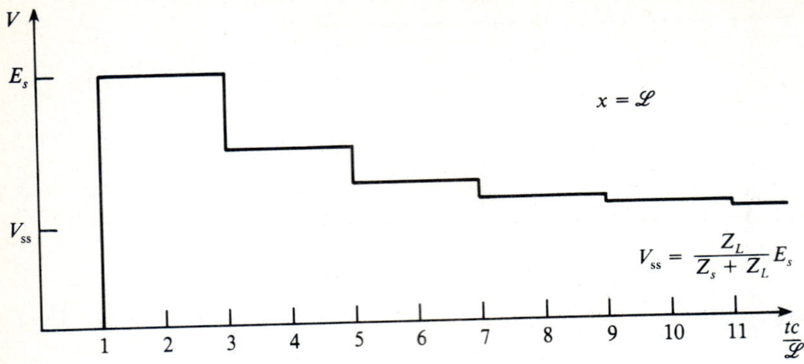

FIGURE 3.16
Voltage as a function of normalized time during the transition period at the position $x = \mathscr{L}$.

The voltage at the end of the line during the transition period is shown in Fig. 3.16 and Table 3.1.

The voltage on the line may be approximated with an exponential relation of the form

$$V = V_{ss} + (V_f - V_{ss}) e^{t/\tau} \tag{3.57}$$

where the time constant τ is given by

$$\tau = (2\mathscr{L}/c)/\ln(C_{Vs} C_{VL}) \tag{3.58}$$

Note, that the time constant τ will be negative since $C_{Vs} C_{VL} < 1$.

These results indicate that an improperly terminated transmission line requires a significant amount of time before the voltage stabilizes. This transition time represents a degradation of the line since the voltage should be stabilized at the load end in order to insure the delivery of a specified voltage pulse to the gate of a transistor. If the line were properly terminated with $Z_L = Z_s = Z_0$, then the

TABLE 3.1
Voltage at $x = 0$ and $x = 1$ during the transition period

Reflection number	Reflection load	Voltage (V_r / V_0) source	Total voltage V_n / V_0	Location $x / \mathscr{L}$
1	$C_{V\mathscr{L}}$	—	$1 + V_{r1}$	1
2	—	$C_{Vs} C_{V\mathscr{L}}$	$1 + V_{r1} + V_{r2}$	0
3	$C_{Vs} C_{V\mathscr{L}}^2$	—	$1 + V_{r1} + V_{r2} + V_{r3}$	1
4	—	$C_{Vs}^2 C_{V\mathscr{L}}^2$	$1 + \Sigma V_{rn}$	0
5	$C_{Vs}^2 C_{V\mathscr{L}}^3$	—	$1 + \Sigma V_{rn}$	1

time required for a stabilized signal to reach the load is the propagation delay $t_p = \mathscr{L}/c$, which is much less than 3τ.

3.10 REFLECTIONS FROM DISCONTINUITIES

In some circuits discrete components, such as resistors, capacitors and inductors, are placed in the line as illustrated in Fig. 3.17. These discrete components represent discontinuities placed in the line and affect the signal that is propagating along the line. Consider the resistor located in the line at position $x = \mathscr{L}_1$ as shown in Fig. 3.17 and note that the input voltage V_{in} to the resistor is

$$V_{\text{in}} = V_f + V_b \tag{a}$$

where V_b is the pulse reflected from the resistor. The voltage out of the resistor V_0 is given by Eq. (3.43) as

$$V_0 = Z_0(I_f + I_b) \tag{b}$$

and the voltage drop across the resistor V_R is

$$V_R = R(I_f + I_b) \tag{c}$$

Combining Eqs. (a), (b) and (c) and noting that $V_f = E_s/2$ when $Z_s = Z_0$, leads to the relation for the reflected pulse:

$$V_b = (E_s/2)[R/(2Z_0 + R)] \tag{3.59}$$

Adding the incident V_f and the reflected V_b pulses gives a total voltage

$$V = E_s[(Z_0 + R)/(2Z_0 + R)] \tag{3.60}$$

which appears at the source after a delay time $t_{\text{pd}} = 2\mathscr{L}_1/c$ as indicated in Fig. 3.18*a*.

If a capacitor is placed in the line at position $x = \mathscr{L}_1$, in place of the resistor shown in Fig. 3.18*b*, the signal is initially grounded and the capacitor

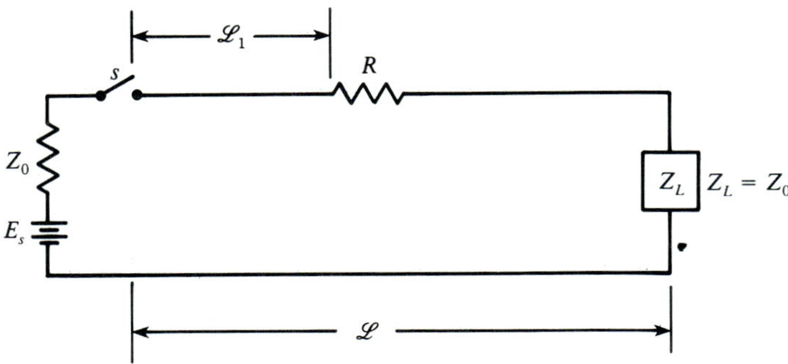

FIGURE 3.17
Placement of a discrete resistor in a transmission line.

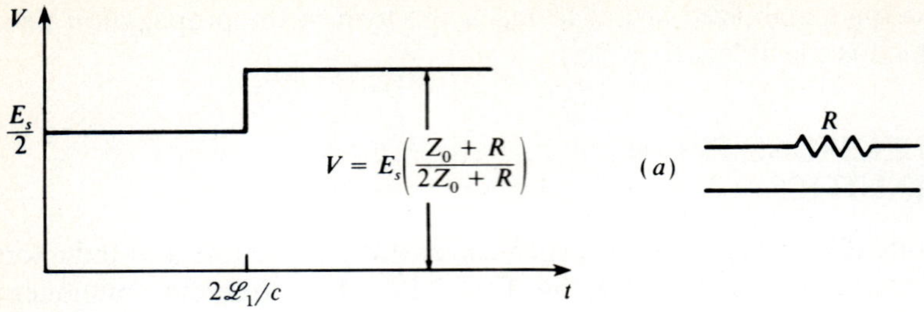

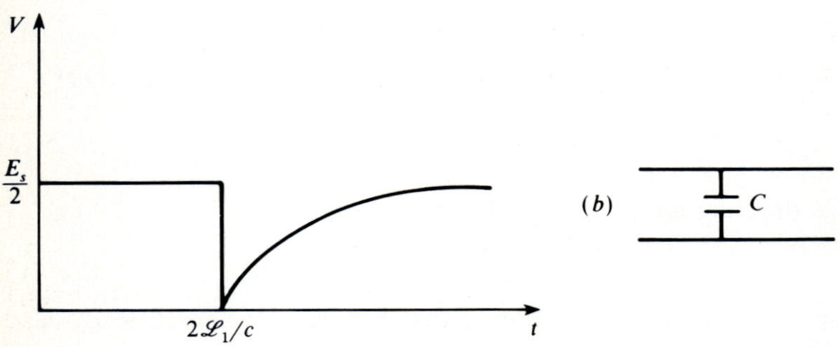

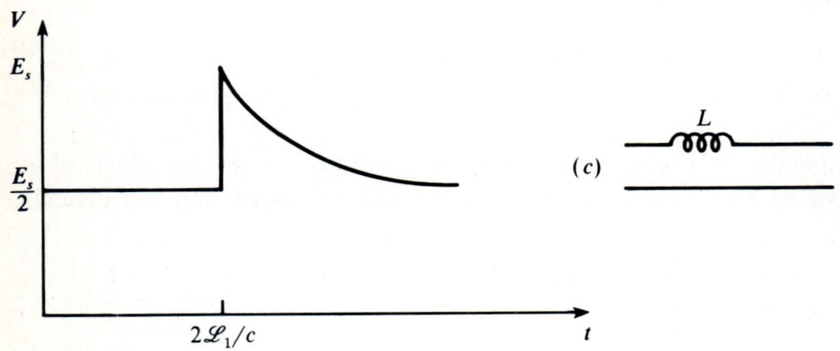

FIGURE 3.18
Voltage time traces at the source showing the influence of discrete components placed at $x = \mathscr{L}_1$.

begins to develop a charge when the forward pulse encounters the component. A reflected current pulse I_b is produced to provide the charging current where

$$I_b = (E_s/2Z_0)e^{-t_1/Z_0C} \tag{d}$$

with t_1 representing the time after the forward pulse reaches the capacitor. The sum of the forward and reflected voltage pulses give the voltage V as

$$V = V_f - Z_0 I_b \tag{e}$$

which reduces to

$$V = (E_s/2)[1 - e^{-t_1/(Z_0 C)}] \tag{3.61}$$

This voltage signal arrives at the source after a delay time of $t_{\text{pd}} = 2\mathscr{L}_1/c$ as illustrated in Fig. 3.18*b*.

Finally, an inductor in the line at $x = \mathscr{L}_1$ as shown in Fig. 3.18*c* produces a reflected voltage pulse

$$V_b = L(dI/dt) \tag{f}$$

Noting that the current flow through the inductor is given by

$$I = (E_s/2Z_0)[1 - e^{-Z_0 t_1/L}] \tag{g}$$

and substituting Eq. (g) into Eq. (f) yields:

$$V_b = (E_s/2)e^{-Z_0 t_1/L} \tag{h}$$

Adding the forward and reflected pulses gives the voltage

$$V = (E_s/2)[1 + e^{-Z_0 t_1/L}] \tag{3.62}$$

The voltage varies with time at the supply as illustrated in Fig. 3.18*c*.

3.11 CHARACTERISTIC IMPEDANCE OF CONDUCTORS

A transmission line consists of two conductors separated by an insulating material. The characteristic impedance of the line depends on the geometry of the conductors, their spacing and the field properties, permittivity ε and permeability μ, of the insulating material. Coaxial cable is commonly employed as a transmission line because its construction leads to a carefully controlled characteristic impedance. A coaxial cable, shown in Fig. 3.19, consists of a central circular conductor of radius a and a shield of radius b. Insulating material fills the annular area between the two conductors. The characteristic impedance of the coaxial cable is

$$Z_0 = (1/2\pi)\sqrt{\mu/\varepsilon}\,\ln(b/a) \tag{3.63}$$

A second insulating material that covers the shield does not affect the characteristic impedance. The outer insulation is used only to isolate the shield and to

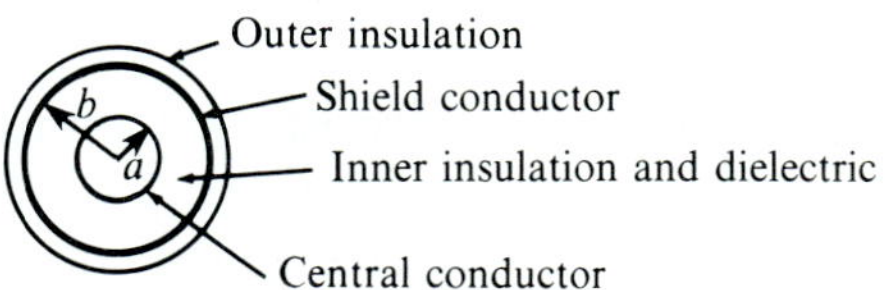

FIGURE 3.19
Cross section of a coaxial cable.

TABLE 3.2
Types of coaxial cable and their characteristics

Type	Center conductor AWG	OD (in.)	Z_0 (Ω)
RG-8/U	11	0.405	50
RG-58A/U	20	0.195	50
RG-174/U	26	0.101	50
RG-58/U	20	0.195	53
RG-59/U	20	0.242	75
RG-59/U	22	0.242	80
RG-62/U	22	0.242	93
RG-62B/U	24	0.242	93
RG-63B/U	—	0.415	125

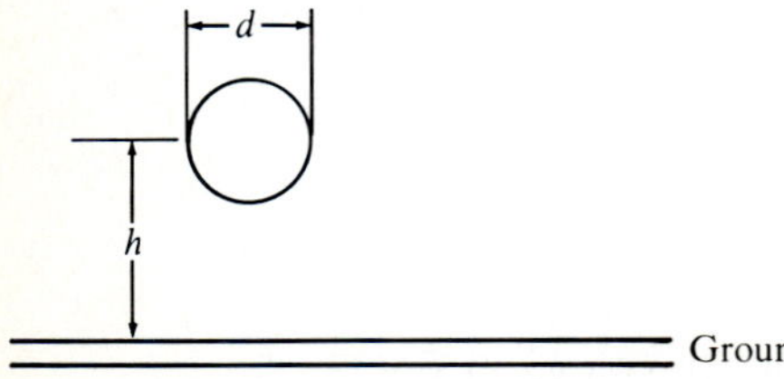

FIGURE 3.20
Transmission line formed by a wire in close proximity to a ground plane.

protect the cable assembly. Many different types of coaxial cable are commercially available and Table 3.2 provides a select listing of cables with Z_0 ranging from 50–125 Ω.

A transmission line can also be fabricated from standard hookup wire (AWG 24–28) by twisting two wires together. A twisted pair with 30 turns per foot of cable length will exhibit a $Z_0 \simeq 110$ Ω.

A transmission line is also formed by passing a single wire over a ground plane as indicated in Fig. 3.20. The characteristic impedance is given by

$$Z_0 = \left(60/\sqrt{e_r}\right)\ln(4h/d) \tag{3.64}$$

where e_r is the effective dielectric constant of the material surrounding the wire. When wires are used for connections in a back plane they are often placed in close proximity to a ground plane and each wire acts as a transmission line. The characteristic impedance of these wires cannot be accurately controlled and Z_0 will range from 75–150 Ω depending upon h and the proximity of adjacent wires.

3.12 TRANSMISSION LINES ON CIRCUIT BOARDS

Wiring lines are produced on circuit boards by an etching process that forms conductors with a rectangular cross section on an insulating laminate as illustrated in Fig. 3.21. If the copper cladding on the other side of the laminate is

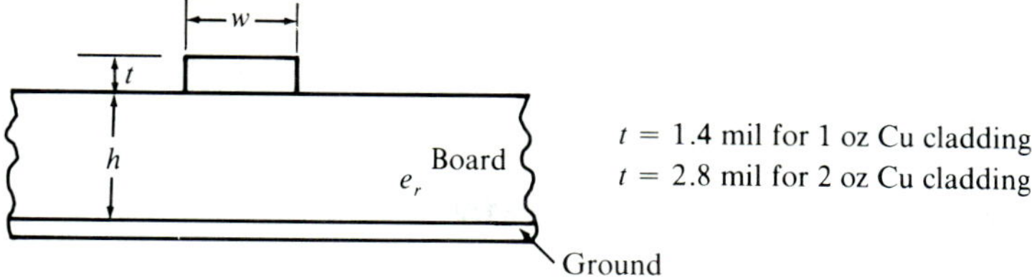

FIGURE 3.21
Cross section of a microstrip line etched on a circuit board.

maintained as a ground plane separated from the signal plane by the board material, then each of the etched conductors represents a transmission line. This type of conductor is termed a microstrip line and its characteristic impedance is given by

$$Z_0 = \left(87/\sqrt{\varepsilon_r + 1.41}\right)\ln[5.98h/(0.8w + t)] \tag{3.65}$$

where ε_r is the relative dielectric constant of the board material (4.7 for the common glass–epoxy laminate G10). h, w and t are dimensions defined in Fig. 3.21.

The capacitance for microstrip wiring traces is a function of the line width w and the thickness h of the board. Capacitance per foot C' for a microstrip line formed on a glass–epoxy board is given in Fig. 3.22. Note, that a trace on a board with h = 60 mil and w = 15 mil will exhibit a capacitance C' = 15 pF/ft.

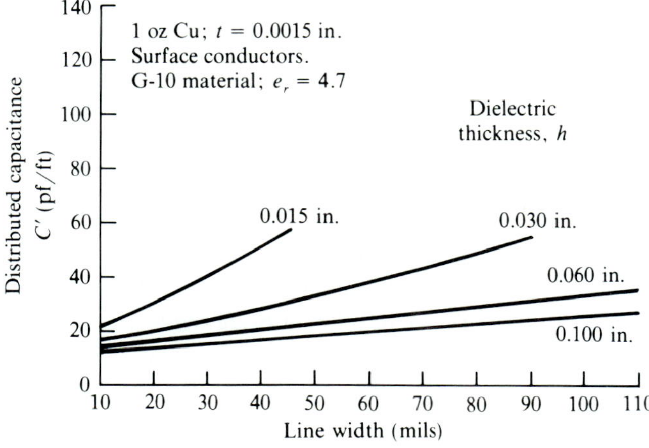

FIGURE 3.22
Distributed capacitance C' as a function of board thickness and line width for epoxy–glass (e_r = 4.7) laminates with microstrip traces. (*Courtesy of Motorola Semiconductor Products, Inc.*)

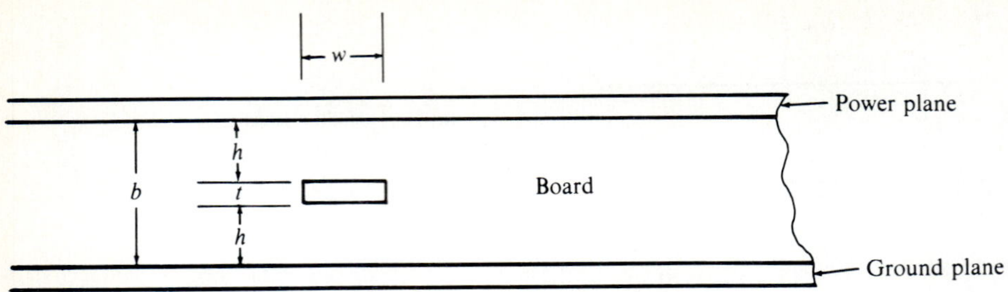

FIGURE 3.23
Strip line construction in a multilayer board.

The inductance L' is usually computed from the capacitance by using Eq. (3.31) to obtain

$$L' = Z_0^2 C' \tag{3.66}$$

The velocity of the pulse moving along the microstrip line is given by $c = 1/\sqrt{L'C'}$ and the propagation delay is

$$t_{\text{pd}} = 1/c = \sqrt{L'C'} \tag{3.67}$$

For a microstrip line,

$$t_{\text{pd}} = 1.017\sqrt{0.475e_r + 0.67} \tag{3.68}$$

where t_{pd} is given in nanoseconds per foot. It should be noted that the time delay depends only on the dielectric constant of the board material and is independent of the geometry of the circuit traces. For glass–epoxy laminate with $\varepsilon_r = 4.7$, the time to propagate a pulse a distance of 1 ft is 1.73 ns.

When circuit boards are fabricated with many layers, the signal planes are often sandwiched between a ground plane and a power plane as shown in Fig. 3.23. In this case the wiring trace is called a strip line and it acts as a transmission line with Z_0 given by

$$Z_0 = \left(60/\sqrt{\varepsilon_r}\right)\ln[4b/[0.67\pi w(0.8 + t/w)]] \tag{3.69}$$

where the dimensions b, w and t are defined in Fig. 3.23.

The capacitance per foot C' for strip lines is dependent upon the line width and the spacing b between the power and ground planes as shown in the graph of Fig. 3.24. The inductance is determined from Eq. (3.66). The delay time t_{pd} is determined in nanoseconds per foot from

$$t_{\text{pd}} = 1.017\sqrt{\varepsilon_r} \tag{3.70}$$

It is clear that t_{pd} depends only on the relative dielectric constant of the board material. For glass–epoxy boards $t_{\text{pd}} = 2.26$ ns/ft with the strip line-type traces.

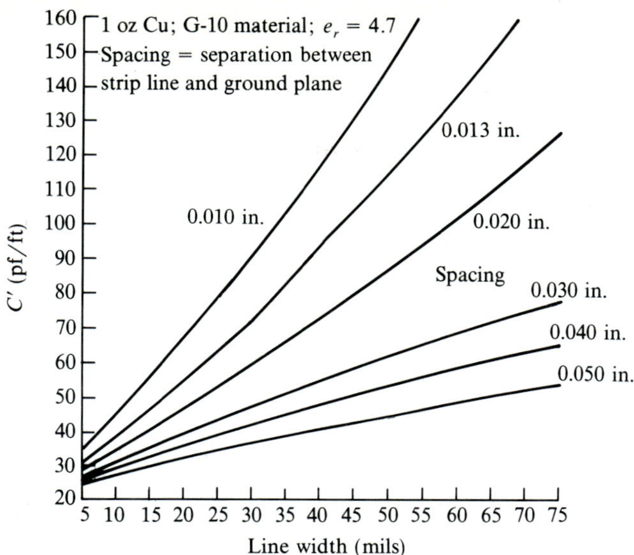

FIGURE 3.24
Distributed capacitance C' for a strip line trace as a function of line width and spacing of the ground and power planes. (*Courtesy of Motorola Semiconductor Products, Inc.*)

3.13 RESISTANCE OF PRINTED CIRCUIT LINES

The resistance R of a line of length L on a printed circuit board may be determined from Eq. (2.1) as

$$R = \rho L/wt \tag{3.71}$$

where w and t are the line width and thickness, respectively, and $\rho = 1.724 \times 10^{-6}$ Ω-cm is the resistivity of copper. The thickness of the copper cladding is usually specified in the number of ounces of copper per square foot of board area with 1 or 2 oz representing the common values corresponding to thicknesses of 0.00135 or 0.0027 in., respectively. Substituting these values of t into Eq. (3.71) gives

$$R/L = 0.503/w \quad \text{for 1 oz copper} \tag{3.72a}$$

$$R/L = 0.251/w \quad \text{for 2 oz copper} \tag{3.72b}$$

where the line width is expressed in mils and the length in inches.

For a line width of $w = 10$ mil, the resistance per inch of a 1 oz copper line is about 50 mΩ/in. The line width required for digital signal transmission is not critical because the power in signal transmission is quite small (a few milliwatts) and the currents are of the order of a few milliamperes. In these cases the line width is selected to ease the manufacturing process and to give higher yields with w ranging from 5–10 mil. Circuit lines carrying power to the chips are much

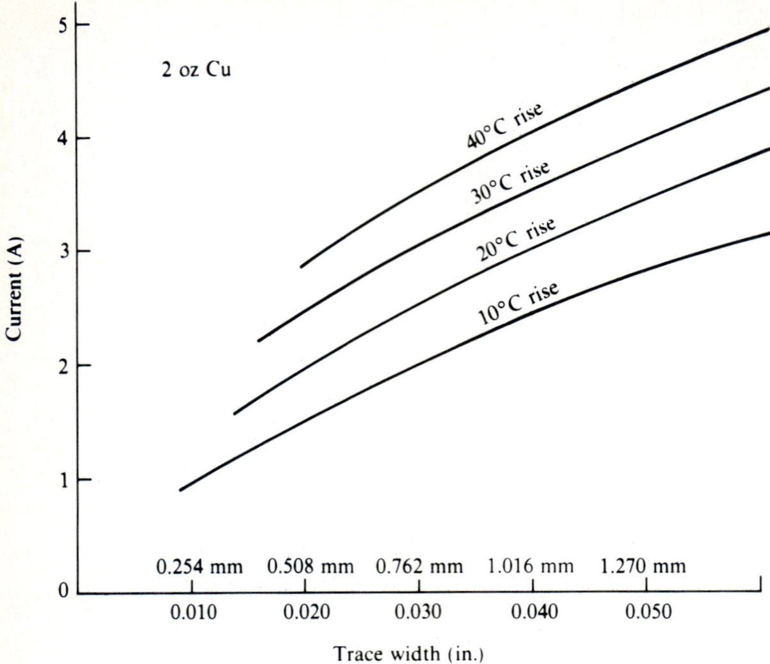

FIGURE 3.25
Allowable current for surface traces as a function of trace width and temperature increase.

wider and the value of w is selected to provide adequate current capacity. The current capacity depends on the allowable temperature increase in the line. Results showing current capacity for microstrip lines as a function of line width are presented in Fig. 3.25 for conductors etched in 2 oz copper. If the lines are etched in 1 oz cladding, the current capacity is about 60% of that shown in Fig. 3.25.

3.14 LOGIC GATE CHARACTERISTICS

The basic logic gates described previously in Section 2.9 are fabricated from resistors, diodes and transistors as integrated circuits. Several different technologies are used to provide a family of devices that can be connected together to provide digital functions and digital systems. The technologies commonly employed include TTL (transistor–transistor logic), NMOS, CMOS and ECL (emitter-coupled logic) and depend on the details of fabricating transistors as described in Chapter 2. Of particular importance is the power delay product relationship that is shown in Fig. 2.18. Gate delay on the chip is due to the RC constant associated with the gate circuit, where R includes the load resistance and the switching resistance and C includes the parasitic capacitance due to the circuit lines on the chip. When a gate is switched from the low state to the high state, the output pulse exhibits a propagation delay and a rise time as indicated in Fig. 3.1. The rise time can vary from about 1–100 ns depending on the feature size of the components, the design of the gates and the type of technology used.

To illustrate typical characteristics of a family of logic chips, consider MECL-10KH, which is an emitter-coupled logic design introduced in 1981. This

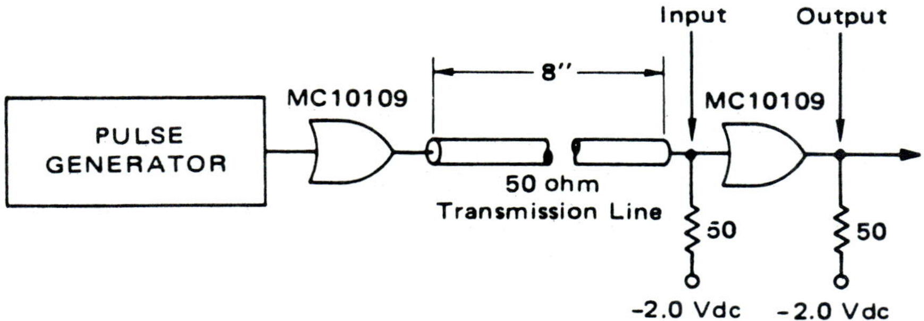

(a) Test Configuration

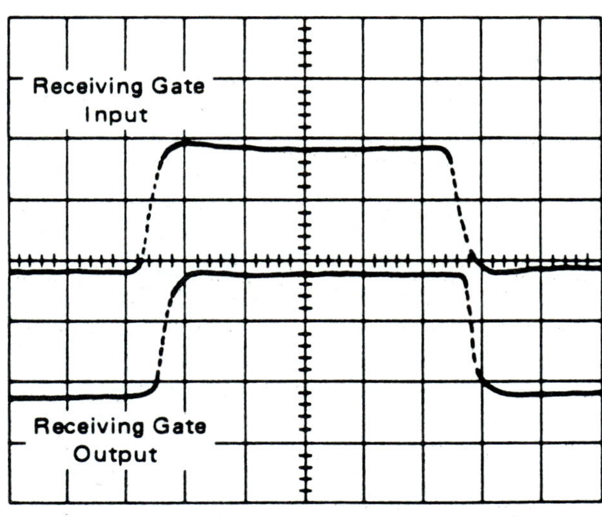

Vertical Scale = 400 mV/cm
Horizontal Scale = 10 ns/cm

(b) Input and Output Waveforms

FIGURE 3.26
Input and output pulses from a high speed OR gate with matched termination at both ends of the gate. (*Courtesy of Motorola Semiconductor Products, Inc.*)

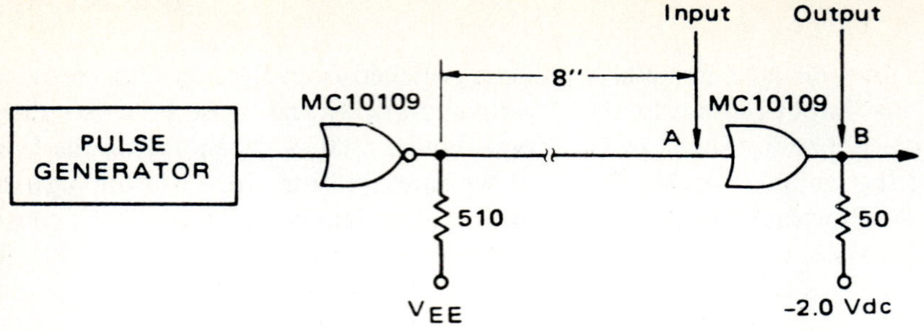

(a) Test Arrangement

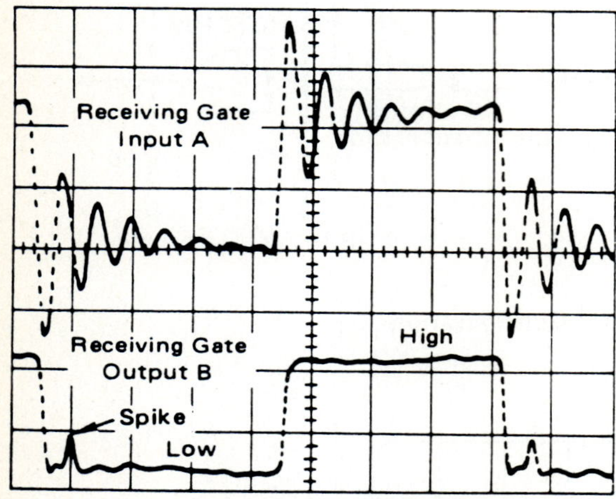

(b) Ground Plane Not Used

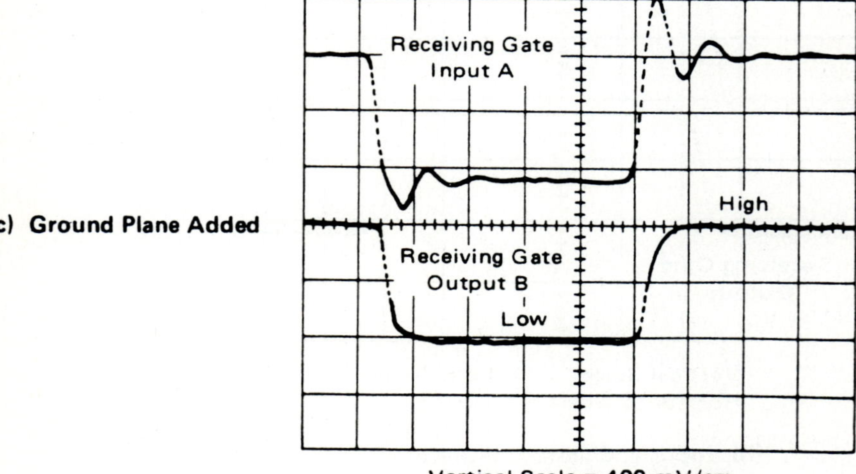

(c) Ground Plane Added

FIGURE 3.27
Input and output pulses for a high speed OR gate with and without impedance controlled input lines. (*Courtesy of Motorola Semiconductor Products, Inc.*)

family of gates exhibits a propagation delay of 1 ns, a gate edge speed (10–90% of the rise time) of 1.8 ns and a power delay product of 25 pJ. The propagation delay is measured from the 50% amplitude point on the input signal to the same point on the output signal. The test configuration and the oscilloscope records, which depict the input and output signals from an OR gate, are shown in Fig. 3.26. It should be noted that a transmission line was used to transmit the input signal to the receiving gate and that both the input and output pulses show well defined voltages associated with the high and low states.

If the same experiment is repeated with the transmission line (coaxial cable) replaced with a relatively long segment of a circuit board trace without an adjacent ground plane and if the terminating resistance is eliminated, then the results obtained are markedly different. Reference to Fig. 3.27 shows that the input signal oscillates with wide voltage variations as the voltage is switched from one state to another. These voltage fluctuations affect the output of the receiving gate as the output signal exhibits a spike as it is switched from the high to low state. Addition of a ground plane beneath the circuit trace converts the trace into a microstrip-type transmission line and even with the absence of a terminating resistor, the voltage oscillations are reduced in magnitude and duration. From this example, it is clear that this gate can tolerate voltage variations in the input signal and provide a clean ramp function in switching with the output pulse relatively free of oscillations.

The pulse shape of the input signal clearly affects the behavior of a gate as illustrated in Fig. 3.27 and as the switching times of the gate are reduced, the tendency for voltage oscillations increase. To determine the effect of the oscillations of the waveform on the performance of the gate, the input signal is analyzed to measure the overshoot and undershoot as indicated in Fig. 3.28. Of particular importance is the undershoot voltage in relation to the input voltage required to switch the gate from low to high. The relation between the input and output voltages for a typical gate presented in Fig. 3.29 shows that the output voltage changes from low to high as the input voltage changes from V_{IL} to V_{IH}. The midpoint of this voltage range is defined as the switching threshold V_{BB}. For proper switching it is important that the input voltage remain above V_{IH} to prevent parasitic switching and oscillations in the output voltage.

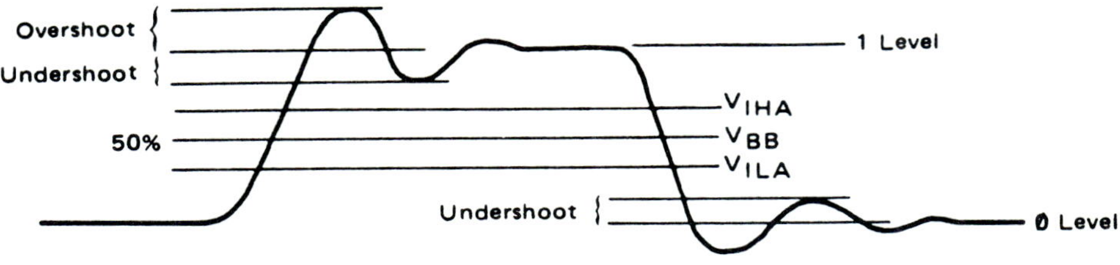

FIGURE 3.28
Analysis of the input signal pulse shape to determine undershoot and overshoot. (*Courtesy of Motorola Semiconductor Products, Inc.*)

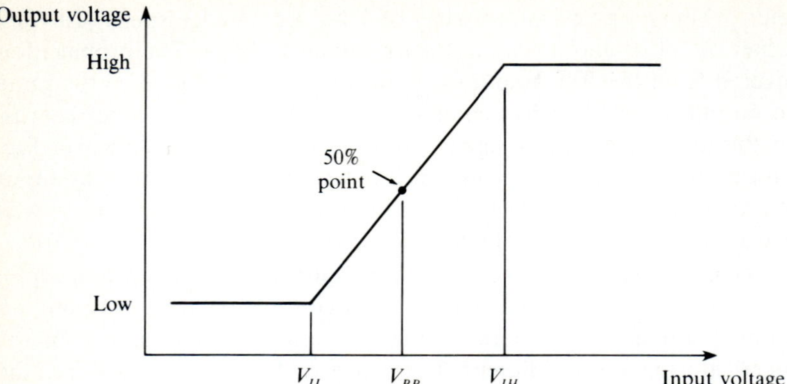

FIGURE 3.29
Typical transfer curve for a gate showing the input voltage required to switch from low to high.

The magnitude of the oscillations that result in overshoot and undershoot depend on circuit design, the printed circuit board layout and the type of wiring used for interconnections. As the switching times approach 1 ns or less, extreme care is necessary in maintaining the fidelity of the signals throughout the system. Interconnect line lengths must be minimized and the use of transmission lines with termination resistors at both the source and the load is essential.

REFERENCES

1. Hildebrand, F. B.: *Advanced Calculus for Engineers*, Prentice-Hall, New York, 1949.
2. Churchill, R. V.: *Introduction to Complex Variables and Applications*, McGraw-Hill, New York, 1948.
3. Dally, J. W., W. F. Riley and K. G. McConnell: *Instrumentation for Engineering Measurements*, Wiley, New York, 1984.
4. Hayt, W. H., Jr. and J. E. Kemmerly: *Engineering Circuit Analysis*, 4th ed., McGraw-Hill, New York, 1986.
5. Ramo, S., J. R. Whinnery and T. Van Duzer: *Fields and Waves in Communication Electronics*, Wiley, New York, 1965.
6. Tugal, D. and O. Tugal: *Data Transmission*, 2nd ed., McGraw-Hill, New York, 1988.
7. Blood, W. R., Jr.: *MECL System Design Handbook*, 4th ed. Motorola, 1983.

EXERCISES

3.1. If the high and low voltages shown in Fig. 3.1 are 1 and 0 V, respectively, find the coefficients A_n and B_n in Eq. (3.1) for $n = 0–3$. Take t_r as 100, 10, 1, 0.1 and 0.01 ns. Show how the frequencies required to represent the pulse increase as the rise time decreases.

3.2. Repeat Exercise 3.1 but find the coefficients C_n in Eq. (3.1) instead of A_n and B_n.

3.3. For the circuit shown in Fig. E3.3 determine:
(*a*) The impedance.
(*b*) The voltage drop across the inductance.
(*c*) The voltage drop across the resistance.
(*d*) The voltage drop across the capacitor.

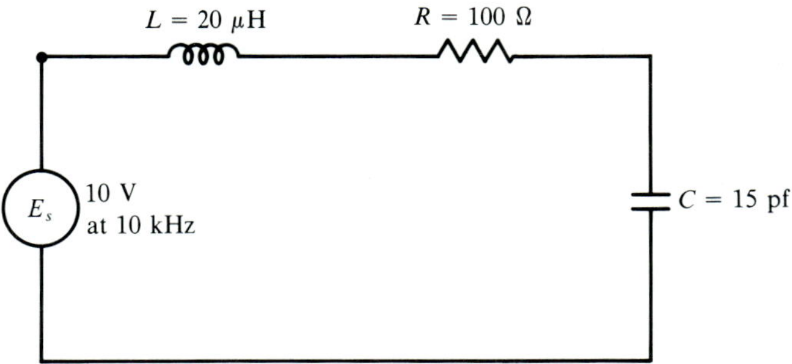

FIGURE E3.3

3.4. Find the phase angle ϕ for the circuit defined in Fig. E3.3. Also, determine the current amplitude I_a.

3.5. Prepare a graph similar to that presented in Fig. 3.3 that indicates the intercepts on the real and imaginary axes, the phase angle and the voltage E_a.

3.6. Use Eqs. (3.16) and (3.18) with Eq. (3.18) to verify the telegrapher's equation.

3.7. List the features illustrated in the circuit element that represent a segment of a transmission line shown in Fig. 3.4. Note in particular the differences between the circuits represented in Figs. 3.2 and 3.4.

3.8. Determine the characteristic impedance of a low loss transmission line if the distributed inductance is 12 fH/ft and the distributed capacitance is 2 pF/ft.

3.9. For an open-ended low loss transmission line with a distributed capacitance of 20 pF/ft, determine the impedance of the line if it is 30 ft long and carrying a signal with a frequency of 500 kHz.

3.10. Determine the current flow in the open-ended transmission line described in Exercise 3.9 if the voltage amplitude of the sinusoidal signal is 4 V. What is the basic cause of current flow in an open-ended line?

3.11. Consider a transmission line with a shorted end as illustrated in Fig. 3.9. If the line is 14 in. long with $L' = 5$ pH/ft and $R' = 2$ μΩ/ft, determine the line impedance. For a source operating at 3 V with a frequency of 10 MHz, find the current flow from the source to ground.

3.12. Determine the velocity of the electrical signal propagating in the transmission line described in Exercise 3.8.

3.13. Consider the transmission line illustrated in Fig. 3.9 with $Z_s = Z_0 = 50$ Ω and $E_s = 3$ V. Determine the forward voltage and current when the switch is initially closed. Find the magnitude of the backward voltage and current pulse if the end of the line is terminated with a 30 Ω resistor.

3.14. If the transmission line of Exercise 3.13 is 28 in. long and the pulse is propagating at a reciprocal velocity of 2 ns/ft, construct a diagram similar to that presented in Fig. 3.12, which shows development of the voltage and current along the line for:
(a) $0 \leq t \leq 2\mathscr{L}/c$.
(b) $2\mathscr{L}/c \leq t \leq 4\mathscr{L}/c$.

3.15. Repeat Exercise 3.13, replacing the 30 Ω terminating resistor with a 70 Ω resistor.

3.16. Repeat Exercise 3.14 using the 70 Ω terminating resistor.

3.17. Find the length along a transmission line to accommodate the linear ramp that will occur when the voltage E_s is switched with a transistor that has a rise time $t_r = 10$ ns. Take the propagation velocity as $1/c = 2$ ns/ft. If the transmission line is only 8 in. long, find the time required to totally charge the line. Comment on the implications of this result.

3.18. Repeat Exercise 3.17 but replace the transistor with a high speed unit that switches in 1 ns.

3.19. Construct a diagram similar to that shown in Fig. 3.14 for $t_r = 3t_p$, $Z_s = Z_0$ and $Z_L = 3Z_0$. Determine the time required for the voltage on the line to reach 95% of the steady state voltage $E_s/2$.

3.20. Repeat Exercise 3.20 with $Z_L = Z_0$.

3.21. Determine the time constant τ for a transmission line with $Z_s = Z_0/3$ and $Z_L = 3Z_0$. Consider a line with $\mathscr{L} = 18$ in. and $1/c = 2.5$ ns/ft.

3.22. If the supply voltage $E_s = 5$ V is applied to the line described in Exercise 3.21, find the voltage on the line as a function of time. Construct a voltage-normalized time diagram similar to that presented in Fig. 3.16 for this case. Compare the voltage predicted by Eq. (3.57) with that shown in your diagram.

3.23. A discrete resistor with $R = 200$ Ω is inserted into a transmission line with $Z_0 = Z_s = 50$ Ω. Determine the magnitude of the voltage pulse reflected from the resistor. Find the magnitude of the voltage that appears at the source after a delay time of $t_d = 2\mathscr{L}_1/c$.

3.24. Verify Eq. (3.61).

3.25. A capacitor with $C = 2$ μF is placed across the end of a 72 in. long transmission line. If $Z_s = Z_0 = 90$ Ω with $E_s = 4$ V, write an equation describing the voltage at the source after $t = 2\mathscr{L}/c$.

3.26. Replace the capacitor in Exercise 3.25 with a discrete in-line inductor $L = 120$ μH positioned at $x = \mathscr{L}$. Write an equation describing the voltage at the source after $t = 2\mathscr{L}/c$.

3.27. A bare copper wire is used as a telegraph line connecting two locations 100 mi apart. The wire is strung on poles and on the average the wire is 20 ft above the ground. If the ground water is an average of 10 ft below the surface, determine the characteristic impedance of the line. Note that the ground water serves as the second conductor of the transmission line.

3.28. A Teflon insulated wire with $e_r = 2.3$ is supported by a copper ground plane. The wire is 22 gage with $d = 0.032$ in. and the diameter of the Teflon insulation is $D = 0.062$ in. Estimate the characteristic impedance of this wire.

3.29. A microstrip line defined in Fig. 3.21 is etched with a width $w = 10$ mil in copper cladding with a thickness of 2 oz/ft^2. The glass–epoxy laminate separating the line

from the ground plane is 60 mil thick. Determine:
(*a*) The characteristic impedance Z_0.
(*b*) The distributed capacitance C'.
(*c*) The distributed inductance L'.
(*d*) The signal velocity.

3.30. A strip line defined in Fig. 3.23, with $w = 5$ mil is fabricated from 1 oz/ft^2 copper. This line is embedded between two ground planes with a spacing $b = 20$ mil. Determine the same four quantities as indicated in Exercise 3.29.

3.31. If the microstrip line in Exercise 3.29 is 9 in. long, determine its resistance R and its distributed resistance R'.

3.32. If the strip line of Exercise 3.30 is 15 in. long, determine the resistance R and the distributed resistance R'.

3.33. A microstrip line 30 mil wide is used to distribute power to several chips on a PCB. If the maximum current carried by the line is 3 A, estimate the increase in temperature of the line.

3.34. Compare the test configurations illustrated in Figs. 3.26 and 3.27 and the resulting voltage time traces recorded in operation of the two systems. Explain why the voltage time record in Fig. 3.27 exhibits overshoot, undershoot and a damped oscillation.

3.35. Using Fig. 3.29 as a reference, describe the switching behavior of a transistor if the input voltage exhibits a damped oscillation. If this transistor is in a logic gate that is in a binary adder, can the voltage oscillation affect the output of the adder?

PART II

PACKAGING

CHAPTER 4

FIRST LEVEL PACKAGING—THE CHIP CARRIER

4.1 INTRODUCTION

The chip carrier is the housing for the thin and fragile silicon chip. The chip carrier serves three main purposes. First, to protect the chip from the environment and from abusive handling, which may occur in manufacturing and from the forces due to either shock or vibration that may occur in service. Second, it serves to facilitate the interconnection of the circuits on the chip, which are extremely compact, to the more widely spaced interconnecting pads or holes located on the circuit boards. Finally, the chip carrier serves to facilitate the manufacturing process by providing a rugged housing that can be handled with automatic machinery without damaging the chip. The chip carrier also provides pins or pads that serve as bases for solder joints. The pins and/or pads are designed so that the solder joints can be made with automatic soldering processes without danger of solder defects such as solder bridges between leads or voids of significant size in the solder.

The chip carrier is also involved in the heat transfer process. The heat generated on the top surface of the chip must pass through the chip carrier as the first step in the path of heat flow from the source to the sink.

A typical chip carrier is illustrated in Fig. 4.1. Its various parts include the chip as the central element, the case, the leads and lead frame, the chip to

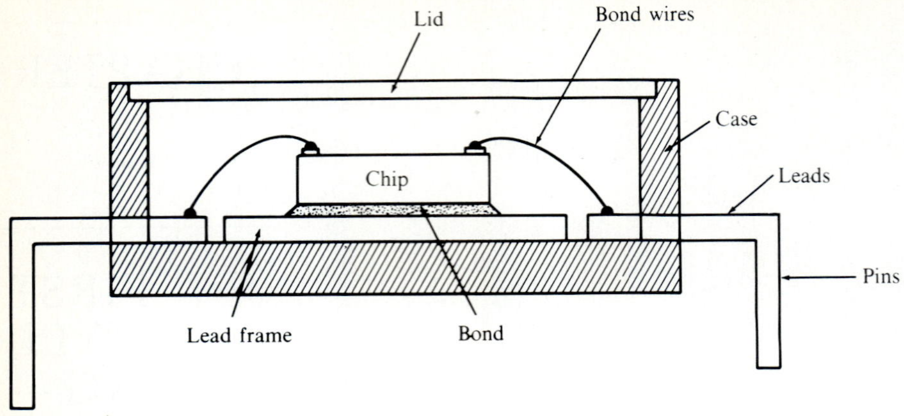

FIGURE 4.1
Typical features of a chip carrier.

package bond, the bonding wires and the lid. Many different types of chip carriers are in use today but with few exceptions they all conform closely in concept to the typical package shown in Fig. 4.1. Unfortunately this concept, while commonly utilized, is deficient from a design viewpoint. The flow of heat from the chip is severely limited because both the top and bottom surfaces are bounded by wiring planes that are thermal insulators. The dissipation of heat from this type of chip carrier is difficult and severe thermal penalties are encountered with chip carriers with wiring schemes that thermally isolate the chip.

Key design features for a chip carrier involve I/O count, heat dissipation or thermal resistance and hermeticity. I/O count is extremely important in housing modern VLSI chips that contain tens of thousands of gates, and access to these chips through a high I/O count is required to fully utilize a high percentage of the resident gates. These modern high gate logic chips dissipate large quantities of heat and reliable operation of these dense chips requires that the chip carrier facilitate rather than impede the transfer of heat from the chip to the heat sink. Hermeticity is required to ensure reliable operation of the chip over very long periods of time. Very small amounts of moisture that gain entry into the chip cavity of the carrier during assembly or by leaks in the package during operation permit corrosion of the metal wiring on the surface of the chip to occur. The presence of corrosion seriously degrades the chip circuits and reduces the life of the component. To eliminate the detrimental effects of moisture-induced corrosion, chip carriers are made of materials that prevent moisture entry by diffusion. Also, all organic materials that will out-gas (release volatiles) with time are prohibited within the chip cavity since the chemicals involved in out-gassing can also induce corrosion.

The common denominator of price and performance, typical of all of design philosophy for electronic systems, holds for chip carriers. Very low cost plastic chip carriers are available that are quite suitable for SSI or MSI chips, which dissipate low powers, require modest I/O count to achieve maximum performance and do not require hermeticity for the product involved. However, for high performance systems operating at very high speed, the low cost plastic chip carrier is not adequate because it often does not have sufficient I/O count, its thermal resistance is much too high to handle the high power level required and its hermeticity is not sufficient to ensure the degree of reliability needed. At this time significant research and development is underway to produce new chip carriers that are capable of housing modern high speed, high powered and high I/O features found in VLSI and ULSI chips that are currently being developed or produced.

4.2 TYPES OF CHIP CARRIERS

Chip carriers can be classified in several different ways. The material used in fabricating the case is one common way to distinguish between different types of chip carriers. Plastic cases formed by injection molding are low cost and provide suitable housings for low and moderate cost product. Ceramic cases are used for products that have stringent hermeticity requirements. These ceramic cases are much more costly than plastic cases and their application is usually limited to military products, which strictly require hermeticity to achieve the high reliabilities that are necessary for strategic systems, and to high performance commercial computers and signal processors, where system availability is essential.

A second major distinction between types of chip carriers pertains to the type of leads used for the I/O emerging from the carrier. Three general types of leads are currently employed.

1. A lead intended for pin in-hole mounting to the circuit board as illustrated in Fig. 4.2.

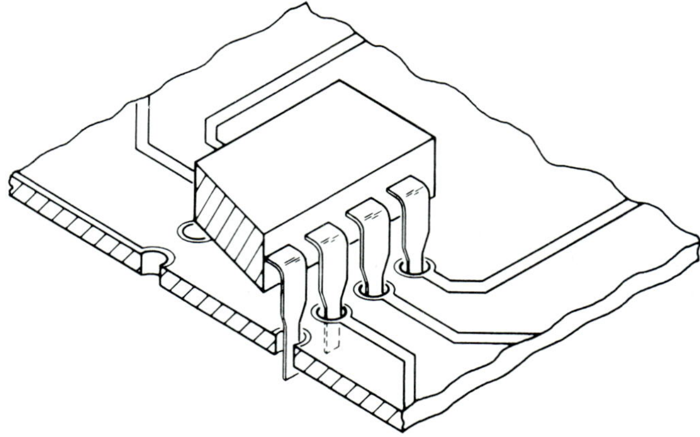

FIGURE 4.2
Chip carriers with leads for pin in-hole mounting. (*Courtesy of Texas Instruments, Inc.*)

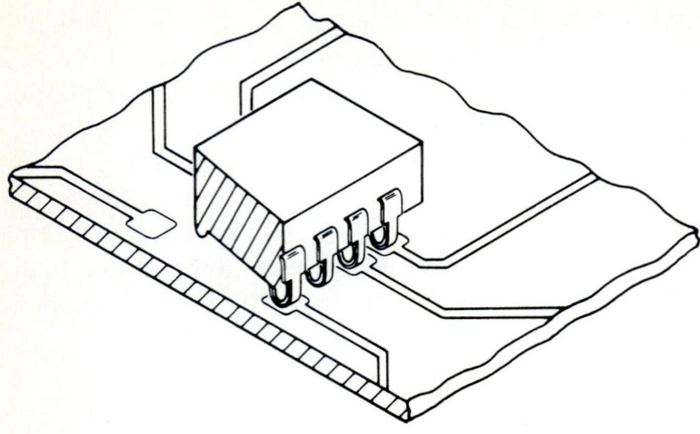

FIGURE 4.3
Chip carrier with *J* leads for surface mounting. (*Courtesy of Texas Instruments, Inc.*)

2. A lead intended for connection to the surface of the circuit board at a solder pad as shown in Fig. 4.3.
3. Leadless, where metallized pads are provided on the chip carrier to provide soldering surfaces for connection to the circuit board. An illustration of this type of chip carrier is presented in Fig. 4.4.

Each of these three types of packages will be described in some detail in the following subsections.

4.2.1 Pin In-Hole Chip Carriers

The most common chip carrier found in electronic packaging today is the dual in-line package (DIP) that is illustrated in Fig. 4.5. The DIP incorporates two rows of pins that are arranged on 100 mil centers along the longer side of the

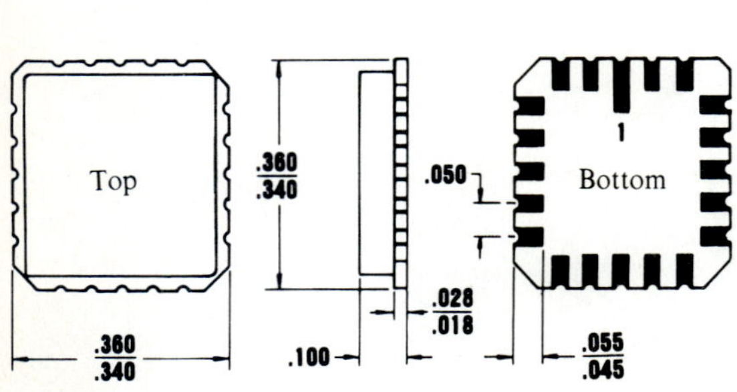

FIGURE 4.4
Leadless chip carrier utilizes metallized pads for solder connection to the circuit board.

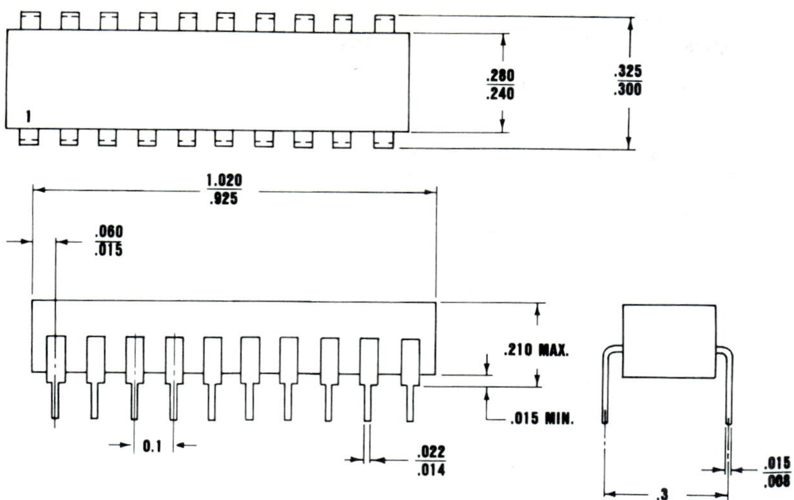

FIGURE 4.5
Dimensions of an 18 pin DIP.

rectangular package. The DIP family of chip carriers is available in different sizes and is often used to package low wattage chips with 8 to 64 bonding pads. The pins are short and relatively large in cross-sectional dimension, which makes them stiff and robust. This robustness is an advantage in automatic assembly as the leads are inserted in holes in the circuit board by pick and place machines. The pins inserted through the circuit board are all soldered from the underside of the board by passing the assembled PCB through a standing wave of molten solder. With this wave soldering process, all of the solder joints on the board are made quickly and efficiently in a single pass through the wave. The width of the pins is increased near the body of the chip carrier to provide a shoulder that serves to seat the DIP a distance of 20–40 mil from the top surface of the circuit card. This standoff is necessary to provide a space under the carrier so that the solder fluxes can be cleaned from the board after assembly is complete. Removal of the solder flux is important to prevent corrosion of the exposed metal on the PCB.

The DIP family has some disadvantages, which include poor area efficiency, limited I/O count and poor wire ability that limit its usefulness in housing modern high density logic chips. The area efficiency is poor because the pins are on 100 mil centers and deployed on only two sides of the package. In more area efficient carriers the pins are on 50 mil centers (or less) and are deployed on all four sides of the package. The maximum I/O count on a standard DIP is limited to 64 and to increase this count without changing the pin spacing would require prohibitively long packages with significant lead lengths within the chip carrier. Wire ability is poor because the pin in-hole mounting requires annular solder pads on the circuit board with an outside diameter of about 60 mil. These pads

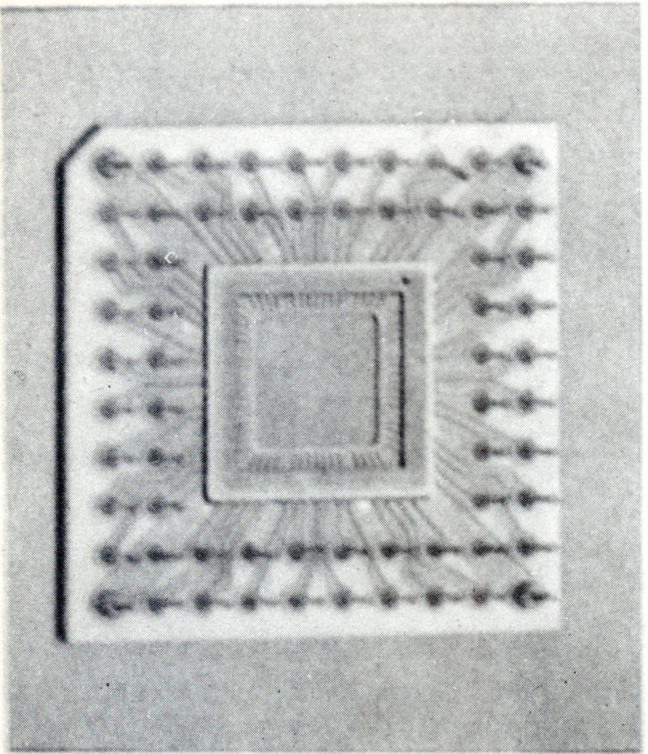

FIGURE 4.6
Pin grid array chip carrier.

leave only 40 mil for the wiring channel between pins, which causes a problem in providing for more than two wiring traces per wiring channel. A detailed description of wiring channels is presented in Chapter 5.

Pin grid array carriers are employed as a substitute for DIPs when additional I/O count is required or if lower thermal resistance is necessary. The pin array carrier, illustrated in Fig. 4.6, shows the general features of this design. The body of the package is fabricated from a ceramic (usually a multilayer alumina) and the body provides a cavity for the chip. A ledge around the cavity is used as the support surface for metallized pads that are the connections for the bonding wires that lead to the chip. Vias (holes containing conductors) lead from these pads to intermediate signal and power planes that carry the wiring traces from the vias to the pin locations. The pins, which are 20–25 mil in diameter, are brazed into the ceramic substrate. After the chip is bonded in the package, a ceramic or Kovar (an alloy with a low coefficient of thermal expansion) lid is placed over the cavity and sealed using inorganic solders that do not out-gas.

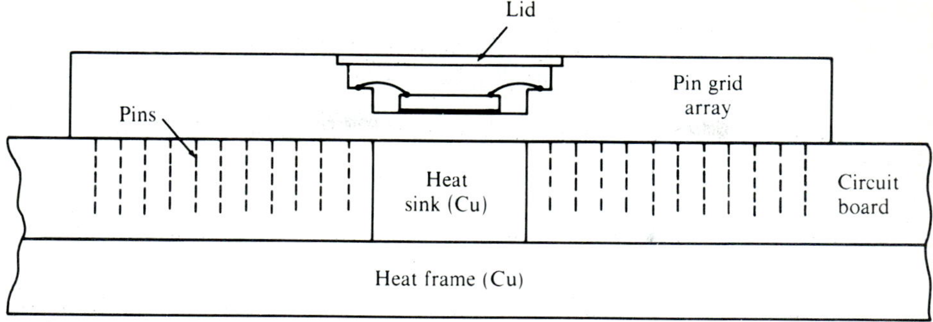

FIGURE 4.7
Improving thermal resistance for a pin grid carrier with direct contact between the chip and the heat sink.

In pin grid arrays, intended for chips with a high power rating, the surface area directly below the chip is not used for pins. Instead, it is reserved for a low resistance thermal path leading to the heat frame. The method of coupling the heat frame to the package is illustrated in Fig. 4.7. While pin grid arrays handle very large I/O count and can be designed to couple directly with a heat frame, they have two disadvantages. First, they are relatively high in cost because the processes involving multilayered ceramics are limited to a very few companies and the market is not competitive. Second, the area on the circuit board occupied

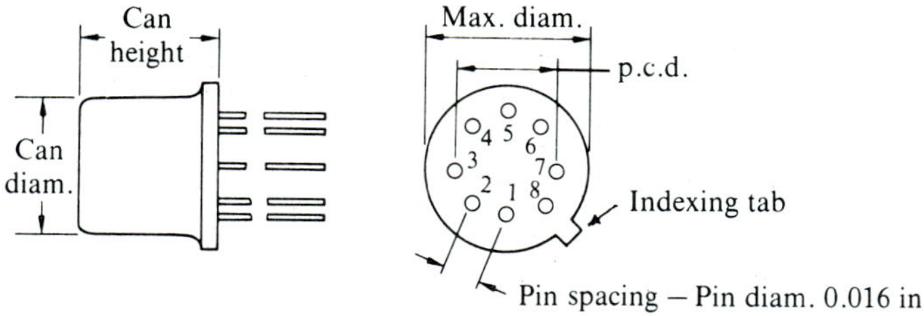

No. of pins	JEDEC No.	p.c.d.	Pin spacing	Height		Max. diam.		Can diam.	
				min.	max.	min.	max.	min.	max.
8	TO 71	0.100 in.	0.038 in.	0.170 in.	0.210 in.	0.209 in.	0.230 in.	0.175 in.	0.195 in.
		2.54 mm	9.96 mm	4.32 mm	5.33 mm	5.31 mm	5.84 mm	4.45 mm	4.95 mm
	TO 70	0.141 in.	0.054 in.	0.065 in.	0.085 in.	0.240 in.	0.270 in.	0.205 in.	0.240 in.
		3.58 mm	1.37 mm	1.65 mm	2.16 mm	6.10 mm	6.86 mm	5.21 mm	6.10 mm
	TO 77	0.200 in.	0.077 in.	0.240 in.	0.260 in.	0.335 in.	0.370 in.	0.305 in.	0.335 in.
		5.08 mm	1.96 mm	6.10 mm	6.60 mm	8.51 mm	9.40 mm	7.75 mm	8.51 mm

FIGURE 4.8
TO cans evolved from vacuum tube bases and are still used today for power transistors.

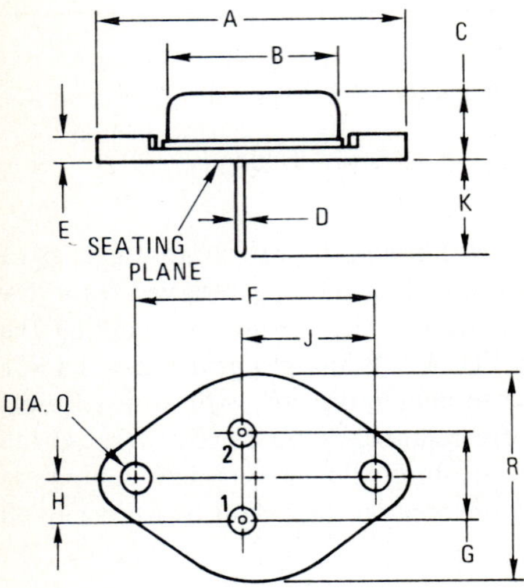

STYLE 1:
PIN 1. BASE
2. EMITTER
CASE: COLLECTOR

DIM	MILLIMETERS		INCHES	
	MIN	MAX	MIN	MAX
A	–	39.37	–	1.550
B	–	21.08	–	0.830
C	6.35	7.62	0.250	0.300
D	0.99	1.09	0.039	0.043
E	–	3.43	–	0.135
F	29.90	30.40	1.177	1.197
G	10.67	11.18	0.420	0.440
H	5.33	5.59	0.210	0.220
J	16.64	17.15	0.655	0.675
K	11.18	12.19	0.440	0.480
Q	3.84	4.09	0.151	0.161
R	--	26.67	–	1.050

Collector connected to case.
CASE 11-01
(TO-3)

FIGURE 4.9
TO can adapted with a flange to enhance heat transfer from the base of the chip carrier.

by the pin grid array is large. The pins are usually on 100 mil centers which gives a pin density of 100/in.^{2}. This is a relatively low density for leads and more modern concepts to be discussed later provide higher densities and more efficient use of circuit board real estate.

The TO can is also a pin in-hole-type chip carrier. This package is the oldest type of chip carrier and its shape reflects the transition from vacuum tube design to the design of packages to house low I/O count transistors on small silicon chips. The pins emerge from the bottom of the package on a pin circle as illustrated in Fig. 4.8. The TO cans are still used today to house chips intended for power transistors or silicon control rectifiers where power dissipation is very high and the lead count is small. The TO can is often fitted with a flange that serves as a heat sink as shown in Fig. 4.9. The flange provides a large contact area for heat dissipation and is fastened directly to a heat frame to complete a low resistance thermal path from the chip to the heat frame.

4.2.2 Leaded Surface-Mounted Chip Carriers

With surface-mounted chip carriers, the leads from the package are soldered to pads that are formed on the top surface of the circuit card as indicated in Fig. 4.3. The fact that a hole is not necessary to accommodate the mounting of the lead wire permits the leads to be placed on closer centers and a more area efficient chip carrier can be developed. As chip I/O count increases and efficient use of circuit board area becomes more important, the advantages of surface mounting with its smaller package sizes has lead to several new developments in first level packaging.

The oldest surface-mounted chip carrier is the flat pack shown in Fig. 4.10. This chip carrier is similar to the DIP in that it incorporates leads deployed along both sides of the package; however, it differs in that the leads are on 50 mil centers rather than the 100 mil centers found on the DIP. The leads on the flat pack are smaller in cross-sectional area and they extend straight outward from the package. These leads are cut and shaped prior to mounting the flat pack to

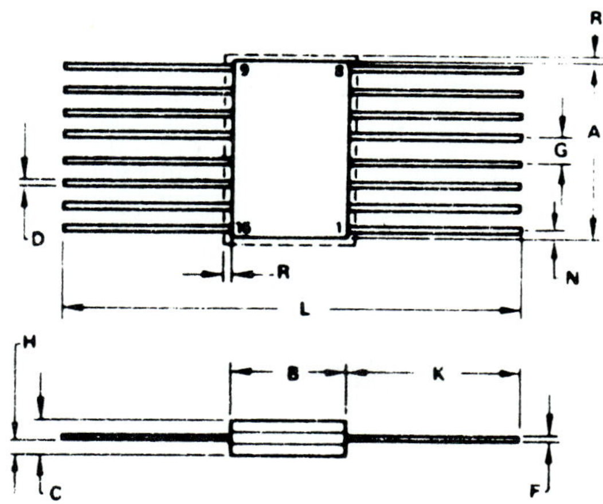

DIM	MILLIMETERS MIN	MILLIMETERS MAX	INCHES MIN	INCHES MAX
A	9.40	10.16	0.370	0.400
B	6.22	7.24	0.245	0.285
C	1.52	2.03	0.060	0.080
D	0.41	0.48	0.016	0.019
F	0.08	0.15	0.003	0.006
G	1.27 BSC		0.050 BSC	
H	0.64	0.89	0.025	0.035
K	6.35	9.40	0.250	0.370
L	18.92	-	0.745	-
N	-	0.51	-	0.020
R	-	0.38	-	0.015

CASE 650

FIGURE 4.10
Geometry of the flat pack-type chip carrier.

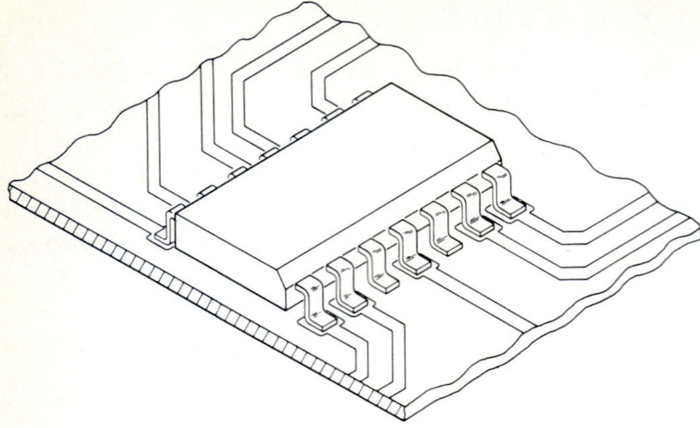

FIGURE 4.11
Flat pack leads formed into a gull wing configuration prior to mounting. (*Courtesy of Texas Instruments, Inc.*)

the circuit board. In the final configuration the leads are shaped like gull wings as indicated in Fig. 4.11.

The major advantage of the flat pack is its relatively small size compared to a DIP with the same I/O count. Since the leads are on closer centers, the flat pack is nearly twice as efficient as the DIP in utilization of circuit board real estate. The flat pack does have disadvantages mostly related to handling. Since the leads are only 3–6 mil thick, they are fragile and the chip carrier requires careful handling during assembly with lead guards used to protect the leads until they are cut and formed. The assembly process is also more difficult because the packages are often positioned and soldered individually. With DIPs the packages are machine-inserted and wave-soldered with all of the connections made simultaneously. Soldering the connections on the flat pack is usually performed with a short hot bar-type of soldering iron with only the leads from one side of the package connected simultaneously. Of course this disadvantage leads to much higher assembly costs that must be compensated for by savings in circuit board area, savings in third level package size and improved performance due to shorter lead length.

A relatively new surface mounted chip carrier is now available that incorporates a *J* lead on 50 mil centers. This chip carrier, presented in Fig. 4.12, utilizes all four sides of the package for leads and hence is more efficient from an area viewpoint than even the flat pack. The *J* lead configuration, which essentially folds the lead into a small pocket on the underside of the package, reduces lead deformations during shipping and handling. The *J* lead is fully configured for assembly. It is possible to position the packages on the board, adhere them with solder paste and to make all the solder joints simultaneously by using a vapor phase reflow soldering process; see Fig. 6.24. The *J* lead with the pocket protecting the lead and the solder joint connecting the lead to the circuit board

NO. OF TERMINALS	A MIN	A MAX	B MIN	B MAX	C MIN	C MAX
20	9.35 (0.368)	10.03 (0.395)	8.89 (0.350)	9.04 (0.356)	8.08 (0.318)	8.38 (0.330)
28	11.89 (0.468)	12.57 (0.495)	11.43 (0.450)	11.58 (0.456)	10.62 (0.418)	10.92 (0.430)
44	16.97 (0.668)	17.65 (0.695)	16.51 (0.650)	16.66 (0.656)	15.70 (0.618)	16.00 (0.630)
52	19.51 (0.768)	20.19 (0.795)	19.05 (0.750)	19.20 (0.756)	18.24 (0.718)	18.54 (0.730)
68	24.59 (0.968)	25.27 (0.995)	24.13 (0.950)	24.28 (0.956)	23.32 (0.918)	23.62 (0.930)

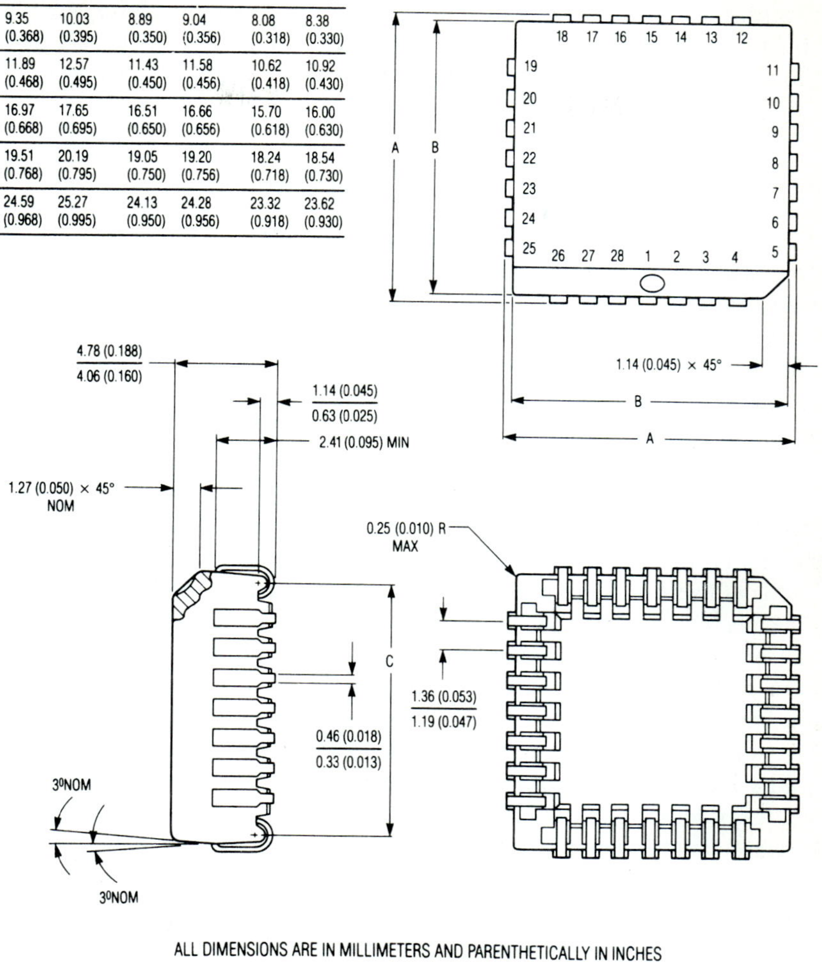

ALL DIMENSIONS ARE IN MILLIMETERS AND PARENTHETICALLY IN INCHES

FIGURE 4.12
J leaded chip carrier. (*Courtesy of Texas Instruments, Inc.*)

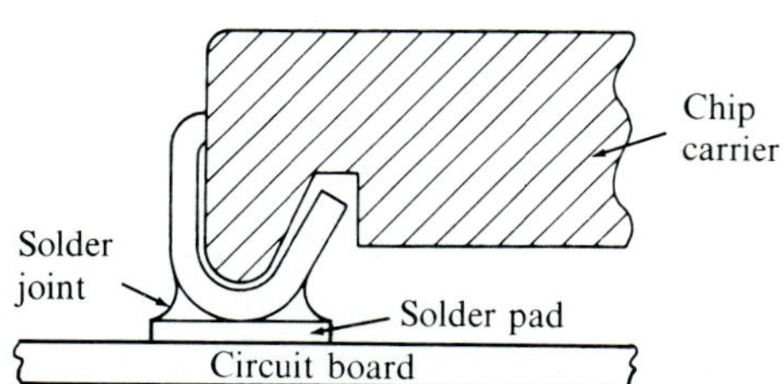

FIGURE 4.13
Cross section showing the pocket to protect the *J* lead and the solder joint connecting the lead to the solder pad.

pad is shown in Fig. 4.13. Clearly the J leaded chip carrier has significant potential to replace the conventional DIP in designs involving high I/O count chips. One disadvantage of this new package is the difficulty in maintaining all of the curved leads in the same plane so that all N leads are seated on the pads of the PCB prior to soldering.

4.2.3 Leadless Surface-Mounted Chip Carriers

Leads from the chip carrier clearly are a nuisance because they are fragile and can easily be deformed in shipping and handling. It is possible to eliminate the leads entirely by replacing them with metallized pads fixed directly on the chip carrier as shown in Fig. 4.4. These metallized pads are usually on 50 mil centers and provide an area efficient chip carrier. The pads on the chip carrier are connected to corresponding pads on the circuit board by using solder paste, which is melted in a vapor phase soldering process described more completely in Chapter 6.

The area efficiency of the leadless chip carrier is mitigated by a very serious disadvantage involving solder joint failure. The solder joints on the circuit board assembly are subjected to large thermally induced strains due to a mismatch of the coefficients of expansion of the chip carrier and the circuit board materials. Thermal cycling may be encountered in operation due to power on–power off, or thermal cycles may be imposed as a manufacturing screening test or in qualification testing. During these thermal cycles the solder joints are exposed to a thermally induced strain of relatively large magnitude. The effect of this strain is to induce fatigue failures of the solder joints. The failures begin with the solder joints located at the corners of the chip carrier and then move inward to more centrally located solder joints. The number of thermal cycles required to initiate fatigue failure depends on the temperature extremes involved in the thermal cycling, the rate of temperature change, the mismatch in the coefficients of thermal expansion and the size of the chip carrier. The data for the shear strain range $\Delta\gamma$, which can be accommodated by a 63/37 Sn–Pb solder joint is given in Fig. 4.14. The shear strain range $\Delta\gamma$ is approximated by the relation

$$\Delta\gamma = (d/h)\left[\alpha_b(T_b - T_a) - \alpha_{cc}(T_{cc} - T_a)\right] \tag{4.1}$$

where d is the distance of the solder joint from the chip carrier (CC) center
h is the height of the solder joint between pads
α_b is the coefficient of expansion of the circuit board
α_{cc} is the coefficient of expansion of the chip carrier
T_b is the maximum temperature of the circuit board
T_{cc} is the maximum temperature of the chip carrier
T_a is the ambient temperature in the power off condition

The use of Eq. (4.1) with the fatigue data given in Fig. 4.14 permits the life of the solder joints on a leadless chip carrier to be estimated. Generally this life is so short for the larger chip carriers that the leadless carriers are limited to an I/O

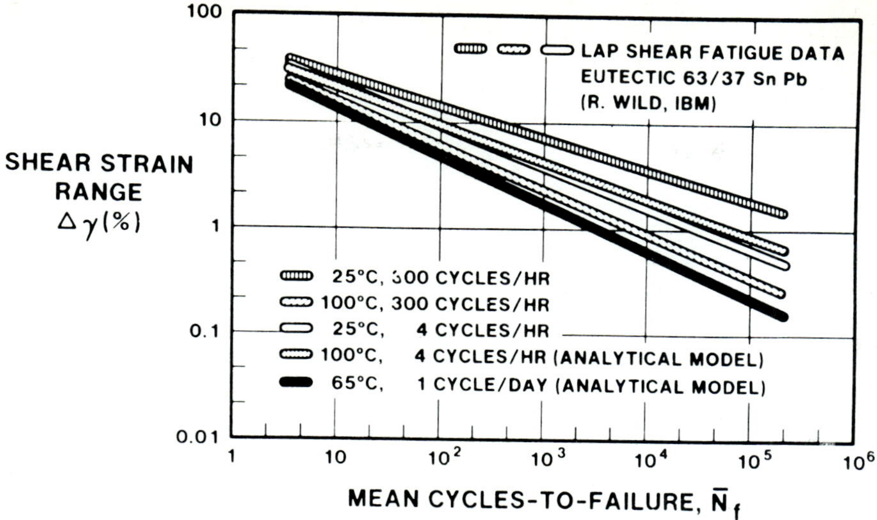

FIGURE 4.14
Fatigue diagram showing the shear strain range–life relationship for 63/37 Sn–Pb solder. (*Copyright, 1984, ISHM reprinted by permission.*)

of 20 or less when they are mounted on the common glass–epoxy circuit board. The exception to this limitation is when a ceramic circuit card is used to support the leadless chip carrier and the coefficients of expansion of the circuit card correspond closely to that of the chip carrier.

4.3 CHIP-TO-CHIP CARRIER MOUNTING

The chip is bonded to the chip carrier to prevent any movement of the chip relative to its housing during the life of the product. Bonding methods vary considerably and depend on the quality of the chip carrier. Consider first the high performance chip carrier with its body fabricated from ceramic. This carrier meets stringent hermeticity requirements and it is essential that the bonding material used to attach the chip to the carrier maintain this hermeticity. In these ceramic housings a eutectic solder of gold and silicon with a melting point of 390°C is often used as the bonding material. This solder is inorganic and will not out-gas with time and is ideal in view of the hermeticity requirements. This material also exhibits a high thermal conductivity of 296 W/m °C and aids in the transfer of heat from the chip to the case.

The bonding of the chip to plastic carriers is entirely different because hermeticity is usually not a significant concern. Instead, the problem is with the difference in the expansion coefficient of the silicon chip and the plastic housing. The expansion coefficient of silicon is 4 ppm/°C and that of plastic about 80 ppm/°C. The bonding of the chip directly to the plastic would result in large

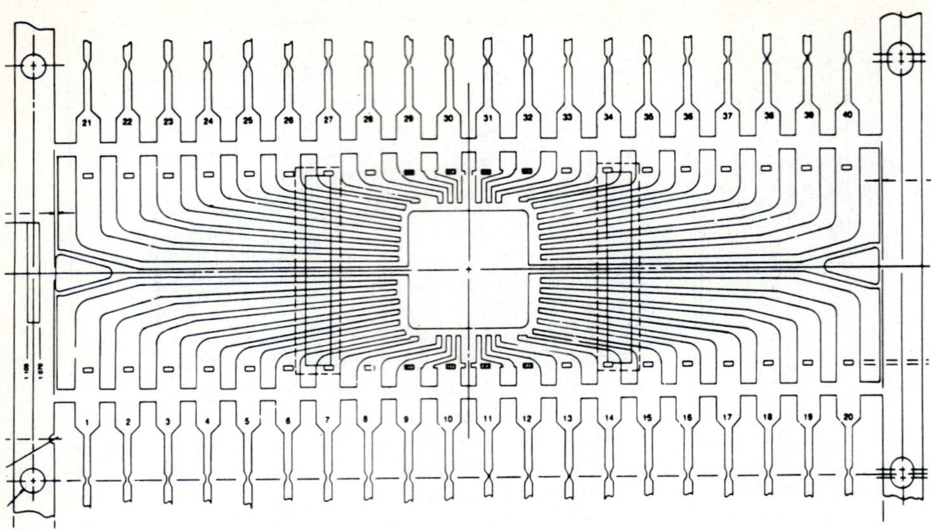

FIGURE 4.15
Typical lead frame used for chip bonding in a plastic DIP housing.

thermal stress induced in the chip due to temperature changes. This problem is circumvented by introducing a lead frame into the housing. The lead frame serves three purposes. First, it is fabricated from a copper alloy that has a coefficient of expansion much more closely matched to silicon and can be used as the bonding surface. Second, it provides the leads that extend out of the package. Finally, the lead frame provides a surface to which the small bonding wires from the chip can be welded. A typical lead wire frame for a 40 lead DIP is shown in Fig. 4.15.

Bonding to the lead frame can be performed with a number of different adhesives. Epoxies are often employed because of their high shear strength and chemical stability over long periods of time. Polyimide adhesives with a metallic filler are also used to extend the temperature capacity of the adhesive joint.

4.4 CHIP-TO-CHIP CARRIER CONNECTIONS

The I/O from the chip consists of a number of bonding pads usually arranged around the top edge of the chip (see Fig. 2.30). The bonding pads are very small, usually 5 mil square on 10 mil centers. On very dense chips the pad size and the spacing is reduced to 3 mil square on 6 mil centers. Connections are made between these bonding pads and the lead wires on the chip carrier. Three different methods, which include automatic wire bonding, tape automated bonding and flip chip bonding, are used in making these connections. Each of these methods will be described in the following subsections.

4.4.1 Automated Wire Bonding

Wire bonding is performed using either a thermocompression bond or an ultrasonic bond. With thermocompression bonding a small diameter wire is fed through a heated capillary and pressed against the bonding pad of a chip as illustrated in Fig. 4.16*a*. The bonding capillary is withdrawn, after a short interval, when the bond has been formed by the diffusion of atoms from the wire and pad materials into each other. The combination of temperature and pressure promotes diffusion, thereby producing a strong welded joint between the bonding wire and the pad.

In ultrasonic bonding the wire is forced against the bonding pad with a head that is vibrated in shear at a frequency of about 60 kHz. The local slip of

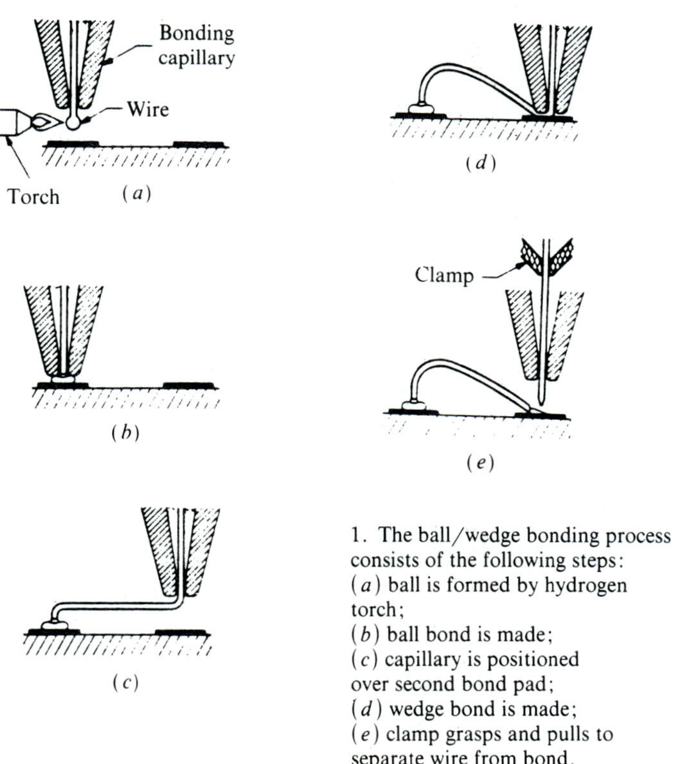

1. The ball/wedge bonding process consists of the following steps:
(*a*) ball is formed by hydrogen torch;
(*b*) ball bond is made;
(*c*) capillary is positioned over second bond pad;
(*d*) wedge bond is made;
(*e*) clamp grasps and pulls to separate wire from bond.

FIGURE 4.16
Steps in producing ball-wedge joints with a thermocompression welding process.

(*a*)

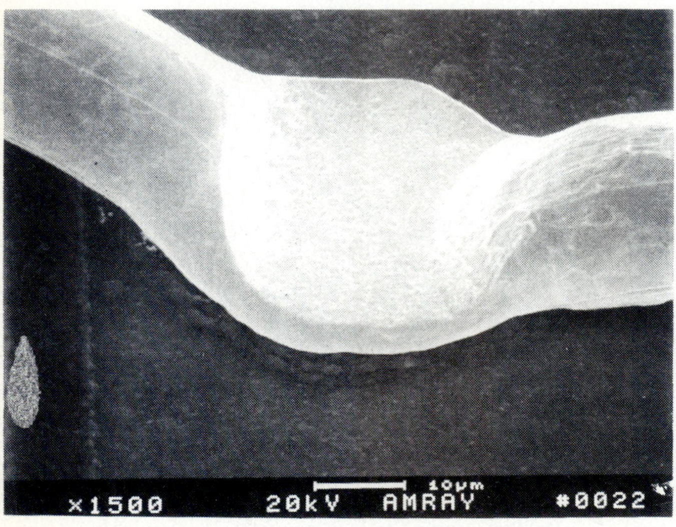

(*b*)

FIGURE 4.17
Typical bonded joints on a chip. (*a*) Gold bond on an aluminum pad with an intermetallic region. (*b*) Gold wedge bond on an aluminum pad.

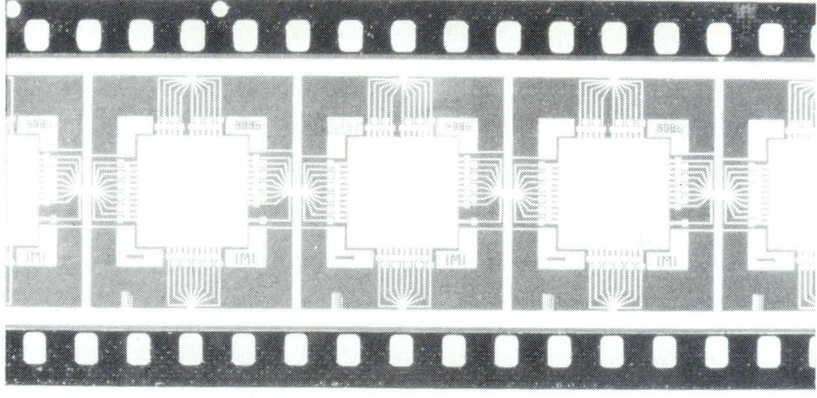

(*a*)

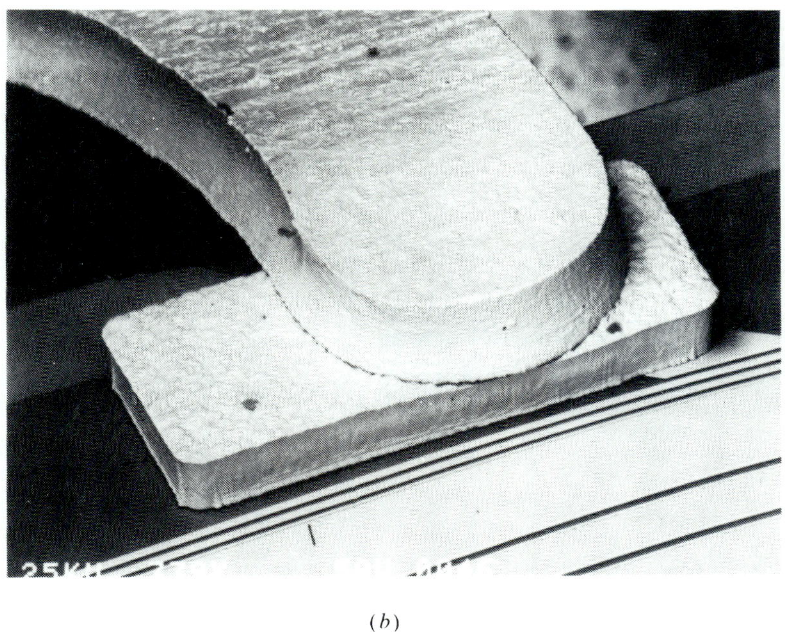

(*b*)

FIGURE 4.18
TAB method of chip-to-chip carrier connections. (*a*) Sprocket holes in the plastic carrier assist in positioning the TAB leads. (*b*) Precision etched foil leads attached to chip bonding pads. (*Courtesy of International Micro Industries, Inc.*)

the wire produces intimate contact of the two materials and friction welding occurs, which produces the joint.

One type of automated machine for wire bonding, illustrated in Fig. 4.16, utilizes the ball-wedge method for making the welded connections. The process starts with a hydrogen torch that heats the end of the wire to form a small ball. Then this ball is joined to the chip site as shown in Fig. 4.16*b*. Next, a loop is formed in the wire as the bonding capillary is moved to the lead wire bonding site as illustrated in Fig. 4.16*c* and *d*. The second thermal compression weld is made at this site and the wire is tensioned to produce failure of the wire adjacent to the joint. This completes the connection of a chip bonding pad to one lead wire on the chip carrier. The process is repeated on each pad until the chip is completely connected to the lead frame on the chip carrier.

The two metals used for the bonding wire are gold or aluminum, and the common diameter of the wire is 1 mil. Aluminum is used primarily in making wedge-type joints with ultrasonic bonding. Gold is easily bonded, using the ball-wedge method illustrated in Fig. 4.16. The ductile properties of gold aid in the welding process and high production rates can be achieved. Also, the gold is a noble metal and resists corrosion that can occur at the joints in plastic chip carriers.

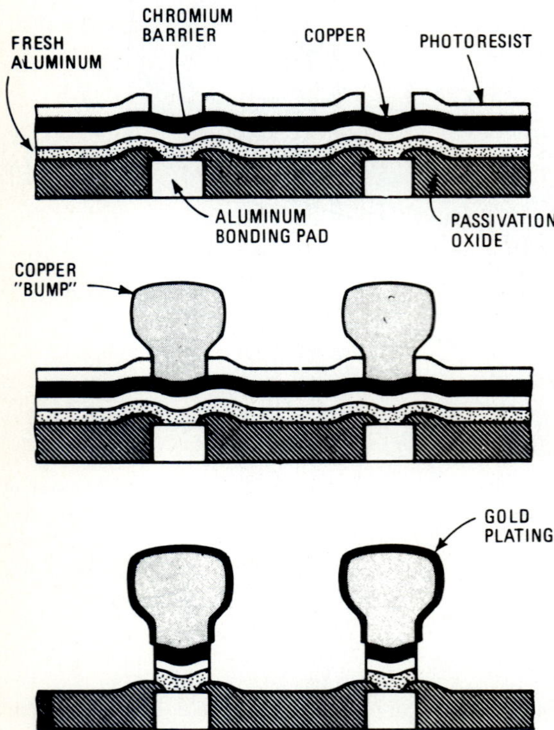

FIGURE 4.19
Chip pad bumps formed to enhance TAB thermocompression bonding.

Examples of the ball-wedge joints made with gold wire on aluminum pads are illustrated in Fig. 4.17.

4.4.2 Tape-Automated Bonding

In tape-automated bonding (TAB) the small diameter wires are replaced with a precisely etched series of foil leads supported on a plastic carrier. An example of typical TAB connecting leads is shown in Fig. 4.18. The plastic carrier is in the form of a roll of film with sprocket holes so that the TAB lead pattern can be automatically positioned over the chip. The lead pattern is etched in very thin copper foil and then plated with gold to protect the foil from oxidation during storage and to facilitate welding to the chip bonding pads. The pattern is positioned over the chip bonding pads and a heated pressure head is lowered over the assembly to simultaneously form a thermocompression bond to all of the chip pads. After this inner lead bond is complete, the chip may be tested to ensure that it is totally functional before completing the assembly of the chip carrier. The final step in the TAB process is making the connections to the pads on the chip carrier. This series of bonds is made in the same way with a larger pressure head designed to conform to the pad array on the chip carrier.

The welded joints formed with TAB require added processing of the chip bonding pads. This added processing is performed on the wafer before it is separated into chips. The aluminum bonding pads are plated with chromium and copper to develop a bump that raises the pad well above the surface of the chip as indicated in Fig. 4.19. The bumps are then gold-plated to facilitate thermocompression welding. The chromium layer serves to limit migration of the gold and aluminum atoms and inhibits the formation of Al–Au intermetallics that can degrade the joint.

The advantage of TAB over wire bonding is reduced labor content in completing the first level package, the ability to test the chip after completing the inner lead joints and the ability to handle higher I/O counts by placing smaller bonding pads on closer centers. As VLSI and ULSI chips become more common, TAB-type connections will become more widely used.

4.4.3 Flip-Chip Connections to the Chip Carrier

In standard chip carriers, the chip is positioned on its back surface and electrical connections are made to the bonding pads located about the perimeter on its top surface. In chip carriers using flip-chip connections this standard orientation of the chip is reversed. The chip is placed face downward and the back side of the chip is oriented upward. This flip-chip orientation has a significant advantage in that it concentrates the electrical functions on the underside of the chip, leaving the top side free for use in developing a highly efficient heat transfer design.

The flip-chip method of connecting the bonding pads to the chip carrier was originally developed by IBM and was used exclusively by that company for many years in its high performance computers. The original process was identi-

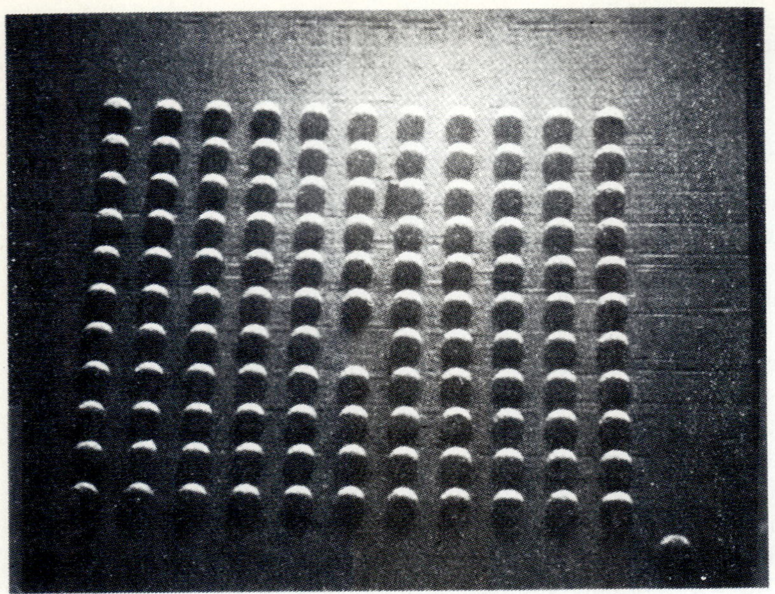

(a)

(b)

FIGURE 4.20
Chip and ceramic substrate for flip-chip bonding. (*a*) Solder spheres located on an area array of chip bonding pads. (*b*) Lead fan out and chip bonding pads on a 240 pin ceramic substrate.

fied as C4 by IBM to denote controlled collapsible chip connection. In this process the chip bonding pads are deployed in an area array over the surface of the chip in contrast to the development of these pads around the perimeter of the chip. By using these area arrays of bonding pads, a higher I/O count is possible. These bonding pads are 5 mil in diameter on 10 mil centers. Matching bonding pads are produced on a ceramic substrate so that the pads on the chip and the ceramic coincide. Spheres of solder 5 mil in diameter are placed on the ceramic substrate pads, as illustrated in Fig. 4.20*a*, and the chip is positioned and aligned relative to the substrate. The assembly is heated until the solder spheres begin to soften and a controlled collapse of the sphere takes place as the solder simultaneously wets both pads. Surface tension of the solder controls the geometry of the solder joints until the solder solidifies. Solder flow from the joints along the circuit traces on either the chip or the substrate is prevented by plating these traces with metals like chromium, which inhibit solder flow. This treatment controls the exact quantity of solder at the joint and provides for solder pillars about 3 mil high between the chip and the substrate. Precise control of the quantity of solder also prevents the formation of solder bridges between the closely spaced leads during the reflow process. An example of the solder balls on the chip pads is presented in Fig. 4.20*a*. A mechanical fan out of the lead pattern from the chip site to the pin holes is shown in Fig. 4.20*b*. This pattern is produced on a ceramic substrate that accepts a very dense chip packaged as a 240 pin grid array.

4.5 MULTICHIP PACKAGES

In most cases, electronic systems are assembled using a large number of components that are packaged as separate items. These individual components are mounted on a circuit card or cards and connected with PCB wiring, connectors and cabling. In some cases it is a disadvantage to package all of the components individually because of the need for added space for the connections and the loss of performance due to wiring length-induced delays. It is often advantageous to house several components that form a circuit function in a single multichip package.

4.5.1 Hybrids

Hybrid packages are first level carriers that house several chips and other passive components together and provide the wiring necessary for circuit connections. The hybrid assembly usually has an I/O count of less than 100. An example of a ceramic hybrid chip carrier is presented in Fig. 4.21. In a sense the hybrid is a combination of a first and second level package. It provides the protection function of a first level package while providing the wiring function of the second level package. The hybrids offer three advantages. In grouping several relatively small chips, the hybrid package reduces the area of the board required when compared to individual packages for each chip. Second, they reduce the complex-

FIGURE 4.21
Hybrid chip carriers. (*Courtesy of Aeroflex Laboratories, Inc.*)

ity in wiring the circuit board since the wiring between the chips is completed in the hybrid. Finally, the on-board lead lengths are short and often of closely controlled impedance to facilitate timing in digital systems and/or circuit matching in analog systems.

Hybrid chip carriers will continue to be an important first level packaging concept until custom design costs for special purpose chips are reduced to the point where a custom chip economically replaces the several small chips used in the hybrid. Replacement of the hybrid with a single customized chip housed in more standard chip carrier will provide a lower cost assembly, a more efficient design and a higher performance circuit.

4.5.2 Thermal Conduction Module

The thermal conduction module (TCM) is a multichip package designed and produced by IBM for use in its high performance computers. In the past few years several other firms have adapted and modified this concept in the design of the CPU for main frame and super minicomputers. The multichip carrier will be discussed in two parts of this text because it represents a major step forward in efficient packaging of electronic devices. Here we will consider the TCM as a very efficient (area and wiring length) chip carrier where the functions of the first and second level packages have been combined. In Chapter 8 we will consider the merits of this package in dissipating very large quantities of heat with a small thermal penalty.

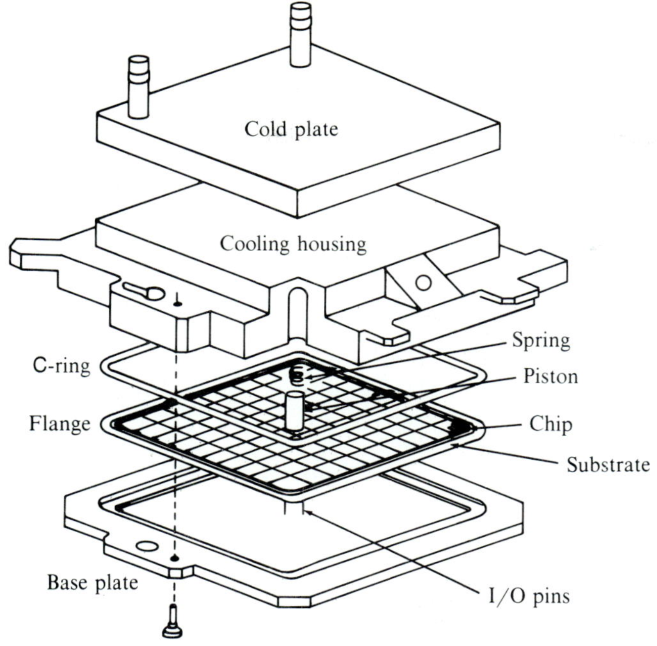

(*a*) Exploded view showing components

(*b*)

FIGURE 4.22
Construction details of the TCM. (*a*) Exploded view showing components. (*b*) Cut away view showing design features. (*Copyright 1982 by International Business Machines Corporation; reprinted with permission.*)

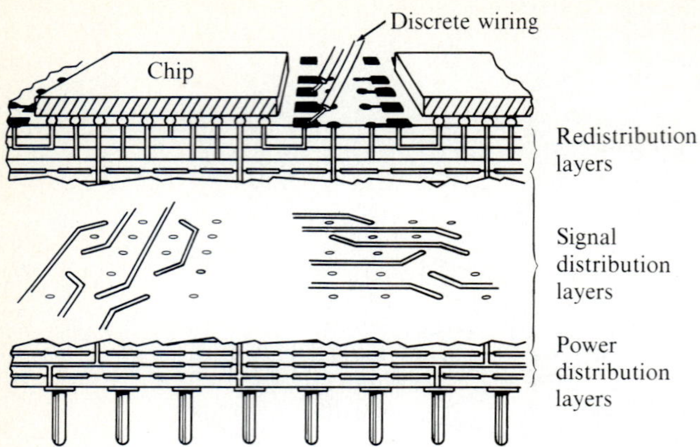

FIGURE 4.23
Construction details for the multilayered ceramic substrate for the TCM. (*Copyright 1982 by International Business Machines Corporation; reprinted with permission.*)

The components involved in assembly of a TCM are shown in Fig. 4.22. The principal element in the assembly is a 90 × 90 mm multilayered ceramic substrate, which provides the power distribution and the signal traces for at least 100 chips, which are simultaneously flip-chip bonded to this substrate. The substrate is made of 33 layers. Since each layer has a wiring plane on both surfaces, there are 66 wiring planes that support the power and signal wiring (impedance controlled) to 12,000 bonding pads (about 120 per chip). The substrate contains 1800 pins that serve as the I/O for the entire 100 chip module. The pins are inserted into connectors mounted on a back panel that connects several modules. The complexity of the substrate is evidenced by the presence of 16 wiring planes for signals, 350,000 vias for layer-to-layer connections and 130 m of *x-y* wiring. The 130 m of wiring represents a very small amount of wiring length required for interconnections to the 12,000 chip I/O associated with each module. The detail of the substrate is illustrated by the section view presented in Fig. 4.23.

The protection of the chips is provided by the 5 mm thick ceramic substrate on the bottom and the substantial aluminum housing on the sides and top. Hermeticity is achieved by employing a C-type metal seal about the perimeter and pressurizing the inside cavity with helium. The helium serves to keep the water vapor from entering the module and enhances the thermal coupling between the chips and the adjacent pistons. The heat transfer is accomplished by contacting spring-loaded aluminum pistons to the back of each chip. These pistons carry the heat by conduction to the water-cooled heat sink at the top of the package.

The very close positioning of 100 chips in a very small and densely wired chip carrier resulted in significant improvement in performance. Relative improvement in cycle time associated with two different types of packaging of high

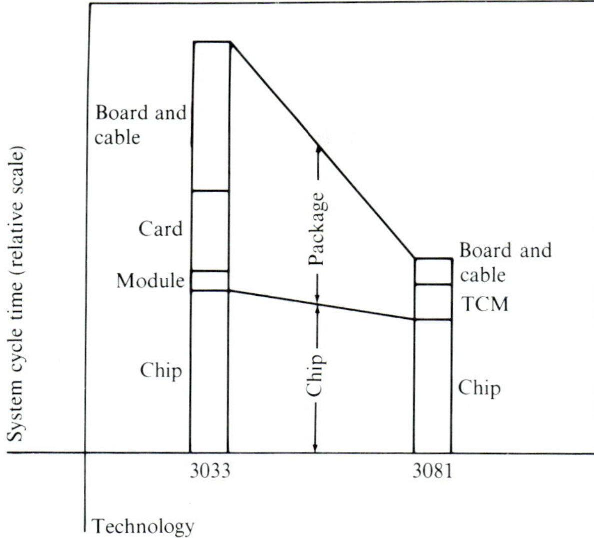

FIGURE 4.24
Improvement in cycle time for high performance computer systems due to enhanced efficiency of packaging. (*Copyright 1982 by International Business Machines Corporation; reprinted with permission.*)

end computing systems is shown in Fig. 4.24. It is important to note from this figure that improvements in system performance due to packaging by the TCM and its back panel were more significant than improvements in logic chip technology that occurred in the late 1970s.

4.6 FIRST LEVEL PACKAGING FOR OTHER COMPONENTS

While the emphasis in this chapter was on chip carriers, it should be recognized that first level packages are also important in protecting other devices like resistors, inductors, capacitors and diodes. Because each of these components is somewhat different from a packaging viewpoint, they will be covered individually.

4.6.1 Resistors

Resistors are available in a very wide assortment of package types as illustrated in Fig. 4.25. There are several reasons for the lack of standardization in the types of packages employed. First, the range of product where resistors are employed is quite large and the amount of power dissipated varies considerably. Second, the importance of circuit density changes significantly with the type of product. Next, the importance of temperature differs considerably with the application. Finally, the assembly process used for the board may influence the choice of package. To avoid having to pass the board through another line to assemble the resistors, the package for the resistors is often selected to match the carrier for the chips.

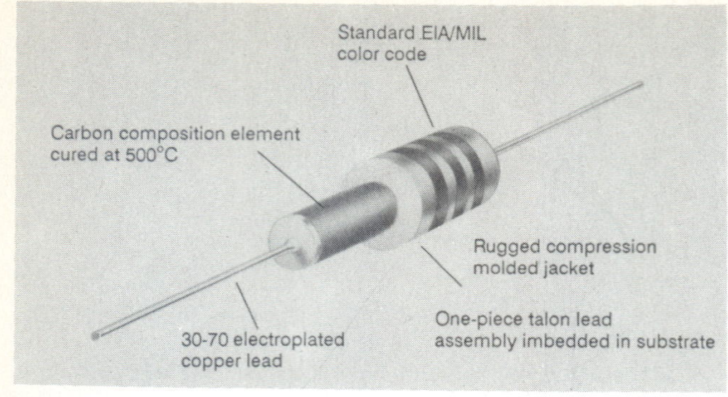

(*a*)

Ultra stable Tantalum
Nitride resistor film system
Tin-lead coated
wrap around
termination
Non-leaching
Nickel barrier

(*b*)

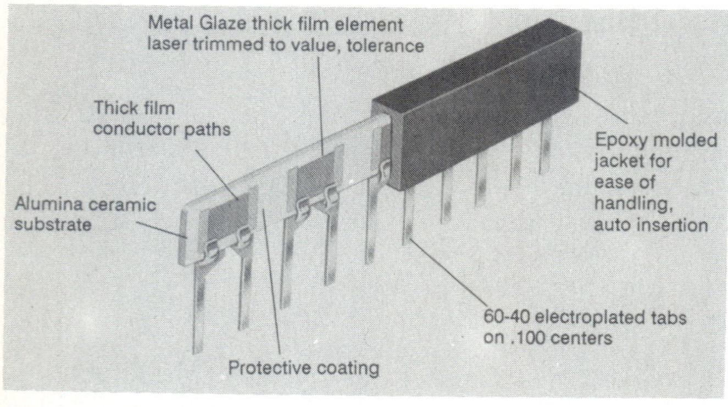

(*c*)

FIGURE 4.25
Representative first level packages for resistors. (*a*) Axial lead, (*b*) lead-less, (*c*) single in-line package (SIP), (*d*) dual in-line package (DIP), (*e*) flat pack network (*f*) LCC network and (*g*) heat sinked power resistor. (*Courtesy of International Resistive Company, Inc. and Dale Electronics, Inc., a Vishay Company.*)

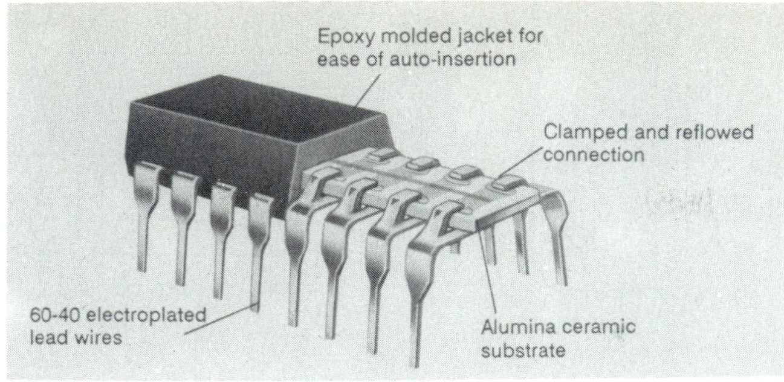

(*d*)

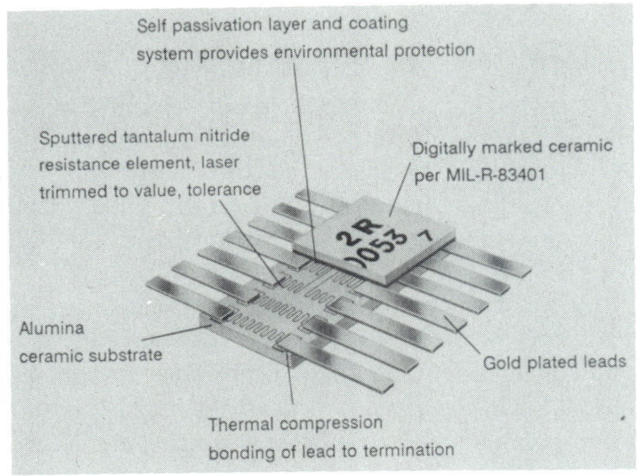

(*e*)

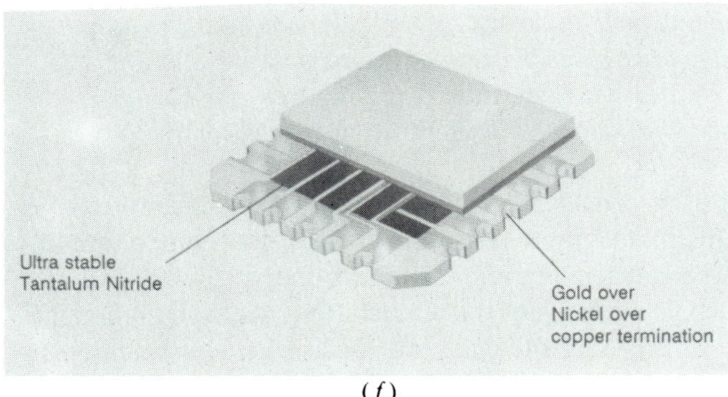

(*f*)

(*g*)

FIGURE 4.25
(cont.)

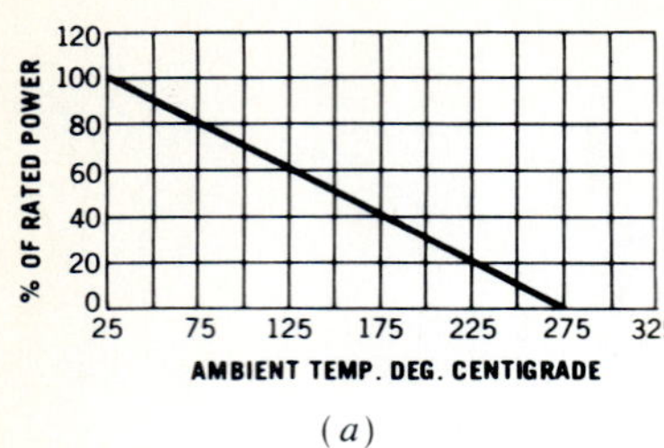

(a)

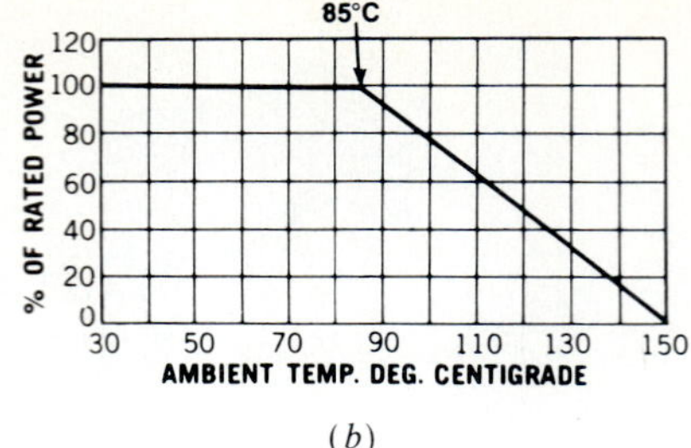

(b)

FIGURE 4.26
Derating curves for two types of resistors. (*a*) Wire wound power resistor. (*b*) Precision thin film resistor.

The most common resistor package is the axial lead type (see Fig. 4.25*a*) since it is easy to assemble into pin in-hole-type circuit boards. The lead-less package shown in Fig. 4.25*b* utilizes less board space and dissipates heat better than the cylindrical axial leaded package. The SIPS and DIPS shown in fig. 4.25*c* and *d* are used to save board space and to conform with assembly procedures involving boards containing only DIP-type chip carriers. Most of these resistor packages are the pin in-hole type and can be assembled with DIP chip carriers and connected to the PCBs with a wave soldering process.

The flat pack and the LCC-type packages also shown in Figs. 4.25*e* and *f* are surface-mounted and are used to conform with the surface mount chip carriers that house the ICs placed on the circuit board. The power resistor shown in Fig. 4.25*g* incorporates an aluminum heat sink. This addition is made to improve the amount of heat dissipation from the resistor.

Temperature plays an important role in the life and performance of resistors. The resistors are rated for a specific power dissipation and this rating is based on an ambient temperature T_a, usually 25°C. If the resistors are used at a temperature higher than T_a, they must be derated (used at a power level lower than the rated value) to avoid earlier than anticipated failures or a change in the specified value of resistance with time. Typical derating curves for two types of resistors are shown in Fig. 4.26.

Self-heating must also be considered in determining the temperature of a resistor. As the wattage dissipated by a resistor increases, its temperature increases relative to the ambient temperature as illustrated in Fig. 4.27. The thermal resistance that controls the slope of the temperature–wattage relation is a function of the size of the resistor and the materials used in the construction. In this case the resistors were cylindrical with axial leads. The resistors were fabricated by depositing a nickel–chrome alloy in spiral form on a high purity ceramic core. Packaging consisted of molding an epoxy case about the core to provide some protection against moisture. It should be noted that the thermal resistance penalty associated with this type of encapsulation is quite high, with

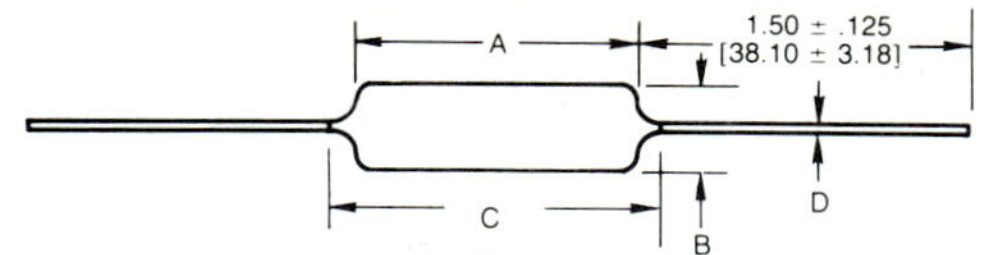

TYPE	DIM. A	DIM. B	DIM. C (Max)	DIM. D
CPF-1	.240±.020 [6.10±.508]	.090±.008 [2.29±.203]	.310 [7.87]	.025±.002 [.635±.051]
CPF-2	.344±.031 [8.74±.787]	.145±.015 [3.68±.381]	.425 [10.80]	.032±.002 [.813±.051]
CPF-3	.555±.041 [14.10±1.04]	.180±.015 [4.57±.381]	.650 [16.51]	.032±.002 [.813±.051]

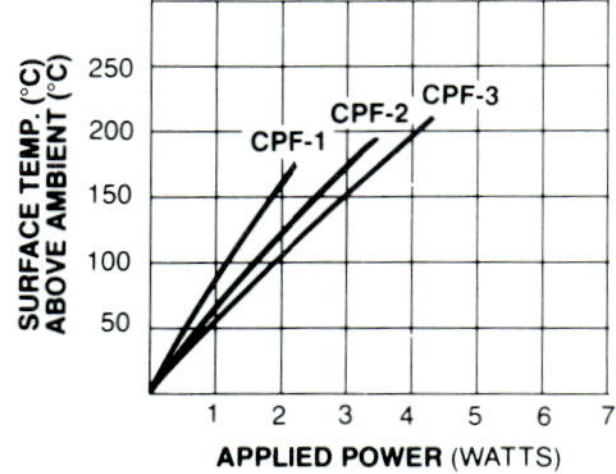

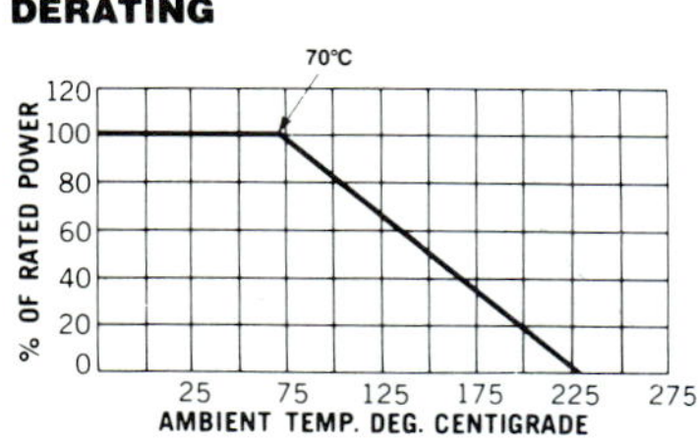

FIGURE 4.27
Self-heating in an axial lead-type resistor.

values ranging from 135–195°C/W. A description of thermal resistance and its effect on component temperature is presented in Chapter 8.

4.6.2 Capacitors

Capacitors, like resistors, are packaged in a wide variety of cases, which must be selected to be consistent with the packaging of the other devices on the circuit board. The PCB usually provides the support and the wiring for the capacitor; however, in some instances the capacitor is too large and it must be mounted on the chassis. Capacitors are generally divided into three categories, namely, aluminum electrolytic, tantalum and film. The aluminum electrolytic utilizes aluminum foil as the conductor and aluminum oxide as the dielectric. Because the electrolyte is liquid, the capacitor is usually packaged in a can, as shown in Fig. 4.28*a*, and sealed. Circuit board mounting with the capacitor leads being inserted into plated through holes is illustrated in Fig. 4.28*b*. The aluminum can is usually insulated by a vinyl chloride sleeve, which also serves for marking the part number, capacitance, working voltage and polarity. The lead wires emerging from the can are a copper-based alloy to provide a material that can be soldered with conventional methods. An aluminum to copper weld is made inside the can because the lead emerging from the aluminum foil is aluminum. Lead forces of about 3–4 lb can be supported. In using electrolytic capacitors it is important that polarity be established and confirmed. If the capacitor is used in reverse

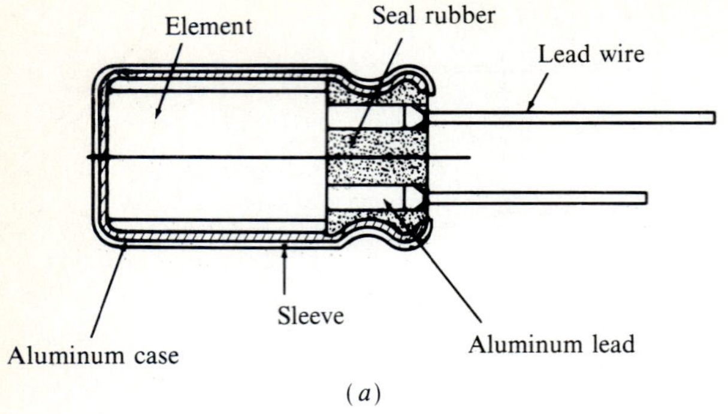

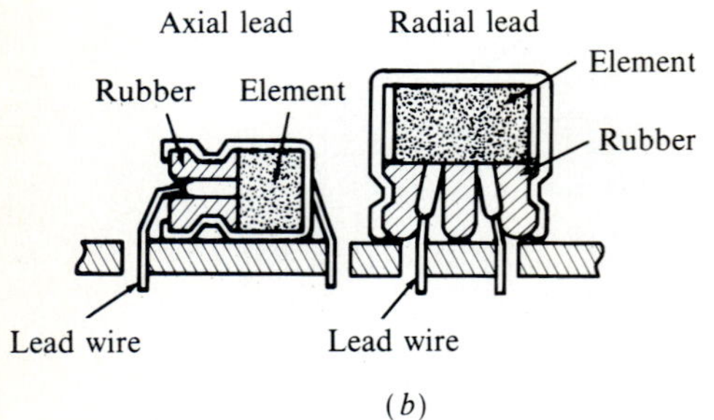

FIGURE 4.28
(*a*) Construction details for an aluminum electrolytic capacitor. (*b*) Mounting details for radial and axial leaded cans. (*Courtesy of Mepco Electra.*)

polarity, its life is reduced in the best of circumstances or the capacitor is destroyed in the worst case.

Tantalum capacitors are also of the electrolytic type and if the electrolyte is a liquid, the tantalum capacitor is sealed in a manner similar to that shown in Fig. 4.28*a*. The advantage of using tantalum in place of aluminum is in the dielectric constant of tantalum oxide ($\varepsilon_r = 24$) as compared to aluminum oxide ($\varepsilon_r = 8$). This property gives the tantalum capacitor the maximum capacitance at a given working voltage for a given volume. Since capacitor size is extremely important, the tantalum capacitor has gained a large share of the capacitor market.

Packaging these tantalum capacitors with a hermetic seal is more difficult than the rubber sealing method described in Fig. 4.28*a*. When hermeticity is required, a glass-to-tantalum seal, as illustrated in Fig. 4.29, is necessary. Note that the combination sealing illustrated in this figure involves an elastomer O ring with a deformed case to seal off the electrolyte and the glass-to-tantalum seal. This glass-to-metal seal was made with a laser weld about the tantalum case, which localized the heat and prevented damage to the elastomer seal. Finally, a solder-coated nickel lead wire is welded to the tantalum feed through tube to complete the container.

Film capacitors are fabricated from polymeric films, either metallized directly or used in conjunction with metallic foils. These capacitors utilize several different film materials including polyester, polycarbonate, polystyrene,

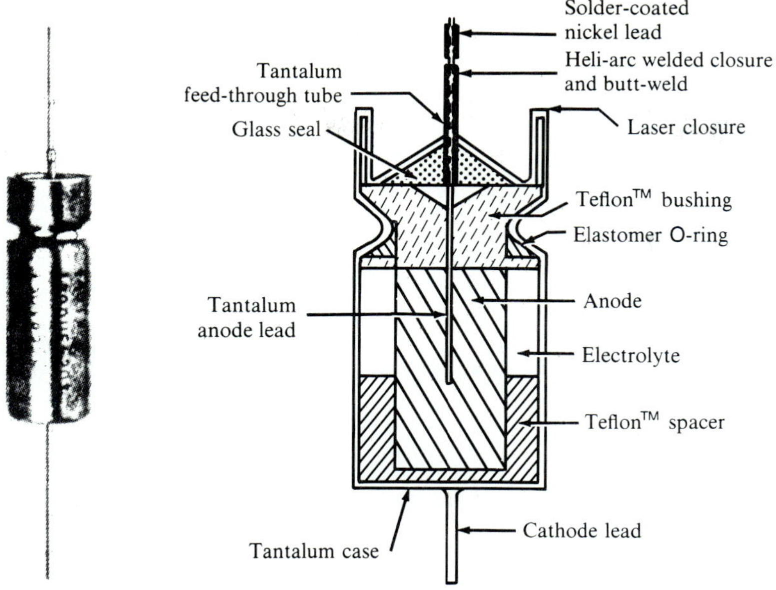

FIGURE 4.29
Hermetic sealing of a tantalum capacitor. (*Courtesy of Mepco Electra.*)

polypropylene, polyester and paper. These capacitors have the advantage of very high reliability and an extended temperature range of operation −55–125°C. However, they are relatively large in volume in comparison to the electrolytics described previously.

Capacitors, like resistors, are affected detrimentally by increasing temperature. Increasing the operational temperature decreases life and decreases the maximum allowable working voltage.

4.6.3 Diodes

Packaging for diodes depends largely on their purpose. Signal diodes either pass or block signals representing low voltages and low currents. These diodes are usually housed in small diameter glass or plastic cylinders with axial leads. They are mounted to a circuit board using pin in-hole technology the same way that an axial lead resistor is mounted to a board. The only difference is that the diode has a polarity and must be inserted in the correct direction for the circuit to function properly.

Diodes used in power supplies to rectify ac voltages are packaged in a wide assortment of housings as illustrated in Fig. 4.30. Because the lead count is very low, the emphasis in packaging the diode is on dissipating large amounts of heat from the device to the heat sink with a small thermal penalty. Many of these power diode packages incorporate a base or a flange that can be fastened directly

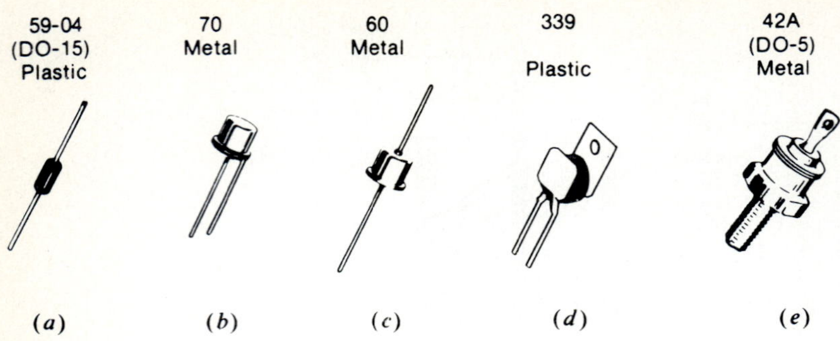

FIGURE 4.30
Diode carriers used in high current switching applications. (*Courtesy of Motorola Semiconductor Products, Inc.*)

to the heat sink to minimize the thermal resistance between the diode carrier and the sink.

REFERENCES

1. Hinch, S. W. and D. R. Cropper: "Evaluating Wire Bond Quality," *Semiconductor International*, vol. 2, pp. 93–107, 1981.
2. Harmon, G. G.: "Microelectronic Ultrasonic Bonding," NBS Special Publication 400-2, 1974.
3. Oswald, R. G. and J. W. Kanz: "Bumped Tape Automated Bonding: Past, Present and Future," *Int. Hybrid Microelectronic Symp.*, Minneapolis, Minn., 1978.
4. Lyman, J.: *Microelectronics Interconnection and Packaging*, McGraw-Hill, New York, 1980.
5. Mullen, J.: "How to Use Surface Mount Technology," Texas Instruments, Inc. Dallas, Tex., 1984.
6. Whalen, B. H.: "Technical Requirements for Advanced Interconnect Systems," *Proc. VHSIC Packaging Workshop*, NSWC, Silver Spring, Md. pp. 1–11, 1985.
7. Pound, R.: "IC Surface Mounting Gains with Standard Packages," *Electronic Packaging and Production*, vol. 4, p. 48, 1987.
8. Blodgett, A. J. and D. R. Barbour: "Thermal Conduction Module: A High Performance Multi layer Ceramic Package," *IBM J. Res. Develop.*, vol. 26-1, pp. 30–36, 1982.
9. Burger, W. G. and C. W. Weigel: "Multi-Layer Ceramics Manufacturing," *IBM J. Res. Develop.*, vol. 27-1, pp. 11–19, 1983.
10. Engelmaier, W.: "Functional Cycles and Surface Mounting Attachment Reliability," in *Surface Mount Technology*, ISHM Technical Monograph series 6984-002, pp. 87–114, 1984.
11. Markstein, H. W.: "Packaging Approaches for Mainframes and Superminis," *Electronic Packaging and Production*, pp. 28–31, October 1988.
12. Lyman, J.: "Fine-Pitch Assembly Affects Every Phase of Surface Mounting," *Electronic Design*, pp. 60–74, August 25, 1988.

EXERCISES

4.1. Sketch three types of solder defects that can occur when a chip carrier is mounted to a circuit board.

4.2. Prepare a sketch similar to Fig. 4.1 showing a chip positioned in the chip carrier. However, in your arrangement position the chip so that only one surface is utilized for wiring while the other surface is used for heat transfer.

4.3. Epoxy is well recognized as a polymer with excellent adhesive properties. Explain why epoxy is not employed to bond chips to ceramic chip carriers.

4.4. Describe the three common methods for designing the I/O connections on chip carriers. Propose another method for the lead arrangements that may increase the I/O count.

4.5. Describe the general characteristics of the dual in-line package (DIP) that is the most commonly employed chip carrier in use today.

4.6. Determine the area efficiency of a DIP that is used to house a chip that is 4 × 6 mm in size. Consider the number of pins on the DIP to be an open variable and graph the area efficiency as a function of I/O count.

4.7. Develop an equation showing the area of a pin grid array as a function of pin count. Prepare a graph showing this relation.

4.8. Locate a power supply preferably 100 W or larger. Remove the cover and locate the power transistors used in the rectifiers. Describe the chip carrier and the hardware used in heat sinking these components.

4.9. Repeat Exercise 4.6 replacing the DIP package with a flat pack.

4.10. Write an equation describing the area of a *J* leaded chip carrier as a function of I/O count and prepare a graph showing this relation.

4.11. Determine the shear strain range for a corner solder joint in a square leadless chip carrier with a side dimension of 0.75 in. The temperatures of the assembly as defined in Eq. (4.1) are $T_{cc} = 80$, $T_b = 30$ and $T_a = 20°C$. The coefficients of thermal expansion are $\alpha_b = 16$ and $\alpha_{cc} = 4$ ppm/°C. The solder joint thickness is 4 mil.

4.12. For the solder joint described in Exercise 4.11, estimate the fatigue life in cycles if the solder joints are made from 63/37 Sn–Pb-type solder.

4.13. Repeat Exercise 4.11 but replace the glass–epoxy circuit board with a ceramic board that exhibits $\alpha_b = 4$ ppm/°C. Find the fatigue life for this new design.

4.14. For the assembly described in Exercise 4.11, find the life of a solder joint located on one of the center lines of the package.

4.15. Describe the effect of the rate of thermal cycling on the fatigue resistance of solder joints. Briefly explain why the fatigue strength of solder is rate dependent.

4.16. Repeat Exercise 4.11 replacing the chip carrier with a rectangular one with dimensions of 0.3 × 0.4 in. Estimate the fatigue life of the corner solder joint and both of the center line solder joints.

4.17. Write an equation for the number of wire bonding pads that can be placed about the perimeter of a chip. Prepare a graph of the I/O count as a function of perimeter length for three different pad sizes and spacings.

4.18. Describe why the bonding of the chip carrier connections is usually performed in developing third world countries.

4.19. Describe the primary advantages and disadvantages of the TAB process for making the connections between the chip and the carrier.

4.20. Calculate the cost of the gold used in the bonding wires for a 40 pin chip carrier. Take the cost of gold at $420 per troy ounce. Comment on the cost trade-off between gold and aluminum bonding wires.

4.21. Note the rectangular array of solder balls positioned on the bonding pads, shown in Fig. 4.20*a*. Sketch the design of a fixture that can be used to collect the array of solder balls, to arrange them in position and to transport the array to the substrate. Recall that the solder balls are 5 mil in diameter and are spaced on 10 mil centers.

4.22. Consider a simple square 5 × 5 array of bonding pads 5 × 5 mil in size located on 10 mil centers. Sketch a lead pattern that will connect these pads to an adjacent array of pins that are mounted on 60 mil diameter pads located on 100 mil centers.

4.23. What is a hybrid chip carrier? List the primary advantages and disadvantages of using a hybrid.

4.24. Based on your reading of recent developments in the personal computer field identify a few custom designed chips that enhance performance of these systems without increasing their size or price.

4.25. Sketch the mounting of a hybrid chip carrier onto a circuit card using both the DIP and the SIP approaches. Discuss the relative merits of both approaches using packaging area and volume as the basis for your comparison.

4.26. Describe the advantages of the multichip module known as the TCM.

4.27. If the TCM represents such a marked improvement in packaging design, why is its use restricted to relatively high performance products?

4.28. Refer to Fig. 4.23 and describe the function of the redistribution layers, the signal layers and the power distribution layers in the ceramic substrate for the TCM. In your description recognize that the TCM is a 300–400 W module with an operating voltage of say 4 V.

4.29. What was the single most important reason for the marked improvement in cycle time associated with the TCM?

4.30. A 1 kΩ axial lead resistor of the PTF type is rated at $\frac{1}{4}$ W. This resistor is used in an application requiring a voltage drop across the resistor of 5 V. Determine the temperature increase due to self-heating.

4.31. Repeat Exercise 4.30 changing the resistance to 100 Ω. Comment on the power rating selected for this resistor.

4.32. Consider a power resistor with a self-heating thermal resistance $R_s = 120°\text{C/W}$ operating in an ambient temperature of 100°C. Determine the percent of rated power at which the resistor may be operated.

4.33. Repeat Exercise 4.32 changing R_s to 50°C/W and the ambient temperature to 125°C.

4.34. Explain why it was not necessary to consider self-heating in the discussion pertaining to the packaging of capacitors. Consider capacitors in both ac and dc circuits in the explanation.

4.35. Examine a circuit board from an available PC and identify the passive components, such as the capacitors, resistors and inductors. Describe some features of the board and the first level packages worth noting as design techniques.

CHAPTER 5

SECOND LEVEL PACKAGING— CIRCUIT BOARDS

5.1 INTRODUCTION

The circuit board is a major element in the mechanical design of an electronic system because it is the primary field replaceable unit that embodies many very important functions which include:

1. A mounting surface for most of the components.
2. Soldering pads to facilitate first to second level and second to third level connections. These pads also provide the mounting areas for attaching the components to the board.
3. Wiring channels to provide conduits for chip-to-chip connections.
4. A test bed to provide accessible and organized points which are probed in making circuit checks.
5. A marking surface to assist in identification of the components placed on the board in assembly. Other board markings support manufacturing operations.
6. A controlled profile on one side of the card with additional markings to support field maintenance and service.

Each of these functions of the circuit board will be discussed individually. The circuit board serves as a mounting plane for the various components such as ICs, resistors, capacitors and diodes. The board provides a method of fixation, either pin in-hole or surface mount, for the component attachment to the board. This attachment essentially fixes the position of each component for the life of the product. The board acting as a planar mounting structure also mitigates shock and vibration forces that are transmitted to the components through the board. In this sense the PCB, like the chip carrier, acts to protect the components. As the circuit boards are usually relatively thin planes with large lateral dimensions, the weight of the components that can be supported is limited. Usually heavy components, like transformers and displays, are mounted directly on the chassis since they can impart forces large enough to damage the board in a shock or vibratory environment.

The circuit board is usually a laminate with a copper cladding on one or both sides. The cladding is configured by an etching process to any shape required to provide attachment pads and the wiring for the electronic components. Footprints etched from the copper cladding, as illustrated in Fig. 5.1, provide soldering surfaces that are precisely aligned to match the corresponding

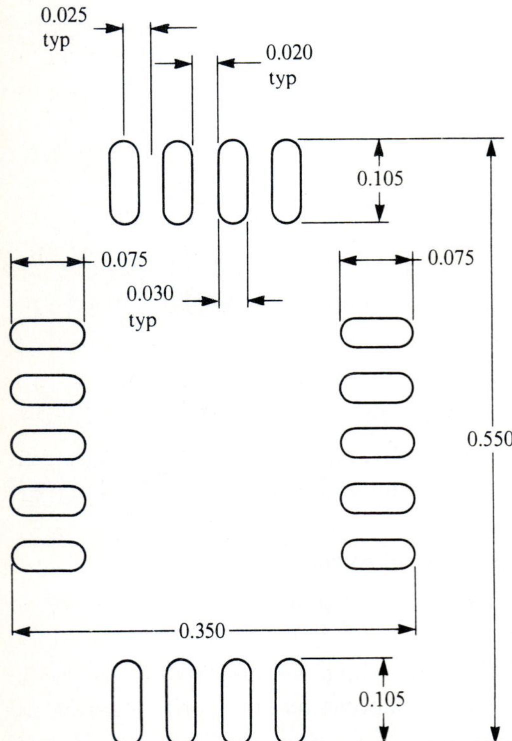

FIGURE 5.1
Footprint showing etched copper pads positioned on the circuit board to provide soldering surfaces for an 18 *J* leaded chip carrier. (*Courtesy of Texas Instruments, Inc.*)

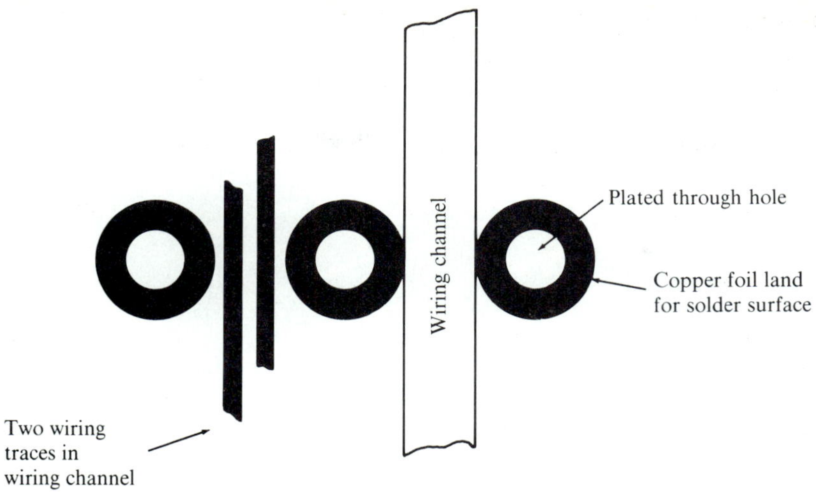

FIGURE 5.2
Wiring channels are formed between etched copper leads.

leads or pads from the components. The size and shape of these soldering pads control to a large degree the formation and geometry of the solder joints connecting the chip carrier leads to the board. Other solder pads are provided to attach to the leads for the I/O cables or connectors that carry power or signals to other PCBs in the system.

The circuit board also supports the wiring traces that lead between the components located on the board. These wiring traces are arranged in wiring channels as illustrated in Fig. 5.2. The wiring traces are etched from the copper cladding and a typical circuit board may support 1000 to 10,000 individual traces leading from one wiring point to another. Wiring traces include conductors for the signal, which are small in cross-sectional area because of the low currents involved in switching the logic gates. Traces for conducting the power to the components are much larger in width to accommodate the higher currents required to operate each device. For some circuit boards that contain several different layers, the wiring planes are embedded on interior layers. The power and ground wiring (V^+ and V^-), in multilayer construction, is provided with nearly solid copper planes dedicated to supplying the required current. The power and ground planes also serve to control the characteristic impedance of the signal lines located on adjacent signal planes.

When the circuit board is completely assembled, as shown in Fig. 5.3, it contains many different types of components and perhaps 100 or more devices. Because failure of at least one of these devices over the life of the product is probable, it is important that the circuit board be field replaceable. The defective card should be removable and easy to repair with equipment found at a service

FIGURE 5.3
An assembled circuit board with a wide variety of components that can be probed in testing. (*Courtesy of Eldre Components, Inc.*)

center. In repairing the board, it is necessary to test many different locations on the board with a voltmeter or an oscilloscope probe to ascertain which, if any, of the components have failed. To facilitate probing and subsequent repair, the devices should have lead structures that are accessible to a voltage probe. Also, the leads should be accessible for specialized soldering tools that can melt the solder at all of the chip carrier joints simultaneously to permit rapid removal of the devices without damage to the board and the remaining devices.

Since the copper cladding on the circuit board material can be etched to any configuration, it is easy to mark the circuit card to facilitate the manufacturing process. The example presented in Fig. 5.4*a* shows markings identifying the location various ICs employed in assembling this board. The markings illustrated in Fig. 5.4*b* indicate the part number, revisions and the date of circuit board production. It is clearly easy to mark the board with component outlines, part numbers, production dates, lot numbers or any other information that will assist in production control, component assembly and inventory control.

To facilitate maintenance, the shape of the card is modified to prevent improper substitution in field substitution and replacement. The profile of the circuit card along the edge where the connector pads are located is usually slotted or notched (see Fig. 5.4*c*). These slots act as keys and permit the correct board to

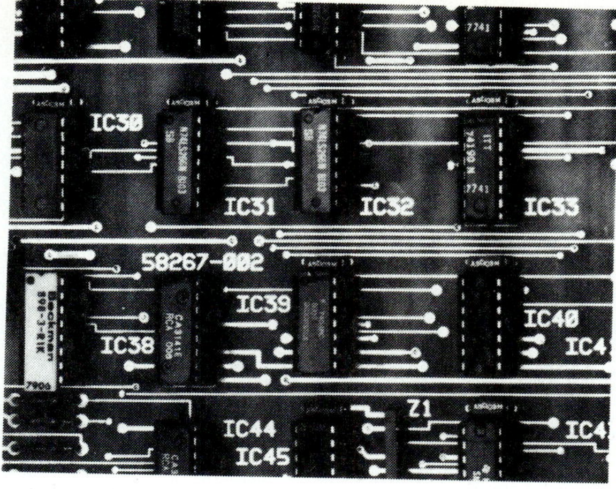

(a)

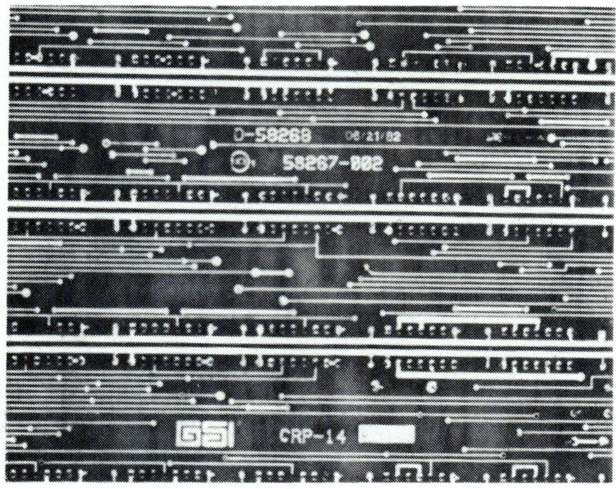

(b)

(c)

FIGURE 5.4
Circuit board marking and profiling to assist in manufacturing and maintenance. (*a*) Markings show IC numbers and locations. (*b*) Markings showing part number and date of production. (*c*) Profiled edge of the card forms a key insuring insertion of only the correct card. (*Courtesy of Gerber Scientific Instrument Co.*)

be inserted into the edge connector but prohibit the incorrect substitution of the wrong part number. Markings of the part number and the probe sites with reference voltages can also be placed on the card to assist maintenance personnel in servicing the card either in the field or in the repair facility.

5.2 TYPES OF PRINTED CIRCUIT BOARDS

There are three common types of printed circuit boards in use today including the single-sided, double-sided and multilayer board. These three types of boards can be fabricated from polymeric composites or from ceramics. The selection of a specific type of board depends to a large degree on circuit density and the number of connections that must be made between the components mounted on the board. For low density product where only 10–20 components with limited I/O are involved, low cost, polymeric composite, single-sided boards, which have circuit lines on only one side, are often used. Holes are drilled or punched into these laminates to support pins from the components; however, these holes are not plated and the solder joint is limited in size to the region between the pin and the top surface soldering pad.

For intermediate density product, double-sided boards are employed, which incorporate circuit lines on both the top and bottom surface of the laminate. The circuits on the opposite sides of the board are connected as necessary by plated through holes (PTHs) that serve as transverse conductors (called vias) through the thickness of the board. An example of a plated through hole is shown in Fig. 5.5. The PTH is also used to contain the pins for components that use pin in-hole-type mounting. In these cases the pin and the hole wall are soldered together and the solder joint encompasses the top and bottom solder pads and fills the entire hole.

Multilayer boards (MLBs) are fabricated from polymeric composites containing several layers of laminates, each layer having two wiring planes. Typical construction of the multilayer board is shown in Fig. 5.6. The layers are prepared in a sandwich with a pre-impregnated epoxy–glass sheet, known as prepreg, positioned between each layer. The epoxy in the prepreg at this stage is only

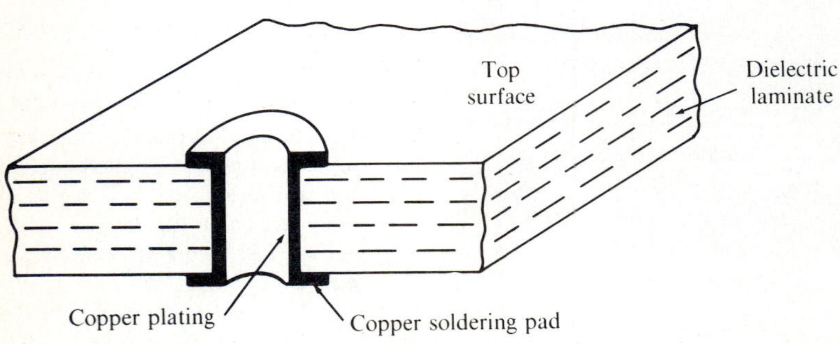

FIGURE 5.5
Plated through hole connecting the top and bottom surfaces of a double-sided circuit board.

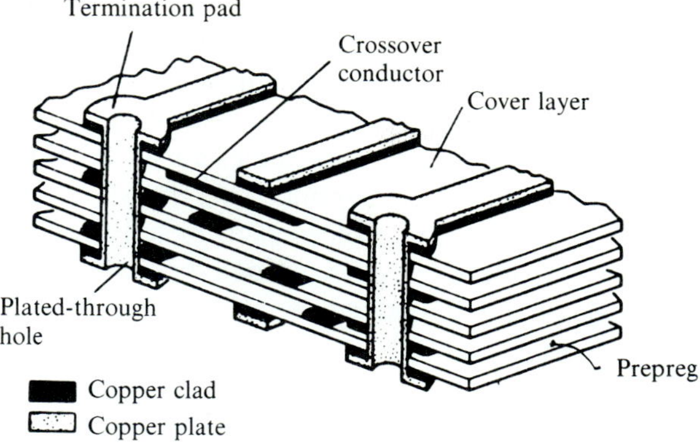

FIGURE 5.6
Multilayer circuit board using plated through holes to connect the top and bottom surfaces with interior wiring planes.

partially polymerized. After the stack of layers and prepreg are assembled, the entire assembly is placed in a hot platen press or an autoclave to complete the polymerization of the prepreg. This lamination process permanently bonds all of the layers together. It is important that the registration of all n layers be maintained for subsequent drilling of through thickness vias. The MLB consists of a single multilayer laminate with $2n$ wiring planes. Plated through holes act as vias to connect wiring in the interior planes with the leads for the chip carriers, which are soldered to the top surface. The plated through holes are used to support pin in-hole chip carriers and they become part of the soldered joints as in the preceding description. The plated through holes may also be filled with solder and used for thermal vias to enhance the conduction of heat through the thickness of the laminate.

Each of these three types of construction has its set of advantages and disadvantages. The single-sided board is the lowest cost and it is easy to maintain. However, it supports only low density circuits, electrical parameter control is not possible and crossovers of wiring traces must be accommodated with jumpers. Double-sided boards permit much higher density, some degree of impedance control is possible and crossovers are eliminated with through thickness vias. However, these advantages are gained with an increase in cost. Multilayer boards permit very high density circuits to be connected but at a significant increase in cost. MLBs also have the advantages of excellent impedance control of the signal lines and power switching through the use of ground and power planes capable of conducting very high currents. Disadvantages, in addition to cost, include lower manufacturing yields with an increasing number of layers and difficulty in repairing shorts or opens that may develop at interior wiring planes.

5.3 CIRCUIT BOARD MATERIALS

A large number of materials are available for use in fabricating circuit boards. Three factors are involved in the material selection, namely cost, electrical characteristics and mechanical properties. Cost factors are extremely important because the market for circuit boards is so large. In 1985 the printed circuit boards produced in the United States had an estimated value of \$3.63 billion, which involved about 75 million square feet of laminate. The immense volume of the market implies that cost will be a major consideration. Mechanical properties of importance include coefficient of thermal expansion, dimensional stability, flexural strength, modulus of elasticity, thermal resistance and resistance to chemicals and solvents. The electrical characteristics of concern include surface and volume resistivity, dielectric constant, dielectric strength, and dissipation and loss factors. These electrical properties are illustrated by describing the test methods used to determine each property.

Surface resistivity is measured by using the specimen shown in Fig. 5.7. A voltage of 500 V is applied across the center and ring electrodes and the bottom electrode is grounded to provide a guard, which eliminates the effects of currents that pass through the thickness of the laminate. The surface resistivity ρ_s is determined from

$$\rho_s = R_s p/s \tag{5.1}$$

where R_s is the surface resistance
p is the perimeter of the ring electrode
s is the spacing between the center and ring electrodes

The units for ρ_s are given as ohms per square and usually range from 10^7–10^{10} Ω/square for polymers. Note, that units are not given to the word square. This word is inserted to remind one that we are dealing with a surface resistivity.

Volume resistivity ρ_v is also measured with the specimen illustrated in Fig. 5.7; however, the voltage is applied between the center and bottom electrodes and the ring electrode is grounded to serve as a guard. The relation for ρ_v is

$$\rho_v = R_v A/t \tag{5.2}$$

where R_v is the volume resistance
A is the area of the center electrode
t is the thickness of the laminate

Most polymers exhibit $\rho_v > 10^{15}$ Ω cm.

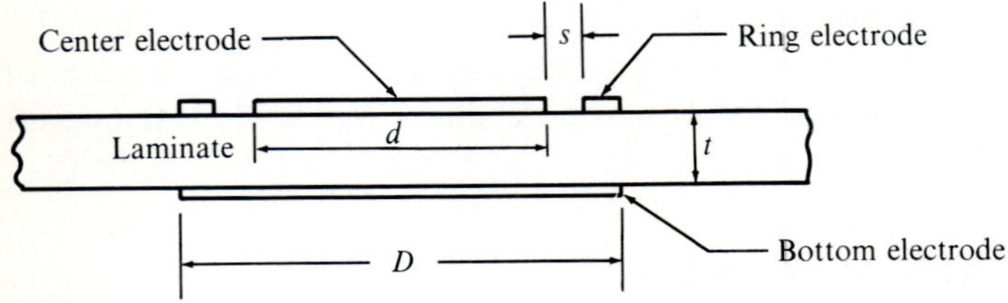

FIGURE 5.7
Test specimen used to measure surface resistivity and volume resistivity (ASTM D257).

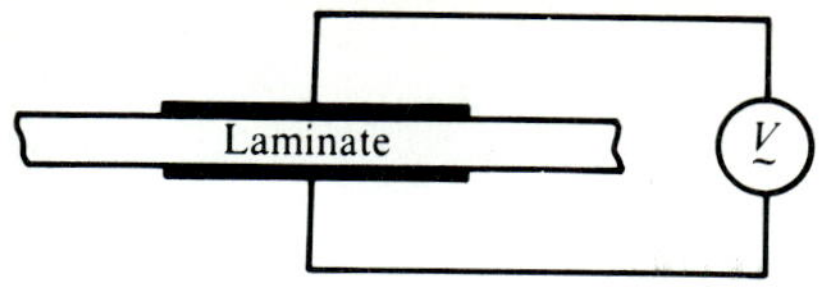

(*a*) Dielectric strength

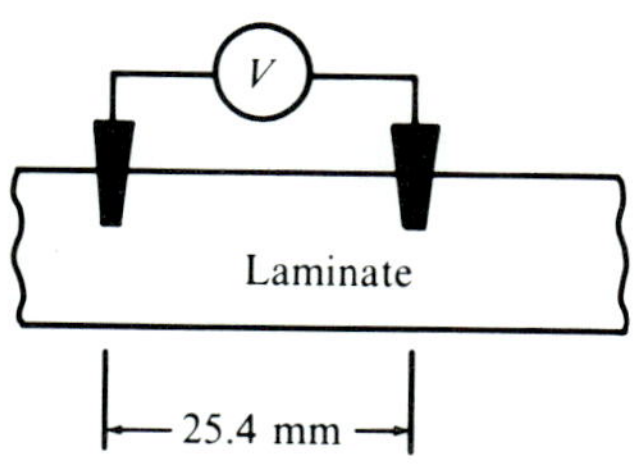

(*b*) Dielectric breakdown

FIGURE 5.8
Test arrangement for determining dielectric strength and breakdown ASTM D149. Both specimens are immersed in oil to eliminate surface conduction.

The dielectric constant ε_r of a material is the ratio of the capacitance C_r of a capacitor with that material used as the dielectric to the capacitance C_v of the same capacitor with vacuum as the dielectric. This definition is expressed as

$$\varepsilon_r = C_r/C_v \tag{5.3}$$

A low dielectric constant is necessary for circuit boards used in high speed signal processing where distributed capacitance is important. Most polymers exhibit ε_r that range from 2–4; however, when these polymers are reinforced with glass fibers that have $\varepsilon_r \simeq 6$, the effective dielectric constant for the laminate is increased to $4 < \varepsilon_r < 5$.

The dielectric strength is the voltage that can be applied per unit thickness before electrical breakdown occurs. The specimen used in this measurement is described in Fig. 5.8*a*. The voltage applied (usually at 60 Hz) across the electrodes is increased until failure of the insulation occurs. The dielectric strength of most polymeric materials exceeds 500 V/mil.

Dielectric breakdown is similar to dielectric strength except that the measurement is made with the potential applied parallel to the surface of the laminate. As indicated in Fig. 5.8*b*, the voltage is applied to tapered pins inserted into the surface of the specimen on 25 mm centers. To avoid surface conduction, which occurs at high voltages, the specimen is immersed in oil. Typical results for dielectric breakdown for polymers usually exceed 10 kV.

The dissipation factor is a measure of the loss of power due to the insulating material given by the ratio of the reactance to the resistance. This ratio is related to the loss angle δ by

$$\tan\delta = [\omega L - (1/\omega C)]/R \tag{5.4}$$

The loss factor F_L is related to the dissipation factor since

$$F_L = \varepsilon_r \tan\delta \tag{5.5}$$

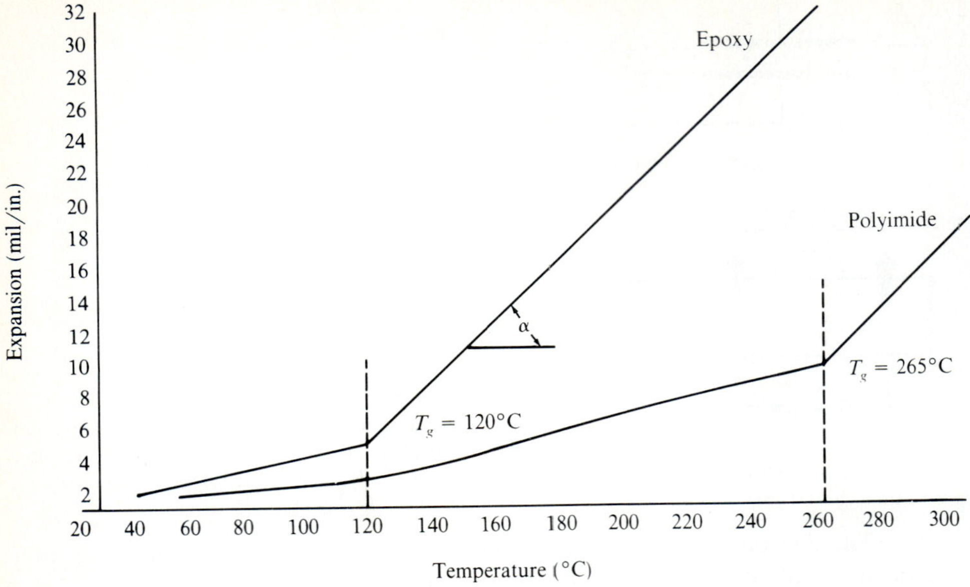

FIGURE 5.9
Expansion–temperature relation for epoxy and polyimide showing the glass transition temperature T_g. (The slope α is the temperature coefficient of expansion.)

The dissipation factor for polymers usually decreases as the frequency increases. This property, usually reported for a frequency of 1 MHz, ranges from 10^{-2}–10^{-4}. To minimize losses it is important to select materials with low dissipation factors.

Mechanical properties, such as flexural strength and modulus of elasticity, are well understood and will not be described here. However, the dimensional stability and the coefficient of expansion require further explanation. Dimension stability is a major concern at elevated temperatures associated with soldering the circuit boards. If the laminate material undergoes large dimensional changes at soldering temperatures, product yield will be adversely affected. The temperature coefficient of expansion is an excellent indicator of the dimensional stability of the polymers used in circuit board fabrication. Consider the expansion temperature curves for epoxy and polyimide shown in Fig. 5.9. The slopes of these curves give the temperature coefficient of expansion α. Note that there is a sharp discontinuity in each of the expansion curves at a temperature labelled as T_g. This term T_g is the glass transition temperature where the material changes from glass-like to rubbery. For temperatures higher than T_g the material is too soft to maintain dimensions if forces are applied in processing. Also, the coefficient of expansion in the direction normal to the laminate α_z is markedly affected when the temperature of the laminate exceeds the transition temperature. A typical example showing α_z as a function of temperature for an epoxy–glass laminate is presented in Fig. 5.10. As α_z increases, the expansion of the thickness of the

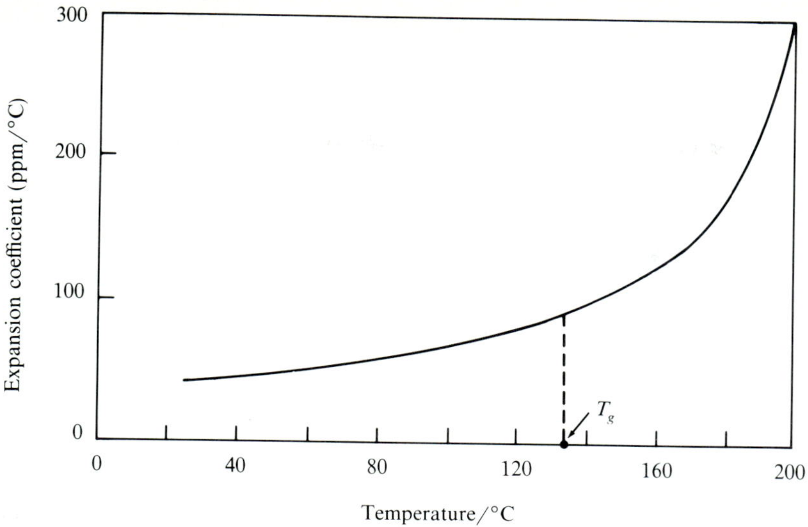

FIGURE 5.10
Coefficient of expansion α_z normal to an epoxy–glass laminate.

laminate increases, which produces very high stresses on the copper-plated through holes. The result, particularly if the temperature is cycled, is cracking of the copper plating in the via holes due to fatigue. Clearly, polyimide with its higher T_g exhibits superior dimensional stability at elevated temperatures.

The operating environment is also important in selecting materials. Humidity resistance is critical because the water absorption by the polymers affects their electrical properties and can seriously degrade performance. The maximum operating temperature is also important because both mechanical and electrical properties degrade rapidly when the glass transition temperature T_g for the polymer is exceeded.

The majority of circuit board laminates are fabricated using polymers as a matrix for a glass cloth, cotton fabric or paper reinforcement. These laminates are often known as rigid boards and they use thermosetting polymers, such as phenolics, epoxies or polyimides, in their production. Flexible circuits are also produced and these circuits employ plastic films without reinforcement as the substrate material to support the circuit traces.

The number of polymers used for the matrix and the number of papers and fabrics available for the reinforcement are both large and many laminate combinations are possible. We will review here only a few of the more common ones that are widely used in the United States.

XXXP is the most widely used grade of phenolic-paper laminate. It is low in cost and offers high insulation resistance under humid conditions. A similar laminate XXXPC affords the opportunity to form the holes by cold punching, which saves on drilling costs for single-sided boards. Another product FR-2 is identical to XXXP except that the phenolic matrix has additives that make it flame resistive.

FR-4 is the most commonly employed epoxy–glass laminate. It consists of a general purpose epoxy as the matrix and several layers of woven glass cloth as the reinforcement. The glass cloth reinforcement gives this laminate a high modulus 3–4 $\times 10^6$ psi, excellent dimensional stability and an implane coefficient of thermal expansion that is closely matched to that of copper, namely 16 ppm/°C. The electrical properties are also excellent although the glass transition temperature for this material is low with $T_g = 120°C$. Additives to the epoxy inhibit the spread of flames and this laminate is considered fire resistant. A similar product, called G-10 is produced without the fire retardant chemical. This laminate is used in most military applications where the out-gassing of fire retardant chemicals is considered more dangerous than the possibility of circuit board fires.

GI laminates incorporate polyimide as the matrix material and glass cloth as the reinforcement. The substitution of polyimide for the epoxy adds considerably to the cost of the laminate; however, the operating temperature is increased

TABLE 5.1
Properties of laminates used for circuit boards

Property	FR4 epoxy 30–35% resin	Keramid 601 polyimide 22–25% resin	FR2 phenolic 40–50% resin	Teflon glass fabric
Flexural strength ($psi \times 10^{-3}$) 25°C	80	71	20	15
299°C	30	57		
Flexural modulus ($psi \times 10^{-6}$)	2.5	4.0	1.2	
Interlaminar bond strength	2.5	2.1	1.0	0.8
Dielectric constant (1 MHz, 25°C)	4.3	4.6	4.0	2.35
Dissipation factor (1 MHz, 25°C)	0.019	0.018	0.03	0.0005
Volume resistivity ($\times 10^{-15}$ Ω cm)	0.2	0.15	0.01	0.2
Electric strength (kV/mm)	22	20	25	28
Deflection temperature (°C) (264 psi)	120	240	60	250
Thermal expansion (XY plane, ppm/°C)	12	11	18	20
Water absorption (%) (24 h)	0.1	0.3	0.6	0.02
Maximum continuous temperature (°C)	150	250	105	250

because $T_g = 265°C$. The high temperature properties of the polyimide also extend the strength and rigidity to operating temperatures well beyond the capabilities of the FR-4 laminate.

Polyimide films are used without reinforcement for flexible circuits. Flexible circuits are fabricated from foils of extremely ductile copper, which is adhesively bonded to a thin insulating film. The polyimide film known as Kapton (a trade name for a Du Pont product) is widely used as the insulating carrier in thickness

TABLE 5.2
Properties of polyimide (Kapton) films

	Type *H*	Type *V*	Type *F* (011)
	Mechanical		
Initial tear, lb/mil	0.75	0.75	0.75
kg/mm	13.4	13.4	13.4
Propagating tear, lb/mil	0.02	0.02	0.06
kg/mm	0.36	0.36	1.30
Tensile strength, psi	25×10^3	25×10^3	14×10^3
kg/mm^2	1.58×10^3	1.58×10^3	0.885×10^3
Tensile modulus, 10^5 psi	4.3	4.3	2.5
kg/mm^2	27.2×10^3	27.2×10^3	15.8×10^3
Elongation, %	70	70	75
	Electrical		
Dielectric constant	3.5	3.5	2.7
Dielectric strength, V/mil	4×10^3	4×10^3	2×10^3
V/mm	160	160	80
Dissipation factor	0.003	0.003	0.001
	Thermal		
Zero strength temperature, °C	815	815	815
Coefficient of thermal expansion (in./in./°C)	2.0×10^{-5}	$\sim 2.5 \times 10^{-5}$	
Flammability		Will not propagate flame; no afterflame	
Limiting oxygen index	35	35	35
Smoke generation (optical index, D_s)	1	1	
Shrinkage, mil/in. (after 1 hr at 200°C)	1–2.5	0.5	1–2.5
	Chemical		
Chemical resistance		Excellent (except for strong bases)	
Moisture absorption, % at 50% *RH*	1.3	1.3	0.4

TABLE 5.3
Properties of polymers

	Volume resistivity (Ω cm $\times 10^{-15}$)	Dielectric constant (1 kHz)	Dissipation factor (1 kHz $\times 10^3$)	Heat distortion temperature (°C)	Maximum continuous service temperature (°C)	Water absorbance (24 h) (%)	Chemical and solvent resistances	Price range per pound†
Thermoplastics								
Acetal	0.1	3.7–3.8	4–5	125	85	0.25	Fair	M
ABS	1–100	2.5–4.5	4–7	80–120	70–115	0.2–0.5	Poor	L
Acrylic	0.1–10	3.0–3.5	40–60	70–100	60–95	0.3	Poor	L–M
Nylon 6	0.01–0.1	3.6–4.0	14–20	65–105	120	1.3	Good	M
Polyamide-imide	100	3.5	1	275	290	0.28	Good	H
Polyarylether	15	2.9	4	150	120	0.25	Excellent	M
Polycarbonate	10	2.9–3.0	20	130–140	120	0.15	Fair	MH–H
Polyethylene HD	$10–10^3$	2.3	0.2–0.5	45–50	80–105	< 0.01	Good	L
PET	$> 10^3$	3.0–3.1	5	230	150	0.6	Good	L
PPO	100	2.6	0.4	100–130	80–105	0.06	Good	MH
PPS	50	3.1	0.4	245	260	0.02–0.03	Good	H
Polypropylene	0.01–100	2.2–2.6	0.5–1.8	50–60	105–125	< 0.01–0.03	Good	L
Polystyrene	> 100	2.4–2.7	0.1–0.3	80–105	60–115	0.03–0.05	Poor	L
Polysulphone	50	3.1	10	175	150–175	0.2–0.3	Poor	MH
PTFE	$> 10^3$	2.1	0.2	55	285	0.01	Excellent	H
PVC	10^{-3}	4–8	80–150	60–80	75	0.15–0.75	Fair	L
SAN	1	2.6–3.3	7–10	90–105	95–105	0.23–0.28	Fair	L
Thermosets								
Diallyl phthalate	> 100	3.3	8	240	150–175	0.15	Good	MH
Epoxy	0.1–10	3–3.7	2–20	45–285	120–285	0.02–0.03	Excellent	M
Melamine	0.01	6.6	0.02–0.03	135	100–200	0.6	Good	L
Phenolic	10^{-4}–0.01	4–5.5	10–50	150–200	150–250	0.2–1	Good	L
Polyimide	10–100	3.4	5	240–360	260	0.32	Excellent	H
Polyurethane	10^{-1}–50	3.4–3.5	5	90–95	95	0.1–0.2	Poor	L
Silicone	0.1–1	2.6–2.7	1–2	—	260	< 0.1	Good	MH

†Price range: L = below \$.50; M = \$.50–1; MH = \$1–2; H = above \$3.

ranging from 1 to 5 mil. The high temperature properties of polyimide described in the preceding paragraph make it the most suitable of the film polymers for the flexible circuit applications. A much lower cost polyester film can be employed for flexible circuits; however, the polyester films are limited to a maximum temperature of 150°C. This temperature is too low for soldering using normal materials and processes. As a result, the utilization of polyester films is limited to applications where connections can be made without soldering.

Properties of some of the boards and films used in hard and flexible circuits are given in the Tables 5.1 to 5.3.

5.3.1 Copper Foil

Copper foil is commonly used as a cladding material on one or both sides of the laminate employed in fabricating a printed circuit board. Because copper sheets were used in the construction industry for roofing applications prior to their application on PCBs, the thickness is measured in ounces per square foot. Converting 1 oz/ft^2 to a thickness gives 1.35 mil for the 1 oz cladding and 2.7 mil for the 2 oz cladding. Copper foil is produced either by an electrodeposition process or by rolling.

Electrodeposited copper is manufactured in a specially configured electrolytic plating tank where the copper from anodes is deposited onto the cathode, which is in the form of a highly polished stainless steel drum. By controlling the speed of rotation of the drum and the plating cell current, the thickness of the copper deposit can be controlled. As the drum rotates, the foil is removed to provide a product that is very smooth on the side in contact with the drum and rough on the other side. The roughness of the foil on one side is considered an advantage because this rough surface provides a mechanical connection when it is adhesively bonded to the circuit board in the lamination process.

The electrodeposited copper exhibits a grain structure with elongation of the grains in the direction perpendicular to the plating surface. This grain structure accounts for the rough surface and while the rough structure improves adhesion, it is not conducive to high strength or flexibility of the foils. In normal hard board applications this lack of flexibility is not a problem because the rigidity of the circuit board limits the strain imposed on the copper foil. However, in flexible circuit applications, the conductors are often subjected to repeated flexings that impose many cycles of relatively high strain on the copper foil. In these applications electrodeposited foils are not satisfactory and it is necessary to specify rolled and annealed foils.

Rolled copper foil is produced by melting purified copper and casting it in ingot molds. These ingots are then reduced in size by progressive rolling and annealing. The final reduction in thickness is usually performed in a Sendzimer mill, which is capable of producing foils of uniform thickness down to 0.5 mil. The strength and elongation characteristics of the rolled copper foil is superior to the electrodeposited foils because of the grain structure. In rolled foils the grains are elongated and parallel to the surface of the laminate, providing a structure resistant to failure due to cyclic strains.

The copper foils are usually treated before they are laminated to the substrate. This treatment may include up to three different additives on one surface or the other to improve the performance of the foil in subsequent steps in the manufacturing process. A bonding treatment on the rough side enhances the chemical activity of the foil to promote the copper–dielectric adhesion. A very thin layer of electroplated and stable alpha brass may be applied to provide a thermal barrier, which prevents degradation of the copper–epoxy bond during processing at high temperatures. Foil stabilizers are added to the smooth surface to passivate the foil and to prevent the foil from oxidizing or staining during storage.

As the requirements for circuit board density are increased to accommodate more modern high gate count ICs, circuit trace width will decrease and thinner foils will be necessary. One can anticipate further refinements in foil manufacturing methods as foil thicknesses are reduced to $\frac{1}{4}$ oz (0.34 mil) or less.

5.4 FOOTPRINT DESIGN

The media for design is a copper clad laminate, large compared to the circuit board dimensions, which can be cut or contoured as required for the product. The circuit traces and the connecting pads are to be produced by a photolithography and etching process that largely removes most of the traditional constraints from design that are due to the difficulties of machining contours. Essentially we are free to design the soldering surfaces and the circuit traces, which provide the required connectivity and protection in the smallest area possible. The primary constraint is in the etching process, which limits line width and line spacing.

The design of a PCB is usually done in three stages, namely, (1) footprint layout, (2) component placement and (3) routing of the circuit traces. We will cover footprint layout in this section and defer the coverage of placement and routing to Sections 5.5 and 5.6.

The footprint layout is arranged to accommodate a certain type of chip carrier or some other component. As an example, the footprint for a 16 pin DIP is shown in Fig. 5.11. The footprint drawing shows the location of the plated through holes for each of the 16 pins together with the tolerances. Note that the tolerances along the length of the chip carrier are not cumulative. The pins on the DIP are very rigid in this direction and, since all the pins are inserted together, it is necessary that close tolerances be maintained over the length of the chip carrier. The tolerances in the transverse direction are relatively large (10 mil). The pins are flexible in this direction and the insertion device inwardly adjusts the pins to fit the row-to-row spacing while maintaining alignment of the row of pins. The dimensions of the hole diameter d and the solder pad diameter D in Fig. 5.11 are not specified. These dimensions depend to a large degree on the precision that can be maintained in several of the manufacturing processes. A large scale drawing of the maximum size pin from a DIP in a plated through hole is shown

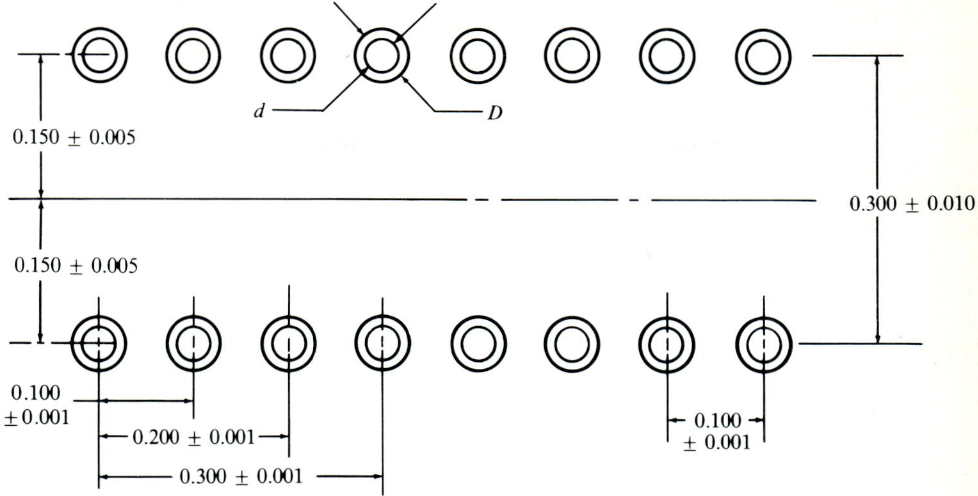

FIGURE 5.11
Footprint for a 16 pin DIP.

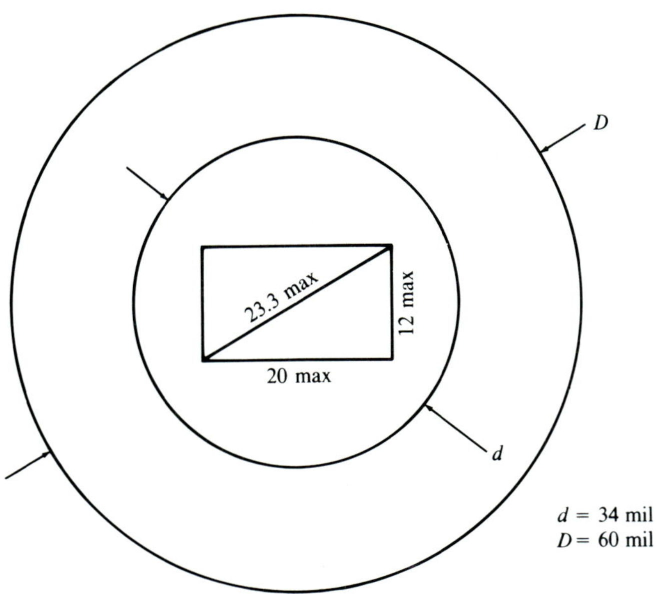

FIGURE 5.12
Scale drawing showing dimensions of a DIP pin in a plated through hole.

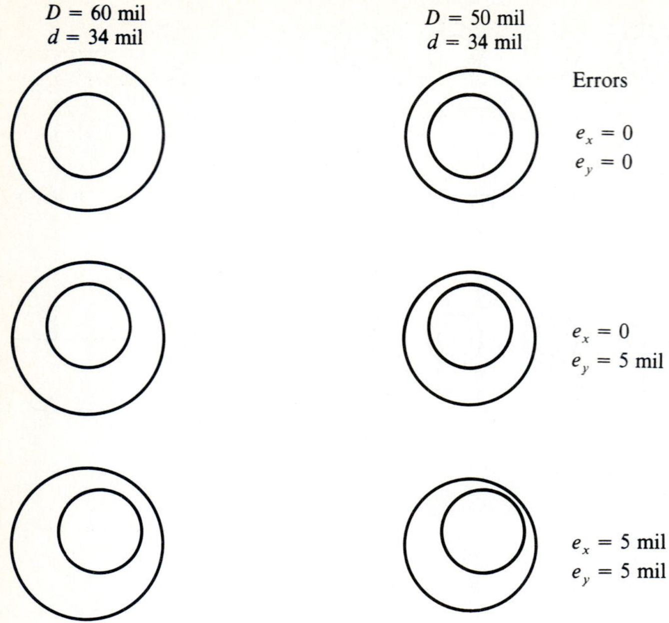

FIGURE 5.13
Influence of drilling registration errors in placement of the plated through hole relative to the soldering pad.

in Fig. 5.12. In this drawing $d = 34$ mil, $D = 60$ mil and the clearance and concentricity of the hole and solder pad are evident. The large diameter solder pad is useful in accommodating registration errors; however, it utilizes valuable board area and reduces the width of the wiring channel.

The influence of registration–drilling errors with solder pads having $D = 60$ and 50 mil is shown in Fig. 5.13. Note that the center of the hole and the center of the pad are located in two different manufacturing processes and differences in registration between these two processes produce errors e_x and e_y in locating the centers for d and D. With $D = 60$ mil it is possible to accommodate $e_x = e_y = 5$ mil while maintaining an annular region suitable for soldering. With $D = 50$ mil the annular region is seriously distorted with the same registration error and effective soldering is impaired. Indeed, when $e_x = e_y = 6$ mil, the hole breaks the edge of the pad and the board is considered a reject.

The footprint for a 12 lead flat pack-type chip carrier is shown in Fig. 5.14. While the leads for the flat pack are on 50 mil centers, the footprint is designed with holes on a 100 mil grid. This practice is common and is used to accommodate drilling machines that are established to drill holes in rectangular arrays on 100 mil centers. Note that the flat pack is a surface-mounted package and the holes shown in Fig. 5.14 are necessary only to serve as vias used to connect flat pack leads to internal wiring planes or as thermal vias to enhance heat transfer

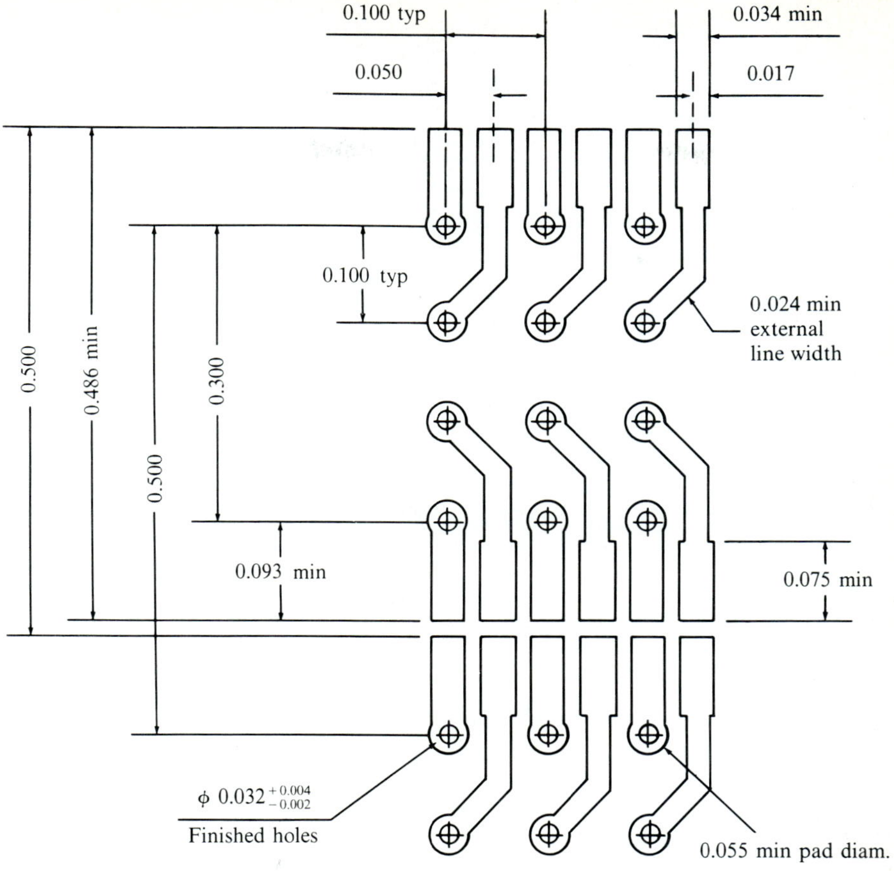

FIGURE 5.14
Footprint for 12 lead flat pack-type chip carrier.

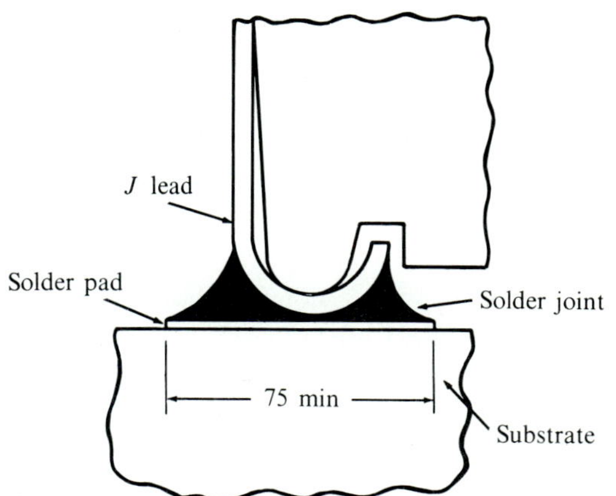

FIGURE 5.15
Length of solder pad to accommodate a relatively long solder joint for a *J* leaded chip carrier.

FIGURE 5.16
Circuit board prior to assembly showing an orderly arrangement of several different footprints.

through the thickness of the board. The diameter of the plated through holes can be reduced, providing that manufacturing has the ability to drill and plate the smaller diameter holes.

The footprint presented in Fig. 5.1 is for a relatively new *J* leaded chip carrier. In this design, the solder pads are on 50 mil centers with a width of 30 mil to accommodate the 18 mil wide *J* lead (see Fig. 4.12). The length 75–105 mil of the solder pad is to provide a long base for generous solder fillets with the large radius of curvature of the *J* lead as shown in Fig. 5.15.

Footprints corresponding to several different chip carriers and components are arranged on a circuit board to accommodate the number and type of devices anticipated in the development. Note that part numbers for ICs may not be specified at this stage in the development, but the type and number of leads for the chip carriers are specified. An example of a completed board prior to assembly, which shows the combination of several footprints arranged in an orderly fashion on a single large circuit board, is presented in Fig. 5.16.

5.5 COMPONENT PLACEMENT

A typical circuit board supports many components and devices and it is often necessary to locate 100 or more components over the available area of the board. The placement (selecting the location) of each of these components on the board is not arbitrary. There are two different design goals influencing the location where each device is placed. The first design goal is wirability, which pertains to minimizing the total length of wire required to connect the devices. Minimizing the wire length reduces transit time of the pulses and alleviates congestion of the circuit traces that results in crossovers and cross talk.

The second design goal is to place the devices at locations that lead to improved reliability of the circuit board. This goal is achieved by placing the more critical (less reliable) components at locations on the circuit board where their temperatures will be minimized. The reliability or wirability design goals often conflict because a placement sequence that ranks well from a wiring viewpoint often ranks poorly when reliability is considered.

Regardless of which design goal is given preference, placement is a difficult problem because of the very large number of unique placement possibilities that exist. The number of placement options k for a board with N components is $k = N!$. Of course, this fact implies that a very simple board with say only eight devices to be placed has $k = 8! = 40{,}320$ combinations to consider if placement is to be performed rigorously. With circuit boards populated by nearly 100 devices, the numbers involved tax very large computers with placement codes seeking to optimize circuit board performance with optimum selection of the location of each of the components.

5.5.1 Placement for Enhanced Wirability

In this discussion, we will separate the two design goals and treat placement to minimize wire length in this section. Because placement techniques that handle very large numbers of components are complex from both mathematical and numerical viewpoints, we will simplify the problem by considering only a small number of components with limited I/O. With this simplified approach it will be possible to understand the concept of placement to enhance circuit board wiring. More rigorous and detailed treatments of this topic are given in references 2 and 3.

Consider a simple example that involves a circuit board with an edge card connector and footprints defining 6 chip carriers as illustrated in Fig. 5.17. The footprints are aligned in two rows labelled ABC and DEF. Each footprint has 8 pins or pads where connections may be made. The edge connector with 14 connecting pads is positioned along the lower edge of the card. Since the connector is a component that we have already placed (along the edge of the card), we consider it seeded and permanently located. A coordinate system for the board is established with its origin at pad No. 1 of the edge connector. The pads are arranged on a grid so that wire length can be determined as grid units. In determining the wire length between two points, the Manhattan distance s is employed, where

$$s = x + y \tag{5.6}$$

The concept of rectilinear measurement of the length between two points, which is illustrated in Fig. 5.18, is the correct measure of distance in circuit boards, since the circuit traces commonly run in either the x or y direction. Diagonals, which yield shorter lengths between two points, are rarely used in routing the circuit lines because the diagonals block wiring channels and cause crossovers.

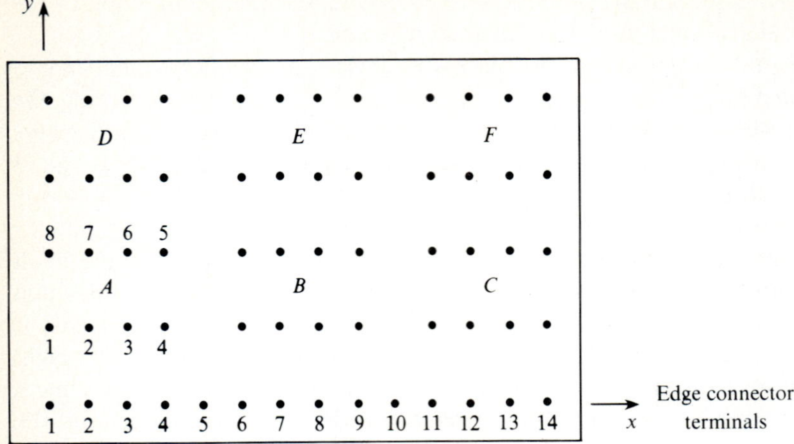

FIGURE 5.17
Circuit board with footprints for 6 chip carriers each with 8 leads and a 14 pin edge connector.

Returning to the placement problem, we are to determine which device from a list of six devices (Nos. 1–6) that are to be placed at the locations *A* to *F* on the circuit board. To make this determination, we refer to a wire list, which gives the connectivity between pins on each device and to the edge connector. An example wire list for this illustration is given in Table 5.4. The wire list is examined to determine the connectivity with the first seeded component (the edge connector). This connectivity, listed in Table 5.5, shows that devices Nos. 2 and 5 each have five connections to the edge connector and all of the other devices have only two leads to pick up the supply voltages. The high connectivity of devices Nos. 2 and 5 indicates that they be placed in the lower row at locations *A*, *B* or *C*. The equal connectivity of these two devices with the edge connector permits us to seed them together. Examining the wire list for device No. 2 shows an average

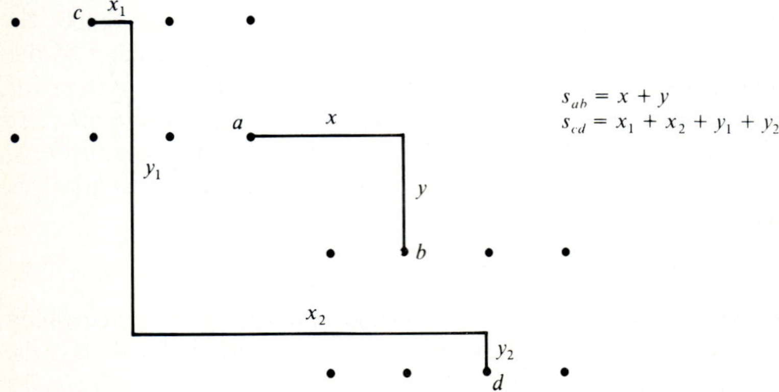

FIGURE 5.18
Manhattan measure of distance between two pairs of points, *a*-*b* and *c*-*d*.

TABLE 5.4
Wire list for the circuit board shown in Fig. 5.17

Device no.	Pin no.	Device no.†	Pin no.
1	1	2	5
	2	2	6
	3	5	5
	4	ec	1
	5	5	6
	6	3	3
	7	3	5
	8	ec	13
2	1	ec	3
	2	ec	4
	3	ec	5
	4	ec	1
	5	1	1
	6	1	2
	7	3	1
	8	ec	13
3	1	2	7
	2	5	7
	3	1	6
	4	ec	1
	5	1	7
	6	4	1
	7	4	2
	8	ec	13
4	1	3	6
	2	3	7
	3	6	6
	4	ec	2
	5	6	7
	6	6	1
	7	6	2
	8	ec	14
5	1	ec	8
	2	ec	9
	3	ec	10
	4	ec	2
	5	1	3
	6	1	5
	7	3	2
	8	ec	14
6	1	4	6
	2	4	7
	3	6	5
	4	ec	2
	5	6	3
	6	4	3
	7	4	5
	8	ec	14

†ec is edge connector.

TABLE 5.5
Connectivity to the first seeded component (edge card connector)

Device no.	Number of connections
1	2
2	5
3	2
4	2
5	5
6	2

pad location on the edge card connector of $x = 5.2$ units. This location corresponds to a position midway between footprints A and B. The average pad location along the edge connector for device No. 5 is $x = 8.6$, which is adjacent to the B footprint. With this information on connectivity along the x axis, it is apparent that device No. 2 be seeded at location A and device No. 5 at location B.

The new placement problem reduces to positioning the remaining four devices. The opening at location C is adjacent to the edge connector and device No. 5. The listing of remaining connectivity, shown in Table 5.6, indicates that device No. 1 has four connections, device No. 3 has three connections and devices Nos. 4 and 6 have only two connections to the seeded components positioned adjacent to location C. Maximizing the connectivity count places device No. 1 in location C.

To position the remaining three components, a connectivity matrix of the type presented in Table 5.7 is prepared. This matrix shows a large connectivity count between devices Nos. 4 and 6 and indicates that devices Nos. 4 and 6 should be placed side by side. Positioning device No. 4 in position E and device No. 6 in position D, gives position F to the remaining device No. 3, where it connects well to devices Nos. 1 and 4. The diagram shown in Fig. 5.19 indicates the final placement and the connectivity to adjacent locations.

This simple example indicates a process for determining the optimum placement of components to enhance the wirability of the board. The importance

TABLE 5.6
Connectivity to the seeded components adjacent to location C—edge card connector and device no. 5

Device	Number of connections		Total
no.	5	ec	no.
1	2	2	4
3	1	2	3
4	0	2	2
6	0	2	2

TABLE 5.7
Connectivity matrix for devices 3, 4 and 6

Devices	1	2	3	4	5	6
3	2	1	0	2	1	0
4	0	0	2	0	0	4
6	0	0	0	4	0	2

of this topic will become more evident in Section 5.6 where board routing is described. In real design situations, where the large number of components to be placed increases the complexity, manual placement techniques require excessive time. The placement process is automated by using extensive computer simulations with placement algorithms in an attempt to arrange the devices at locations that reduce the total length of the circuit traces required to connect the devices.

5.5.2 Placement for Enhanced Reliability

The failure rate λ of an integrated circuit IC is given by

$$\lambda(T_j) = A + Be^{-C[(1/T_j)-(1/298)]} \tag{1.1}$$

where the constants A, B and C depend upon the type of IC and the chip carrier used to house the IC (see reference 4). T_j is the junction temperature of the IC in Kelvin. Note, that the failure rate depends strongly on the junction temperature T_j of each chip. To achieve maximum reliability for a given board, it is necessary to consider the sum of the individual failure rates for each IC placed on the board. The failure rate for a board supporting N components is given by:

$$\lambda_T = \lambda_1 + \lambda_2 + \cdots + \lambda_N \tag{5.7}$$

To optimize board reliability the components must be placed to minimize the total failure rate λ_T. This is a difficult problem to solve in the exact sense because

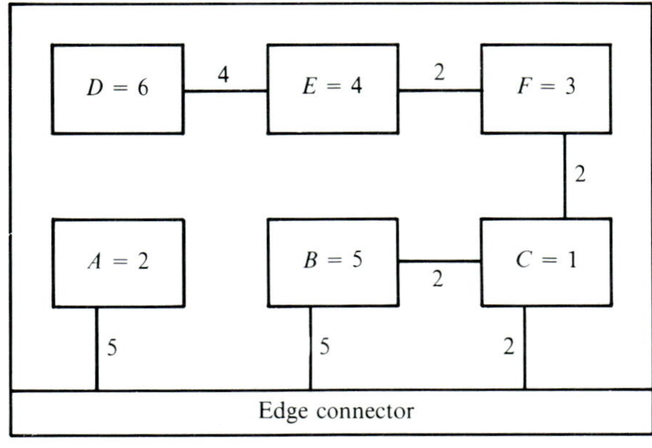

FIGURE 5.19
Placement of six devices showing connectivity to adjacent components.

of the number of computations required. There are $N!$ placement combinations to consider with a different λ_T for each combination. We could consider each combination, compute each junction temperature, determine the individual values of λ_i, then λ_T and finally select the combination corresponding to the minimum value of λ_T. However, with a large N, N factorial $N!$ becomes excessive and this procedure is too involved; therefore, approximate methods are usually employed.

Recently Osterman and Pecht [5] introduced an effective approximate method for placement of ICs on conduction-cooled circuit boards of the type shown in Fig. 5.20. Their approach was first to divide the circuit board into rows as indicated in Fig. 5.20 and then to consider each row independently. This approach assumes that heat flows in only the x direction, which is not exactly true. However, the simplification permits each row of ICs to be placed without considering the effect of adjacent rows. This simplification reduces the number of placement possibilities from $k_1 = N!$ to $k_2 = (N/r)!$, where r is the number of rows. Since $k_2 \ll k_1$, the complexity of the placement procedure for maximum reliability is reduced significantly.

The failure rate of a component clearly depends on its junction temperature and this temperature depends on the location of the device on the board. For conduction-cooled boards, like the one illustrated in Fig. 5.20, the locations adjacent to the heat sink give minimum junction temperatures and locations near the center of the board give maximum junction temperatures. (Procedures for determining junction temperatures for conduction-cooled systems are described

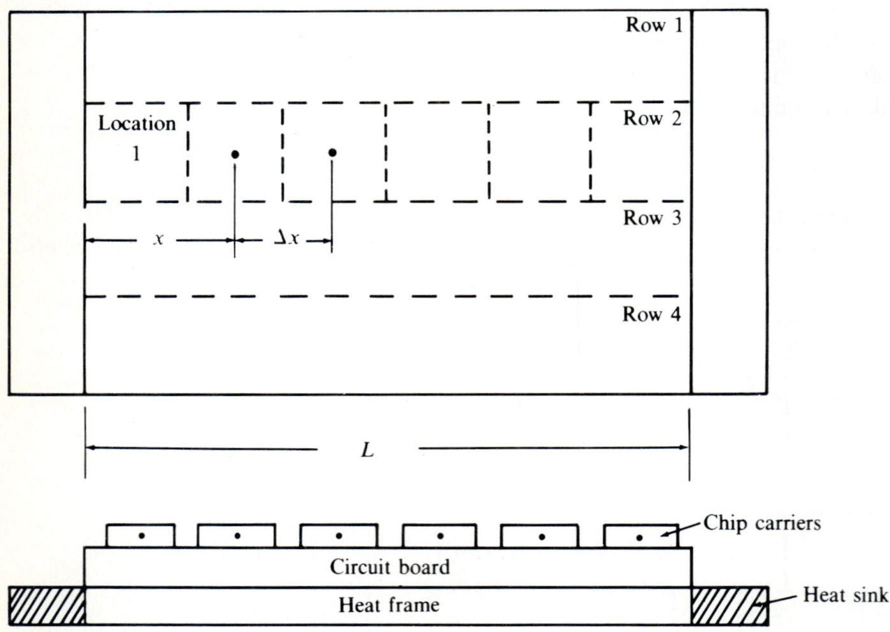

FIGURE 5.20
Conduction-cooled circuit board considered in placement to enhance reliability.

in Chapter 8. In this development we assume the reader to be familiar with analysis methods of heat transfer by conduction.)

The approach is to examine the ICs to be placed in a single row and to select from this group the most critical IC, namely, the component with the highest failure rate. The most critical IC is placed adjacent to the heat sink at the location marked 1 in Fig. 5.20.

Osterman and Pecht have defined a priority ranking number (PRN) for each component i, which is given by

$$\mathrm{PRN}_i = \left[d\lambda_i/dT_i^j\right]q_i \tag{5.8}$$

where q_i is the rate of heat dissipation from the ith component. The subscript $i = 1$ to k_2 is used to identify the components to be placed. The PRN is determined for each component positioned in the row under consideration. The maximum value of PRN identifies the most critical component. This component is positioned at the lowest temperature location 1, which is adjacent to the heat sink.

The value of PRN_i depends on the slope of the failure rate temperature relation which is obtained by differentiating Eq. (1.1) to give

$$d\lambda_i/dT_i^j = B_i C_i \left(T_i^j\right)^2 e^{C_i[(T_i^j - 298)/298T_i^j]} \tag{5.9}$$

The junction temperature T_i^j, which is an unknown in Eq. (5.9), depends on the location being considered for placement. In this approach we start with position located at $x = x_1$ as shown in Fig. 5.21. The junction temperature is determined using the electrical analog illustrated in Fig. 5.22, which permits us to write

$$T_i^j = T_i^f + q_i R_i^{jf} \tag{5.10}$$

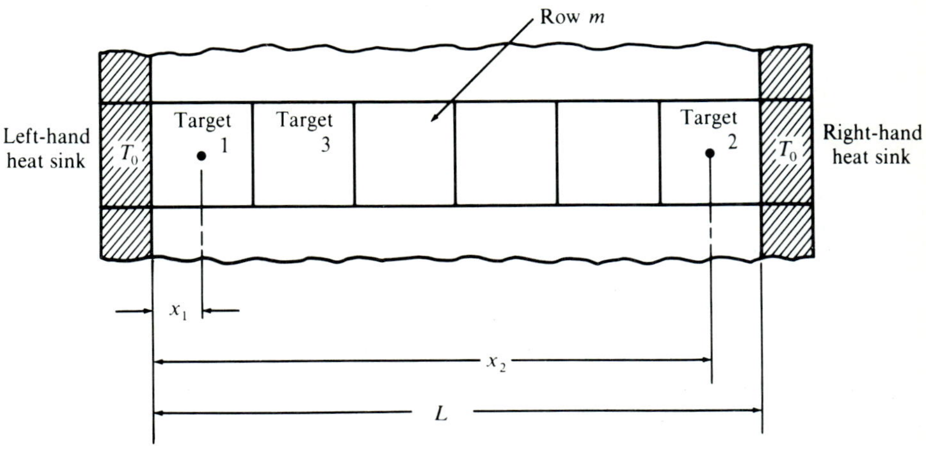

FIGURE 5.21
Locations or target 1 and 2 for placement along row m.

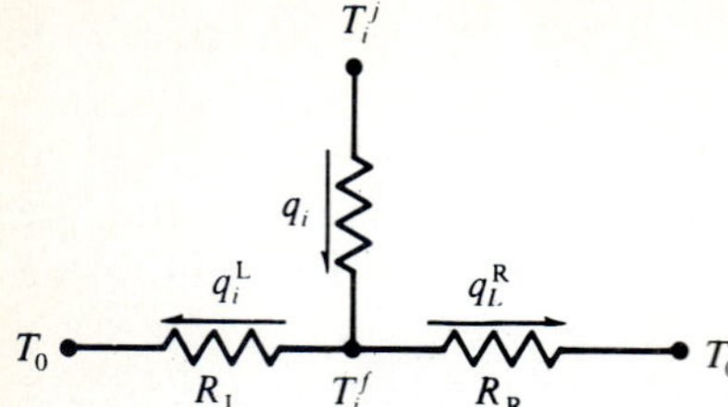

FIGURE 5.22
Resistor network showing the heat flow to heat sinks and the relation between T_i^f and T_i^j.

where T_i^f is the temperature of the heat frame at $x = x_1$ and R_i^{jf} is the combined thermal resistance including the chip, chip carrier and circuit board to the heat frame.

It should be noted that placement controls T_i^f and hence T_i^j. The term $q_i R_i^{jf}$ is important but this term is a constant that is independent of the placement location. The resistor network shown in Fig. 5.23*a* could be used to precisely solve for T_i^f; however, for large k_2 the calculations require extensive amounts of computer time. It is possible to reduce the computational time by considering the approximate model shown in Fig. 5.23*b*.

In this approximate model the ith component is placed over the target location and the remainder of the heat q_c is considered to be dissipated at the center line of the heat frame. We write q_c as

$$q_c = q_t - q_i \tag{5.11}$$

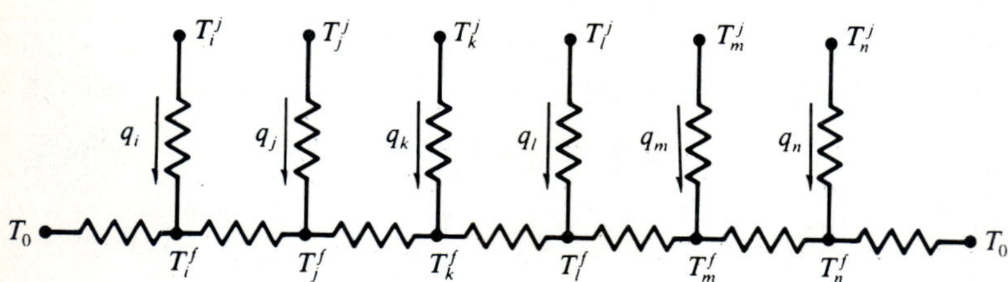

(*a*) A complete resistor network for the *m*th row

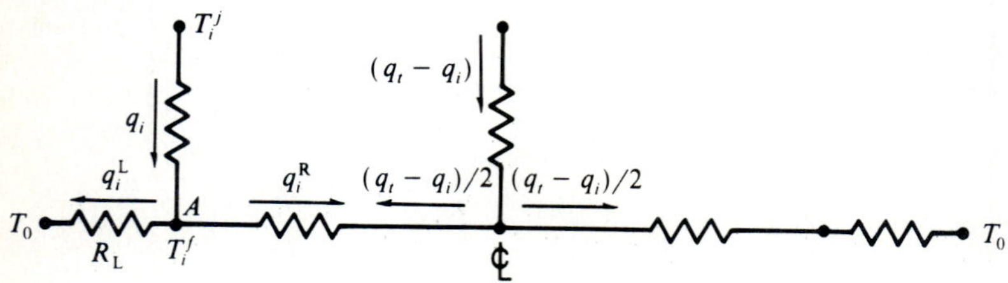

(*b*) An approximate thermal model

FIGURE 5.23
Exact and approximate thermal resistance models used to determine T_i^f and T_i^j.

where q_t is the sum of the heat dissipated by all the components to be placed in the row. To account for branching of the flow of heat at node A in Fig. 5.23b, we write

$$q_i = q_i^{\mathrm{L}} + q_i^{\mathrm{R}} \tag{5.12}$$

The heat flow to the left and to the right may be expressed as

$$q_i^{\mathrm{L}} = (T_i^f - T_0)/R_{\mathrm{L}} = (T_i^f - T_0)kA/x_1 \tag{5.13a}$$

$$q_i^{\mathrm{R}} = (T_i^f - T_0)/R_{\mathrm{R}} = (T_i^f - T_0)kA/(L - x_1) \tag{5.13b}$$

where k is the coefficient of thermal conductivity for the heat frame material. A is the cross-sectional area of the heat frame corresponding to the row height. L and x_1 are dimensions defined in Fig. 5.21. R_{L} and R_{R} are thermal resistances of the heat frame. Combining Eqs. (5.12) and (5.13) gives the temperature T_i^f due only to q_i as

$$T_i^f - T_0 = x_1(L - x_1)q_i/kAL \tag{5.14}$$

Note, that one-half of q_c flows to the left side heat sink and produces an increase in T_i^f due to R_{L}, which is

$$T_i^f - T_0 = x_1(q_t - q_i)/2kA \tag{5.15}$$

Using superposition and combining Eqs. (5.14) and (5.15) gives the frame temperature as

$$T_i^f = x_1(L - x_1)q_i/kAL + x_1(q_t - q_i)/2kA + T_0 \tag{5.16}$$

Next, use Eqs. (5.10) and (5.16) to obtain the junction temperature

$$T_i^j = x_1(L - x_1)q/kAL + x_1(q_t - q_i)/2kA + T_0 + q_iR_i^{jf} \tag{5.17}$$

Finally, substitution of numerical results from Eq. (5.17) into Eq. (5.9) gives $d\lambda_i/dT_i^j$, which is then substituted into Eq. (5.8) to obtain the priority ranking number PRN_i. This process is repeated for each of the k_2 components to be placed in the subject row. Selection of the maximum value of PRN_i permits the placement of the first component at the most favorable location $x = x_1$.

The procedure is repeated with the remaining components and the target for placement of the second component is location 2 at $x = x_2$ (see Fig. 5.21) adjacent to the right-hand heat sink. Determination of T_i^j follows the form of Eq. (5.17) except that the heat flow is in the opposite direction to the right-hand heat sink. We can accommodate this mirror image flow by letting

$$x_1 = L - x_2 \quad \text{and} \quad L - x_1 = x_2 \tag{5.18}$$

Substitution of Eq. (5.18) into Eq. (5.17) permits T_i^j to be determined at location 2. Again PRN_i is determined for the remaining components, the maximum value is selected and the second most critical component is identified and placed at $x = x_2$.

Placement of the remaining $(k_2 - 2)$ components follows the same process with T_i^f determined at location $x = x_3$ from the resistance network presented in

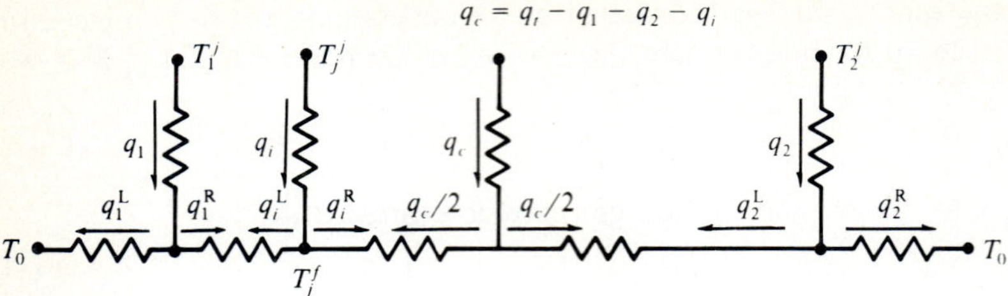

FIGURE 5.24
Resistor network for the temperature determination needed to place the third and fourth components.

Fig. 5.24. Examples of this placement procedure for maximizing reliability are given in the exercises following Chapter 8, because of the conduction theory of heat transfer used in this placement method.

5.6 ROUTING METHODS

Routing a circuit board pertains to the layout of the traces from one point to another on the circuit board. This layout is based on a wire list, as illustrated in Table 5.4, that describes the required connectivity between the I/O for the components placed on a given board. Routing appears simple. We need to connect prescribed pins or pads together according to the wire list, without crossovers that produce unspecified and detrimental connections. Also, we must avoid long, parallel and closely spaced paths that produce cross talk between signal traces. While simple in concept, actual routing is quite complex because of the density of the components placed on a board and the large number of pads that must be connected together within a limited circuit board area. Indeed, manual placement of the traces on a dense board with 1000 or more pads requires 10–100 hr and is prohibitively expensive. A more effective approach is to employ an automatic routing program that utilizes wiring algorithms to route most of the traces. Manual intervention at the end of the automatic routing is usually necessary to clear the crossovers and to place the wires that could not be placed automatically with the controlling algorithms. The combination of the automatic and manual approaches usually reduces the time required in routing the board by a factor of 3–5.

In this description of routing, we will cover four important topics, namely, surface organization, design rules that constrain routing options, automatic routing programs and classical wiring algorithms.

5.6.1 Surface Organization

To lay out circuit traces in a logical and systematic process, the signal planes are organized to provide defined pathways for the traces. This process is analogous to providing right-of-ways for the streets in planning and allocating space in the

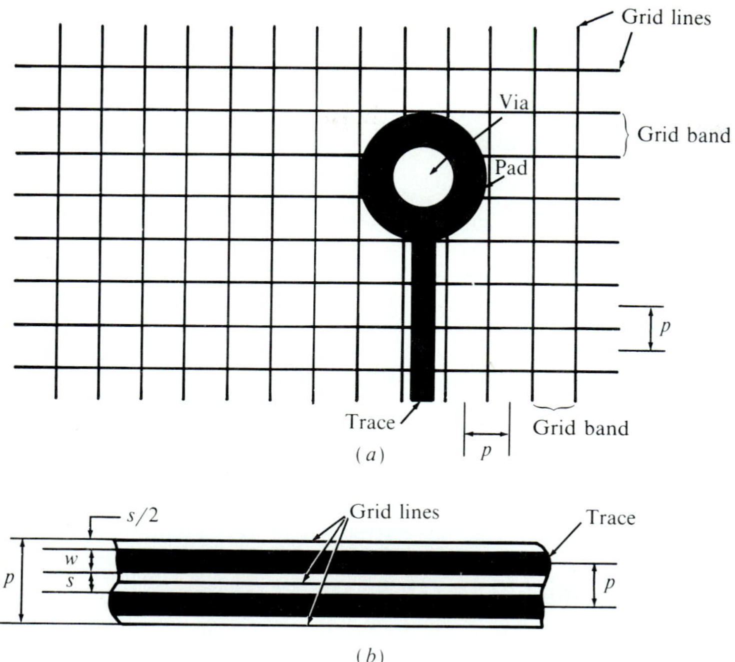

FIGURE 5.25
Grid method of surface organization. (*a*) Grid layout showing pitch, a pad, a via and a trace. (*b*) Space *s*, half space $s/2$, trace width *w* and pitch *p* for a grid layout.

design of a new city. There are four common methods used for surface organization, which include gridded, gridless, plastic grid and channel. The grid method divides the entire surface into a uniform grid with a fixed pitch *p* in both the *x* and *y* directions. Traces and half-spaces may be placed on any grid band that does not interfere with a pad, component lead, via or reserved area. Vias may be placed at intersections of the *x* and *y* grid bands. The grid method of surface organization is illustrated in Fig. 5.25. Reference to this illustration shows the grid lines forming the grid bands. In each grid band one or more circuit traces, each of width *w*, may be etched. A space *s* is required between traces and from the geometry shown in Fig. 5.25*b*, it is evident that

$$w + s = p \tag{5.19}$$

Commonly employed dimensions for *p*, *s* and *w* are listed in Table 5.8. The particular pitch selected depends on the center-to-center dimension of the leads on the chip carriers and the capability of manufacturing to produce high quality boards with fine line etching. For example the pitch $p = 20$ mil is effective with pin in-hole chip carriers where the center-to-center dimension is 100 mil. Two traces may be fitted into the space between the pads. A pitch $p = 10$ mil would

TABLE 5.8
Common dimensions (mils) for trace geometry

p	w	s
25	13	12
20	10	10
16.67	8.67	8
12.5	6.5	6
10	5	5

be better, since four traces could be placed between the pads; however, the trace width would only be 5 mil. This width is small enough to be considered as a fine line. Many of the firms producing circuit boards do not control their etching process well enough to insure adequate yields with fine line traces. The grid method is an excellent approach to surface organization when all of the components on the board are similar and regular patterns can be extended over the entire surface of the board.

The gridless surface permits the use of traces where w, s and p can be varied at select regions on the board in order to avoid interference with obstacles such as vias or pins. This approach permits better utilization of board space but increases the complexity of the automatic routing process significantly. It also requires fine line etching capabilities in production, since in crowded areas of the board the pitch actually used is small. In some cases where a two-sided board is used in place of a multilayered board, the added cost of gridless routing is justified.

The plastic grid surface organization is similar to the gridless method, because this technique permits each routing grid to be of a different pitch. This feature assists in routing those odd components that do not fit the fixed grid, which is suitable for a majority of the components. The plastic grid method is a compromise between the grid and gridless methods. It usually reduces the need

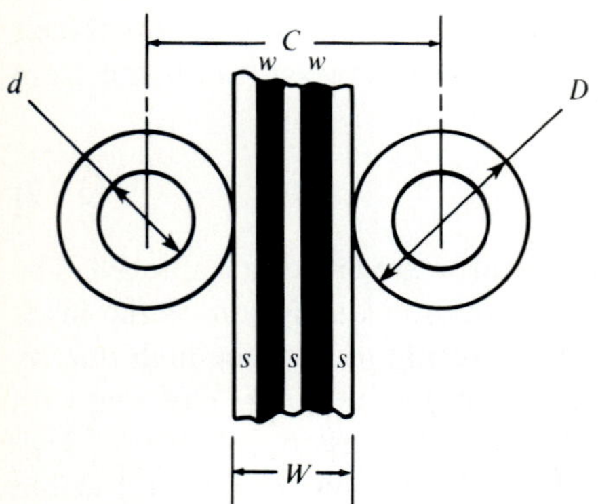

FIGURE 5.26
Two trace wiring channel between pin in-hole-type pads with controlling dimensions shown.

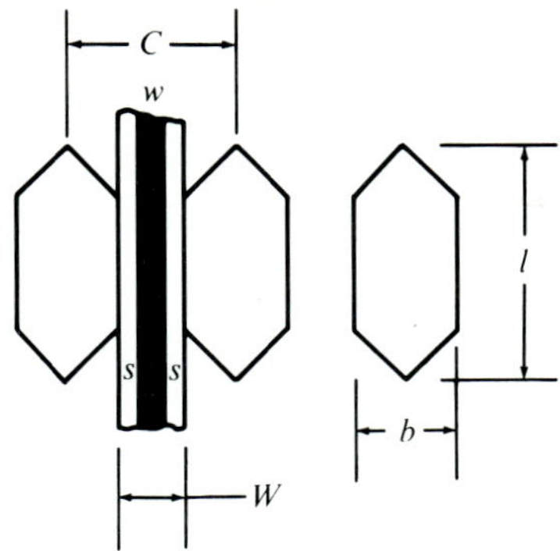

FIGURE 5.27
Single trace wiring channel between SMT pads with controlling dimensions shown.

for manual routing of many of the off-grid components that occur when the fixed grid method is employed. It also reduces the computer costs when compared with the gridless technique.

The channel method organizes the area of the board into channels that pass between the pads of the components as illustrated in Fig. 5.26. Vias are placed between the channels using the same centers as the component I/O. This procedure reserves space for both the vias and the traces so that the two

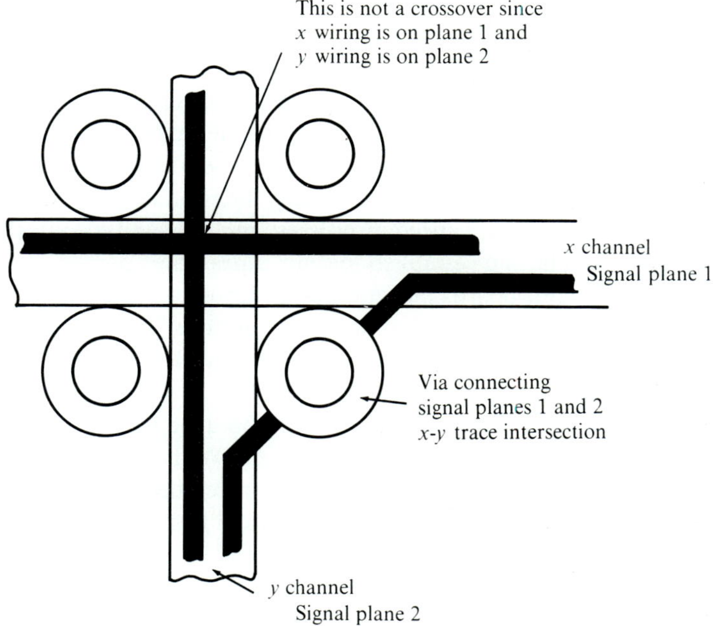

FIGURE 5.28
Trace turning in a multilayer board using the channel method of surface organization for each signal plane.

functions do not compete for board area. This advanced reservation of board area is important in routing very dense boards that require multilayers to resolve all of the wiring assignments.

The width W of the wiring channel is illustrated in Figs. 5.26 and 5.27 for pin in-hole and surface mount boards, respectively. The channel width W is determined by the footprint as

$$W = C - D \quad \text{or} \quad W = C - b \tag{5.20}$$

The wiring channel usually contains one to three wiring traces, depending on the available width W and the degree of control of the etching process, which dictates the minimum trace width w. Note that the dimensions of the traces in the wiring channel shown in Figs. 5.26 and 5.27 are given by

$$W = w + 2s \tag{5.21a}$$

$$W = 2w + 3s \tag{5.21b}$$

$$W = 3w + 4s \tag{5.21c}$$

Wiring channels extend in both the x and y directions and intersections of x and y channels allow for the traces to be turned on a given signal plane. Trace turning is also accomplished by intersecting, say, the x trace on one signal plane with a via that connects to the y trace on another signal plane. An example of trace turning with the channel method of surface organization on a multilayered board is presented in Fig. 5.28.

5.6.2 Design Rules for Routing

Design rules are developed by mutual agreement between the design and manufacturing functions. These rules are imposed to force the designer to lay out circuit boards that employ only those features that can be fabricated with processes and equipment that are available for fabricating the product. The firms with the best equipment, the most advanced processes and a quality control method that insures high yields can establish design rules that permit the development of high density circuit boards. Several design features, including via hole diameter, via pad diameter, trace and spacing dimensions, layered vias, dynamic vias, bus structure and buried resistors, should be considered in writing the design rules that govern the board layout and routing.

Via hole diameter is extremely important in routing because it is a significant factor in controlling the real estate available for wiring. With PCBs for pin in-hole devices there is little choice in specifying this diameter because the via serves to fasten the pin as well as to connect one signal layer to another. Reference to Figs. 5.12 and 5.13 indicates that $d = 34$ mil and $D = 50$ or 60 mil are common design rules for pin in-hole layout. For surface mount devices the diameter of the via hole is dependent on the fabrication processes because electrical functions can be accomplished with holes that are only a few mils in diameter. Today many fabricators prefer a 20 mil hole drilled in a 40 mil pad for SMT. However, with leads spaced at 50 mil centers on SMT chip carriers, the available space for a wiring grid or channel located between the pads is extremely

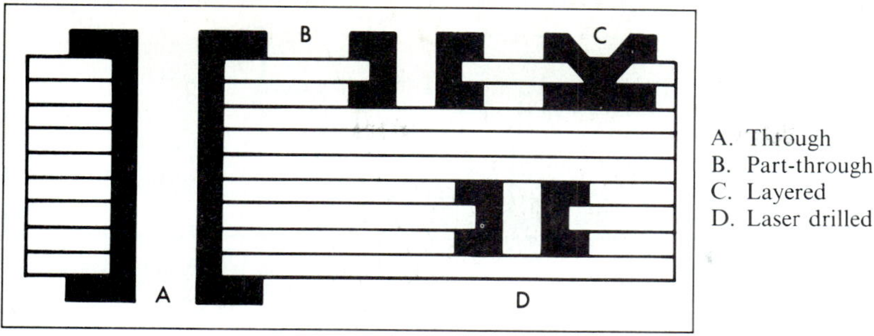

FIGURE 5.29
Different via structures.

limited (only 10 mil). Better drilling methods permits the via hole to be reduced to 15 mil and the via pad width b to 25 mil, allowing for a wiring channel 25 mil wide between the SMT pads.

Trace and space width are also specified as design rules that are extremely important in organizing the board surface areas. The copper cladding is etched away to produce a space of width s and a circuit trace of width w. The photolithography and etching processes to be described in Chapter 6 control the precision and smoothness of the traces and the yield of the boards being processed. Fabricators today are producing lines and spaces that usually range from $s = w = 5\text{–}10$ mil with 10 mil considered standard and 5 mil considered fine line technology. As fine line technology becomes more commonplace, the density of the circuit boards that can be wired will increase and the number of layers required to wire the boards will decrease.

Via structure must also be considered in design rules, particularly with multilayered boards. Many fabricators restrict design and permit only through vias on a fixed grid like those shown in Fig. 5.6. These vias often are not required to support the chip carrier and they effectively destroy wiring area on the interior signal planes. Layered or buried vias, which are illustrated in Fig. 5.29, provide the most efficient structure for wiring. The via is employed on only those layers where interconnections are required. This structure conserves area on all the wiring planes not affected by the via and provides additional area for wiring channels. Of course the use of buried layers requires drilling and plating the vias on a layer-by-layer basis, which adds significantly to the fabrication difficulty and cost.

Dynamic vias are plated through holes of the type illustrated in Fig. 5.6. They are termed dynamic because they are used only when it is necessary to make a connection with one or more interior layers. This is in contrast to a fixed via structure commonly employed where the vias are drilled on 100 mil centers, whether the via is required or not. Dynamic vias are placed when routing SMT boards and only those vias required for connectivity are used. This procedure conserves area on the interior layers and permits more dense circuits to be wired on fewer layers.

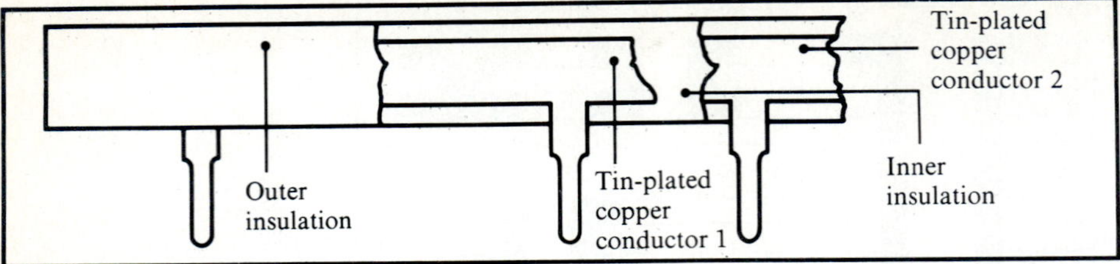

FIGURE 5.30
Bus bar for pin in-hole mounting on PCBs. (*Courtesy of Rogers Corporation, Bus Products Division.*)

Bus structure refers to power distribution, which often utilizes a significant amount of wiring area. Signal traces can be narrow with a small cross-sectional area because of the low currents associated with the signal pulses; however, wiring for the supply voltage can require, with high power chips, large cross-sectional areas. There are three commonly utilized bus structures, depending on circuit density. The first is the power trace, which is identical to the signal trace except that the width w is about 60 mil to provide the cross-sectional area required to handle the current. The power trace is commonly used in lower density circuits mounted on two-sided boards. The wider traces shown in Fig. 5.4b carry the supply voltage to the ICs mounted on this two-sided board.

For moderate density circuits with high switching speeds, small bus bars mounted on the circuit board provide the conductors to distribute the power. These bus bars, illustrated in Fig. 5.30, carry the V^+ and the V^- to a row of ICs as indicated in Fig. 5.31. The bus bars conserve board space, add rigidity to the board and provide a conductor with low inductance and low resistance consistent with the requirements imposed to limit noise generation in high speed switching.

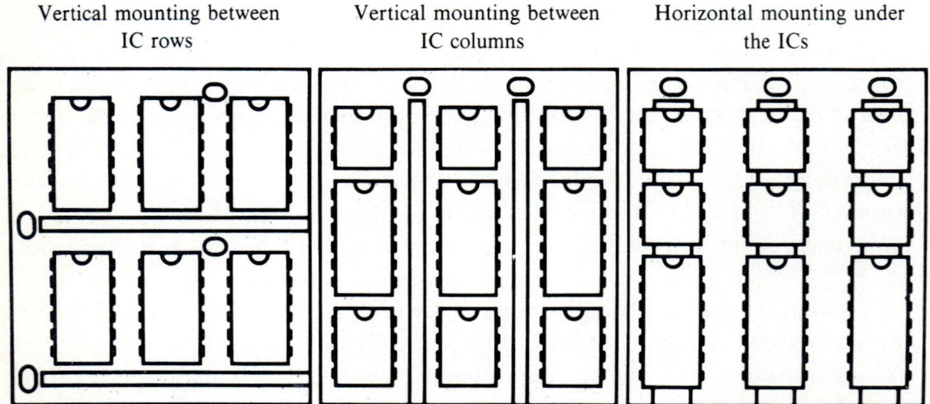

FIGURE 5.31
Power distribution to rows and columns of chip carriers with bus bars. (*Courtesy of Rogers Corporation, Bus Products Division.*)

This approach for distributing the power is most useful in two-sided boards where high inductance in the power distribution conductors can cause large voltage oscillations when 32 or 64 drivers are switched simultaneously in 1 or 2 ns.

The third method for power distribution is used in multilayer construction. With MLBs the V^+ is supplied over an entire plane, making up one side of a layer in the MLB. This approach has several advantages. First, the power plane

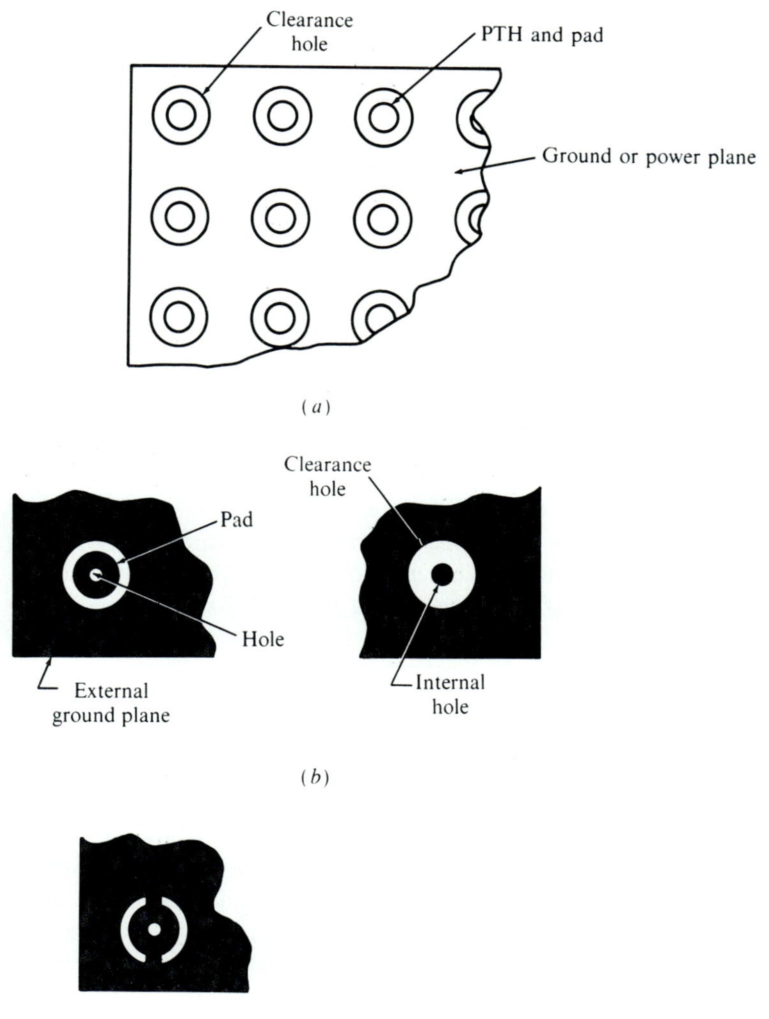

FIGURE 5.32
Ground or power plane design for multilayer board. (*a*) Perforated copper cladding forms the ground or power plane. (*b*) Detail showing the isolation of a pad that does not require a connection. (*c*) Detail showing connection of a pad to the ground or power plane.

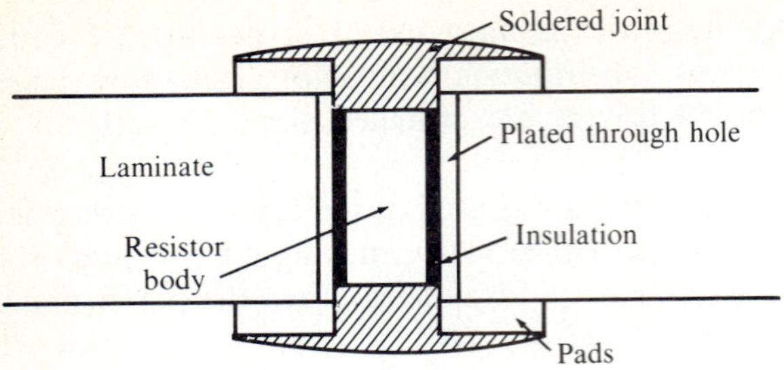

FIGURE 5.33
Buried resistors—z direction mounting in a plated through hole.

provides a low inductance and low resistance conductor to supply voltage to all of the ICs on the board. Second, the power plane in close proximity to the signal plane (the other side of the layer) permits construction of strip lines and microstrip lines to give controlled impedance of the signal wiring. Finally, the continuity of the power plane on one side of the layer gives added in-plane rigidity to the layer during the lamination process. This rigidity is important because it prevents movement of the pads located on interior layers at the elevated temperatures when the epoxy softens and the laminate becomes viscous. An example of a power or ground (V^-) plane is shown in Fig. 5.32. Observe here that the power plane is not continuous because holes are etched in the cladding either to isolate signal pins or to connect the power–ground pins. Details of the clearance hole design for isolation are given in Fig. 5.32*b*. The clearance hole detail for the case when the pin is connected to the power plane is illustrated in Fig. 5.32*c*. Note that two short traces extend from the power plane to the pad and the annular clearance area is maintained. This practice prevents high rates of heat transfer from the PTH to the copper power plane during the soldering operation. If the PTH is connected directly to the power plane, adequate temperature of the assembly is difficult to maintain and a cold-soldered joint may be produced.

Buried resistors are a relatively new structure in PCBs. They use the z direction of the board rather than the surface area of the board to mount resistors. Vias may be placed in the board to house low wattage resistors between two circuit traces as shown in Fig. 5.33.

In addition to the constraints on the design imposed by the limitations in the manufacturing process, we have additional constraints due to the limitations of the automatic routing programs. This topic is changing too rapidly to warrant detailed description at this time. The effort by software firms to develop automatic routing programs is intense. Any limitations identified this year will probably be resolved next year. Some of the more fundamental aspects of automatic routing programs will be discussed in the next section.

5.6.3 Automatic Routing Programs

The task of deciding all of the routes for a thousand or more wires on a typical circuit board is simple in concept but incredibility difficult and tedious in practice. While circuit boards can still be routed with the most efficient use of board area by manual methods, the time required to accomplish this task is not conducive to the early introduction of a new product or to the maintenance of a satisfied work force. To reduce the tedium of manual routing, software firms have developed automatic wiring routines where at least most of the connections are made with CAD methods. The software incorporates one or more of the wiring algorithms and accommodates the design rules described in the preceding section.

A recent survey of CAD programs for PCB design (see reference 7) indicated availability of 28 different products ranging in cost from $1250 to $40,000. In most cases these programs run on microcomputers and the cost of computer hardware is not significant. In addition to vigorous developments by the software firms, the larger electronic companies have developed their own routing programs, which are usually proprietary. While it is not within the scope of this text to describe these automatic routing programs, it is possible to list features that should be included in the CAD programs:

1. Automatic routing for boards containing both pin in-hole and surface mounted devices.
2. Variable grid routing.
3. Daisy chain connection of nets.
4. Routing the shortest connection first.
5. Routing high speed time critical circuits before the noncritical circuits.
6. Automatic placement and connection of line termination resistors.
7. Preassignment of critical pins for priority routing.
8. Report of traces that exceed a specified length.
9. Report of parallel length for traces.
10. Assignment of traces to specific layers.
11. Assignment of trace width and space width by layer.
12. Control of via usage per net or circuit.
13. Control of stub length.
14. Matching trace length on time critical circuits.
15. Listing of crossovers and incomplete circuits.

These features are not all available in every CAD system available today and the development of a complete code may require several more years. However, as both hardware and software improve, more of these features will be available, permitting better circuit board design at lower design cost.

In addition to the listed features, the CAD system should interface with other facilities necessary in the production of the board. Most important is the interface between the CAD system and the photoplotter that is used to prepare the masters for the phototools. This interface can be a modem that allows design information to be transmitted to the photoplotter over a phone line. Alternatively, a tape can be prepared with routing data that is delivered to the photoplotter. Of course, the format of the design data must be consistent with the format required by the photoplotter. Next, the CAD system should produce a drilling tape that can be used with the numerically controlled drills with the coordinates of all of the via holes. Finally, the CAD system should provide the design data necessary for the solder mask. The form of this data depends on the solder mask process, screened or dry film, and includes the locations and sizes of the areas to be exposed to the solder.

5.6.4 Wiring Algorithms

The problem of routing a circuit board can be divided into four parts, which include:

1. Preparing the wire list.
2. Assigning traces to specified layers.
3. Assigning the order for making the connections.
4. Laying out the traces.

In this treatment we will not attempt to completely describe all of the methods to accomplish these four tasks, because to do so would require a separate textbook. Instead, we will introduce each task with one or two techniques that illustrate the concepts involved. A more detailed treatment is presented in reference 10.

The wire list (see Table 5.4) shows the connections between the leads of the various devices on the circuit board and is related to the circuits being placed on the circuit board. With digital circuits, many common points occur in the circuit where the voltage will be either high or low. A connection of these common points is called a net and the wire list should connect these points together while minimizing the length of the traces used to make the specified connections. One approach to determining the most suitable wire list comes from the classic travelling salesman problem. This problem involves a salesman who is to travel to N cities and return to the starting point (home) following the shortest route. To illustrate this approach, consider the four points shown in Fig. 5.34 and labelled A to D. The scheme is to start from home and travel to a different city to determine the minimum travel distance. In this simple example, there are only three unique routes (remember we are not concerned with the direction of travel) as illustrated in Fig. 5.34*b*, *c* and *d*. The result in Fig. 5.34*b* shows that the minimum distance travelled is 10.5 units as compared to 12.1 and 12.4 for the other two unique routes. The wire list for this net would specify the sequence, A, B, D, C, A.

This example is quite simple because we selected $N = 4$; however, the complexity increases rapidly as N is increased. In general, the number of routes k

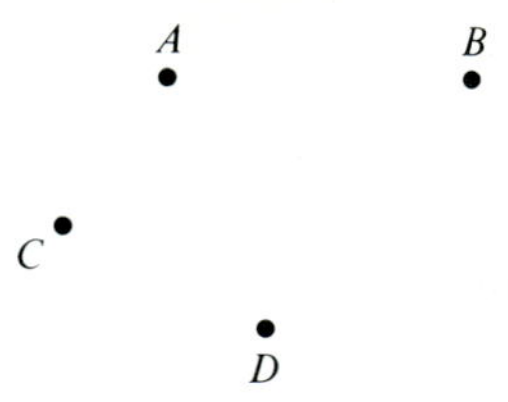

(*a*) Four cities *A*–*D* with home at *A*

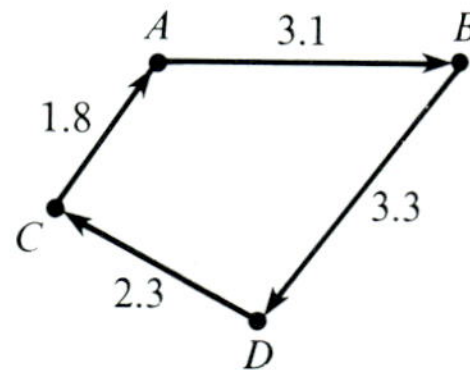

(*b*) Route initiating with *A* to *B*, $L = 10.5$ units

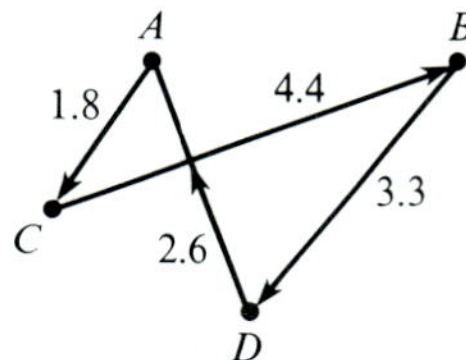

(*c*) Route initiating with *A* to *C*, $L = 12.1$ units

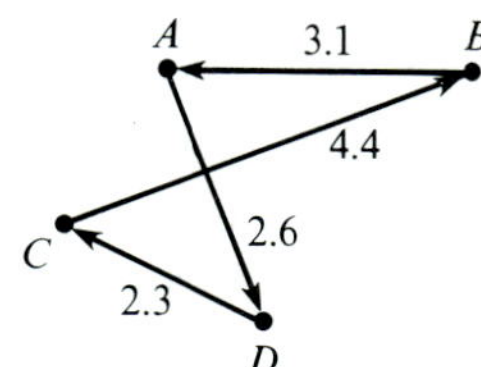

(*d*) Route initiating with *A* to *D*, $L = 12.4$ units

FIGURE 5.34
Illustration of the travelling salesman problem.

to consider in travelling to N cities is

$$k = (N - 1)!/2 \tag{5.22}$$

For $N = 8$ there are 2520 unique routes that can be considered in minimizing trace length. Clearly, algorithms and computers are required for this task if wire length is to be minimized with large nets.

Layering involves the assignment of specified traces to one of the signal planes in a multilayered circuit board. Of course, in one-sided boards layering is not possible. In two-sided boards layering is not necessary since one side is essentially reserved for the x traces and the other side for the y traces. The need for layering and the use of a multilayered board to facilitate routing become evident if all of the traces specified on the wire list cannot be placed on a two-sided board. Additional wiring planes are added to accommodate traces that cause the most crossovers. The addition of wiring planes adds significant cost to the PCBs and, consequently, each interior plane that is added should significantly enhance wirability.

In commercial products where wiring of the external planes is common practice, the first step in layering is to remove the power and ground traces from the external planes and place them on a single internal plane. This clears area on the exterior planes to allow for additional signal traces. If additional wiring

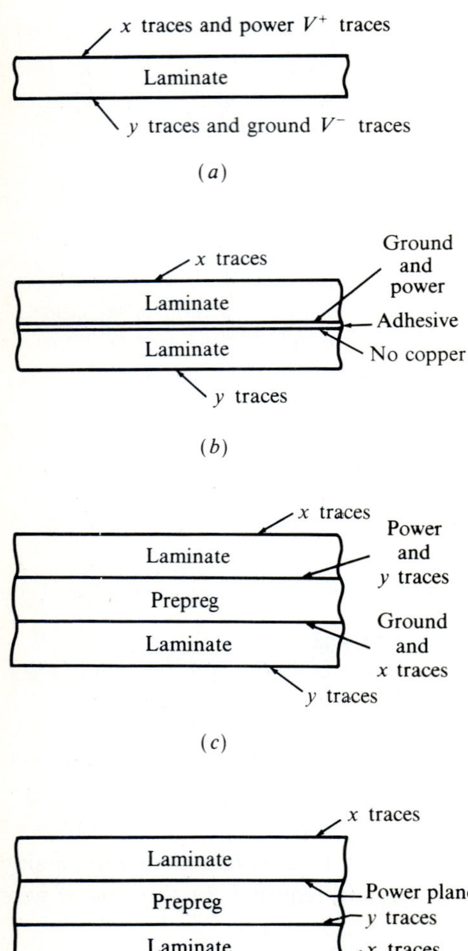

FIGURE 5.35
Multilayer construction techniques used in commercial products to provide additional wiring layers. (*a*) Two-sided board. (*b*) Two layer adhesively bonded three plane board. (*c*) Two layer prepreg bonded four plane board. (*d*) Three layer prepreg bonded six plane board.

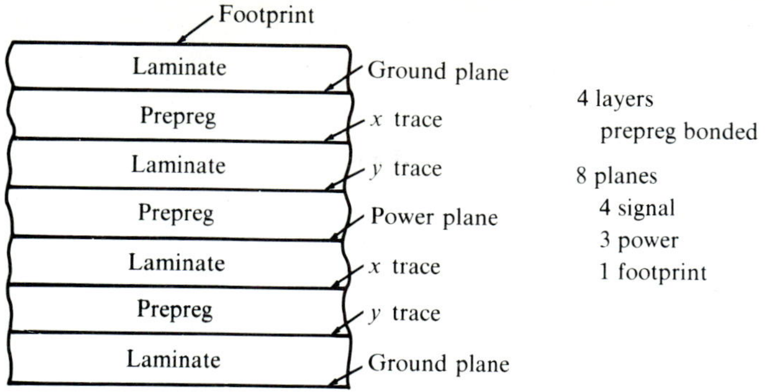

FIGURE 5.36
Multilayer construction techniques used in military products to provide additional wiring layers.

planes are required, more layers are added to the circuit board as illustrated in Fig. 5.35. Assignment of the circuits or nets to the various layers is based primarily on those connections that produce crossovers. These crossover connections are assigned to an interior plane until crossovers become excessive on these planes. At this point, additional planes are added or the remaining traces are manually routed.

With military systems, signal traces on the exterior planes are often prohibited. All wiring is performed on the interior planes. It is common practice to alternate between signal and power or ground planes as indicated in Fig. 5.36. Assignment of connections to the signal planes is again based on crossovers. All of the traces are initially placed on the outermost signal planes and those routes producing a large number of crossovers are progressively moved to the more interior layers.

The order of the routing of the traces is important since the placement of one wire can block the pathway for several other wires. A very simple rule, which is followed in manual placement, is to route the shortest leads first. This practice places the maximum number of connections and utilizes the minimum channel area. Another method used to order the placement of connections is the Aker's rectangle approach that is demonstrated in Fig. 5.37. The five pairs of points to be connected are shown in Fig. 5.37*a*. The first step in the process is to construct rectangles with the points to be connected forming the corners across the diagonal of the rectangle. The number of wiring points contained in each rectangle is counted and tabulated as indicated in Fig. 5.37*b*. The ordering is based on connecting the points on the rectangle that enclose the minimum number of points. In this example the ordering priority is *E-E*, *B-B*, *C-C*, *D-D* and *A-A*. Finally, the layout of the traces is performed by using two edges of the rectangle. The edges selected for elimination remove the crossovers evident in Fig. 5.37*b*. The resulting layout of the connecting traces is presented in Fig. 5.37*c*.

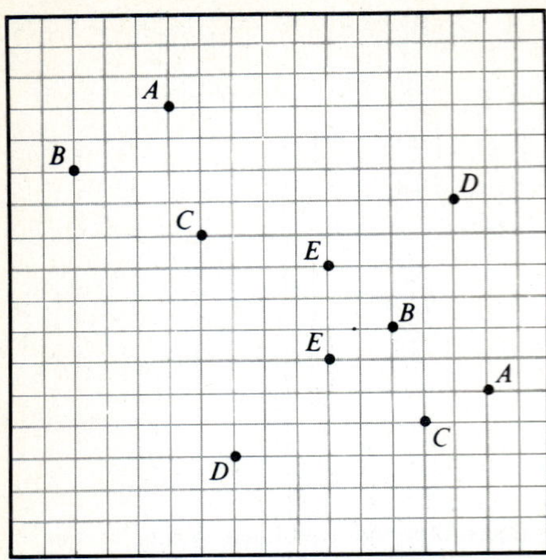

(a) Points to be connected

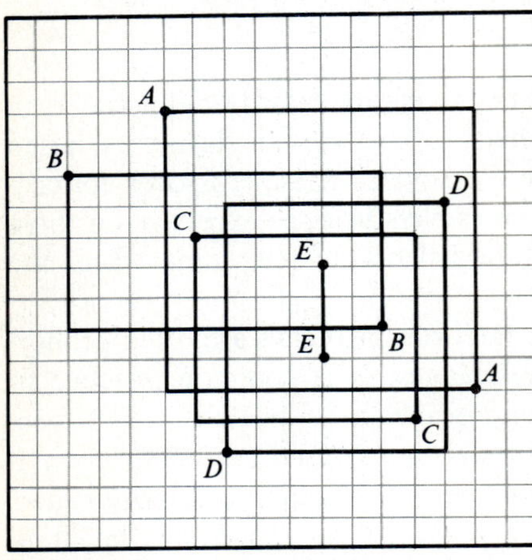

(b) Rectangles formed by connected points

Rectangle	Points enclosed	Ordering
A	5	5th
B	2	2nd
C	3	3rd
D	4	4th
E	0	1st

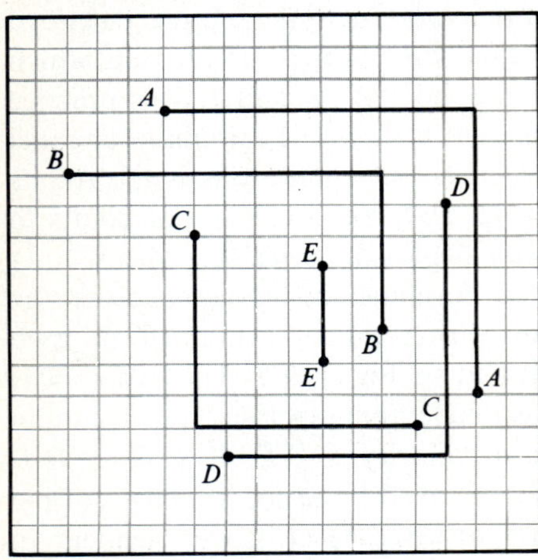

(c) Connections

Connections	Trace length	Ordering
A-A	19	5th
B-B	15	3rd
C-C	13	2nd
D-D	15	4th
E-E	3	1st

FIGURE 5.37
Aker's method for ordering the placement of connections.

The minimum length rule for ordering was also used for this example. The Manhattan length for each trace is listed in Fig. 5.37*c*. The ordering according to the minimum length rule is *E-E*, *C-C*, *B-B*, *D-D* and *A-A*.

The purpose of the final component in a wiring algorithm is to determine the route of the trace from one connection point to another. In the simple example illustrated in Fig. 5.37, we were able to determine clear paths that provided a direct Manhattan connection with a single right turn. In more realistic examples obstructions are encountered and it is difficult to find clear pathways that permit such direct connections. There are two different algorithms used to establish routes across a board with a series of obstacles. An obstacle refers to a region on the board where a wiring channel or grid may already exist and is analogous to the black regions on a crossword puzzle where letters are not permitted.

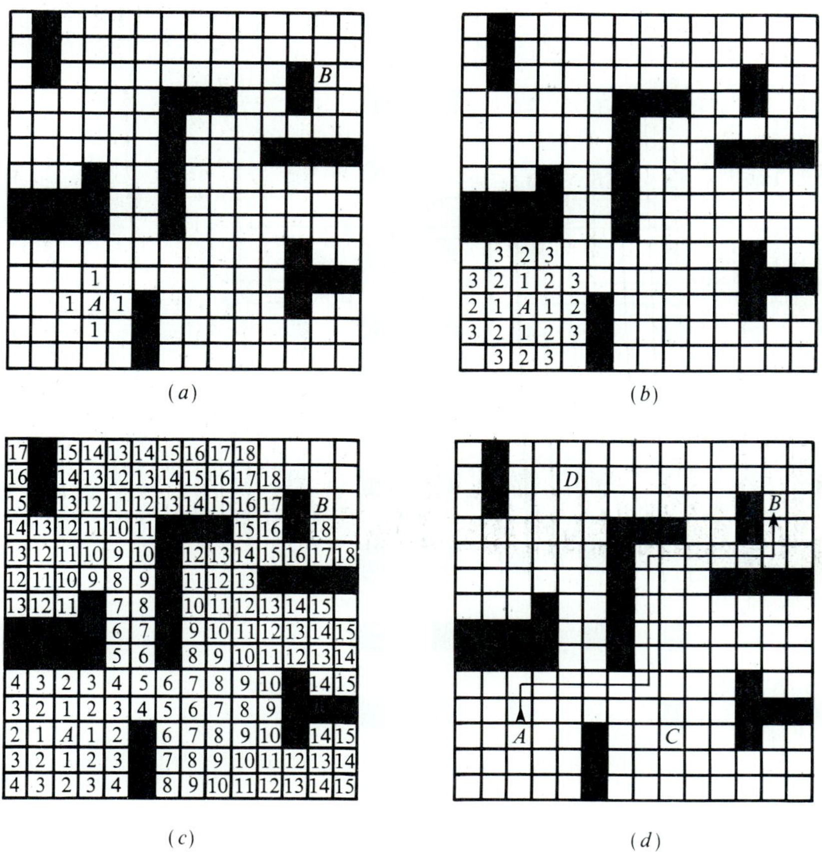

FIGURE 5.38
Lee's algorithm for determining minimum length connection between point *A* and point *B*.

Consider first Lee's algorithm, which is demonstrated in Fig. 5.38 by connecting a trace between points *A* and *B*. The board is divided into a grid and the obstructed areas that cannot be crossed are blackened. Beginning at point *A*, we number adjacent grid squares 1 as indicated in Fig. 5.38*a*. Since Manhattan wiring is employed, the grid squares on the diagonal cannot be numbered. The numbering process is continued by placing 2 adjacent to each 1 and then 3 adjacent to each 2 as illustrated in Fig. 5.38*b*. The numbering process is completed in Fig. 5.38*c*, when the number 18 is at the grid square adjacent to the *B* location. The final step is the construction of the route from *A* to *B* as shown in Fig. 5.38*d*. It should be noted that there are several routes, each 18 units long, that connect *A* and *B*. All of these routes are evident from an inspection of Fig. 5.38*d*. Other routes longer than 18 units can also be determined.

The graphic illustration of Lee's algorithm is used only to describe the method. In practice, the method is implemented with a computer, which stores the locations of the entire grid structure. The numbering advances automatically from a single point. A route is established when the square adjacent to the second point is filled with a number. The advantage of Lee's algorithm is that it will

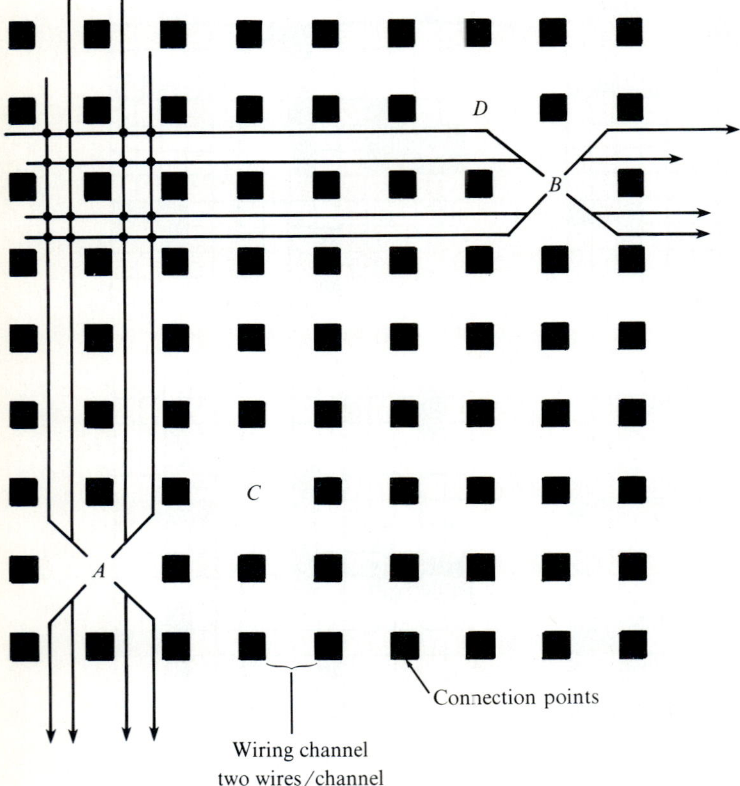

FIGURE 5.39
Graphic illustration of Hightower's fast line search.

always find a route between two points if one exists, and this route will be of minimum length. The disadvantage is that the method requires a very large amount of memory and significant CPU time.

The second approach is the Hightower [11] fast line search that is illustrated in Fig. 5.39. The field of the circuit board is arranged with wiring points and wiring channels running in the x and y directions between the connection points. We locate points A and B to be connected on the board. The Hightower search routine begins simultaneously at both connection points. From point A it connects open grid squares in the $\pm y$ direction of the wiring channel. From point B it connects the available grid squares in the $\pm x$ direction. The locations of the intersections (16 in this example) are recorded as an intersection identifies a possible route. The length of each route is computed and the shortest route is selected from the solution set. The grid squares used in establishing this route are eliminated from the wiring channel (this area now becomes an obstacle) and the program goes on to routing the next pair of wiring points.

The advantage of the Hightower approach is in reduced computer memory and time required to place most of the wires on a board. Also, it is usually possible to place more wires per layer with this approach than with Lee's algorithm. The disadvantages are that the routes are not always minimum length and that manual routing of the last 5–10% of the connections is usually necessary.

REFERENCES

1. Bucci, G. D. and R. D. Savage: "Copper Foil Technology," *Printed Circuit Fabrication*, vol. 8-7, pp. 22–33, 1986.
2. Hanan, M. and J. Kurtzberg: "Placement Techniques," in *Design Automation of Digital Systems: Theory and Techniques*, M. A. Breuer, ed., Prentice-Hall, Englewood Cliffs, N.J., Chap. 5, 1972.
3. Quinn, N. and M. Breuer: "A Force Directed Component Placement Procedure for Printed Circuit Boards," *IEEE Trans. Circuits and Systems*, vol. CAS-26, No. 6, pp. 377–388, 1979.
4. "Reliability Prediction of Electronic Equipment," Military Handbook MIL-HDBK-217E, Rome Air Development Center, February 1987.
5. Osterman, M. D. and M. Pecht: "Placement of Integrated Circuits for Reliability on Conductively Cooled Printed Wiring Boards," ASME paper 87-WA/EEP-8, December 1987.
6. Stredde, G. H.: "Printed Circuit Board Bus Bars," *Western Electric Engineer*, pp. 1–6, January 1979.
7. Lahore, H.: "CAD for PC Design: A Software Survey," *Printed Circuit Design*, vol. 4, no. 11, pp. 19–23, 1987.
8. Ritchey, L. W.: "A Comparison of PCB Routing Methods," *Printed Circuit Design*, vol. 4, no. 11, pp. 9–18, 1987.
9. Aranoff, S. and Y. Abulaffio: "Routing Printed Circuit Boards," *IEEE 18th Design Automation Conference:* pp. 130–136, 1981.
10. Breuer, A. M.: *Design Automation of Digital Systems*, vol. 1, Prentice-Hall, Englewood Cliffs, N.J., 1972.
11. Hightower, D. W.: "A Solution to Line Routing Problems on the Continuous Plane," *Proc. 6th Design Automation Workshop*, pp. 1–24, 1969.
12. Lee, C. Y.: "An Algorithm for Path Connections and Its Applications," *IRE Trans. Electronic Computers*, vol. EC-10, no. 3, pp. 346–365, 1961.

EXERCISES

5.1. Consider a J leaded chip carrier soldered to a circuit board with the attachment areas described in the footprint shown in Fig. 5.1. Determine the force necessary to shear the chip carrier from the board if the shear strength of the solder used to attach the carrier to the board is 3500 psi. Comment on the implications of your result.

5.2. A transformer weighing 2 lb is attached in the center of a circuit board with dimensions of $0.060 \times 5 \times 12$ in. The short edges of the board are simply supported and the long edges are free. If the product is subjected to a shock load of 9 G, determine the stresses imposed on the PCB by the transformer. Comment on the significance of your results.

5.3. Prepare a list of information that can be etched onto a PCB to assist in the manufacturing process. What is the cost of providing this information on the circuit card.

5.4. Examine a circuit card from an available PC and classify the board as one-sided, two-sided or multilayered. Describe the features you used in the classification process.

5.5. For the circuit card of Exercise 5.4, identify the plated through holes and describe the solder joint that is formed. Describe the shape of the solder pads and indicate the manner in which these pads affected the shape of the solder joint.

5.6. For the circuit card of Exercise 5.4, examine the edge of the card that is inserted into the connector. Describe the key code used to avoid accidental insertion of this card into the wrong slot.

5.7. Suppose an electronic system has eight different types of cards located in a single row and all connected to the same back panel. Design a key code for the PCBs that will insure insertion of the cards in one and only one slot. How many part numbers will you have in implementing your design? Discuss the implication of significantly increasing the number of part numbers involved in the construction of a system.

5.8. List the three common circuit boards in use today. Describe the advantages and disadvantages associated with each type of board. Classify by product type the usual application of these three circuit boards.

5.9. A multilayer circuit board is fabricated from six layers. How many wiring planes could be incorporated in this board? What method is used to connect a wire on one signal plane to a wire on another plane? Some wiring planes are reserved for the power V^+ and the ground. Describe the features of copper conductors used for these planes. Describe the features usually found on the wiring planes used to transmit the signals.

5.10. Define the following terms as used to describe wiring.
(*a*) Crossovers.
(*b*) Shorts.
(*c*) Opens.

5.11. The surface resistivity of a laminate with two parallel conductors is 10^8 Ω/square. If the two conductors are 8 in. long and spaced 6 mil apart, determine the current flow along the surface from one conductor to the other. The voltage difference between the traces is 4 V.

5.12. Determine the resistance of a laminate 0.060 in. thick, if the conductor on the top surface is 14 in. long and 10 mil wide. A ground plane covers the bottom surface of the laminate. Note that $\rho_v = 10^{15}$ Ω cm.

5.13. For the conductor and the laminate described in Exercise 5.12, find the total and distributed capacitance of the line if the dielectric constant $\varepsilon_r = 4.7$. Find the time required to charge this line if it is open at one end.

5.14. Write the specifications for a flexure test designed to measure the strength and modulus of a typical circuit board laminate. In the specification give the specimen dimensions, points of measurement, transducer suggestions and requirements, and the formulas to be used in the data analysis.

5.15. From the expansion–temperature curves shown in Fig. 5.9, determine the coefficients of expansion for epoxy and polyimide.

5.16. Estimate the thermal stress imposed on a copper PTH 0.100 in. long that is in a circuit board made from an FR-4 laminate. The circuit board is temperature-cycled from 20 to 100°C.

5.17. List the polymers commonly used as matrix materials for circuit boards. Also list the reinforcing materials.

5.18. Compare FR-2 and FR-4 circuit board materials for use in a very high volume competitively priced product intended for a normal office environment.

5.19. Construct a graph showing the thickness of copper foil in units of inches and micrometers as a function of ounces per square foot.

5.20. Discuss the advantages of copper foils produced by the electrolytic plating process. Compare this type of foil with the copper foils produced by rolling.

5.21. A 2 oz copper foil is bonded to a polyimide film 3 mil thick and etched to form conductors for a flexible circuit. The circuit is flexed repeatedly, forming a loop with a 2 in. radius of curvature. Determine the strain and stresses in both the polyimide film and in the copper.

5.22. Lay out the footprint for a standard eight pin DIP showing dimensions of the holes. Include in the layout, some of the holes associated with the four neighboring DIPs. Define the pad diameters $D = 60$ mil and determine the width of the wiring channels between the pins. What controls the spacing of the neighboring DIPs.

5.23. Repeat Exercise 5.22, changing D to 50 mil. If the conductor traces are 10 mil wide and the spaces are 10 mil wide, how many wires can be placed in a wiring channel?

5.24. Repeat Exercise 5.22, changing D to 30 mil. If the traces are 6 mil wide and the spaces are 8 mil wide, how many wires can be placed in a wiring channel.

5.25. Write an equation that relates pin spacing, pad diameter, hole diameter, annulus width, trace width, number of traces and space width that can be used to analyze a footprint for a pin in-hole chip carrier.

5.26. A 20 mil diameter hole is to be drilled in a 30 mil pad. Prepare a drawing showing the effect of registration errors in drilling. Consider registration errors of 1, 2 and 3 mil in the x direction, in the y direction and simultaneously in both directions. If each error has a 5% probability of occurring, what is the probability of drilling a satisfactory hole through the land.

5.27. A circuit board has 1000 holes of the type described in Exercise 5.26. For the same drilling errors and probabilities, determine the percentage yield in producing the board. What can the designer do to improve the yield of the board in production? What can the manufacturing engineer do to improve the yield?

5.28. A new chip carrier is designed with a U lead intended for surface mount solder attachment. The radius of the U is 10 mil. Design a solder pad that will provide a solder joint with generous fillets. Show a scale drawing of the solder joint.

5.29. Describe the terms crossover and cross talk as used to describe problems in circuit board layout.

5.30. Using the wire list shown in Table E5.30, prepare a connectivity table to the first seeded component that is the edge connector.

5.31. Place the bottom row of components if the board format shown in Fig. 5.17 is used with the wiring list of Table E5.30.

5.32. Prepare a connectivity matrix similar to the one shown in Table 5.7 for the components remaining after the solution of Exercise 5.31. Then place the remaining three components to minimize wire length.

5.33. Using the results from Exercise 5.33, prepare a drawing showing the placement of the six components and the edge connector. On this drawing draw lines connecting the pins according to the wiring list of Table E5.30.

5.34. Fifteen components are to be placed on a circuit board with component locations selected to minimize the failure rate of the circuit board. Determine the number of possible placement patterns for the components. If the circuit board is arranged with three rows of five components and each row is placed separately, determine the number of possible placement patterns. Estimate the savings in computational effort afforded by the placement-by-rows approach.

5.35. Consider the grid and channel methods of surface organization and sketch the intersection of a circuit trace with a via pad for both methods.

5.36. For the circuit board shown in Fig. 5.17 lay out the surface of the board using the grid method. Use a grid pitch of 20 mil and take 100 mil for the pin and pad centers. Note that the layout should be based on using a two-sided board and include both the top and bottom layers.

5.37. Repeat Exercise 5.36 but using the channel method for surface organization.

5.38. Using the wire list shown in Table 5.4, the placement described in the text for this wire list and the surface organization of Exercise 5.37, manually route the six component board.

5.39. Repeat Exercise 5.38 but use the surface organization method of Exercise 5.36.

5.40. Use the grid method to organize the surface of the board that is to be used with a flat pack with surface mounted leads on 50 mil centers. Lay out a footprint showing four 8 leaded chip carriers. On the layout show typical wiring dimensions indicating p, s and w.

5.41. Repeat Exercise 5.40 but use the channel method of surface organization.

5.42. If d is the diameter of the via hole and D is the diameter of the pad, prepare a table showing the number of conductors in a wiring channel between the pads as a function of d and the trace width w. Let $D = 2d$ and $s = w$. Consider pin–pad spacings of 100, 50, 40 and 20 mil. Comment on your results.

5.43. Describe four different types of vias. Sketch each type.

TABLE E5.30
Wire list for the circuit board shown in Fig. 5.17

Device no.	Pin no.	Device no.†	Pin no.
1	1	3	6
	2	5	1
	3	ec	11
	4	ec	7
	5	ec	12
	6	ec	13
	7	ec	14
	8	ec	8
2	1	6	3
	2	6	5
	3	2	7
	4	ec	7
	5	5	3
	6	5	6
	7	2	3
	8	ec	8
3	1	ec	1
	2	ec	2
	3	ec	3
	4	ec	7
	5	ec	4
	6	1	1
	7	5	2
	8	ec	8
4	1	ec	5
	2	ec	6
	3	ec	9
	4	ec	7
	5	ec	10
	6	6	6
	7	6	7
	8	ec	8
5	1	1	2
	2	3	7
	3	2	5
	4	ec	7
	5	6	1
	6	2	6
	7	6	2
	8	ec	8
6	1	5	5
	2	5	7
	3	2	1
	4	ec	7
	5	2	2
	6	4	6
	7	4	7
	8	ec	8

†ec is edge connector

5.44. Locate several circuit boards from a number of different products. Examine these boards and identify the method used to supply the power to the various components. Identify the power bus structure.

5.45. In routing a PCB, why should the shortest connections be made first? Explain the implication of making a daisy chain connection of logic nets.

5.46. Why is it important in automatic routing to have a program that lists the crossovers and incomplete circuits at the conclusion of the run? What do you do with this list?

5.47. Why should one give priority to routing the I/O of critical components.

5.48. The TCM was described as a combined first and second level package in Chapter 4. Referring to this description, estimate the number of pins that were connected in routing a typical substrate. Can this routing be performed without error? How is the possibility of error accommodated in the design process?

5.49. Why should we be concerned with matching the trace length on time critical components?

5.50. Why should we be concerned with the via usage per net or circuit?

5.51. For the four points shown in Fig. E5.51, use the travelling salesman approach to determine the minimum length for the trace connecting the points.

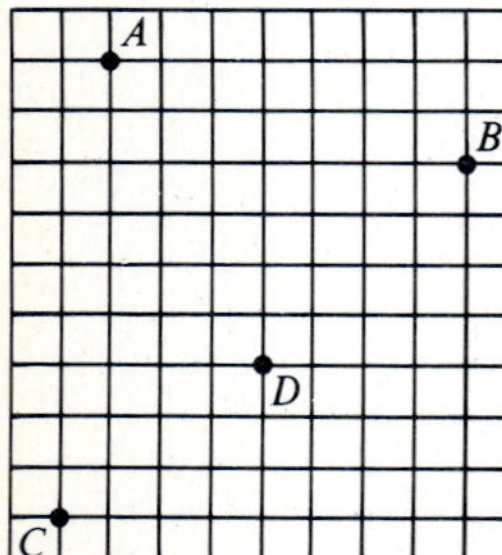

FIGURE E5.51
Four points representing cities for the travelling salesman.

5.52. Prepare a table showing the number of possibilities that must be considered in using the travelling salesman approach to determining the minimum route of a net as the number of points in the net varies from 3 to 10. Comment on these results.

5.53. As a class exercise, determine the minimum connection length for the five points shown in Fig. E5.53.

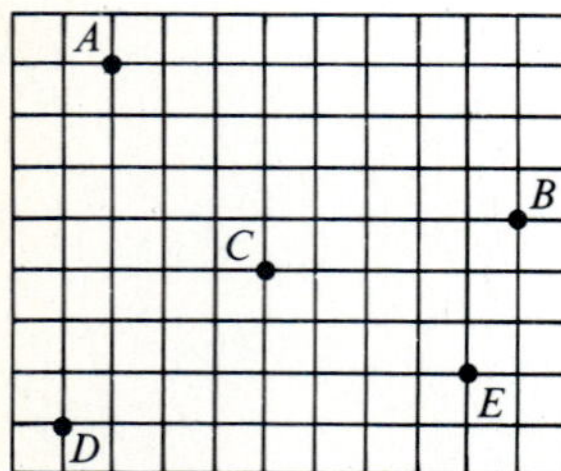

FIGURE E5.53
Five points representing cities for the travelling salesman.

5.54. Use Aker's rectangle method to determine the wiring order for the six pairs of points shown in Fig. E5.54.

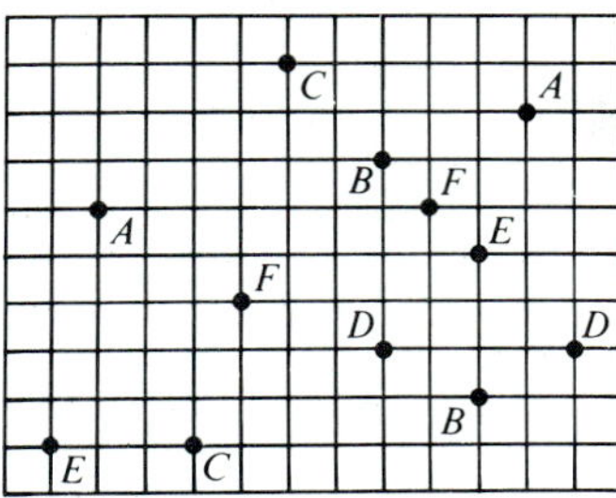

FIGURE E5.54
Connect the point pairs with Manhattan lines.

5.55. Use the minimum length rule to determine the order of wiring for the six pairs of points shown in Fig. E5.54.

5.56. Describe the difference between Manhattan distance and Euclidean distance. Write the equations for both of the distances. Give two reasons for the use of Manhattan distances in design of PCBs.

5.57. Use Lee's algorithm to determine the route from point A to B in Fig. E5.57.

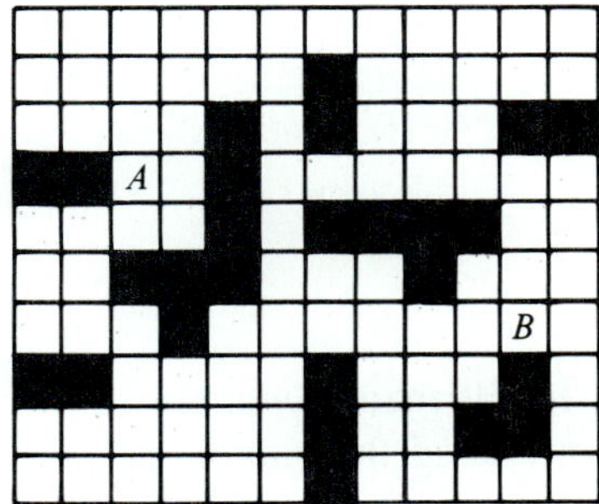

FIGURE E5.57
Connect A and B.

5.58. For the example illustrated in Fig. 5.38, determine the number of unique routes between A and B that are 18 units long. Lay out these routes on a grid drawing.

5.59. Suppose that the 18 unit routes to point B shown in Fig. 5.38 are excluded. Find the next shortest route.

5.60. Use Lee's algorithm to wire point pairs A-B and C-D.

Hint: Use the minimum length rule to establish the order of the routing.

5.61. Consider the 16 intersections obtained with the Hightower line search illustrated in Fig. 5.39. Determine the minimum line length and block out the grid squares associated with this route.

5.62. After completing Exercise 5.61 and blocking the grid squares for the route from point A to point B, determine the route from point C to point D. List the minimum length of the available paths and block the grid squares used for this route. Can you now place a pair of points on the diagram of Fig. 5.39 that are impossible to wire.

5.63. Suppose you have only a few wires that cannot be wired without adding an interior wiring plane. Because of the costs you clearly do not want to add another wiring plane. How can you resolve the problem?

CHAPTER 6

PRODUCTION OF PRINTED CIRCUIT BOARDS

6.1 OVERVIEW OF THE PRODUCTION PROCESS

A printed circuit board, as described in Chapter 5, fulfills many functions in the basic art of packaging. Indeed, the PCB is often considered the most important level of the packaging task because its design involves a significant portion of the engineering time when developing a new electronic system. Also, the production of PCBs is a large and important business. About 100 million square feet of reinforced polymeric circuit board are produced annually in the United States, and the market is growing.

Due to the technical and economic importance of the PCB, it is critical that this component be designed not only to accommodate the electrical and mechanical requirements but also be designed so that it can be manufactured with both high quality and competitive costs. Good design implies an understanding of the manufacturing processes so that design requirements are in close conformity with production capabilities. We will describe here the basic methods, procedures and processes that are used to produce a circuit board. We will then cover assembly and soldering operations that are followed in mounting the components to the board.

As the name "printed" circuit boards indicates, the production of the PCB involves many processes common to the printing or lithography industries. Indeed, it is the close correspondence of the two production systems that has led

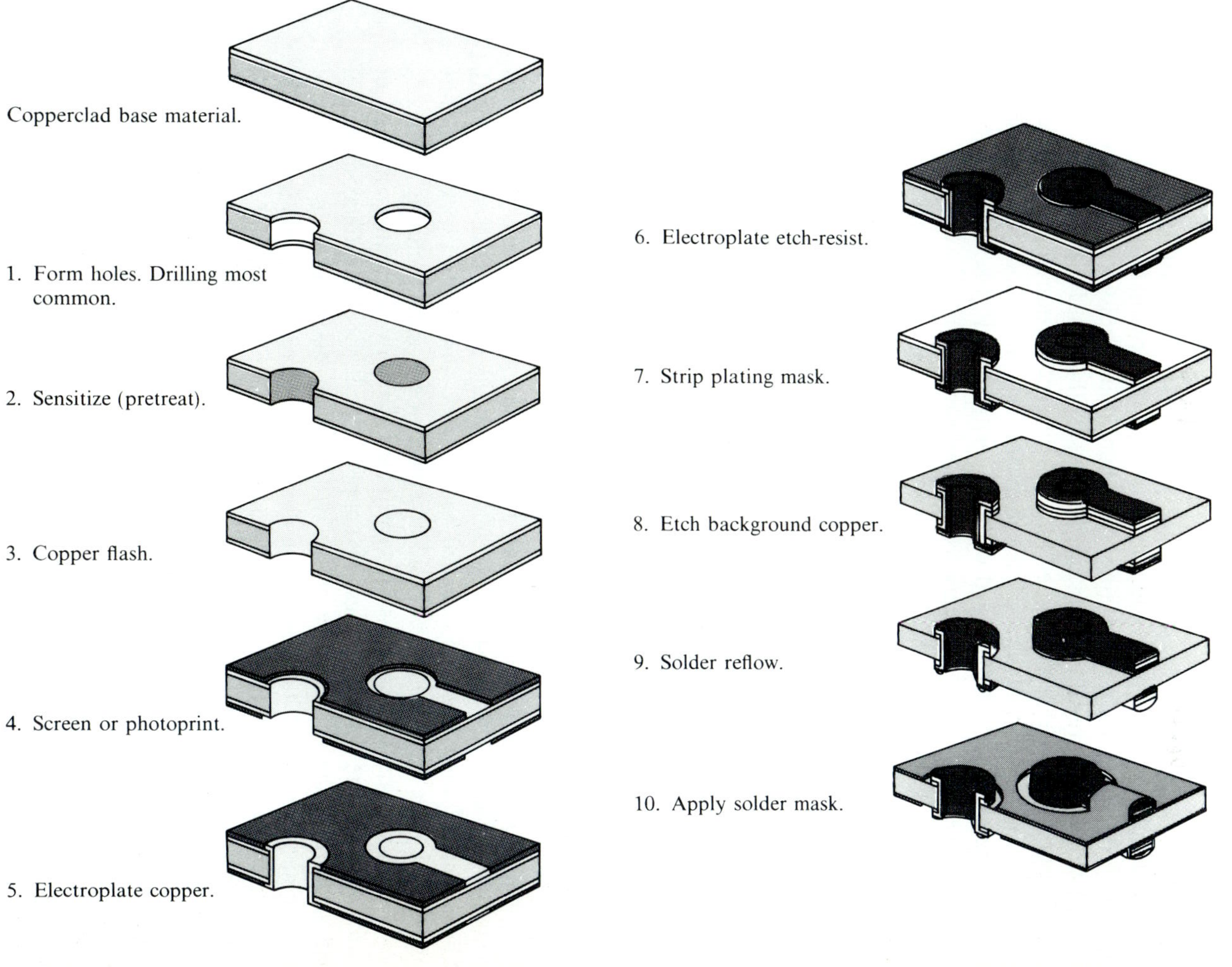

FIGURE 6.1
Major manufacturing processes used in the production of printed circuit boards.

to the rapid development of a highly efficient method for producing specialized circuit boards in a short period of time at remarkably low cost. The entire process is illustrated in Fig. 6.1 where the primary manufacturing steps are shown. The interface between design and manufacturing functions is the layout of the circuit board. This layout is represented by a master copy of the detail design which is used to prepare a phototool. The phototool is a pattern that is copied to each copper-clad board with a printing operation that employs a photosensitive coating called photoresist. The surface features of the board are formed with an etching operation where the copper foil, which is not required for the pads or traces, is chemically removed. Holes used for vias and mounting are drilled through the board and then plated with copper to form PTHs. Solder plating is also employed to protect the exposed copper surfaces from oxidation and to pretin select soldering surfaces. Solder masks are applied to the board to control the flow of the solder during the automatic soldering processes. Components are assembled at the proper locations on the board and the leads are soldered to the mounting points. Following soldering the board is cleaned and then stored in inventory until it is used in a new product or as a field replaceable unit in the repair of an existing product.

We will cover each of these steps in detail in the following sections. We will also introduce a brief description of ceramic circuit cards that have a more limited application in commercial electronic systems.

6.2 PREPARATION OF MASTER LAYOUTS

The interface between the design and manufacturing functions is at the completed design layout, which shows the features of each plane in the circuit board. The layout of these features is called the artwork and this artwork is usually prepared by the design engineers and accepted by the manufacturing engineers. Two methods are in general usage to produce the artwork, namely manual or automated. Before describing these two approaches, it should be understood that the artwork is the master copy of the detail design that is used in a series of photographic processes to produce the detail features of each plane in the PCB. The requirement for high precision and quality in the preparation of the artwork is self-evident, since the master copy represents the first step in a long and involved series of production processes used in fabricating the circuit board.

6.2.1 Manual Preparation of the Artwork

The artwork displays those design features that are to be reproduced in the "printing" process as features in the copper cladding that form the surface of the laminate. This pattern includes all of the circuit traces, the solder pads, component identification numbers and outlines, part numbers, polarity symbols, dates and other information helpful in manufacturing and maintenance. The manual

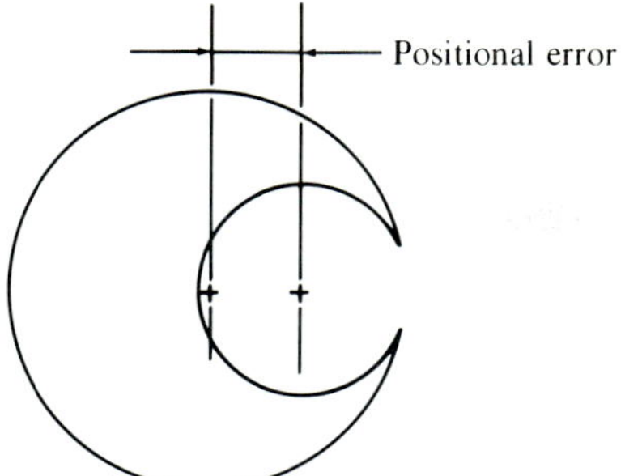

FIGURE 6.2
Breakout on an annular solder pad due to position errors in pad layout.

preparation of the artwork is usually performed at four times the actual size to facilitate accuracy in the layout of the features.

Today most manual artwork is prepared by using tape to lay out the circuit traces and adhesive decals of the correct size and shape to represent the solder pads. The material used as the base to support the layout is polyester film (7 mil thick), which is selected for its dimensional stability. Specifications on stability indicate that a typical film will exhibit changes in length of 0.0011 and 0.0017% for a change in humidity of 1% and temperature of 1°C. Furthermore, dimensional changes due to aging are about 0.02% in 5 yr. These specifications indicate that the artwork will retain the dimensional features required over reasonable product life if the polyester films are stored in record rooms where reasonable temperature and humidity are maintained.

The layout is aided by a rectangular grid on the film. If a clear film is used, the grid can be placed on the surface of the light table that is employed to backlight the pattern. The spacing of the grid is selected to correspond to the spacing of the solder pads and the circuit traces. Precision in pad placement is important to avoid broken pads when the circuit board is drilled. An example of pad breakout due to positioning error is shown in Fig. 6.2. Accuracy in positioning the pads and traces depends on the skill of the layout person and usually ranges from 6–12 mil. With a scale factor of 4 : 1 between the artwork and the actual board dimensions, these tolerances are reduced to $\pm$1.5–3 mil on the board.

The final step with the artwork is the production of a master copy that is a photographic negative (or positive) that has been reduced by the scaling factor of 4. This negative shows the exact features at actual size on a sheet of polyester film 5–7 mil thick. Several copies of this master are produced by contact printing. These contact copies of the master are called phototools. They are used in the photoresist process to transfer details of the features to the PCB. The master is filed in a secure location at a controlled temperature and humidity so that it can be used to produce additional phototools when necessary. The original artwork is also filed so it can be used at a later date for design revisions.

A complete description of manual procedures for preparing artwork is given in reference 1.

6.2.2 Automated Artwork by Photoplotting

Manual layout of circuit board features is being replaced in many firms with automated procedures where the board outline and the board footprint are designed using a CAD system. The CAD system provides a digital record of all of the feature sizes, locations and dimensions. The board is then wired using a CAE system, which automatically routes the circuit traces and places and records all the vias. An extensive description of the usage of CAD/CAE in the PCB design process is given in reference 2.

New software for layout, placement and routing is under continuous development. Today over 20 different programs are available with a wide range of costs and capabilities. The general trend of new development is toward the availability of lower cost programs that run effectively on personal computers or small work stations.

While the use of computer-assisted design procedures has significantly reduced the time and tedious aspects of board layout, some manual intervention is usually required. Usually some of the wires are routed with crossovers. The output from the program will usually locate the crossovers, but the program does not correct them. These crossovers are eliminated manually with an operator selecting new routes for the offending traces.

The digital output of the CAD/CAE system is usually in the form of a tape or a series of tapes. One tape, which contains a record of the coordinates of the via holes, called the drill tape, is used as input for numerically controlled drilling machines. Another tape, which contains the feature information, is used to drive a photoplotter. The photoplotters are used to produce the master "artwork," which consists of a photographic negative of the board features with a 1 : 1 scale on either a glass or film format.

There are two major technologies followed in the design of photoplotters, namely vector and raster. Vector plotters are like *x-y* pen plotters that use a light beam on film instead of a pen on paper. Raster plotters using laser light sources are like dot matrix printers and produce images by turning dots on and off. Vector photoplotters often use a flat bed to hold the film and move this bed as an *x-y* table under a beam of light. They image one feature at a time by moving the table so that the beam of light essentially paints the feature on the film plane. The size and shape of the beam of light is controlled by an aperture positioned in the photohead. The aperture is selected to correspond with the geometric characteristics of the feature being imaged.

As vector-type photoplotters must move the flat bed to the location of each feature to be imaged, their productive capacity is directly proportional to the flat bed speed. The time required to plot each signal plane depends on the number of features and the efficiency of the CAD/CAE system in organizing the plotting instructions. The accuracy for positioning which can be achieved ranges from ± 2 mil for the lower cost plotters to a very precise $\pm 20\ \mu$in. for the higher cost and higher quality instruments.

Raster photoplotters form images by connecting pixels together to develop a pattern. As the size of the pixel decreases, the quality of the image increases. Generally feature dimensions must exceed the pixel size by a factor of 6 or more to produce images with acceptable quality. Raster photoplotters also move a flat bed with the film under a beam of light. However, the raster plotter does not image a single feature at a time. Instead, the raster plotter sweeps across the film with the light beam switching on or off at each pixel location to form the selected elements of all of the features located along the sweep line. The table is indexed and the line sweep is repeated until the entire area of the plane is plotted. Because of the raster plotting process, the time required to cover the area of the plane is independent of the number and the complexity of the features. The computer that interfaces with the plotter must read and process the CAD/CAE tape before initiating the plotting, and the computer processing time is dependent on the signal plane complexity but the plotting time is not.

Generally raster photoplotters are faster and more expensive than vector plotters. Vector plotters usually exhibit better positional accuracy and better image quality, but the throughput is much lower. For those firms with a limited number of applications, the use of service bureaus that offer photoplotting services with very short turnaround times is probably advisable. The cost of most quality photoplotters is in the range of $250,000–500,000 and a significant use factor is required to justify this capital expenditure.

The output from the photoplotter is a negative or positive, with a 1 : 1 scale, which serves as the master for the phototools. This master can be on a film or on a glass plate, depending on the precision required. It is always safely stored and used only to make the phototools. Other output from the photoplotters are the negatives that show the features of the solder masks.

6.3 LITHOGRAPHY

Lithography and intaglio printing have been employed by artists for more than a century for making reproductions of their artwork. The lithographic processes developed in the printing industries for making printing plates were adapted to the manufacture of printed circuit boards in the early 1940s when printed circuit boards were first introduced. Two specific processes were adapted from the photolithographic industry: First, the process to transfer the image of the circuit board features from the phototool to the copper-clad laminate using photoresist techniques; second, the etching of the copper cladding to chemically machine all of the individual features required on a specific signal or power plane. These two processes are described in the following subsections.

6.3.1 Photoresist Printing

Photoresists are polymeric coatings that are sensitive to light and are similar in many respects to black and white photographic emulsions used in standard

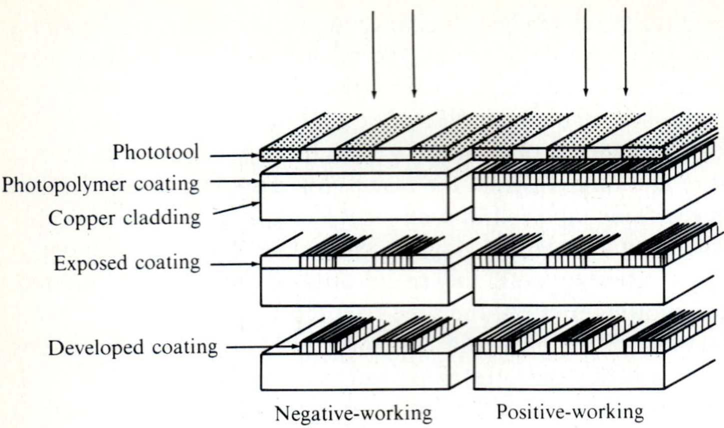

FIGURE 6.3
Effect of exposure and development in image transfer with photoresist coatings.

photography. In addition, the photopolymers must be resistant to select chemicals, adhere well to metallic surfaces, particularly copper, be water soluble and form thin films. These photopolymers are applied to the copper-clad laminates to produce a continuous, light sensitive film over the copper surface. The phototool is placed over this coated surface and brought into close contact with the laminate. A vacuum frame is employed to remove the layer of air separating the phototool from the coated surface. The photopolymer is then subjected to an ultraviolet (UV) light source to expose those regions of the coating under the transparent regions of the phototool. The exposed regions of the photopolymer undergo chemical and physical changes that make it insoluble to many chemical solutions. The regions not exposed to light are not affected and the coating remains soluble. After exposure, the laminate is placed in a developing solution that strips those regions of the coating that have not been exposed. This developing action does not affect the exposed regions where the coating is insoluble. The resulting image, which is the negative of the phototool, can be hardened by baking. A graphic describing the steps in the negative photoresist process is shown in Fig. 6.3.

Positive-type photopolymers are also available, which act in the opposite sense to the negative resists. The positive-type resists give coatings that are insoluble prior to exposure. Those areas exposed to the light are change chemically and physically and become soluble.

Commercial photoresists are available as a liquid or as a dry film. The liquid resists may be applied to the copper-clad laminates by spraying, rolling or dipping. The dry films are applied by laminating in a rubber roller laminating press. The dry films are usually preferred because the film resists are ready to be exposed immediately after lamination. The liquid resists must be dried and

FIGURE 6.4
SEM microphotograph showing a 3 mil trace in 2 mil thick photoresist. Note the relatively smooth edges and the excellent contact at the surface. (*Courtesy of Advanced Systems, Inc.*)

baked, and this requirement delays the process and adds to the floor space necessary to accommodate the production line.

The exposure of the resist becomes critical as the line width decreases. For normal width lines, with $w \geq 8\text{–}10$ mil, mercury arc lights arranged in a reflector to give a uniform intensity of light over the field of the phototool provide an exposure adequate to resolve the lines. For fine lines with $w < 5$ mil, the definition of the line becomes more difficult to achieve. It is necessary to employ collimated light in the plate maker, thin resist films and very close control of the exposure and developing times. When the proper equipment is used with precise process control, the fidelity of the reproduction of the lines and spaces in the photoresist is excellent. An example of the quality of the photoresist pattern that can be achieved is indicated by the SEM microphotograph presented in Fig. 6.4.

Transfer of images by the photoresist method is easier for single-sided boards than for double-sided or multilayer boards. With single-sided boards the registration issue does not arise because only one pattern is employed for wiring. With double-sided boards both sides of the board have an image and the board must be exposed in two separate operations. Since there are many vias connecting one side with the other, it is imperative that these two images be in precise registration. This registration is usually accomplished by drilling three tooling holes in the laminate and positioning the laminate in the printing frame over registration pins that pass through these alignment holes.

6.3.2 Etching

Etching is a chemical machining process that strips the copper cladding away from those regions of the board not protected by an etchant resistant coating. The copper that remains after the etching is completed serves to define the board features such as circuit traces and solder pads. There are two common processes in use today in etching the circuit features. The first employs a photoresist coating to protect the copper cladding from the etchant. The second utilizes a

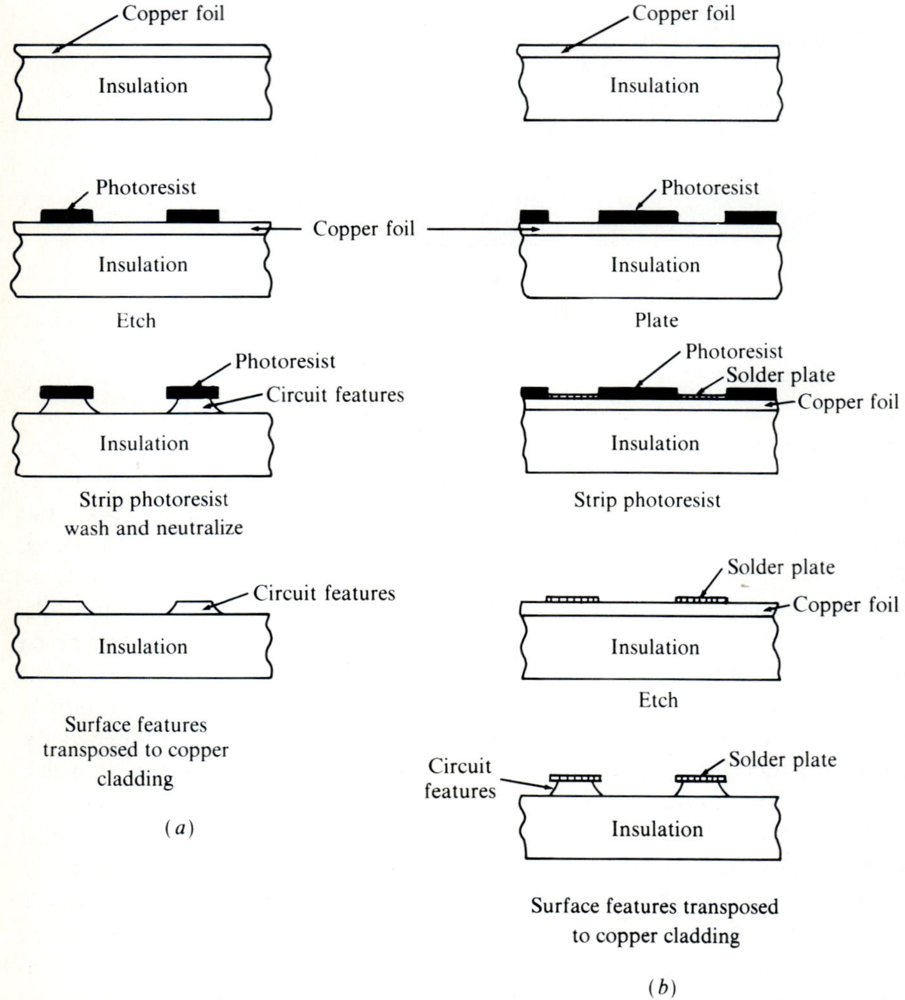

FIGURE 6.5
Two methods for placing etch resistant coatings. (*a*) Direct photoresist coating to protect surface features. (*b*) Solder plating to protect surface features.

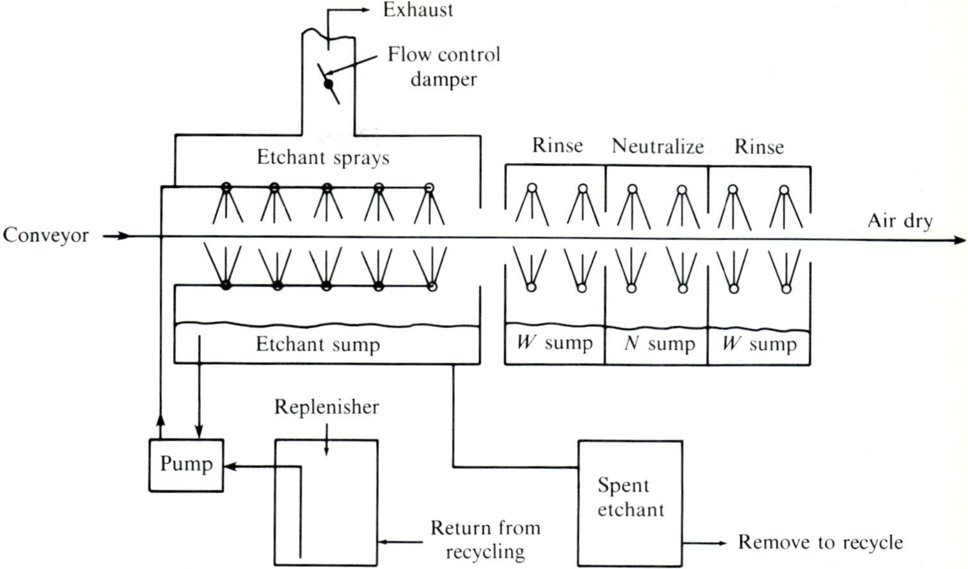

FIGURE 6.6
Schematic illustration of an automated etching line.

solder plate (60Sn-40Pb) 0.3–1 mil thick as the protective coating. These two processes are illustrated in Fig. 6.5.

The etching is performed in automated etching lines, illustrated in Fig. 6.6, that incorporate sprays that deliver the etchant to both sides of the board at high velocity. Alkaline ammonia (NH_4OH), which can be used in continuously operated lines with automatic solution replenishment, is the most commonly employed etchant. Ammonium hydroxide and ammonium salts combine with copper ions to form cupric ammonium complex ions [$Cu(NH_3)_4^{+2}$] that hold the dissolved copper in solution. The concentration of dissolved copper is usually in the range from 18–30 oz/gal. The spent etchant with high concentrations of dissolved copper can be recycled and the copper recovered. This recycling is very important because it eliminates the need to dispose of large amounts of toxic chemicals. After etching, the board is washed to thoroughly remove the etchant and then treated with a mild acid to neutralize the surfaces.

The etching process is not selective. This fact implies that material will be etched away in all directions at nearly the same rate. The spraying of the etchant onto the surface of the board does introduce a modest degree of selectivity into the process because fresh etchant is always being delivered to the bottom of the channels where the copper is being dissolved. However, the lack of selectivity in etching produces an undercutting of the etchant resistant coating as shown in Fig. 6.7. This undercutting is extremely important. The board features as etched are a different dimension than the features described in the phototool and the

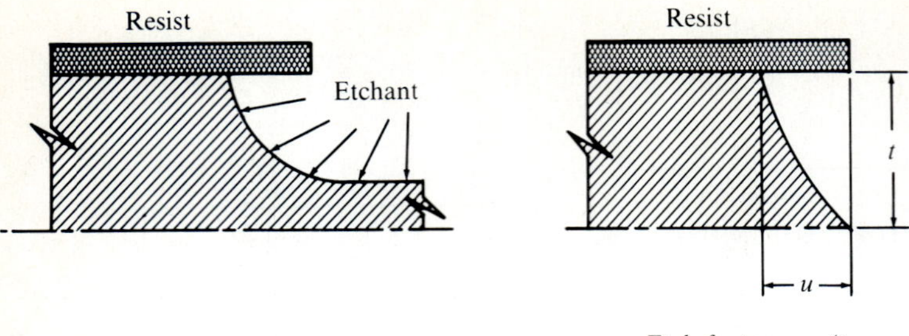

FIGURE 6.7
Undercutting the resist protection and the definition of the etch factor.

photoresist. The top surface of the copper cladding is undersized on each edge by a distance u. The magnitude of the undercut dimension u is determined by the etch factor F_e, which is defined as

$$F_e = u/t \tag{6.1}$$

where t is the thickness of the copper foil. It is important in the layout of the circuit features to take into account the etch factor, which requires one to increase the line width on the layout. For example, consider a circuit trace that is to be 8 mil wide and etched into a foil with a thickness equivalent to 1 oz of copper with an etch factor of 0.8. The circuit trace on the layout should have a width

$$w = 8 + 0.8 \times 1.35 \times 2 = 10.1 \text{ mil}$$

to accommodate for the dimensional changes anticipated by underetching.

The relation used to account for undercutting the resist is

$$w_L = w_f + 2F_e t \tag{6.2}$$

where subscripts L and f refer to the width at layout and the final width after etching.

6.4 DRILLING AND PUNCHING

Circuit boards contain an extremely large number of holes to either accommodate pins from components or to serve as vias that are used for connections between layers or for heat conduction. The holes are formed either by drilling or by punching, depending on the type of laminate used in constructing the board. Punching the holes is limited to single-sided PCBs fabricated from a paper reinforced phenolic material (XXXPC), which is a punchable laminate. Because the quality of the walls of the punched holes is not suitable, it is not possible to plate the through holes and the PCB is limited to single-sided applications. The

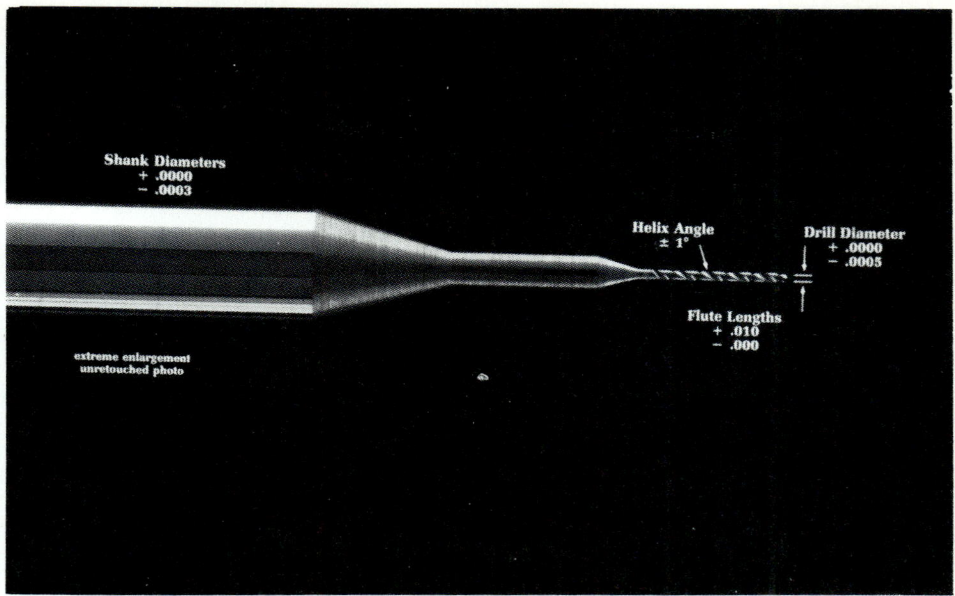

FIGURE 6.8
Detail of a solid carbide drill for producing small diameter holes in PCBs. (*Courtesy of International Carbide Tool.*)

exception to this statement is when the insulating material is ceramic green sheet. These ceramic sheets are similar to paper in thickness and texture and can be punched and preformed with precision prior to firing and sintering the ceramic particles.

Mechanical drilling is performed mainly with numerically controlled drilling machines with multiple heads, using small diameter twist drill bits fabricated from carbide as shown in Fig. 6.8. Manual drilling procedures are not usually employed. Location of the center of the holes to within ± 2–3 mil and the need to reduce labor costs usually require using multiple spindle numerically controlled drilling machines. With modern numerically controlled drilling stations using four drill spindles, it is possible to drill 200 to 400 holes per minute with positional accuracy of ± 0.2 mil and diameter tolerance of ± 1 mil.

Two different types of holes are usually drilled in the copper-clad laminates. The first, tooling holes, are of relatively large diameter, 125–250 mil. The second, via holes, are small diameter holes that accommodate pins and circuit feed throughs in the body of the circuit board. The via holes can be divided into two categories, namely those for pin in-hole-type chip carriers and those for surface mount-type chip carriers. To accept pins, holes are usually drilled with a diameter of 40–45 mil so that clearance for the pin exists after the hole is plated and its diameter is reduced to 32–36 mil. The holes for surface mount components may be much smaller in diameter because these holes are used only to connect one

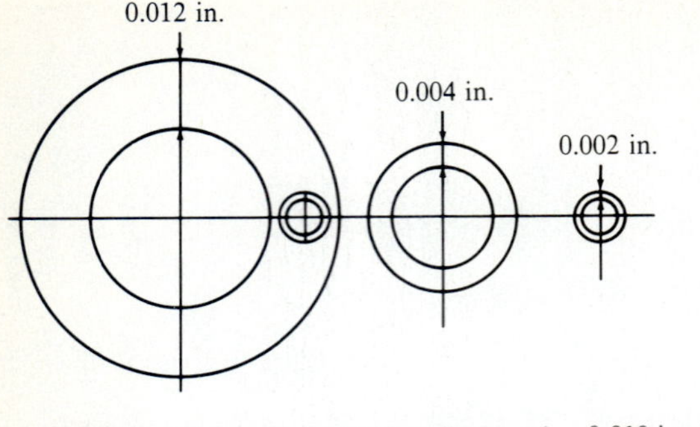

0.060 in. diam. pad	0.026 in. diam. pad	0.010 in. diam. pad
0.035 in. diam. hole	0.018 in. diam. hole	0.006 in. diam. hole
0.012 annular ring	0.004 annular ring	0.002 annular ring

FIGURE 6.9
A scale drawing illustrating the importance of registration, position drilling and diameter drilling tolerances as the hole diameter decreases.

signal layer to another. Indeed, the minimum diameter of a via for circuit connections is established by drilling technology. When the via hole diameter is reduced below 20 mil drill breakage increases markedly. Currently, most firms will not attempt to drill holes less than 10–15 mil in diameter in production.

With a reduction in the diameter of the via hole, the registration of the hole with respect to the pad becomes more critical. A common dimension for the annular ring formed by drilling a hole through a circular pad is 1/3 of the diameter of the plated through hole. For example, a 30 mil hole has an annular pad with $r_1 - r_2 = 10$ mil. This sizing rule gives a pad with an outside diameter of 50 mil. These tolerances give high yields in production even when the circuit board contains 1000 or more holes.

For surface mount chip carriers, the hole diameter can be much smaller. This fact implies that the tolerances on the position of the holes and on the diameter will be smaller and will require more precision in drilling. The scale drawing presented in Fig. 6.9 shows the problem of maintaining concentricity in pad positioning and drilling. Remember that the position of the pad and the position of the center of the hole are established in two totally independent manufacturing processes. Errors in either one or both can result in pad breakout, illustrated in Fig. 6.2, and rejection of the board.

Usually laminates are stacked in a pile, with registration maintained by the tooling holes, and several boards are drilled. The number of boards that can be drilled together depends on the board thickness and the aspect ratio that can be handled at the drilling station without excessive drill breakage. The aspect ratio, the depth of the stack of boards divided by the diameter of the drill, that can be achieved usually ranges from about 6–8. To avoid drill deflection upon entry and the formation of burrs upon exit, the stack of circuit boards is usually sandwiched between entry and backup layers. These layers are fabricated from

FIGURE 6.10
Cross section of a two-high stack of multilayer board with entry over lay and foil board backup to assist drilling burr-free holes. The drill size shown here is No. 57 (43 mil). (*Courtesy of L.C.O.A., Laminating Company of America.*)

materials that facilitate a low force entry and a burr-free exit. An illustration of drilling a two-high stack of multilayer boards, presented in Fig. 6.10, shows the entry and exit layers.

6.5 LAMINATION

Lamination is the first step in the production of a typical single- or double-sided board. The thin copper foils that form the clad surfaces of a laminate are spread over a polished steel sheet to control the flatness of the foil. Next, several sheets of prepreg are placed over the copper foil until the specified thickness of the board is achieved. Prepreg is a sheet material formed by impregnating glass fabric with liquid epoxy (stage A). The epoxy resin is partially cured and becomes rubbery (stage B), permitting the prepreg sheets to be cut and handled easily. The second foil of copper is then placed on top of the prepreg lay-up. Finally, another polished steel sheet (called a press pan) is placed over the assembly. Mold release is used with the press pans to insure that the laminates will separate easily after the polymerization of the epoxy is completed. This process is repeated until a stack of about 100 boards is developed. The sheets are pressed together in a large hydraulic platen press. The press develops about 1000 psi pressure over the area of the sheets, typically 36 × 72 in. in size. In addition the platens of the press are heated to raise the temperature of the stack. As the temperature of the partially

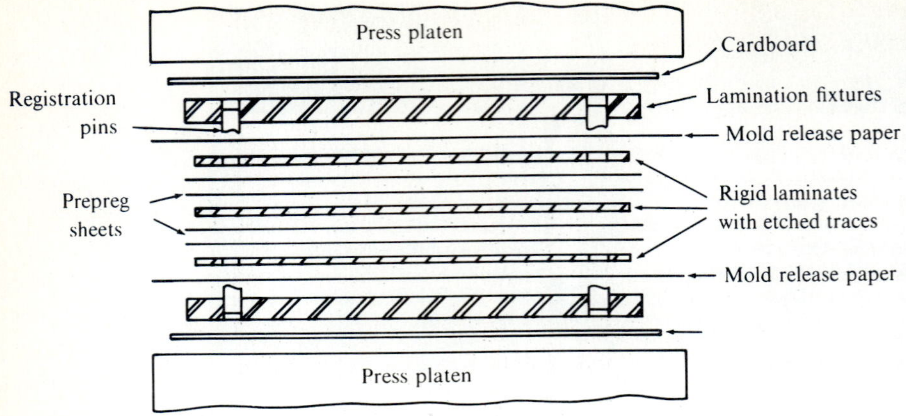

FIGURE 6.11
Multilayer board lamination with lamination fixtures to insure registration.

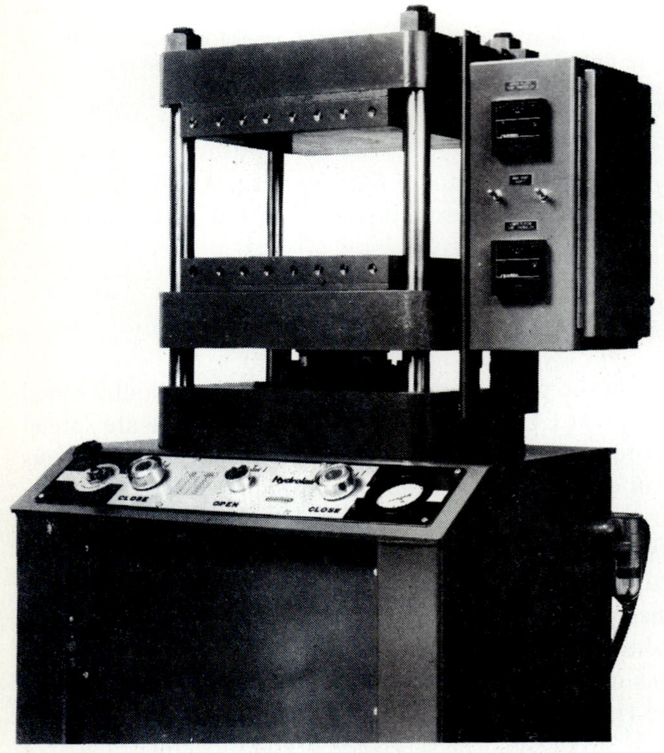

FIGURE 6.12
Details of an electrically heated platen press used in laminating multilayer circuit boards. (*Courtesy of Greenerd Press and Machine Company, Inc.*)

cured (B stage) epoxy increases, the resin flows, eliminating voids. With time and temperature the polymerization of the epoxy continues until the epoxy becomes rigid (C stage). The laminates are separated from the stack and are heated in an oven at a temperature slightly above the heat distortion temperature to relieve any residual stresses developed in the curing epoxy.

Lamination is also used as a step in producing multilayer boards. In this instance, several layers, each representing a two-sided circuit board with previously etched power and signal planes, are laminated together. The process is similar to the preceding description in that prepreg is used to separate the layers. The thickness of the prepreg is controlled to properly space the signal and power planes. However, the process is more difficult since registration of the layers is very important in the buildup of a multilayer board. Remember that these layers are connected electrically after lamination by drilling and plating through holes. The registration in lamination is achieved by using fixtures that incorporate two or more registration pins that pass through tooling holes drilled in each layer. Because of the need for close control over registration, the size of the multilayer panel is smaller (20 × 20 in. is common). Also, the stack height is limited to less than 1 in. to better control the heat transfer through the stack. A typical laminate lay-up with the laminating fixture is shown in Fig. 6.11 and a typical lamination press is shown in Fig. 6.12.

6.6 PLATING

Plating is the electrochemical deposition of thin layers of various metals on the PCB. Plating is employed in several steps in manufacturing a circuit board, including finishing plated through holes, providing an etch resistant surface, applying a gold coating for connector contacts and forming thin tin or solder coatings for oxidation resistance.

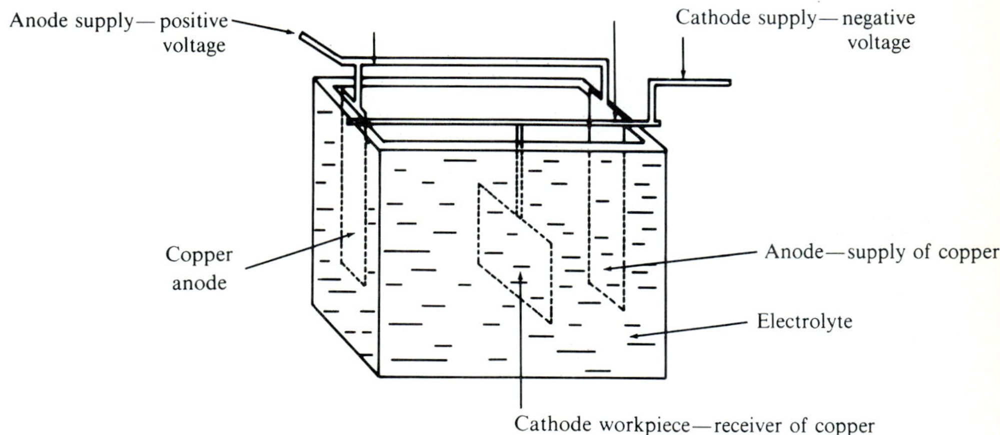

FIGURE 6.13
Electrolytic plating bath showing anode and cathode positions with copper as the plating material.

The plating occurs in a bath, as illustrated in Fig. 6.13, where the workpiece (the PCB) is connected to the negative side of a current supply, making it the cathode in the dc circuit. The metal to be plated, copper in the illustration, is connected to the positive side of the supply and becomes the anode. The anode and cathode are coupled together with an electrolyte, which serves to transport positively charged copper ions from the anode to the cathode. The rate of deposition is controlled by the cell voltage, which is adjusted to give current densities consistent with anode-to-cathode spacing and the specified deposition rates.

6.6.1 Plated Through Holes

After the holes have been drilled, it is necessary to connect the circuits that exist on both sides of the board. This is usually accomplished by depositing a layer of copper over the wall of the hole to form a plated through hole shown in Fig. 5.5. The process for plating the inside surface of a hole is divided into two steps. First, the formation of a very thin conducting layer using electroless copper and, second, the deposition of a thicker layer of electrolytic-plated copper.

The electroless plating of copper is a chemical deposition process that makes use of special activators to sensitize the walls of the holes. These activators are colloidal solutions of stannous and palladium ions. The board is immersed in solutions containing these ions. The stannous ions are adsorbed onto the surface to be catalyzed and subsequently they reduce the palladium ions, converting them to the metallic state. This series of reactions creates the precious metal sites necessary to initiate the electroless copper reaction.

The board is then transferred to an electroless copper bath where copper ions are reduced to metallic copper that is deposited onto the palladium sites. The deposition rates are quite low, 1–2 μin./min. To conserve process time, only a very thin layer (20 μin.) of copper is deposited in the electroless bath. The board is then transferred to the electrolytic plating bath where 0.2–0.5 mil of copper are flash plated onto the walls to give a more rugged film of copper. Additional copper is added in subsequent plating operations to bring the copper barrel to a wall thickness that ranges from 1–3 mil.

6.6.2 Panel and Pattern Plating

Two different approaches are followed in developing the full thickness of the copper plate in the through hole, namely, panel and pattern plating. In panel plating the entire surface of the board, including the inside surface of all of the holes, is plated immediately following the application of the electroless copper. Clearly this panel plating operation is independent of the image transfer process. Pattern plating, on the other hand, is selective since the copper is deposited only on the surfaces of the holes, the pads and the circuit traces. Of course, pattern

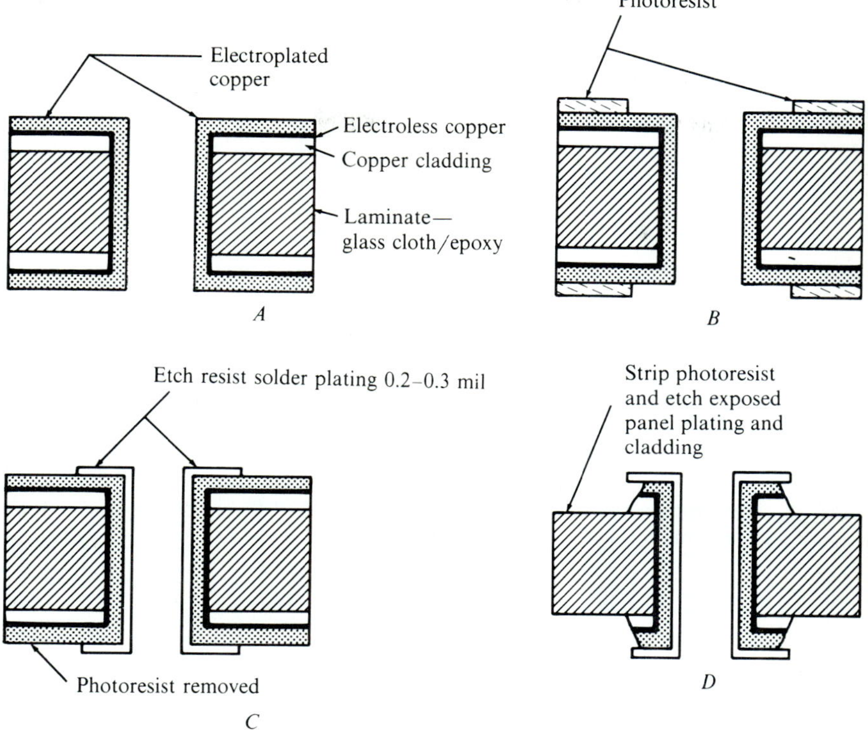

FIGURE 6.14
Four production steps in producing PCBs using panel plating methods.

plating requires completion of the image transfer process before plating begins. The presence of the resist patterns influences the distribution of the current density, making it difficult to achieve plated coatings that are uniform in thickness over the entire area of the board.

The panel plating approach is illustrated in a four step sequence in Fig. 6.14. In the first step, a 1 mil thick layer of copper is deposited over the wall of the through hole and the entire surface of the cladding. In the second step, photoresist is applied and the image is transferred using techniques described in Section 6.3.1. In the third step, a thin layer of Sn-Pb is plated over the areas of the board that are free of resist. In the final step the resist is stripped, exposing copper. This exposed copper is subsequently removed by etching. The Sn-Pb plating serves as the etch resistant coating, protecting the copper used to form the surface features from the chemical attack of the etchant.

The pattern plating approach is described with a three step sequence of operations shown in Fig. 6.15. The first step involves the application of photoresist and the image transfer. Next, copper and then Sn-Pb platings are deposited

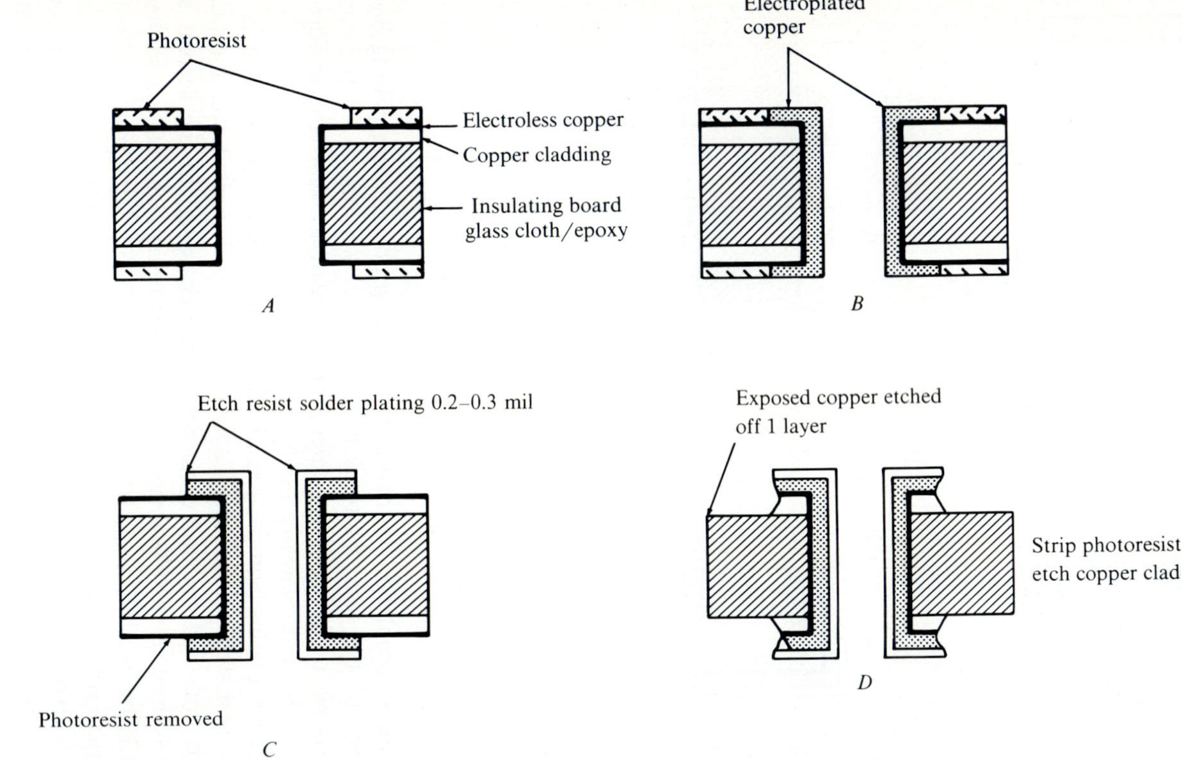

FIGURE 6.15
Production sequence employed with the pattern plating approach to form plated through holes and the circuit board features.

over those board features not protected by the resist. Finally, the resist is stripped and the exposed copper is etched away to produce the complete pattern.

6.6.3 Tin-Lead or Solder Plating

Solder plate consists of a composition of 60% tin and 40% lead and serves as an etch resistant coating. The coating protects the circuit details and plated through holes as the excess copper cladding is being etched away. The solder plate also serves to protect the copper surfaces on the pads and traces from oxidation during the storage period prior to assembly. When the board is ready to be soldered, the solder plate can be reflowed. In the liquid state the solder acts to wet the surfaces to be joined. Because the composition of the solder plate is near the eutectic composition (63% Sn and 37% Pb), it melts at a relatively low temperature of 190°C. The thickness of solder plate is about 0.3 mil, which is sufficiently thick to resist the chemical attack of the etchant and to flow under the action of surface tension when heated above the eutectic temperature.

The solder plate produces one problem in etching, which is related to undercutting. The undercutting produces overhanging ledges of solder plate as shown in Fig. 6.16*a*. These ledges tend to break off and produce small needle-like slivers that are extremely difficult to detect and to remove from the surface of the board. The slivers are also a serious detriment because they can cause short circuits between adjacent conductors. To reduce the effects of the overhanging

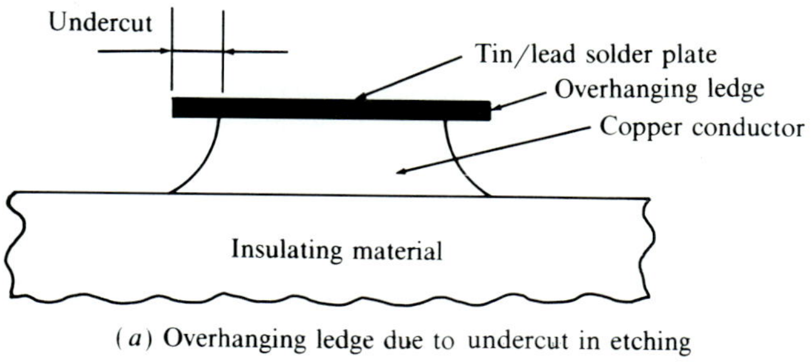

(*a*) Overhanging ledge due to undercut in etching

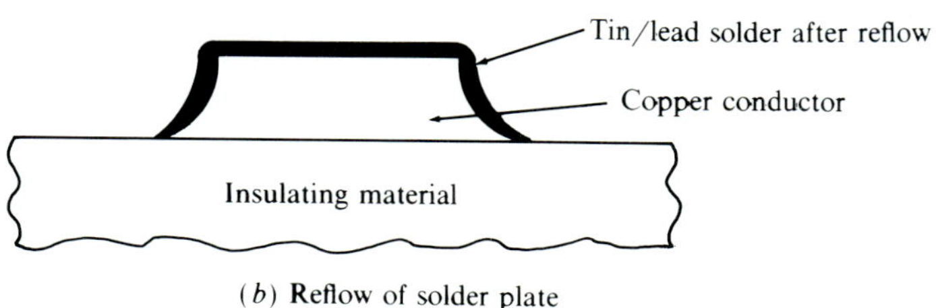

(*b*) Reflow of solder plate

FIGURE 6.16
Effect of undercutting during etch and postetch reflow on the geometry of the tin-lead plating.

ledges, the boards are heated above the eutectic temperature. The solder plate melts and the solder in the ledges is drawn over the surface of the conductor as indicated in Fig. 6.16*b*.

6.6.4 Gold Plating

Gold is a noble metal and as such it is very resistant to oxidation and corrosion. It is also a very good conductor with a resistivity of only 2.44×10^{-6} Ω-cm. These two properties indicate that gold plate is an ideal coating for connector parts that come into sliding contact. The cost of gold is quite high, ranging from $400–500 per ounce depending upon the fragility of the financial community. It is clearly important to minimize the thickness of the gold coating that is required. MIL specs call for gold coatings with a minimum thickness of 80 μin.; however, in commercial applications, thinner coatings are often employed. Most gold plate is deposited on top of a layer of 0.2 mil thick nickel plate. The nickel plate serves to smooth the surface, producing an improved base for applying gold plate that is free of pinholes and porosity. The nickel plate also is chemically stable and acts together with the gold to produce a corrosion resistant surface.

In connector applications, where two contacts are forced together and designed to slide as the contacts are seated, wear of the surfaces is important. When wear is a criterion, the two surfaces (the pin and the contact) are usually plated with gold of differing hardness. Hard gold, an alloy of gold and nickel, is used on the pin and pure soft gold is used on the female contacts. With proper plating procedures, the life of connectors often exceeds 500 cycles before wear of the plating becomes excessive.

Very thin gold plate is sometimes used on expensive components to protect the leads of the chip carrier from oxidation during storage. When these components are soldered to the circuit board the gold plate is stripped by the molten solder and the solder joint is essentially composed of tin and lead. Problems can arise if the gold plate is too thick and is not removed by the molten solder. In these cases, the gold becomes an important alloying element in the solder contained in the joint. Indeed, if the concentration of gold is excessive, a gold-tin intermetallic forms at the bond line of the solder joint. This intermetallic is very brittle and it impairs the strength of the solder joint.

6.7 SOLDER MASKS

After the circuit board has been drilled, plated, imaged, etched, neutralized and cleaned, it is ready for the solder mask to be applied to the exposed surfaces. The role of the solder mask is to control the placement of the solder during the automated soldering process. The mask also provides a protective barrier against surface contamination during the operating life of the PCB. The solder mask is a thin polymeric coating that is applied over all areas of the board that will not be soldered during the assembly process. The solder mask can be applied in two different approaches, namely, by screening or by film lamination.

The film lamination process is almost identical to the photoresist process in that a photopolymer, which serves as the mask, is roller laminated to the board. The photopolymer is then imaged using a phototool, which permits exposure of those areas where solder joints are to be made. The exposed area is removed by stripping and the remaining film forms the solder mask. The mask is placed with the precision of the original photoresist process. The resolution of the solder masks formed by film lamination is excellent; however, the cost is higher than that incurred in the screen printing process that is described next.

The screening process, illustrated in Fig. 6.17, utilizes a fine mesh screen, usually fabricated from nylon fibers or stainless steel wire. A stencil film covering only those areas on the PCB to be soldered is bonded to the screen. The screen is mounted in a frame that keeps it stretched and tight. The screen is lowered in

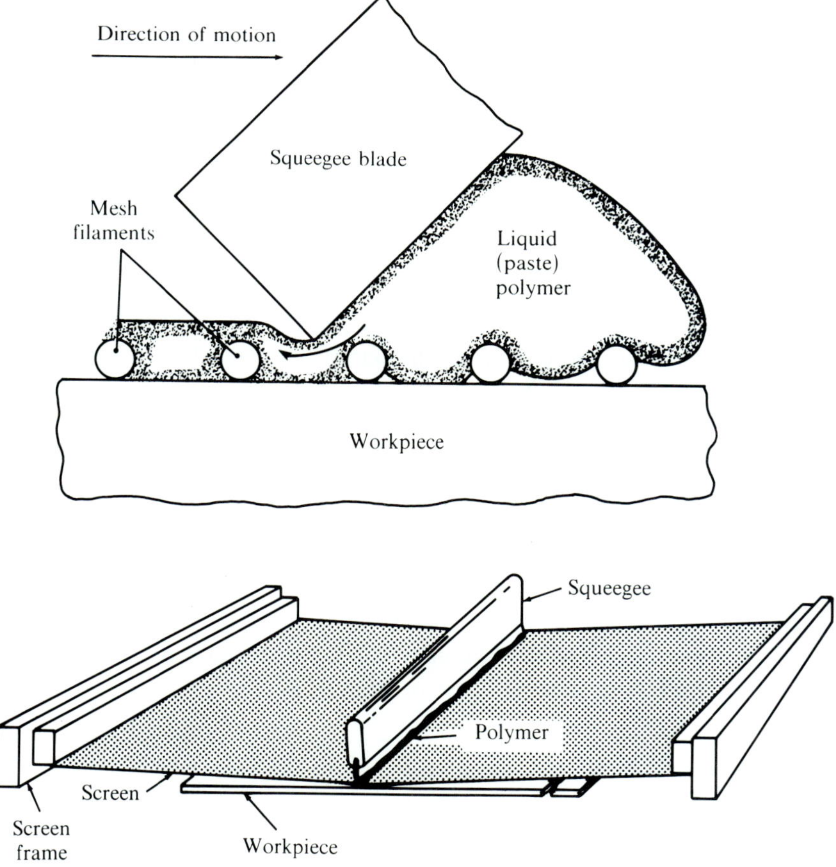

FIGURE 6.17
Illustrations showing features of the screening process to transfer images from a stencil to a workpiece.

registration over the circuit board and liquid polymer is forced through the screen with a wiper tool called a squeegee. The area pattern of the polymer transferred to the board is controlled by the stencil and the thickness of the polymer is maintained by the combined thickness of the stencil and the screen. After screen printing, the coating of polymer, is permitted to dry, forming the solder mask. While lower in cost, registration and resolution of the solder mask applied with screen printing may not be sufficient for dense circuits with pads on close centers.

In addition to solder control during assembly and corrosion control over the life of the product, the solder mask serves a third important function: It provides an insulating coating that encases the circuit traces over major areas of the board. This insulation prevents voltage induced dendritic growth of copper crystals that occurs with time between the closely spaced traces. The dendritic crystals grow on free surfaces and can produce short circuits when they contact adjacent traces.

The solder mask also protects the surface of the board from detrimental effects of moisture over extended periods of time.

6.8 PRETINNING

The PCB is pretinned with the solder plate applied as described in Section 6.6.3. However, the components that are to be soldered to the board should have their leads pretinned. Pretinning is the application of a thin layer of solder to the leads a short time before assembly. Pretinning is usually performed in a solder pot, which contains a molten bath of solder and an overlay of a solder flux. When the leads of the component are inserted into the bath, the flux acts to strip away the layers of surface oxidation and chemically prepare the surface for a metallurgical bond to the solder. The pretinning of components can be accomplished in automatic stations where pick and place devices handle the entire process.

6.9 ASSEMBLY

Assembly involves the placement of devices and components on the circuit board with a temporary means of fastening to hold them in place. The devices are fixed permanently when the solder joints are made.

There are three main problems that the designer must consider in circuit board layout to facilitate assembly. These include:

1. Type of first level package.
2. Footprint design.
3. Machine placement of the components.

We will discuss each of these issues from the viewpoint of automated assembly.

We noted in Chapter 4 that electronic components are available in a wide variety of packages and that these packages are divided into two broad categories

—pin in-hole and surface mount. Assembly of components using pin in-hole packages involves trimming the leads to the correct length, bending them to the proper shape, pretinning, insertion and clinching. All of these operations can be performed with a high degree of automation. However, if several different types of components are specified for the same board, such as both axial lead and radial lead resistors and capacitors in addition to DIPS, the assembly becomes more difficult. A design rule to facilitate assembly minimizes the number of different package types used on a given PCB. Also, mixing pin in-hole- and surface mount-type packages on one board causes major problems. The difficulties arise due to the use of different assembly equipment and the need to employ different soldering methods.

Surface mount packages are more difficult to handle in assembly because they cannot be aligned by the insertion operation. They must be aligned by placing pad on pad or lead over pad. Also, once placed the surface mounted package is subject to movement since the leads are not temporarily fastened by insertion into the board. Assembly of the surface mounted packages depends on the type of surface mount technology employed, and mixing two different types of surface mount packages on the same board can cause serious and costly assembly problems. For example, the flat pack leads (gull wing-type) are usually positioned over the pads of the circuit board and soldered immediately with a hot bar soldering tool. The flat packs are soldered individually because it is not possible to hold them in position if the board is moved prior to soldering. On the other hand, *J* lead packages and leadless chip carriers are positioned over solder pads and are held in position with solder paste. This paste acts like an adhesive as it dries and it holds the chip carrier until it is heated and melted to form a permanent solder joint.

Clearly significant assembly problems arise when packages of different types are required on the same PCB. In some instances the use of different packages cannot be avoided because some components are only available in certain package types. However, it should be recognized that mixing package types adds significantly to the assembly time and cost.

Circuit board layout for assembly is equally important in minimizing costs. For automatic placement of the components, dimensional accuracy is critical, since all of the holes must line up with the pins on each chip carrier. Note that errors in automatic insertion accumulate from four sources, which include variations in the lead diameter, the table position, the insertion head position and the accuracy of the position of the drilled hole. We often accommodate for the relatively large cumulative error by enlarging the plated through hole diameter D according to

$$D = d + e_1 + e_2 \tag{6.3}$$

where d is the diameter of the lead wire
e_1 is twice the positional error due to the insertion machine
e_2 is twice the positional error of the plated through hole

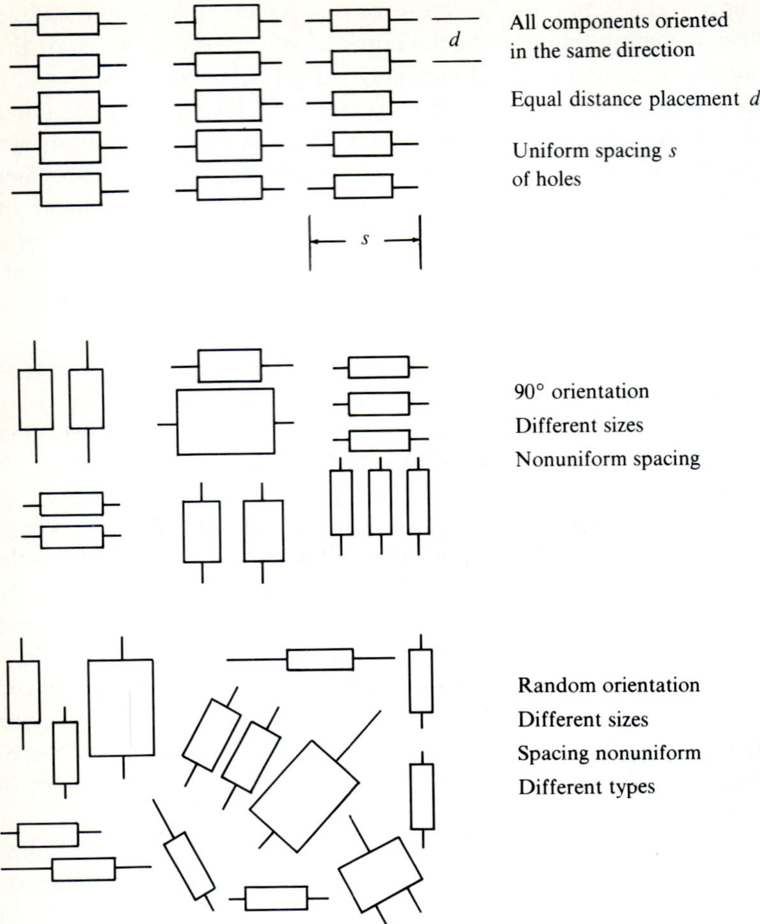

FIGURE 6.18
Examples of component organization that can affect easy assembly.

Values of D = 35–40 mil are common with d = 25 mil. An expression similar to Eq. (6.3) can be developed for the length and width of the pads to accommodate positioning errors in placing surface mounted chip carriers.

Component organization and component spacing on the board also strongly affect ease of assembly. Consistent orientation and constant spacing of components over the board reduce the requirements imposed on the pick and place machinery as well as the number of passes through the line that are required to complete the assembly. An illustration showing examples of component organization ranging from good to terrible is presented in Fig. 6.18.

The equipment for automatic assembly is currently being improved. As the cost of computer controlled machinery decreases, the assembly systems are being developed with markedly improved controls. Basically the pick and place machine is becoming more flexible as the enhanced controls permit changes in the machine functions by programming. The pick and place systems used in component assembly consist of four main subsystems including the conveyor, the positioning mechanism, the feeder and the heads. The flexibility of the system is usually described by the number of degrees of freedom of motion that the entire system has in controlling the position of the component relative to the circuit board.

A very simple 3 degree of freedom system, shown in Fig. 6.19, indicates the principle of operation of a pick and place machine. The circuit board is placed in a holder on a conveyor belt and the movement of the belt controls the x position of the board. The pick and place mechanism with two prime movers arranged in a right angle configuration controls the y and z positions. In these simple pick and place machines limit switches are adjusted to control the position x_1, y_1 and z_1 necessary to pick up the component from the feeder. A second set of limit switches is adjusted to define the coordinates x_2, y_2 and z_2 used in locating the position for inserting the component into the board. For more flexible machines with computer control, these manually positioned limit switches are eliminated and programmable automatic controls permit selection of the x, y and z coordinates associated with pickup at the feeder and placement at the board.

Frequently a number of these pick and place machines are arranged in a line with each machine dedicated to the insertion of a single component. With a programmable machine, it is possible to place several components acquired from different feeders and positioned at different locations on the board with a single machine.

The head on a pick and place machine serves to grip the components and to hold them during transit from the feeder to the board. In some applications the head is also used to control the position of the leads. For example, in inserting

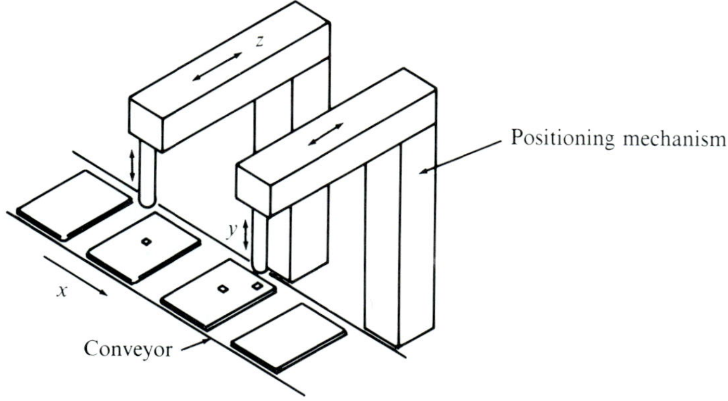

FIGURE 6.19
Simple pick and place machines with 3 degrees of freedom.

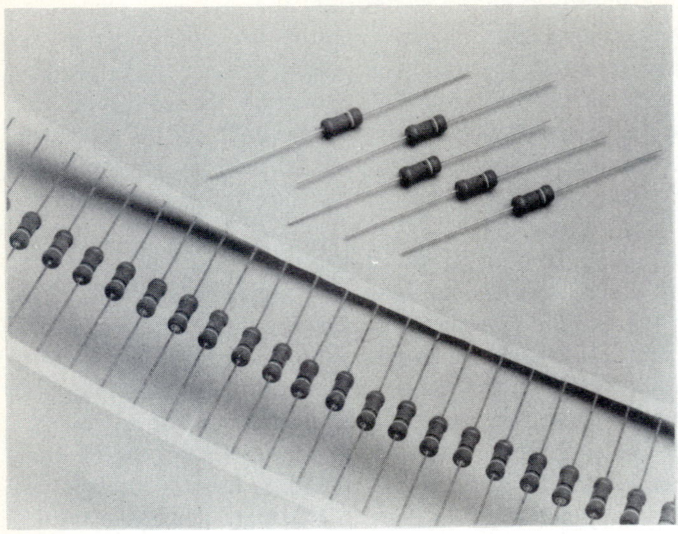

FIGURE 6.20
Axial lead resistor with a belt carrier for feeding in automatic assembly machines. (*Courtesy of IRC, International Resistive Company, Inc.*)

DIPs the leads are initially spread wider than the center-to-center distance on the two rows of holes. The head grips the DIP, squeezes the leads to the proper center distance and supports the component as it inserts the pins in the holes. Most heads employ one or two movable jaws for gripping; however, since the components to be placed are very light, vacuum chucks can also be used effectively. Regardless of the detail design of the head, it is important to space the components far enough apart to give clearance for the head during the insertion phase in the assembly process.

Feeders are a critical component in any automatic placement system. The supply of parts to be mounted on the board must be presented to the head at a prespecified location with the correct orientation. Methods to accomplish this usually entail placement of chip carriers in magazines that contain several hundred parts of the same type. The parts are indexed through the magazine with a spring force, which moves a new part to the target location to replace that part removed by the head. Axial leaded components are usually handled in belt feeders, with the resistors forming cross members in the belts as illustrated in Fig. 6.20.

6.10 SOLDER AND SOLDER FLUXES

Solder is an alloy of tin and lead that is widely employed to make the connections between the chip carrier and the circuit board. A large number of different solder alloys with different compositions, as indicated in Table 6.1, are available.

TABLE 6.1
Solder alloys with melting range and strength

Alloy	Melting range (°C)	Tensile strength (lb / in²)		Shear strength (psi)		
		20°C	100°C	− 130°C	25°C	150°C
62 Sn-36 Pb-2 Ag	179					
63 Sn-37 Pb	183	6120	2700	12700	4130	1165
60 In-40 Pb	174–185					
60 Sn-40 Pb	183–188	2700	580			
50 In-50 Pb	180–209					
50 Pb-50 Sn	183–212			11100	3515	1100
96.5 Sn-3.5 Ag	221	5260		16600	4650	1510
95 Sn-5 Pb	183–222					
99 Sn-1 Sb	235			15100	2900	1125
95 Sn-5 Ag	221–240					
95 Sn-5 Sb	232–240	4410	2900	18150	4625	1880
80 Au-20 Sn	280					
10 Sn-90 Pb	275–302	2850	1160	7300	2800	1500
97.5 Pb-2.5 Ag	303			5300	2590	1440
97.5 Pb-1.5 Ag-1 Sn	309	4980		5900	3045	1520
90 Pb-5 In-5 Ag	290–310			6220	3470	1755
95 Pb-5 Sn	310–314					
88 Au-12 Ge	356					

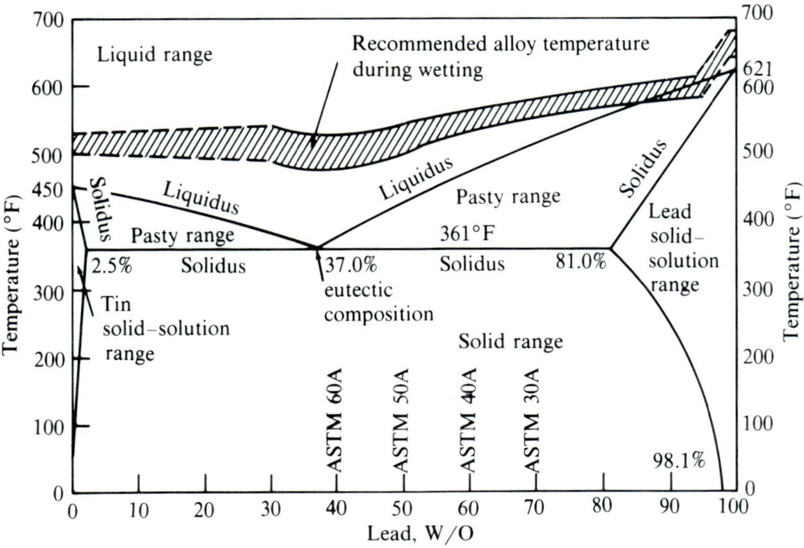

FIGURE 6.21
Tin-lead phase diagram.

TABLE 6.2
Rosin fluxes for soldering PCBs†

Rosin superactivated	Rosin or resin with strong activators
Activated (RA)	Rosin or resin with activator
Mildly activated (RMA)	Water white rosin with activator
Nonactivated (water white rosin) (R)	Water white rosin only

†The vehicle for all four rosin fluxes is a mixture of alcohols, organic solvents and glycols.

The most common solder alloy is the eutectic mixture with 63% tin and 37% lead. The phase diagram for tin-lead, shown in Fig. 6.21, gives the melting temperature for the entire composition range.

A flux is necessary to chemically prepare the surfaces so that the melted solder will form a metallurgical bond at the solder joint. The flux removes surface contaminates, which are usually oxides, and leaves the surfaces clean and wettable. The rosin fluxes commonly employed in electronic applications are defined in Table 6.2. Usually the fluxes that contain only rosin or rosin with mild activators are employed on circuit boards. These fluxes must be removed after soldering and the rosins with mild activators are more easily removed by cleaning than the more active fluxes. The activators employed are usually halogenated organic compounds that dissociate above their activation temperature to produce agents like HCl (hydrochloric acid), which effectively remove surface oxides. Since the acids are present after soldering, they must be removed by cleaning to prevent corrosion from degrading the circuit board.

6.11 SOLDER METHODS

The number of solder joints produced annually in the United States is approaching 10^{12}. With a production operation that is repeated this often each year, it is imperative to maximize the efficiency and minimize defects that occur. To give an example of the importance of defects, it should be noted that some typical soldering processes have defect rates of 1000 ppm (parts per million). The fact that a common circuit board has on the order of 1000 solder joints implies that a defective solder joint will probably occur on each board. Hopefully, the defective joint will be identified during quality control inspections and repaired before the board is incorporated in a product and shipped. The cost of a defective solder joint escalates as a function of time after the soldering process is complete. After the board is placed in the system, the defective board must be located and then the defective joint must be found before repair can be initiated. After the product is shipped, the customer must deal with a product that does not function properly, which is clearly not acceptable under any circumstances.

A sharp reduction in the number of defective solder joints can be achieved with markedly improved process control. Defect rates of 10–30 ppm are possible. However, achieving these rates requires diligent attention to the details of cleaning prior to soldering, elimination of holes and voids in the solder plate and rigorous control of every aspect of the soldering process. There are several soldering methods employed in the industry depending on the type of solder joint being fabricated or the type of solder coat being applied. We will describe some of the more common methods used in soldering electronic components in the following subsections.

6.11.1 Solder Pots and Wave Soldering

A solder pot is used in the simplest form of soldering, usually pretinning leads. The pot is normally fabricated from cast iron and is heated with electrical elements arranged over the bottom and up the sides to maintain a uniform temperature distribution in the molten solder bath. The pots range in size and may contain only a few pounds of solder for a pretinning operation to several hundred pounds of solder for larger circuit board applications. The heaters are thermostatically controlled to maintain the temperature of the solder slightly higher than its melting point.

The molten solder will oxidize in the pot with time and form a film, called dross, that floats on the surface. The dross is detrimental to the soldering process and to inhibit the formation of dross, rosin flux is added to the solder pot. The

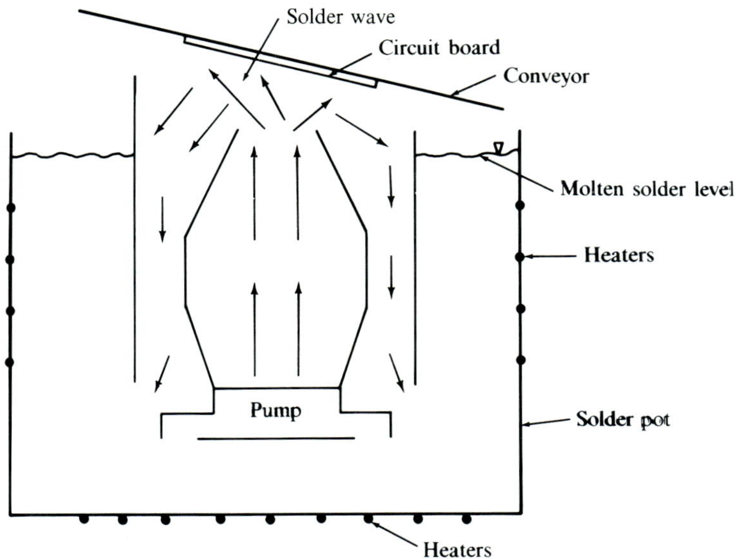

FIGURE 6.22
Cross section of a solder wave facility incorporated in a solder pot.

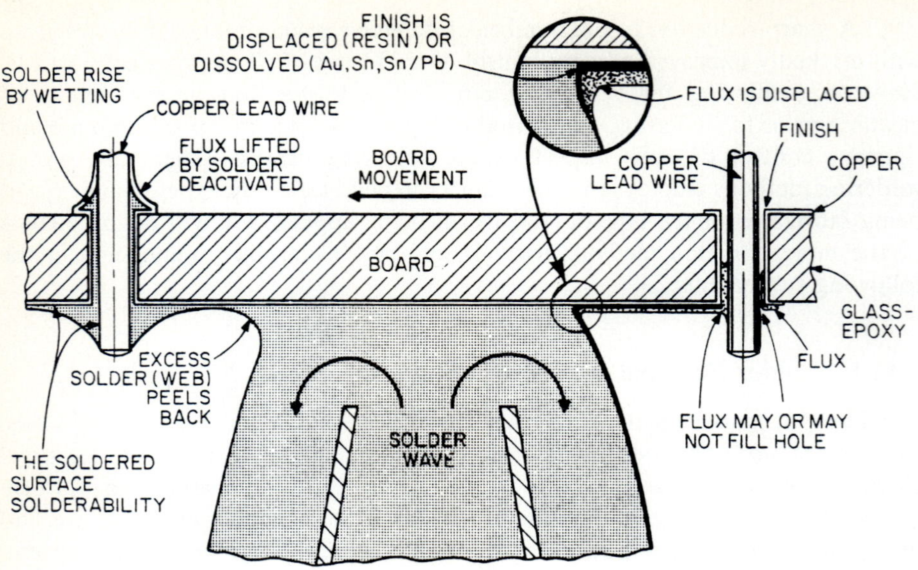

FIGURE 6.23
Interaction of solder wave with pin in-hole-type circuit board.

flux provides a liquid film that acts as a surface blanket and reduces the exposure of the molten solder to air. Another approach is to remove the dross as it accumulates to restore the surface quality of the molten solder.

Components are soldered or tinned by dipping them in the solder pot and holding them in the molten solder until the surface is wetted and coated with solder. For producing solder joints on circuit boards, the pot is often equipped with a pump that produces a solder wave. The solder wave is adjusted to produce a flow of liquid solder that impinges on the underside of the circuit board. The wave simultaneously supplies molten solder to all of the solder joints on the board. An illustration of the pumping arrangement for producing the solder wave is presented in Fig. 6.22. The formation of solder joints by transfer of the liquid solder from the wave to the pin in-hole-type circuit boards is shown in Fig. 6.23.

6.11.2 Vapor Phase Soldering

While the wave soldering process is used extensively for pin in-hole-type joints, it is not effective for surface mounted components where the solder joint is not accessible from the bottom side of the PCB. For surface mounted components other soldering methods are required. One of these methods—vapor phase soldering—is often used with leadless or with *J*-leaded components. These components are placed on the board and held temporarily in place with solder paste, which acts as an adhesive. When the solder paste is heated, it melts and forms a solder joint that solidifies when the board is cooled to room temperature.

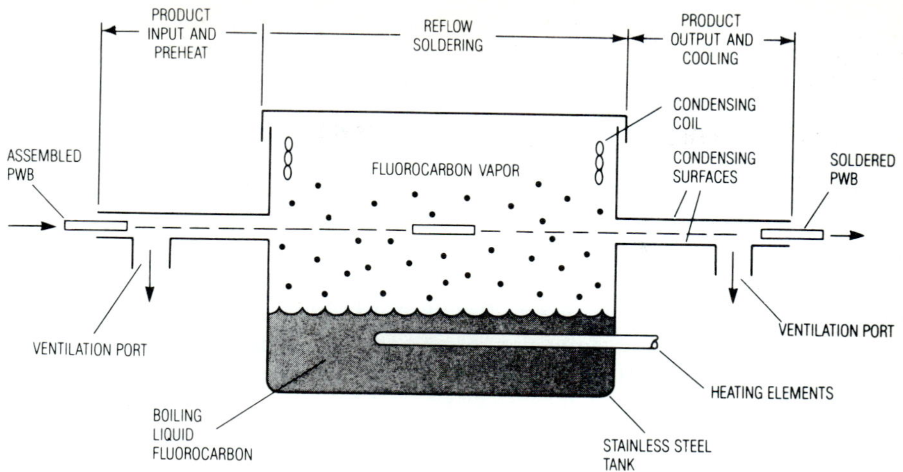

FIGURE 6.24
Vapor phase soldering process for surface mounted components. (*Courtesy of Texas Instruments, Inc.*)

All of the solder to be employed in fabricating the joint is placed with the application of the paste and the pretinning of the component leads or pads. The soldering process in this case requires only that the temperature of the joint be increased until the solder melts and flows. The surface temperature of the mating parts must be above the melting temperature of the solder for the solder to properly wet the surfaces. Note, that the flux necessary to clean the surfaces and remove the oxides is contained in the solder paste.

The vapor phase soldering process, illustrated in Fig. 6.24, is often employed to fabricate the solder joints for surface mounted components. This process achieves excellent temperature control over the entire board and is effective in the rapid transfer of heat to the board and all of the components. The circuit board is preheated to about 100°C before it is moved into the vapor chamber to avoid component damage from thermal shock. When the PCB enters the chamber, it is subjected to the vapor from freon FC-70, which is at a temperature of 215°C. The vaporized freon gives up its latent heat of vaporization as it condenses on the board, components and solder joints. This latent heat produces a rapid increase in the temperature of the entire board and the components. The circuit board is then removed from the vapor to cool. The solder solidifies simultaneously, forming all of the joints on the board. To prevent grain growth in the solder after initial solidification, the joints should be cooled rapidly to room temperature.

6.12 CLEANING AND COATING

After soldering, it is important to thoroughly clean the circuit board to remove residual surface residues that accumulate during the automated soldering processes. Surface contamination is usually divided into two classifications. The first involves polar contaminates, such as plating residues and solder flux activators, that can conduct current and will in time produce corrosion. The second includes nonpolar contaminates, such as oil, grease and rosin fluxes, that produce nonconducting films. These films may deposit on the surfaces of the fingers (pins) associated with the edge connector and produce intermittent opens. Both of these types of surface contamination must be removed to achieve a quality PCB with a long trouble-free life.

Vapor degreasing is the most commonly employed process for removing both types of surface contaminates. A solvent cleaning solution is heated in a sump until it boils. Vapors from the solvent rise and condense onto the circuit board, which is positioned above the sump. The solvents that condense on the board are pure and wash the boards with contamination free cleaners. Rinsing with the pure solvent continues until the board reaches the boiling temperature of the solvent and the condensation process terminates. The condensate containing surface impurities drops to the boiling sump where it is recycled. The impurities that collect in the sump are removed periodically.

When clean, the PCB is often treated with a conformal coating. The purpose of this coating is to minimize the degradation in electrical properties of the circuit board due to the long term effects of humidity. Recall that water vapor absorbed by the polymer in the circuit boards lowers insulation resistance, reduces high voltage breakdown, increases the loss factor and promotes corrosion. Conformal coatings are blends of polymers, solvents and plasticizers. Acrylics, polyurethane and epoxy are commonly used as the base polymer because of their ease in application and relatively low cost. Silicone and polyimide are also used in higher temperature applications; however, they are more difficult to apply and are much higher in price.

This treatment provides the reader with a brief overview of some of the manufacturing methods used in the production of common circuit boards. The methods of manufacturing are very important because the quality, performance and cost of the PCBs are controlled by these many processes. Moreover, as the circuit densities on the chips increase with the implementation of VLSI and the introduction of ULSI, the circuit densities on the circuit board will also increase. These changes will result in smaller spaces between circuit traces, finer lines, smaller pads and extremely small diameter holes with demanding registration requirements. New manufacturing processes will be required to produce the circuit boards that will be found in the electronic systems scheduled for introduction in the early 1990s.

6.13 CERAMIC CIRCUIT BOARDS

Ceramics are employed in the production of chip carriers, hybrid circuit boards that support a relatively small number of components and the larger more traditional circuit boards. In the chip carrier application, the ceramics have the advantage of excellent hermeticity and a relatively low thermal resistance. In circuit board applications, both hybrid and conventional, the ceramics have the advantage of a temperature coefficient of expansion that is closely matched to the silicon chip or the ceramic chip carrier. A second advantage, particularly important for hybrids, is the ability to directly print resistors onto the circuit board. Another advantage is the stability of the ceramic with both time and temperature in harsh environments.

Circuit boards fabricated from ceramics are comprised of three different components, which include the ceramic substrate, sintered metal circuit traces and pads, and glass insulation. We will describe the components individually in subsequent subsections.

6.13.1 Ceramic Substrate Materials and Processes

Some of the ceramic materials employed as substrates are listed in Table 6.3. Alumina-based ceramics, containing 80–90% Al_2O_3, are the most common because of their strength, high electrical resistivity and economy. The substrates are produced in sheet form by dry pressing or sheet casting, followed by sintering. For substrate applications where heat transfer is the most important consideration beryllium oxide (BeO) can be employed because it has the highest thermal conductivity, 7–8 times larger than alumina, of any of the ceramic oxides.

In recent years nitrides and carbides have been considered for substrate applications. These materials are very strong and offer interesting combinations of properties. For example, with both aluminum nitrate and silicon carbide, the thermal conductivity is high and the thermal coefficient of expansion is closely matched to silicon. The problem of low electrical resistivity of silicon carbide was recently solved by Hitachi by producing a composite ceramic incorporating both BeO and SiC. The fabrication process, illustrated in Fig. 6.25, involves mixing powders of the two materials together with a binder and a solvent. This mixture is sprayed dried and then dry pressed to form disks. These disks are stacked between graphite spacers and hot pressed in a vacuum at 2100°C. The BeO segregates at the grain boundaries and the SiC forms as relatively large and very pure grains with a thermal conductivity approaching 1000 W/m °C. The BeO in

TABLE 6.3
Ceramic substrate materials with select properties

Material	Thermal expansion coefficient (20–200°C)	Thermal conductivity (W/m K)	Dielectric constant
Aluminum oxide	6.7×10^{-6}/K	25–35	9.0–9.5
Silicon	2.7	150–160	
Aluminum nitride	3.3	140–170	8.5–10.0
Silicon nitride	2.3	25–35	6.0–10.0
Silicon carbide	3.5	120–250	20–40
BeO	9	220–240	6–7

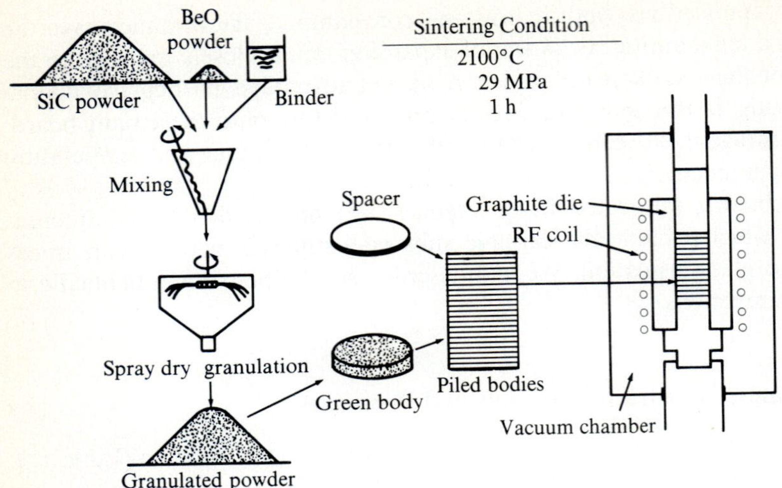

FIGURE 6.25
Hitachi process for producing a composite ceramic (SiC + BeO).

the grain boundaries forms a series of thin high resistance layers between each grain which significantly increases the electrical resistivity of the composite ceramic.

The dry pressing and sintering method of producing thin flat plates for ceramic substrates has been described in Fig. 6.25. A second method for producing thinner and larger ceramic plates involves the development of a "green sheet." To form a green sheet the ceramic and glass powders are blended and then milled in a ball mill to ensure proper particle size. The powders are then mixed with organic binders and a solvent to form a slurry. The slurry is formed into a uniform thickness sheet by casting the mixture onto a continuous plastic carrier, as shown in Fig. 6.26, that supports and transports the sheet as the

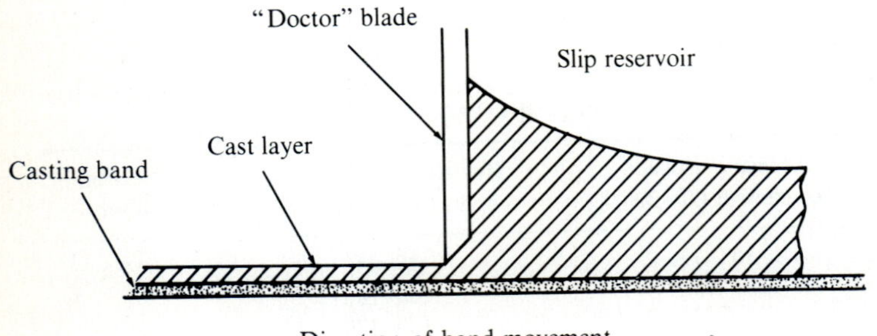

FIGURE 6.26
Continuous casting of ceramic green sheets.

solvents are removed by evaporation. Upon drying, the "green sheet" is similar to a piece of paper. The binder holds the particles of glass and ceramic together, providing a reasonable degree of flexibility and strength. These properties permit the unfired sheet to be handled during subsequent processing operations. In the green state, prior to sintering, the sheets can be cut, holes can be punched and circuit lines and vias formed by printing with metallized inks. The green sheets become ceramic sheets only after they are fired. The firing drives off the organic binders and fuses (sinters) the glass and ceramic particles together. During the sintering process the sheet shrinks by about 15–20%. This change in dimensions must be accommodated in the design layout where lengths between features and hole diameters are increased to account for the shrinkage.

6.13.2 Metallization

There are three methods commonly employed to apply the metal required to form pads and circuit traces on substrates, namely, thin film, thick film and cofired. Thin film technologies have been developed for the past three decades to produce integrated circuits. This technology is now being applied in a few applications of high density packaging on ceramic or silicon circuit boards. The method involves the development of a pattern representing the pads and circuit traces in a layer of photoresist. The metal is then applied in a thin film by evaporation. The metal charge to be evaporated is placed in a crucible and the substrate to be coated is placed above the crucible on a rotating stage. The entire assembly is contained in a bell jar and a high vacuum is used to prevent oxidation of the metal during its vaporization. Very thin pads and circuit traces, usually 1–3 μm thick, are formed using the evaporation technique.

Thick film technology is the most common method used to apply metal features to the ceramic substrate. The metals used in this process are in the form of pastes that are deposited by a screen printing process (see Fig. 6.17). The pastes contain a binder to attach the metal to the substrate and a vehicle to facilitate printing. Control of the viscosity of the paste is important. The paste must be fluid enough to pass through the printing screen yet firm enough to form sharp edges without significant spreading during the drying process. After drying, the printed deposits are fired at elevated temperature. This firing burns off the organic binders and sinters the inorganic binders (glasses) to the substrate. A typical time temperature profile for a belt furnace used in firing is illustrated in Fig. 6.27.

Pastes are formulated to provide circuit traces representing both conductors and resistors. For pastes used to produce a conductor, the metal is in the form of fine particles with an equivalent diameter of 1–3 μm. A small amount of glass frit is added to the paste to control shrinkage during sintering and to improve bonding of the metal to the substrate. Gold particles perform well as a metallized conductor and would be more widely used except for its cost. It can be wire bonded and it is resistant to oxidation at the firing temperatures. The properties of gold can be improved with the addition of either platinum or palladium, which

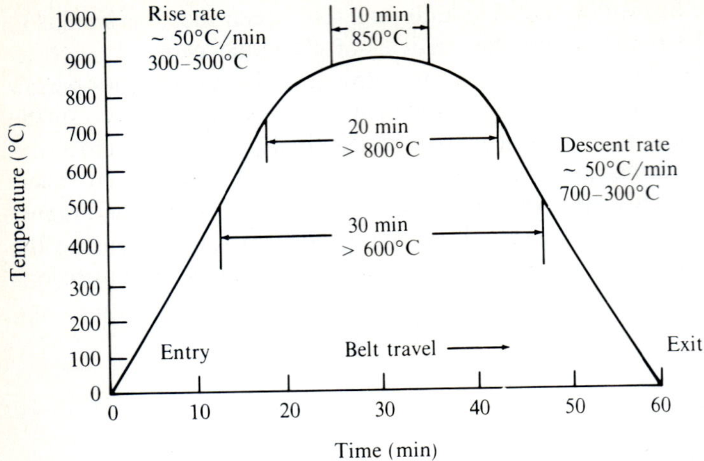

FIGURE 6.27
Typical time temperature profile used in a belt furnace for firing metal pastes.

enhance its solderability and increase its resistance to leaching in solder. Silver is sometimes used because of its reduced cost, but it exhibits migration under voltage with time. Additions of platinum and/or palladium reduce the tendency of silver to migrate and improve its resistance to leaching. Copper can be employed to advantage in pastes since it is resistant to migration, it does not leach in solder, it has a high conductivity and it is available at a low cost. The difficulty in using copper-based pastes is in the firing. The copper will oxidize in air and it is necessary to use furnaces with nitrogen as the cover gas. Of course, the absence of oxygen in the furnace makes it impossible to burn out the organic binders. To alleviate this problem, the binders are oxidized with air as the cover gas in the furnace at low temperature. The furnace is flushed to remove the oxygen and the hydrocarbons from the binder and is then flooded with nitrogen prior to elevating the temperature to sinter the remaining copper particles.

Resistors are formed directly on the substrate by sintering a mixture of metal and oxides together to form a high resistivity conductor. The geometry of the line (width, length and thickness) controls the resistance of the element. Ruthenium mixed with glass frit is used for high resistivity films and silver palladium alloys are compounded with glass for the lower resistivity films. The actual size of the resistor depends on the power dissipation, but minimum sizes are usually about 20 mil wide by 30 mil long. Tolerances on the resistance are usually $\pm 20\%$. Laser trimming of the conductor is often employed to modify the resistances to provide closer control of the resistance value.

Cofired metal conductors and resistors are produced using methods nearly identical to those used with the thick film. Pastes much like printing inks are screened onto the ceramic to form the surface features required. The primary difference is the fact that the ceramic is in the form of a green sheet. The green sheet and the pastes are fired together and both are sintered in the same operation. The use of the cofired process places special requirements on the pastes used for metallization. The firing temperatures that are necessary for sintering the ceramics are quite high and the alloys previously described are not suitable at these higher temperatures. Instead molybdenum, tungsten and other refractory metal alloys are employed, which do not chemically react with the ceramic at these elevated temperatures. The multilayer ceramic substrate for the thermal conduction module, illustrated in Fig. 4.23, was produced using the cofired process.

6.13.3 Glasses as Dielectrics and Seals

Multilayered circuit boards are produced with metallization on ceramic circuit boards. With the cofired process, layers of green sheet with printing on one side only are stacked together in registration. The stack of sheets is sintered as a single unit and remains in registration even with shrinkage of 15–20%. With the thick film process, the production of multilayer boards differs appreciably because several printings and firings are used to fabricate the required number of signal and power planes. The first firing is that of the metal paste on the ceramic substrate as described previously. The next step is to apply, by printing, a layer of

TABLE 6.4
Composition of some glasses used for seals

Component (wt %)	Glass 1	Glass 2	Vitreous cane seal glass	Devitrfying seal glass	Composite glass
PbO	56.0	73.0	66.0	77.0	73.1
B_2O_3	21.5	12.5	14.0	9.4	9.5
SiO_2	12.0	14.3	2.0	1.0	1.9
	1.0	0.2	3.5	2.1	0.5
CaO	5.5	0	0	0	0
MgO	2.0	0	0	0	0
Na_2O	2.0	0	0	0	
Bi_2O_3	0	0	1.5	0	0
ZnO	0	0	10.5	10.5	0
CuO	0	0	2.5	0	0
Beta-eucryptite	0	0	0	0	15.0
Thermal expansion coefficient $\times 10^7/°C$ (RT 300°C)	86.5	84.0	84.0	89.0	77.0
Softening point (°C)	542	473	413	370	350
Seal temperature (°C)	480	435	470	450	425

Source: R. R. Tummala and R. B. Shaw, "Ceramics in Microelectronics," *Ceramics International*, vol. 13, pp. 1–11, 1987.

an insulating dielectric over this layer of metallization. The dielectric is a glass paste that is screen printed with via holes aligned to the first layer of metallization. The glass paste is dried and fired to form a continuous layer of insulation of uniform thickness with via holes at select locations to provide signal plane connections. The layer of glass provides the surface for the application of the second layer of metal paste. The printing and firing processes are repeated many times to fabricate a multilayered ceramic circuit board.

Glass is also widely used for seals in a wide range of packaging applications. Seals may be necessary to hermetically bond components together or to provide mechanical connections and support structures. Combinations of glass, ceramics and metals may be involved in a specific design. The composition of the glass is very important if the seal and/or bond is to be effective. The temperature coefficient of expansions of all of the materials in the joint must be closely matched. If mismatches occur, large differences in the coefficient of expansion result in thermal stresses that can cause cracking of the joint. The usual approach in dealing with mismatch is to use a series of glasses to produce a graded seal. In grading the glasses across the joint, the glasses with a higher coefficient of expansion are placed around the glasses with a lower temperature coefficient of expansion. This procedure produces compressive residual stresses in the graded glass joint and enhances its strength. A select list of glasses that may be considered for seal applications is presented in Table 6.4.

REFERENCES

1. Lindsey, D.: *The Design and Drafting of Printed Circuits*, Bishop Graphics, Inc., Westlake Village, Calif., 1984.
2. Munich, S.: "The Effects of Integrated CAE/CAD on the Design Cycle," *Printed Circuit Design*, vol. 5, no. 2, 1988.
3. Ducas, C.: "Electrical Apparatus and Methods of Manufacturing the Same," US Patent No. 1,563,731, Dec. 1, 1925.
4. DeForest, W. S.: *Photoresists—Materials and Processes*, McGraw Hill, New York, 1975.
5. Coombs, C. F.: *Printed Circuit Handbook*, 2nd ed., McGraw Hill, New York, 1979.
6. Lund, P.: *Printed Circuit Board Precision Art Work Generation and Manufacturing Methods*, Bishop Graphics, Inc., Westlake, Calif., 1986.
7. Bello, D. C.: "Preventing Wave Solder Defects," *Printed Circuit Fabrication*, vol. 10, no. 10, 1987.
8. Mullen, J.: *How to Use Surface Mount Technology*, Texas Instruments, Inc., Dallas, Tex., 1984.
9. Roos-Kozel, B.: "Solder Pastes," *Surface Mount Technology*, ISHM Technical Monograph 6984-002, pp. 115–170, 1984.
10. Romenesko, B. M.: "Cleaning," *Surface Mount Technology*, ISHM Technical Monograph 6984-002, pp. 229–244, 1984.
11. Manko, H.: *Solders and Soldering*, McGraw-Hill, N.Y., 1964.
12. Ikegami, A. and T. Yasuda: "High Thermal Conductive SiC Substrate and Its Applications," *5th European Hybrid Microelectronics Conference*, ISHM, Reston, Va., pp. 465–471, 1985.
13. Tummala, R. R. and R. B. Shaw: "Ceramics in Microelectronics," *Ceramics International*, vol. 13, pp. 1–11, 1987.

14. Holmes, P. J. and R. G. Loasby: *Handbook of Thick Film Technology*, Electrochemical Publications Ltd., Glasgow, Scotland, 1976.
15. Moran, P.: *Hybrid Microelectronic Technology*, Gordon and Breach Science, New York, 1984.

EXERCISES

6.1. Prepare a block diagram showing the key steps in the production of a printed circuit board. Begin with the design engineering layout and track the product flow to inventory.

6.2. Artwork is prepared under temperature and humidity of 20°C and 40% is stored for 3 yr at a temperature of 25°C and humidity of 85%. Determine the errors anticipated in using this artwork for design changes.

6.3. Plot a line using the principles of raster graphics and comment on the quality of the line if it is along the x axis, along the y axis and along a 45° diagonal. Compare this result to the lines that can be constructed by using vector plotting.

6.4. Plot an equilateral triangle using the principles of raster and vector graphics.

6.5. Plot a circle using the principles of raster and vector graphics. Comment on the differences in the quality of the images produced using these two procedures.

6.6. Compare the advantages and disadvantages of photoplotters of the vector and raster types.

6.7. Prepare a cost analysis showing the cost trade-off between buying and operating a photoplotter and using a service bureau to prepare the required artwork. Assume the necessary costs as required to support your analysis.

6.8. Describe the differences between negative-type photoresist and positive-type photoresist.

6.9. Design a frame to hold the laminate in a photoresist printing operation. The frame should incorporate some feature that will permit registration of the image on both sides of the laminate. Discuss the principle of operation for your registration technique.

6.10. An etching process is yielding product with an etch factor $F_e = 0.6$. Determine the width of lines to be used in the phototool if the line width on the board is to be 8 mil. Consider copper cladding of weights 0.5, 1, and 2 oz/ft^2.

6.11. Describe the types of circuit boards that have holes that are punched. Give some characteristics of the products likely to use this type of PCB.

6.12. Prepare a graph showing the diameter of the pad D and the thickness of the annular ring on a PCB as a function of the hole diameter d. The design rule for the layout indicates that the thickness of the annular ring is (a) $\frac{1}{2}d$, (b) $\frac{1}{3}d$ and (c) $\frac{1}{4}d$.

6.13. Write the equations needed to perform a tolerance analysis for hole breakout in a PCB. Before beginning, list all of the factors that contribute to the error accumulation.

6.14. A drilling station is capable of operation with an aspect ratio of 10 without excessive drill breakage. We are planning to drill 20 mil holes in circuit boards that are 40 mil thick. How many boards do we place in the stack to be drilled simultaneously?

6.15. Describe the process used to produce the two-sided epoxy–glass laminate.

6.16. Describe the lamination process used to fabricate multilayer circuit boards. Indicate how registration between the layers in each board is maintained. Discuss control of the stack during the elevated temperature curing cycle.

6.17. Why is the plating process used so frequently in the production of PCBs?

6.18. What factors control the deposition rates in a plating process?

6.19. Describe the electroless plating process, showing the essential differences with the electrolytic plating process. What is the role of the stannous and palladium ions in this process.

6.20. Describe the panel and pattern plating processes. List the advantages and disadvantages of both processes.

6.21. Solder is an alloy containing both tin and lead. How can both of these elements be plated to form a coating that is composed of the proper amounts of the elements?

6.22. Give an example of reflow that is analogous to reflow of the solder plate. Describe wetting or the lack of wetting in this analogy.

6.23. Show the price of gold over time beginning with 1932. What are the most important factors that affect the price of this material.

6.24. What are the two most important properties of gold that warrant its use in select connector applications.

6.25. Describe the advantages of using photopolymer films for solder masks.

6.26. Describe the advantages of using liquid polymers and the screening operation for the application of solder masks.

6.27. List the defects that could occur if a PCB were soldered using the wave soldering process without the benefit of a solder mask.

6.28. Sketch a manufacturing cell designed to pretin the leads on DIPs. Would it be possible to use this cell for pretinning flat packs? Indicate why.

6.29. Why are surface mounted packages more difficult to assemble than pin in-hole packages? What practices are employed to ease the assembly problems with surface mounted components?

6.30. List several design rules to be followed in reducing cost and complexity in assembly of components on circuit boards.

6.31. Determine the diameter D of a plated through hole that is to accommodate a pin with a diameter $d = 25$ mil. The pin is to be placed with a pick and place machine that can locate the x, y, and z positions to ± 3 mil. The tolerance on d is ± 2 mil, on D is ± 3 mil and on the location of the center of the hole ± 4 mil on each coordinate.

6.32. What is the primary difference between a pick and place machine with fixed stops and one with computer control of the motion. Cite advantages and disadvantages of both of the machines. Which would you select if you needed to assemble 500,000 boards of the same type per year? Would your answer change if you produced 500,000 boards per year with 500 different part numbers?

6.33. Sketch the design of a head used to acquire, move and insert a 14 pin DIP.

6.34. Sketch the design of a head used in a pick and place machine to acquire, move and place the 44 J leaded chip carrier illustrated in Fig. 4.12.

6.35. Sketch the design of a feeder to provide a supply of the *J* leaded chip carriers defined in Exercise 6.34.

6.36. You are shipping a product with 16,000 solder joints. If your soldering process has a defect rate of 200 ppm, how many defect joints will occur in an average product. If your inspectors locate 70% of these defects and the repair is effective, what percent of your shipments will contain defects. Comment on the advisability of this procedure.

6.37. What is the melting point of the following solders: (*a*) 60% Sn-40% Pb; (*b*) 50% Sn-50% Pb; (*c*) 40% Sn-60% Pb; (*d*) 5% Sn-95% Pb.

6.38. Write a short paragraph describing the features required of a flux used to facilitate soldering of joints on a printed circuit board.

6.39. Describe the effect of small pieces of dross that are engulfed by the solder wave and impinge on the circuit board to form solder joints.

6.40. With the wave soldering process, how does one avoid the formation of solder bridges between adjacent joints.

6.41. Compare vapor phase soldering with oven soldering where infrared heating elements are used to increase the temperature of the board, the leads and the components.

6.42. Explain why the vapor phase process is often called an isothermal process.

6.43. Describe the problems encountered by the presence of contamination on circuit boards. Classify the contaminates in your description of the problems.

6.44. Describe the vapor degreasing process and give the advantages and disadvantages associated with this process.

6.45. What are the merits of a conformal coating? Cite one important disadvantage of the conformal coating.

6.46. Compare the advantages and disadvantages of printed circuit boards of the organic (glass–epoxy) and ceramic types.

6.47. Discuss the common substrate materials for ceramic PCBs. Include in your discussion the nitrides and carbides.

6.48. Cite the recent development by Hitachi in modifying SiC and indicate the importance of this development.

6.49. What are the three common methods used to form surface features on ceramic and silicon substrates? Briefly describe each of these methods.

6.50. Describe the process for manufacturing green sheets of ceramic. What constituent gives the sheets strength and flexibility?

6.51. Describe the primary difference in firing conductor pastes made from the noble metals and those made from copper.

6.52. Why would you prefer to place your resistances on a circuit board by direct firing rather than by placement of discrete components?

6.53. Describe the manufacturing process for producing a multilayer ceramic circuit board using thick film technology.

6.54. Repeat Exercise 6.53 but use cofired technology.

6.55. Repeat Exercise 6.53 but use thin film technology.

CHAPTER 7

THIRD LEVEL PACKAGING

7.1 INTRODUCTION

Third level packaging includes all of the hardware required to house the circuit boards and peripheral equipment necessary for a complete electronic system. Large systems, with 10^2–10^3 circuit boards, are housed in one or more electronic enclosures. These enclosures support the boards, provide a cooling medium, protect the components from the environment and contain the back panels, bus bars and cabling necessary for the higher level connections. In large systems the design of the third level packaging is a significant effort because many different mechanical components are involved. In smaller systems, the packaging tasks are less complex because the number of circuit boards is reduced. However, with the smaller systems competitive pricing pressures complicate the design effort since cost limitations eliminate some of the design approaches.

In addition to housing the entire electronic system, the third level packaging must permit easy access to service personnel so that the components located within the enclosure can be identified, tested and replaced. For large high performance networks, system availability is critical and rapid diagnostics and repair are essential to ensure timely recovery of a malfunctioning system. Design for accessibility is an essential part of the maintenance strategy. At the same time, the third level packaging must inhibit access by the well meaning and always curious klutzes who may damage themselves and/or the system. It is clear that these conflicting goals for accessibility require ingenious design of fasteners, keys and tools involved in opening and closing the enclosures.

Safety is always a paramount design requirement. One may believe that the voltages associated with a logic circuit are always low and the dangers of shock and electrocution are nonexistent. It is true that low voltages (less than ± 10 V) are common in digital circuits; however, these low voltages are often associated with very high currents. Bus bars servicing a dense back panel can carry several hundred amperes. These bars must be shielded to prevent accidental shorting by service personnel. If the power bus is shorted, severe arcs will result and the possibility of blinding and burning exists. Higher voltages are found in the power supply compartments where the ac supply is converted to the dc voltages required for the digital circuits. Shielding of the higher voltage terminals is essential in meeting safety requirements.

A final factor to consider in the design of the third level package is the operator/system interface. A typical interface includes switches and indicator lights for turning the system on and off, a display for messages and/or graphics and a keyboard or keypad for entering commands. It is essential that this input/output panel be positioned to minimize operator effort in monitoring the system function and in providing commands to direct system activity.

We will cover third level packaging, beginning with the edge card connectors that serve as the third level of connection in the connection hierarchy, and then move to back panels and cables used for still higher levels of connections. We will also describe subassemblies and enclosures and then cover cooling devices, such as fans and cold plates, that are commonly used.

7.2 CONNECTORS

On a typical PCB, we frequently connect 1000 or more pins to form an element of a logic function. The complete logic function requires many more components and connections and it is necessary to use many circuit boards to support the very large number of chips involved in the complete circuit. These boards are connected together using edge card connectors. A typical edge card connector, illustrated in Fig. 7.1, shows four features common to almost all connectors:

1. A series of pins or leads that are soldered to the circuit board to make connections to the I/O traces from the board.
2. Another series of pins that are inserted into a receptacle to carry the signals to the next level of connection.
3. A core, usually fabricated from plastic, that spaces the pins on closely controlled centers and insulates the connector assembly.
4. Two guide pins that serve to align the connector pins with the holes in the receptacle. The two guide pins can also be arranged to serve as a key that prevents insertion of the wrong plug into the receptacle.

There are many different types of edge card connectors, ranging from the simplest male connector, presented in Fig. 7.2, to very dense connectors with pins

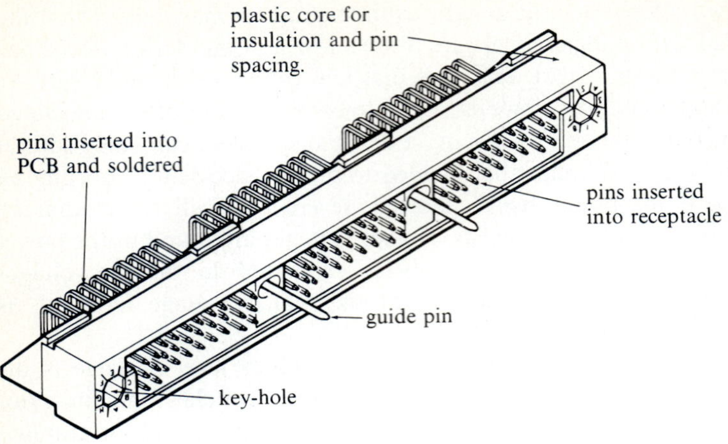

FIGURE 7.1
Typical edge card connector showing the four common features. (*Courtesy of Berg Electronics, A Du Pont Company.*)

on 50 mil centers as shown in Fig. 7.3. The simple male connector of Fig. 7.2 is made by etching blade surfaces from the copper cladding on both sides of a printed circuit board. These finger-like surfaces are plated with gold to provide a suitable surface for electrical contact. The dense connector of Fig. 7.3 has a very high pin count to handle the large I/O requirements typical of a high performance circuit board.

The receptacles contain contacts and accept the pins from the edge card connector. The receptacles may be mounted on a back panel, a wire wrap board, a cable or a chassis. Some examples of commonly employed receptacles are presented in Fig. 7.4 and a typical example of the PCBs (daughter boards) connecting with a back panel (mother board) is illustrated in Fig. 7.5.

FIGURE 7.2
A simple arrangement of gold-plated fingers on the edge of a PCB serves as pins for one type of edge card connector.

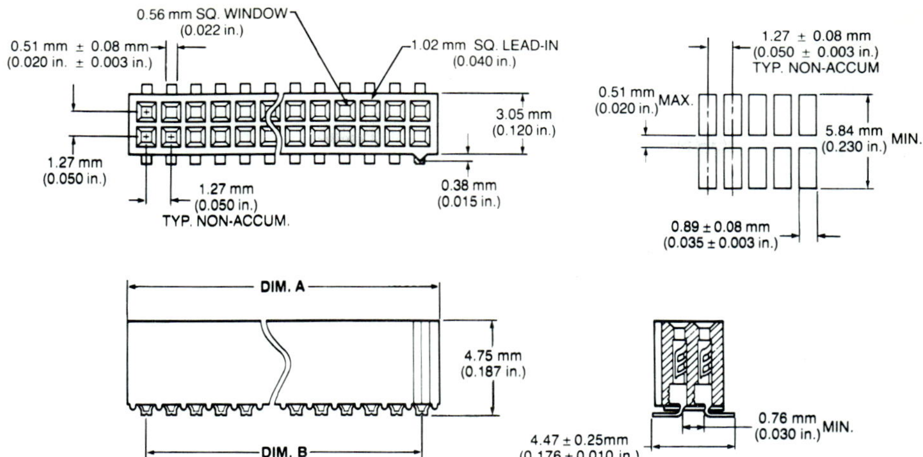

FIGURE 7.3
A dense card connector with contacts on 50 mil centers that is surface mounted. (*Courtesy of Berg Electronics, A Du Pont Company.*)

There are many different designs of connectors; however, the basic principles of design are relatively few in number. We will consider the design of the basic components, including the pin, contact and insert, in the next subsection.

7.2.1 Connector Pins, Contacts and Inserts

The male side of a connector has one or more pins or blades that are inserted into the contacts on the female side of the connector. The pins, illustrated in Fig. 7.6*a*, are cylindrical with a diameter d that usually ranges from 15–85 mil. The nose of the pin is rounded and tapered to assist in the alignment of the pin with the contact during engagement. The center region of the pin is larger in diameter and usually provided with a discontinuity to provide for positive retention in the plastic core of the connector. The other end of the pin is called the tail and it is configured in a number of different ways to accept the incoming leads.

The force that can be carried by the pin during insertion is controlled by buckling. Reference to Euler buckling theory gives the critical load associated with collapse of the pin as

$$P_{cr} = C\pi^2 EI/L^2 \tag{7.1}$$

where $C = 1/4$ for one end fixed and the other free
$C = 2$ for one end fixed and the other end guided
$I = \pi d^4/64$ for a circular cross section
L is the length of the pin
E is the modulus of elasticity

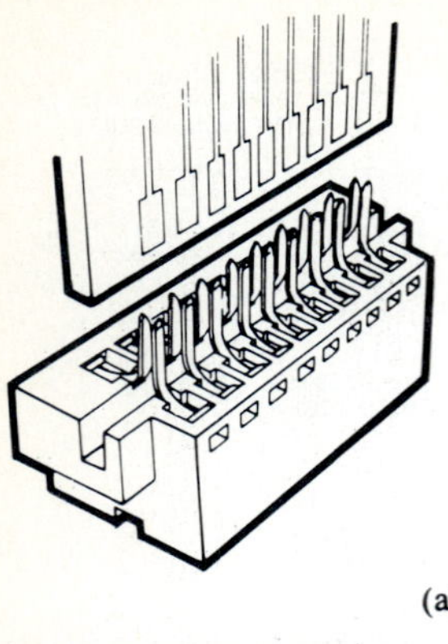

(a)

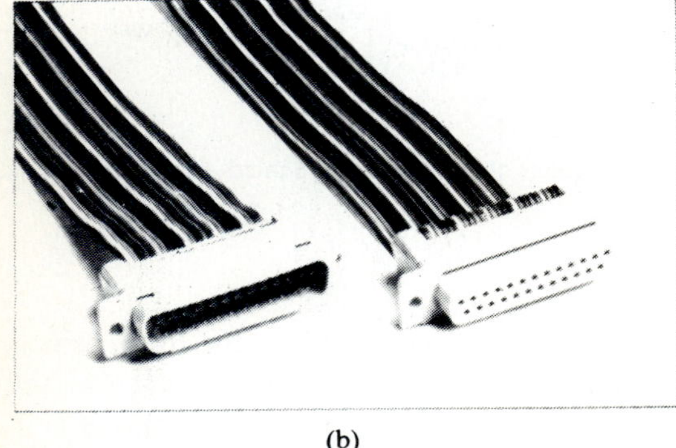

(b)

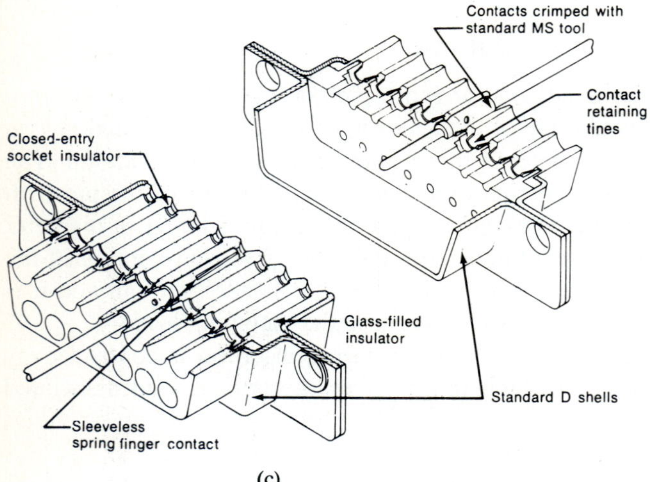

(c)

FIGURE 7.4
Examples of different connector types. (a) Receptacle for card edge. (b) Flat cable connectors–plug and receptacle. (c) Panel *D* shell connectors. (d) Round cable connector. (*Courtesy of ITT Cannon-Components Division*.)

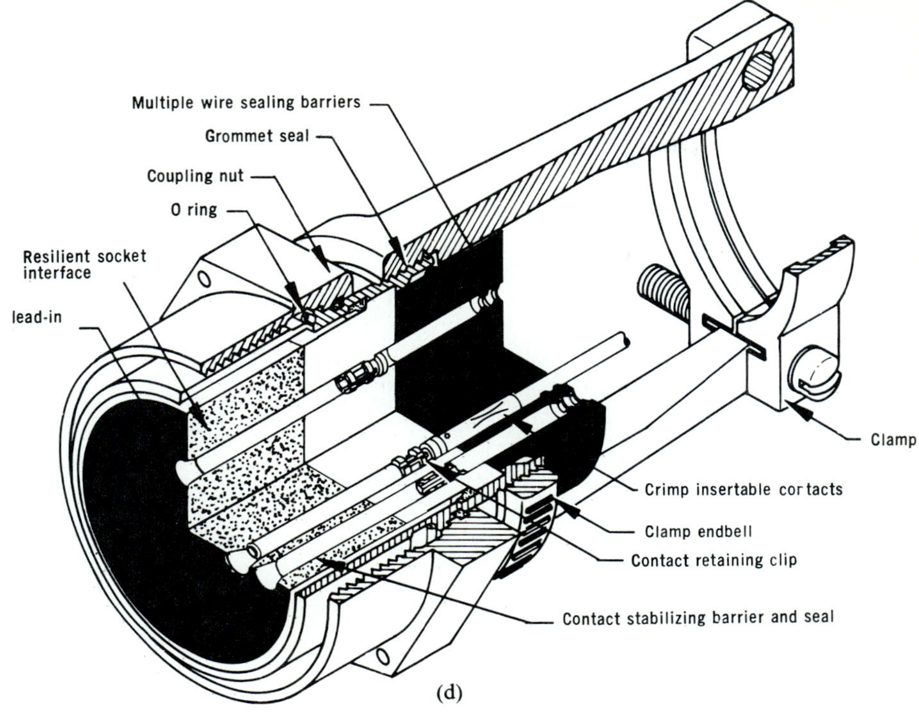

(d)

FIGURE 7.4
(cont.)

Blades, illustrated in Fig. 7.6*b*, are sometimes used in place of pins when the number of connections to be made is small. The moment of inertia for the blade, $I = bh^3/12$, is used in Eq. (7.1) to determine the buckling force.

Contacts are geometrically more complex than the pins because of the need for the contact to first accept and then to clamp the pin. The contact acts as a small spring and exerts a normal force on the pin. This normal force is essential in maintaining a low contact resistance across the connection. Some typical examples of contacts showing different types of spring elements are presented in Fig. 7.6*c*.

The materials used in fabricating the pins and the contacts must exhibit low resistivity, high strength, high modulus of elasticity and excellent wear. These materials must also be sufficiently ductile to permit easy fabrication of connector parts by bending and stamping processes. Copper-based alloys, including brass, beryllium copper, phosphor–bronze and copper–nickel, which are commonly employed for both pins and contacts, are described in Table 7.1. After the contacts and pins have been fabricated, they are plated with relatively soft metals to enhance the contact areas and to further reduce the electrical resistance across the connection.

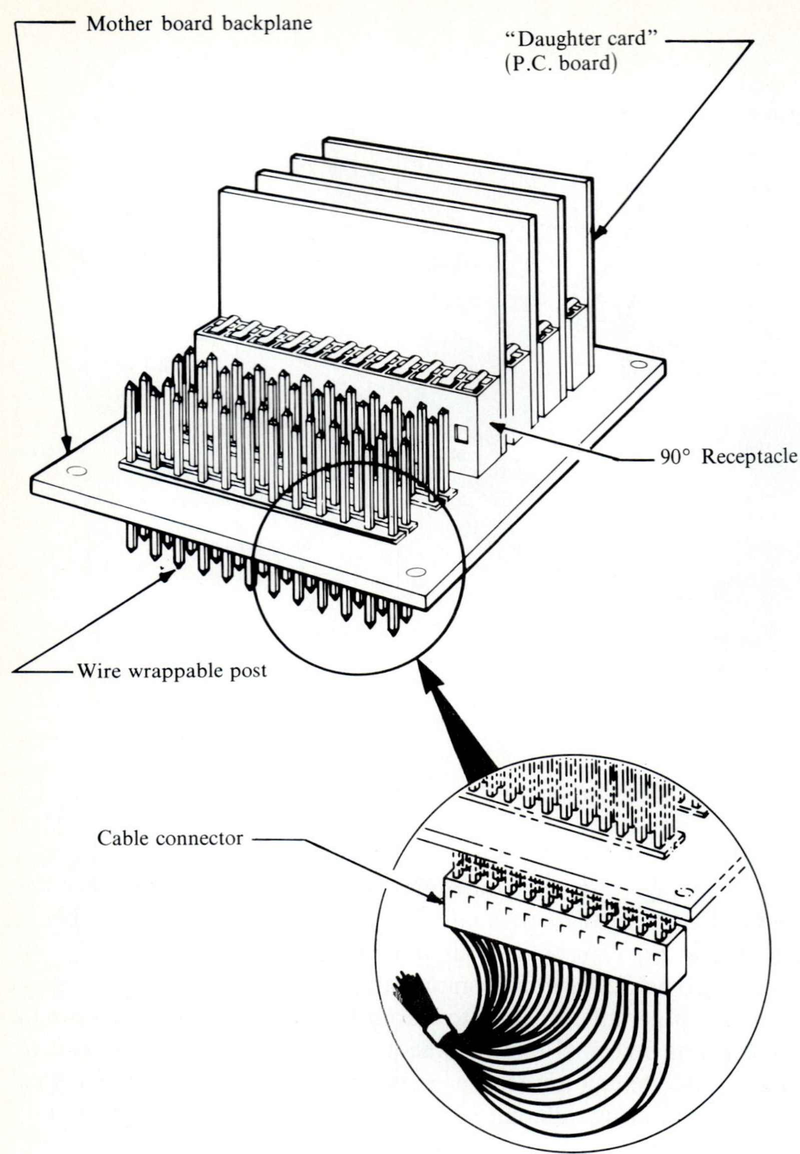

FIGURE 7.5
PCBs connected together with edge card connectors, receptacles and back panel. (*Courtesy of ITT Cannon-Components Division.*)

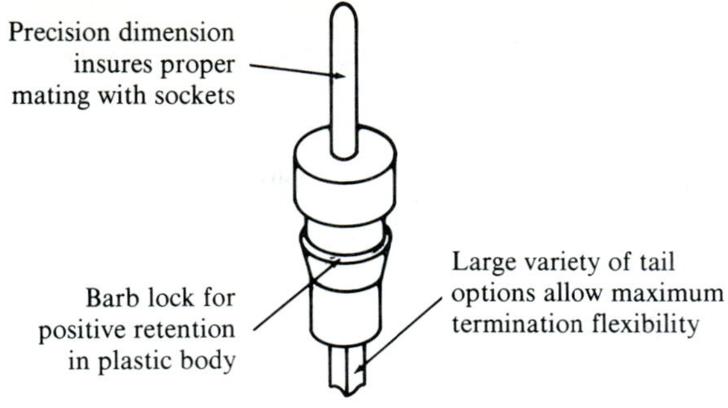

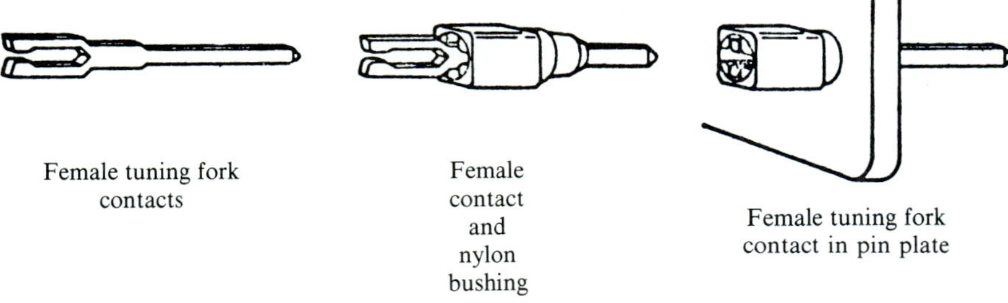

FIGURE 7.6
Examples of pins and contacts used in connectors.

The plastic core or insert of the connector holds the pins in position with close dimensional tolerances and also insulates one pin from another. These inserts are fabricated from either thermoset or thermoplastics. Thermosets polymerize and harden with the application of pressure and/or temperature, forming a three-dimensional cross-linked molecular structure. After hardening, the thermosets can not be remelted although they do turn rubbery at elevated temperature. In forming connector inserts from thermosets, molds are employed and the inserts are essentially cast. The thermosets can be reinforced with chopped glass fibers, and they provide excellent electrical and mechanical properties. The common thermoset materials used today are listed in Fig. 7.7. The advantages of the thermosets are their high strength, modulus and high heat distortion temperatures. The disadvantage is the higher cost of the fabrication process.

Thermoplastics are characterized by the ability to be resoftened several times without degrading their properties. Also, they can be injection molded in high volume production to reduce molding costs. The advantages in fabrication costs are apparent in Table 7.2, which indicates relative costs for the different

TABLE 7.1
Copper based alloys used in fabricating pins and contacts for connectors

Nominal composition	Alloy 260 brass (Cu 70.0 ZN30.0)	Alloy 172 beryllium–copper (cu 98.1 Be 1.9)	Alloy 510 grade A phos.–bronze (Cu 94.81 SN 5.0 P 0.19)	Alloy 638 (Cu 95.0 Al 2.8 Si 1.8 Co 0.4)	Alloy 725 (Cu 88.2 Ni 9.5 Sn 2.3	Alloy 762 nickel–silver (Cu 59.25 Zn 28.75 Ni 12.0)
Electrical conductivity						
Percent IACS at 68°F	28%	22%	15%	10%	11%	9%
megmho-cm at 20°C	(0.163)	(0.128)	(0.087)	(0.058)	(0.064)	(0.050)
Thermal conductivity						
btu/ft^2/ft/hr/°F at 68°F	70.0	62–75	40	23	31	24
cal/cm^2/cm/s/°C at 20°C	(0.29)	(0.26–0.31)	(0.164)	(0.097)	(0.130)	(0.100)
Density						
lb/in.^{3} at 68°F	0.038	0.298	0.320	0.299	0.321	0.314
g/cm^3 at 20°C	(8.54)	(8.26)	(8.86)	(8.29)	(8.89)	(8.70)
Modulus of elasticity						
10^6 psi	16.0	18.5	16.0	16.7	19.0	18.0
10^6 kg/mm^2	(11.2)	(13.0)	(11.2)	(11.7)	(13.5)	(12.7)
Yield strength, 0.2% offset						
Annealed, ksi	10–33	155	22	58–67	25	29
kg/mm^2	(7–22)	(109)	(15)	(41–47)	(18)	(20)
Half hard, ksi	42–60	175	47–68	76–89	57–73	58–82
kg/mm^2	(29–41)	(123)	(33–48)	(53–63)	(40–51)	(41–58)
Hard, ksi	67–78	180	74–88	91–103	74–79	82–97
kg/mm^2	(46–53)	(127)	(52–62)	(64–72)	(52–56)	(58–68)
Spring, ksi	82–91	N.A.	92–108	100–112	78–93	101–110
kg/mm^2	(58–63)		(65–76)	(70–79)	(55–65)	(71–77)
Extra spring, ksi	86–98	N.A.	98–110	107 min.	89–102	102 min.
kg/mm^2	(60–69)		(69–77)	(75 min.)	(63–72)	(72 min.)

polymeric materials. The common thermoplastics used in connector applications are listed in Fig. 7.7. One of the best thermoplastics for connector applications is polyphenylene oxide resin. This resin has the lowest water absorption rate of the engineering thermoplastics, only 0.0066%, and excellent dielectric properties. Mechanically, the material is tough and rigid over a wide range of temperature from −40–300°F. The resin exhibits low shrinkage in injection molding, permitting close control of the dimensions required in fabricating multipin connectors. In general, the thermoplastics have lower modulus, more flexibility, better color, enhanced processing and lower total costs than the thermosets.

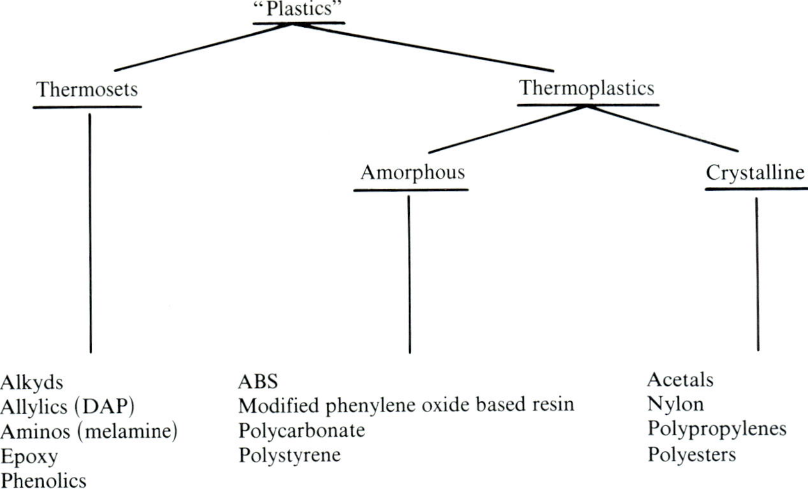

FIGURE 7.7
Classification of polymer families with a listing of the common plastics.

7.2.2 Connector Resistance and Plating Materials

The connection made between the pin and the contact is totally mechanical. We do not have the metallurgical bond of a soldered joint or the chemical bond of an adhesive joint. Yet the electrical resistance across the connection between the pin and the contact must be reduced to a minimum, usually 10–15 mΩ. The connection resistance R_c is due to two factors, namely, a resistance due to constriction effects R_{ct} and a resistance due to oxide films R_f, which occur between the pin and the contact.

TABLE 7.2
Cost factors for materials used in fabricating molded inserts for connectors

Polymer	Cost factor
Diallyl phthalate	1.0
Glass phenolic	0.9
Phenolic	0.6
Polycarbonate	0.6
Nylon	0.5
Polyphenylene oxide	0.5
Polyesters (PBT)	0.4

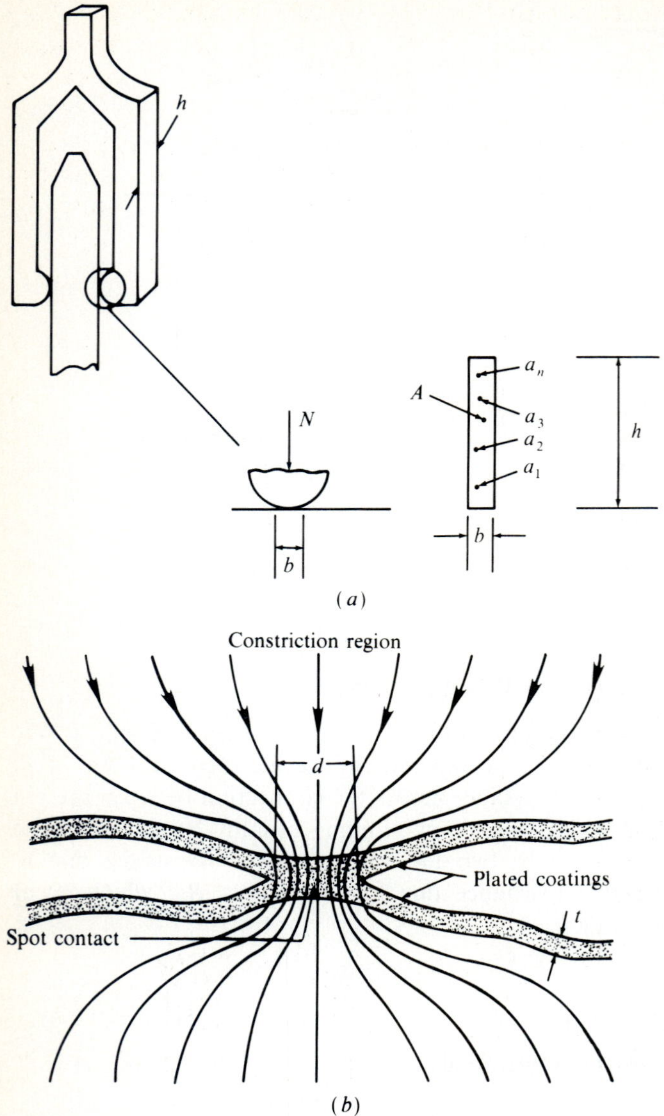

FIGURE 7.8
(a) Apparent contact area $A = bh$ and true contact spots of area $a_1, a_2, \ldots, a_n$. (b) Idealization of current flow through a contact spot in a plated pair of conductors.

Consider the engagement of a blade in a tuning fork contact that appears to give a large area A of contact as shown in Fig. 7.8. However, the surface finish over this area of contact is far from perfect. The surface imperfections, notably the many asperities, produce a large number n of true contact spots, each of which is quite small. When the total area of true contact is determined by summing the areas a_c of the n contact spots, one finds that true contact area is often less than 1% of the apparent contact area. Increasing the normal force N on the contact enlarges the area a_c of the existing contact spots and produces more contact spots, but even with excessively large normal forces the true contact area remains a small fraction of the apparent contact area.

A better approach for increasing the true contact area and decreasing the connector resistance is to plate the surfaces of both the blade and the contact with a relatively soft material. The diameter d of the contact spot formed by bringing the two plated surfaces together is shown in Fig. 7.8*b*. The diameter d is related to the hardness H and the normal force by

$$d \propto \sqrt{N/H} \tag{7.2}$$

where the hardness is approximated as 3 times the yield strength. Next, note that

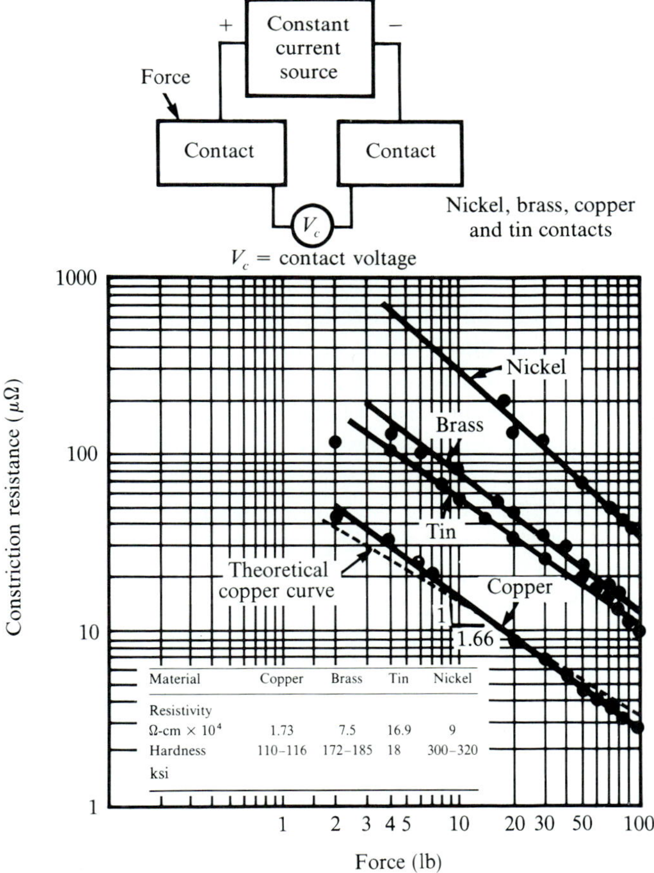

Material	Copper	Brass	Tin	Nickel
Resistivity Ω-cm × 10⁴	1.73	7.5	16.9	9
Hardness ksi	110–116	172–185	18	300–320

FIGURE 7.9
Constriction resistance R_{ct} as a function of normal force N for several different materials. (*Courtesy of AMP Incorporated.*)

the constriction resistance R_{ct} can be written as

$$R_{ct} = \kappa\rho/d \tag{7.3}$$

where ρ is the resistivity of the plating material
κ is a parameter dependent on geometry and type of contact

Combining Eqs. (7.2) and (7.3) gives

$$R_{ct} \propto \kappa\rho\sqrt{H/N} \tag{7.4}$$

The results of this proportionality indicate the importance of both hardness and normal forces in minimizing the constriction resistance. Typical experimental results for a number of different materials showing R_{ct} as a function of normal force N are presented in Fig. 7.9 and confirm the validity of Eq. (7.4).

The effect of the film resistance on the connector resistance is often more important than the constriction resistance. With the exception of gold, all other materials tend to form oxide films that, to some degree, insulate the pin from the contact. Even gold is not a perfect plating material since it can wear away, it erodes under arcing contacts and it can be defeated by corrosion products that develop due to the porosity of the plated coating. There are four factors that must be considered in evaluating the effectiveness of a connector to resist the development of a film resistance. The first is the quality of the plating as indicated by the porosity of the coating. Porosity is a measure of the number of pores that occur in the coating. Clearly the pores are detrimental since they permit migration of base metals that can oxidize to form surface films. Porosity may be controlled by coating thickness as indicated in Fig. 7.10; however, costs of noble metal plating at several hundreds of dollars per ounce dictate extremely thin coatings. A more suitable approach for controlling the effects of porosity is to provide an underplate coating of nickel 1.25 μm thick. The nickel serves to form a passive, very thin, oxide film at the base of the pore sites. It also prevents migration of corrosion products from the edges of the contacts. The use of the nickel underplate reduces the required thickness of the precious metals to 0.5–0.75 μm.

The second factor to consider is wear of the plating material at the contact area as the connector is cycled in service. The cycling can occur due to either the need to repeatedly disconnect and reconnect the system or due to small motions, called fretting, produced by vibration or thermal cycling of the connector. There are several approaches for extending the life of a connector. One of the most common design methods is to use either hard gold plate or a palladium–nickel alloy plate over a nickel underplating. The results of wear testing with these two platings, presented in Fig. 7.11, indicates long life for both coatings. However the superiority of the palladium–nickel alloy is evident for extended life applications.

In some applications tin is used as the plating material in order to reduce costs. Tin oxidizes easily and forms a film that is thin, hard and brittle. The normal forces acting at the contact rupture this film and the metallic tin extrudes through the cracks to form a low resistance connection. With repeated usage or fretting, the contact is repeatedly broken and new oxide films are generated on

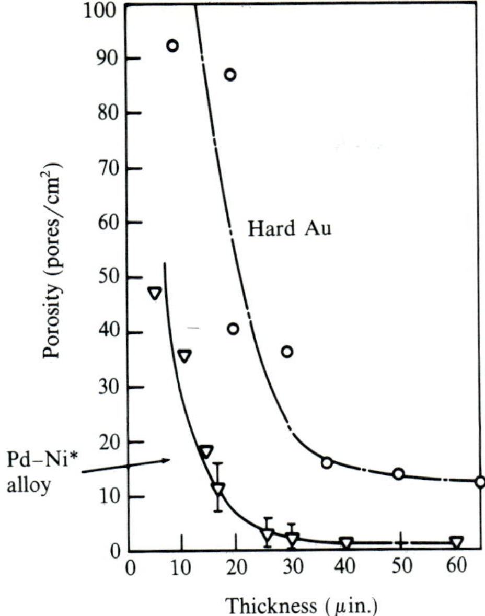

FIGURE 7.10
Porosity of plated coatings as a function of coating thickness.

each cycle. With time tin–tin oxide debris forms, which significantly increases the contact resistance as shown in Fig. 7.12. The life of the tin plated contact can be extended considerably by lubricating the assembly. The lubricants can be passive, such as mineral oil, or active proprietary fluids that attack oxide films. The passive lubricant reduces wear, flushes debris from the contact area and limits the amount of oxygen reaching the exposed metal at the contact. The active lubricant is more effective than the passive lubricant in that it removes the oxide films from the tin.

The third factor that markedly affects connector performance is the normal force acting on the contact area. High normal forces produce lower connector

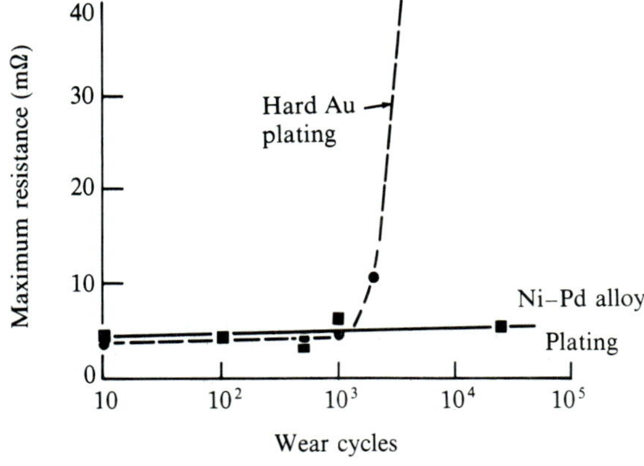

FIGURE 7.11
Resistance of a connection as a function of the number of cycles (N = 300 g).

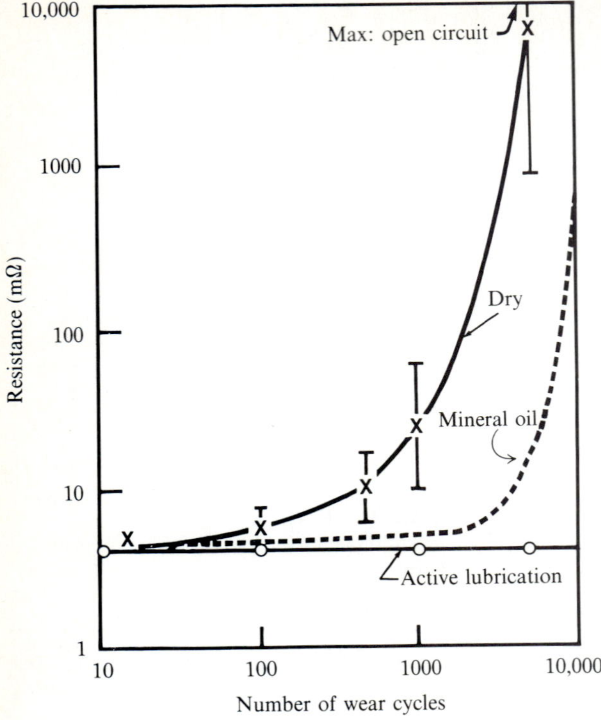

FIGURE 7.12
Resistance as a function of the number of wear cycles for tin plate. (*Courtesy of AMP Incorporated.*)

resistance since they reduce both the constriction and film resistances. The effect of the normal force on constriction resistance is clearly evident in Eq. (7.4). The effect of normal forces on film resistance is less obvious because it depends on the characteristic of the films formed over the contact area. These films are insulators and must be ruptured during the sliding of the pin over the contact, which takes place during engagement. A high normal force insures that the film is ruptured; however, this high normal force increases wear rates and increases both the insertion and extraction forces. Normal forces of 200–300 g are commonly employed in many connectors. With newer connectors that have a higher pin count, a need exists to reduce insertion forces to permit easier assembly. The insertion forces may be reduced if the normal forces are decreased to 100 g or less. In these instances the normal force may not be sufficient to rupture the oxide films and then precious metal platings are mandatory.

The final factor to consider in the development of film resistance is the power transmitted through the connection. The connector should always be engaged without power (i.e., a dry circuit). When the circuit is activated, current flows through the connector and is concentrated at the contact spots shown in Fig. 7.8. The voltage drop across the contact area breaks down the films and the heat from the I^2R losses melts the local high spots to establish a larger more intimate contact. The level of power involved significantly affects behavior of the connection. For very high power, with voltages greater than 100 V and currents greater than 50 A, the conducting characteristics of the base metal are extremely important and plain copper is the preferred material. If the connection is to be made under power, as in a switch, then silver plate is essential to prevent fusing of the contacts due to arcing as the switch closes.

For power applications with currents exceeding 1 A at 10 V, the constriction resistance is the important factor. High normal forces and effective wiping are more important in reducing the constriction resistance than plating materials. For intermediate power levels with current less than 1 A at 10 V both film and constriction resistance are important. In this power range one seeks a balanced design that includes effective plating to inhibit film development and moderate normal forces to lower constriction resistance. The balanced design provides a connection with wear resistance, a long life and stable connection resistance. At the low power levels with the current less than 0.1 A, film resistance becomes the dominant factor and precious metal platings and/or lubrication are design approaches followed in providing reliable connections with a long life.

7.2.3 Connector Forces

There are three different forces that are considered in the engagement and disengagement of a connector, namely, the normal force, the insertion force and the extraction force. We have already discussed the normal force and have noted its importance in minimizing constriction resistance. Insertion and extraction forces are also important because these forces control the ease of assembly of a circuit board or of a set of cables in an electronic system. To determine the insertion force, consider the free body diagram of the blade and contact shown in Fig. 7.13*a*. As the blade is inserted into the contact, the lead on the blade encounters the contact and the normal forces developed are not perpendicular to the axis of the blade. Equilibrium relations give the insertion force F_i as

$$F_i = 2N_i(\sin\theta + \mu\cos\theta) \tag{7.5}$$

where θ is the lead angle
μ is the coefficient of friction
N_i is the normal force

The normal force N_i is controlled by the stiffness of the contact and the amount of interference $\Delta h = h_2 - h_1$ between the blade thickness and the contact opening. If the contact is treated as a cantilever beam, we may write the force deflection equation as

$$P = 3EI\delta/L^3 \tag{7.6}$$

where P is the transverse force
δ is the deflection of the free end

The forces acting on the pressure point at the entrance to the contact, shown in

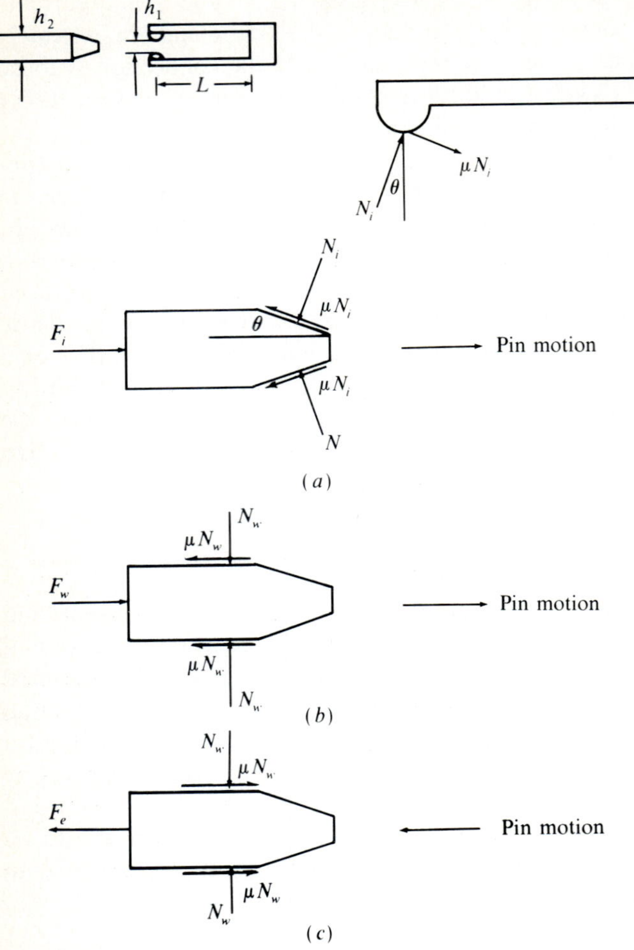

FIGURE 7.13
Free-body diagrams showing forces acting on the connector blade. (*a*) During contact opening as the blade is inserted. (*b*) During wiping as the blade is inserted. (*c*) During extraction.

Fig. 7.13*a*, may be resolved to give the force P transverse to the beam as

$$P = N_i(\cos\theta - \mu \sin\theta) \tag{7.7}$$

Combining Eqs. (7.6) and (7.8) gives

$$N_i = 3EI\delta/\left[L^3(\cos\theta - \mu\sin\theta)\right] \tag{7.8}$$

Examination of this equation shows that the normal force on the connection is a function of the stiffness of the contact, the deflection imposed on the contact during insertion, the lead angle and the coefficient of friction.

Next, substitute Eq. (7.8) into Eq. (7.5) to obtain the insertion force during engagement of the blade in the contact as

$$F_i = 3EI\Delta h(\tan\theta + \mu)/[L^3(1 - \mu\tan\theta)] \tag{7.9}$$

where Δh is the interference given by

$$\Delta h = 2\delta = h_2 - h_1 \tag{7.10}$$

After insertion into the contact beyond the lead, the blade is wiped and the normal force N_w is now perpendicular to the axis of the blade as shown in Fig. 7.13*b*. The force N_w decreases to

$$N_w = P = 3EI\delta/L^3 \tag{7.11}$$

and the insertion force F_w during the wiping part of the engagement reduces to

$$F_w = 3\mu EI\Delta h/L^3 \tag{7.12}$$

Upon extraction, the direction of the friction forces reverses, as indicated in Fig. 7.13*c*, and the extraction force F_e becomes

$$F_e = F_w \tag{7.13}$$

The preceding relations describe the forces required to engage and disengage a single pin of a connector. In most cases the connectors have many pins and it is necessary to multiply the single pin force by the number n of the pins being engaged. Single pin insertion forces usually range from 0.25–0.5 lb and multiple pin connectors often require relatively large insertion and extraction forces. When the total insertion force exceeds about 35 lb, it is necessary to provide mechanical aids to facilitate assembly of the connector. The mechanical aids may be as simple as two screws that are used to draw the connector together or levers that are hinged to the chassis to give a mechanical advantage to the person making the connection.

7.3 BACK PANEL, WIRE WRAP BOARDS AND CABLE CONNECTIONS

The printed circuit cards are connected together with edge connectors that permit the boards to plug into a back panel, a wire wrap board or a cable arrangement. All three of these components serve the same function, namely, to carry power to all of the boards and to carry signals from one board to another. The choice of which component to use in design depends largely on complexity and on the volume of the product produced. We will describe these three methods of connecting the PCBs individually in the following subsections.

7.3.1 Back Panels

A back panel is similar to a multilayer printed circuit board in that it is laminated with alternating layers, providing planes of copper cladding for V^+,

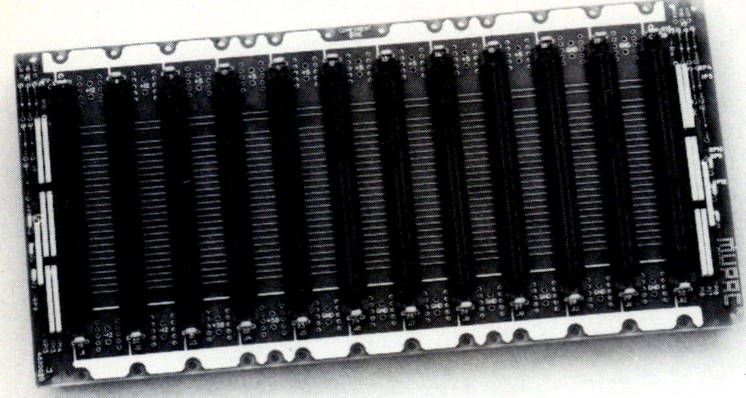

FIGURE 7.14
Back panel with press fitted 96 pin DIN receptacles. (*Courtesy of Mupac Corporation.*)

signal and ground. It is different from a PCB in that it is larger in size, contains more layers and the footprint is designed to accept the contact side of the connector. An example of a relatively small back plane, illustrated in Fig. 7.14, shows the plated through holes that accommodate the pins from the connector. In this design the pins on the reverse side of the receptacle are press fitted in the plated through holes to make connections with the interior wiring planes in the back panel. The nine receptacles shown in this figure each accommodate 96 pins from the edge connectors, giving an I/O capacity of 854. This particular back panel is a six layer design with three layers or six planes devoted to signal. The voltage planes are independent of each other and distribute +5 and ±12 V. The ground planes are connected together with a number of plated through holes to avoid ground loops and to ensure that the ground acts as a single plane. Power

TABLE 7.3
Features of a complex high performance back panel

Size $W \times L \times T$ (mm)	$600 \times 700 \times 4.6$
Circuit length (m)	1,400
No. of connector contacts	19,200
No. of plated through holes	41,644
No. of programmable vias	14,400
No. of card I/O	16,200
No. of cable I/O (signal)	2,160
No. of signal/power planes	$6S/12P$
Trace width (mm)	0.08
Trace thickness (mm)	0.048

blocks are fitted to the back side of the panel to provide for high current connections to the power bus.

Back panels vary in size and complexity, depending on the application. The back panel that supports nine TCMs on high performance IBM computers is much more complex. The measure of this complexity is illustrated by listing several features of the high performance back panel in Table 7.3. Note that the back panel contains 1.4 km of circuitry, accommodates 19,200 I/O pins and in excess of 41,000 plated through holes with an additional 14,400 programmable vias.

Back panels are expensive and are usually used only in large, higher performance processors with moderate to large volume production. For smaller processors with more limited production, wire wrap panels are often more cost effective than back panels.

7.3.2 Wire Wrap Panels

Wire wrap panels, illustrated in Fig. 7.15*a*, are similar to back panels in that they are multilayer boards with an array of plated through holes to accommodate pins. A cutaway view of the board, presented in Fig. 7.15*b*, shows that the power and the ground are distributed to appropriate holes with multiple planes of 2 or 4 oz copper. On the connector side of the board the holes are available to accept component pins, connector pins or socket pins. On the wire wrap side of the board, wire wrap posts extend beyond the surface. The point-to-point connections are made by wrapping the bare ends of an insulated wire in a tight helix about a four-cornered square post. A typical wire wrapped connection is illustrated in Fig. 7.16.

A pin plate provides another method of connection that is similar to the wire wrap panel. The pin plate is usually a sheet of aluminum or plastic with holes drilled in a rectangular array. Insulators and female contacts are inserted into these holes to form the receptacle portion of a connector as shown in Fig. 7.17. The tail on the contact extends beyond the lower surface of the pin plate and serves as the four-cornered post used in wire wrapping.

(a)

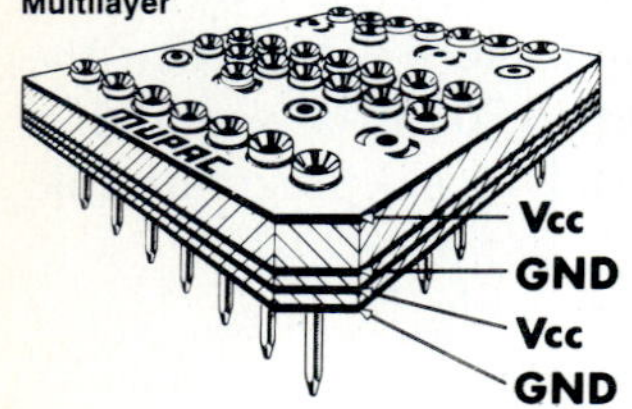

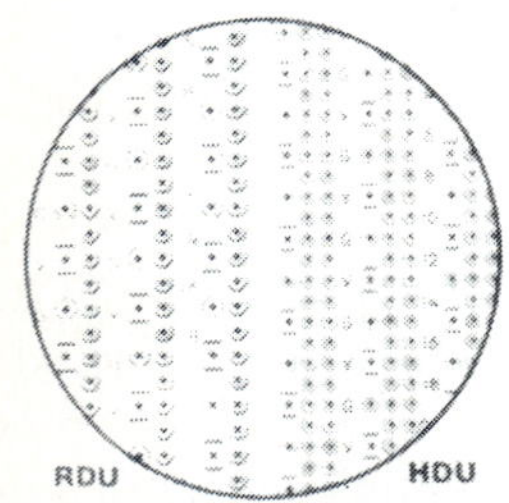

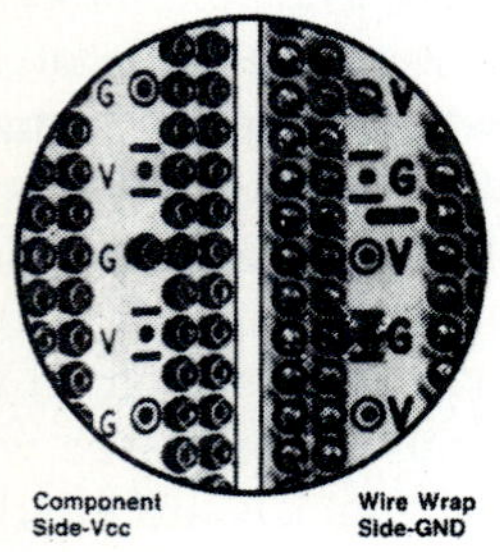

(b)

FIGURE 7.15
(*a*) Component side of a wire wrap panel. (*b*) Enlargements of each side and a cutaway view. (*Courtesy of Mupac Corporation.*)

FIGURE 7.16
A wire wrap connection is a tight helical winding on a four corner post.

Wire wrapping a pin plate or a panel provides a method of connecting together circuit boards with moderate density and size. However, if the volume of the production is large, the labor costs for making the large number of wire wrapped joints becomes prohibitive and the use of a back panel becomes more cost effective. With small volume production and semiautomatic operations, the initial tooling costs are moderate and wire wrapping is often the preferred approach.

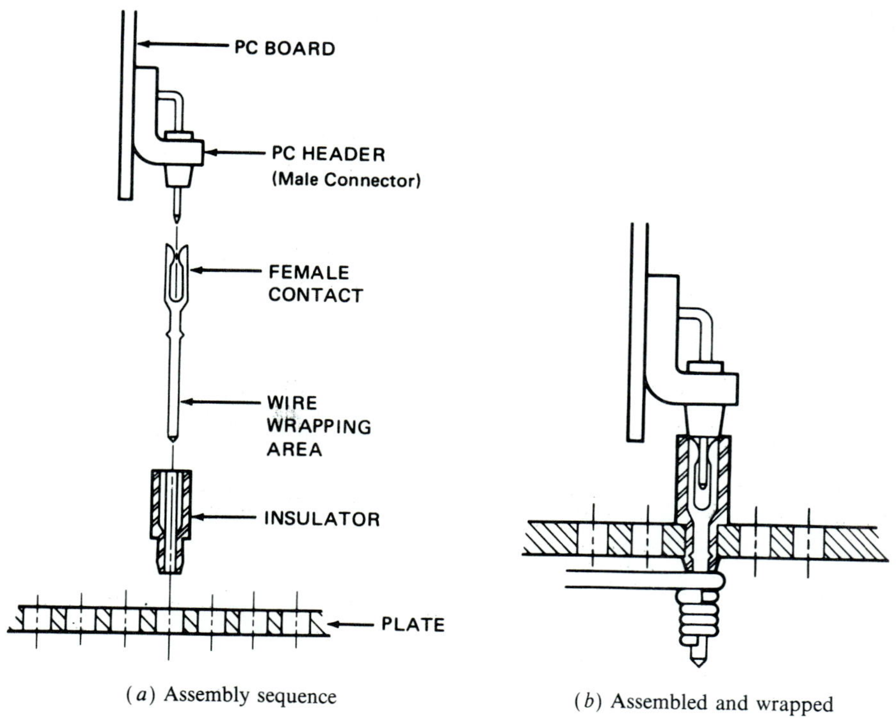

FIGURE 7.17
Pin plate with wire wrapping. (*a*) Assembly sequence. (*b*) Assembled and wrapped.

7.3.3 Cable Connected Boards

A common method for connecting a few circuit boards together, which avoids the complexities and costs of a back panel or a wire wrapped board, uses flat cables or flexible circuits. A flat cable, illustrated in Fig. 7.18, consists of a number of adjacent wires laying in a common plane. The wires are connected together by a web of insulation so that the multiconductor assembly resembles a sheet of

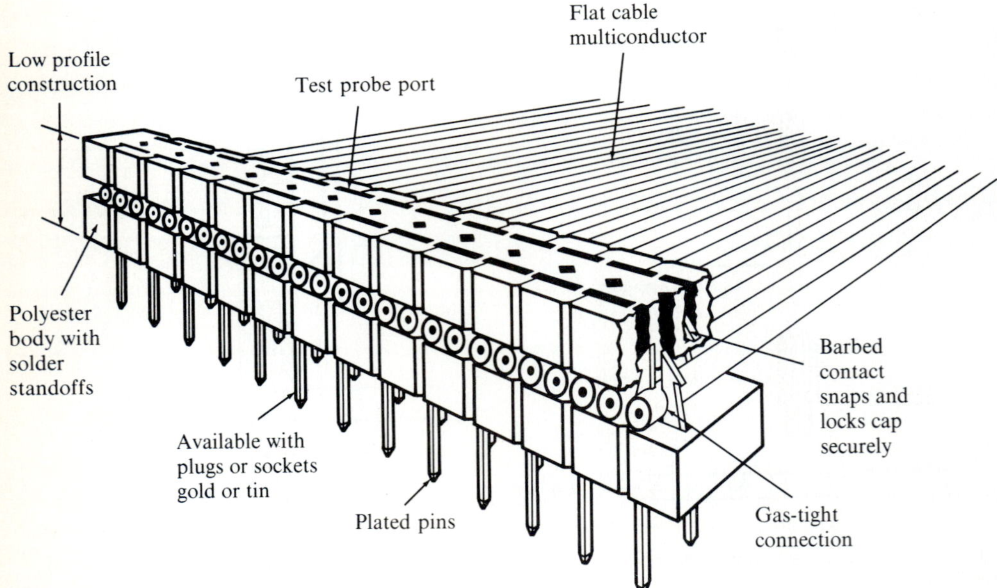

FIGURE 7.18
Flat cable connections. (*Courtesy of Samtec Inc.*)

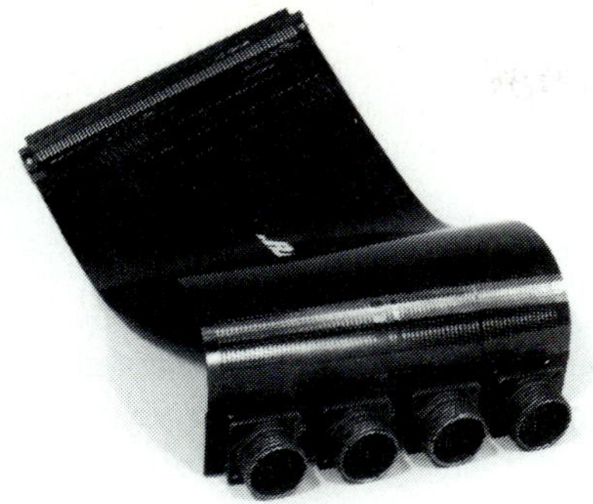

FIGURE 7.19
Flexible circuit used as a multiple conductor cable with a connector termination. (*Courtesy of Buss Systems Inc.*)

parallel conductors. Connectors, either male or female, can be attached to the cables at one or more locations along the length of the flat cable. The conductors in the cable serve the same purpose as the back panel or wire wrap panel in connecting together two or more circuit boards. It is interesting to note, in Fig. 7.18*b*, that the connector shown incorporates a contact that acts as a knife to cut through the cable insulation and then clamps the wire to make an airtight connection. A barb on the top of the contact snaps into the connector cap and permanently locks it in place without fasteners. This type of design facilitates rapid assembly, which significantly reduces labor costs in preparing cable connections.

A similar design concept involves the use of flexible circuits in place of the flat cable as shown in Fig. 7.19. The flexible circuit is the same as a printed circuit board except that a polyimide film is used as the substrate instead of the rigid circuit board materials. The conductors are etched from copper cladding that is bonded to the flexible substrate. A second layer of polyimide film is bonded over the copper conductors to form a sandwich structure that completely insulates the flexible circuit. The flexible circuits can be mass terminated with a connector, as shown in this illustration, or they can be soldered directly to the circuit boards to act as jumpers connecting one board to another.

Either flat cable or flexible circuit connections between PCBs are employed when relatively few connections are necessary. The method provides for a cost effective solution to a limited class of connection problems since the flat cables are much lower in cost than the back panels. The flexible circuits have additional advantages that include:

1. The ability of the conductors to be formed into almost any imaginable configuration and then used to connect randomly spaced points on a given plane.
2. The flexible substrate supports the conductors and provides a connection system capable of large motions that are cyclic. This flexible circuit exhibits a

very high fatigue life. An example of a fatigue resistant connection is the flexible circuit used on most printers to carry signals and power to the printer head.

7.4 POWER SUPPLIES AND BUS BARS

Power supplies and power distribution in electronic systems is not difficult, but it is an important topic. The power required in large systems is significant and is specified in tens of kilowatts. On the other hand, the power required in small systems, your watch, is measured in microwatts. The production of the dc voltages and the distribution of the power are topics of major importance in the design of a wide variety of electronic systems, which may range from a \$1 watch to a $\$10 \times 10^6$ super computer.

7.4.1 Batteries as Power Supplies

Operation of most electronic equipment requires a source of dc power for one or more voltages. For small systems, like your watch or calculator, the dc source is a small battery that is disposable after a life of several years. Recent advances in lithium batteries are particularly important for the low power applications. These batteries exhibit outstanding performance characteristics with high individual cell voltages, very low decay of voltage under load, extremely long shelf life and high energy density. The performance of a lithium-iodine battery, shown in Fig. 7.20, indicates continuous operation with negligible voltage decay for 3–9 yr depending on the current demand. The capacity is 110 V A hr/lb, which is nearly a factor of 7 better than a nickel–cadmium or a lead–calcium battery.

For applications requiring greater power, nickel–cadmium batteries are used because they can be recharged up to 500 times before replacement is

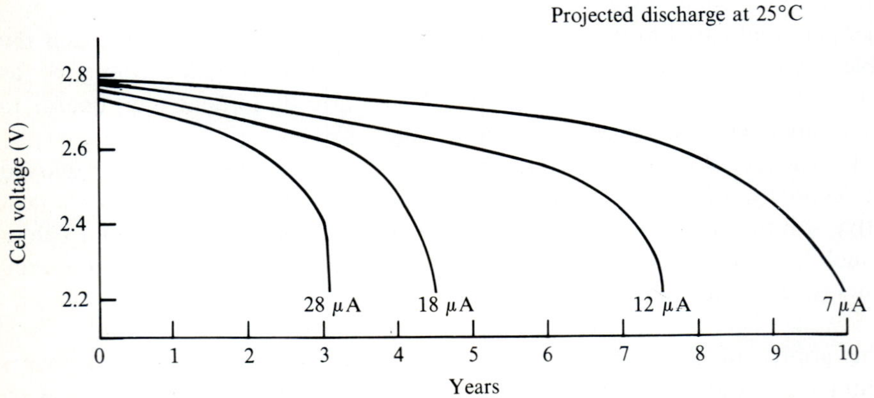

FIGURE 7.20
Voltage-life curves at different current levels for lithium–iodine batteries.

necessary. The individual cell voltage for the Ni–Cd battery is 1.2 V and the capacity at this voltage is 16 V A hr/lb. Cells are often connected in series to increase the voltage. The current–time capacities available for this rechargable battery range from 100–4000 m A hr.

For still larger battery supplies the lead-calcium rechargeable system is employed. The individual cell voltage is 2.5 V, but the batteries are usually supplied with a regulator to give a nominal 2 V per cell. The Pb–Ca batteries are available as sealed units and can be operated in any position and treated as a dry cell. The batteries are available in many different sizes with nominal voltages ranging from 4–12 V. The discharge rate is extremely large at 20 A/hr and the capacity is a function of the size of the cells. At 12 V the Pb–Ca battery has a specific capacity of 16 V A hr/lb, which is the same as the Ni–Cd battery. The units can be recharged about 1000 times if they are recharged from a discharge state of 50%.

7.4.2 Power Supplies

Power supplies are used to convert ac line voltage at 115 or 230 V to a dc source with one or more fixed voltages. The power supplies may be packaged in a number of different ways, as indicated in Fig. 7.21. In some applications the supply is totally enclosed and can be used as a free-standing unit. In other cases, the unit is enclosed in a safety screen and then housed in another electronic enclosure together with other parts of the system. Other power supplies are open and are assembled on a simple sheet metal platform. These open supplies must be enclosed and shielded in the enclosure to insure the safety of service personnel.

There are several features of the power supply that must be considered in incorporating the supply into an enclosure. First, power supplies are large and they require from 1–3 in.^{3} of space per watt depending on the type, capacity and efficiency. Second, they are heavy, because the transformers used in the supplies to convert the ac voltage involve large copper coils wound about heavy laminated steel cores. Typical power weight ratios range from 40–100 W/lb with the lower capacity supplies exhibiting the smaller power weight ratios. Third, the supplies dissipate large amounts of heat and must be given serious attention in the design of the cooling system. An examination of Fig. 7.21 shows cooling fins attached at critical locations to dissipate the heat away from the supply. However, the management of the heat from these fins to the environment is a task handled by those designing the total system enclosure. The large heat dissipation is due to the relatively poor efficiencies of the power conversion process. Typical power supplies operate at efficiencies ranging from 65–85%.

The output from dc power supplies is typically ± 5, ± 10, ± 12 and $+24$ V. Some supplies are designed with a single voltage output and others have multiple voltage outputs. There is no difficulty in specifying any voltage required for the system if custom designed power supplies are procured. With relatively low voltages and high power output, it is important to recognize that the supplies provide large currents. For example, a 1 kW supply at 5 V has a current output of

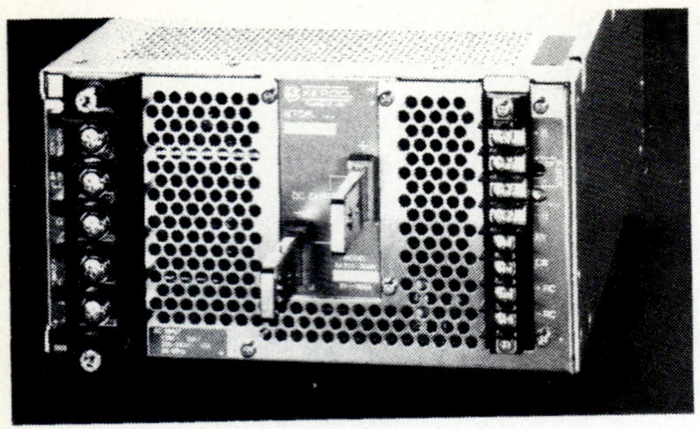

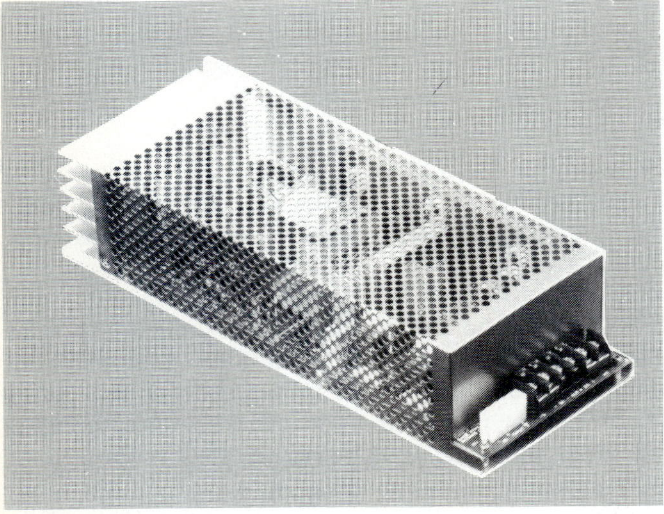

FIGURE 7.21
Power supplies for converting 115–230 V ac line supply voltages to appropriate dc voltages. (*Courtesy of Kepco, Inc.*)

200 A. Distribution of these high currents and the termination of the distribution conductors requires careful consideration to minimize (IR) voltage drops along the distribution line and (I^2R_c) losses at the terminations.

7.4.3 Bus Bars

Bus bars are conductors in the same sense that wire is a conductor and in some instances the bus bar is simply a wire that is tin-plated and not insulated. We refer to bus bars here as the conductors used to distribute high amperage currents from the power supplies to the back panels.

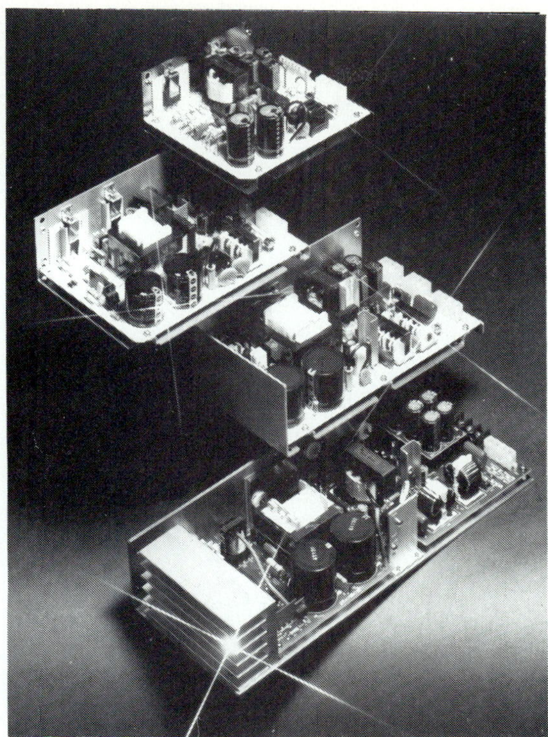

FIGURE 7.21
(cont.)

The bus bars can be heavy gage copper wire with a circular cross section. Copper wire sizes are commonly designated in the American wire gage (AWG) standard, which has numbers that extend from 0000 as the largest conductor to 40 as the smallest. The diameter, resistance, weight and current capacity of copper wire of different gage based on the AWG number is given in Tables 7.4 and 7.5. The current carrying capacity of a given wire number depends on the insulation covering the wire. The current capacity ranges from 343 A for the 0000 wire to 7 A for 18 wire with insulation rated at 75°C. For bus bar applications, heavy conductors are common with wire size of 8 or less employed in normal designs. Current densities of 0.001 A/circular mil are usually employed in large distribution circuits. Note that a circular mil is the area of a circle inscribed in a square 1 mil on each side.

For very high currents, bus bars are often fabricated from sheet copper with several layers of copper insulated with plastic film. The sheet copper is easy to form into long, thin but wide conductors that can be attached with a simple screw clamp to the power and ground planes on the back panel.

Low resistance termination at the junctions to the bus bar is critical to avoid excessive heating of the joints. To illustrate the possibility of heating, consider a termination that conducts 100 A with a connection resistance of 10 mΩ. The power dissipated at this joint is 100 W, which is so large that the

TABLE 7.4
Capacity of single-insulated copper conductors in free air at an ambient temperature $T_a = 40°C$

Size AWG	Temp. rating of insulation		
	60°C	75°C	90°C
18			16†
16			22†
14	24†	30†	35†
12	30†	39†	45†
10	41†	51†	61†
8	55	71	83
6	73	94	109
4	96	124	145
3	112	145	169
2	128	165	192
1	148	191	223
0	171	221	258
00	198	255	298
000	229	295	345
0000	266	343	400
Ambient temp. °C	**For $T_a \neq 40°C$ multiply capacity by the appropriate scale factor**		
21–25	1.32	1.20	1.14
26–30	1.22	1.13	1.10
31–35	1.12	1.07	1.05
36–40	1.00	1.00	1.00
41–45	0.87	0.93	0.95
46–50	0.71	0.85	0.89
51–55	0.50	0.76	0.84
56–60	—	0.65	0.77
61–70	—	0.38	0.63
71–80	—	—	0.45

Note: The overcurrent protection for conductor types marked with a dagger (†) shall not exceed 7 A for 18 AWG, 10 A for 16 AWG, 15 A for 14 AWG, 20 A for 12 AWG and 30 A for 10 AWG copper.

Source: D. G. Fink and H. W. Beatty, *Standard Handbook for Electrical Engineers*, 12th ed. Mc-Graw Hill, NY, 1987.

connection will fail due to oxidation and eventually melting. The connection resistance must be reduced to less than 10–100 $\mu\Omega$ in high current applications to limit power dissipation at the connections to less than 0.1–1 W. This is usually accomplished by making the contact areas large and clamping them together with a screw that permits the application of contact forces of the order of 100 lb or more. These high contact forces fracture any oxide films and produce a true contact area, which is a large fraction of the apparent contact area.

TABLE 7.5
Capacity of multiconductor cables with three or less insulated conductors in free air at an ambient temperature T_a = 40°C

Size AWG	Temp. rating of insulation			
	60°C	75°C	85°C	90°C
18				11†
16				16†
14	18†	21†	24†	25†
12	21†	28†	30†	32†
10	28†	36†	41†	43†
8	39	50	56	59
6	52	68	75	79
4	69	89	100	104
3	81	104	116	121
2	92	118	132	138
1	107	138	154	161
0	124	160	178	186
00	143	184	206	215
000	165	213	238	249
0000	190	245	274	287
Ambient temp. (°C)	**For $T_a \neq$ 40°C multiply capacity by the appropriate scale factor**			
21–25	1.32	1.20	1.15	1.14
26–30	1.22	1.13	1.11	1.10
31–35	1.12	1.07	1.05	1.05
36–40	1.00	1.00	1.00	1.00
41–45	0.87	0.93	0.94	0.95
46–50	0.71	0.85	0.88	0.89
51–55	0.50	0.76	0.82	0.84
56–60	—	0.65	0.75	0.77
61–70	—	0.38	0.58	0.63
71–80	—	—	0.33	0.44

Note: The overcurrent protection for conductor types marked with a dagger (†) shall not exceed 7 A for 18 AWG, 10 A for 16 AWG and 15 A for 14 AWG, 20 A for 12 AWG and 30 A for 10 AWG copper.

Source: D. G. Fink and H. W. Beatty, *Standard Handbook for Electrical Engineers*, 12th ed. McGraw-Hill, NY, 1987.

7.5 CARD RACKS

The printed circuit cards are assembled in orderly rows in a card rack. The card rack supports the lateral edges of the cards during the life of the product, provides guide rails for the initial insertion of the cards during assembly and provides alignment of the cards during substitution involved in servicing. A typical card rack is made by combining several subracks, illustrated in Fig. 7.22, into a vertical assembly. The subracks are usually fabricated from sheet metal

FIGURE 7.22
Subrack structure showing assembly details. (*Courtesy of Mupac Corporation.*)

and extrusions to form a rectangular opening much like a book shelf. The back panel with the receptacle half of the connectors forms the rear wall of the subrack. Card guides are spaced at uniform intervals along the length of the rack to facilitate the alignment of the card with the connector during insertion. The top and bottom surfaces are open to permit air flow between the cards if direct impingement convection cooling is used. The structural rigidity of the subracks varies with the application of the system. For systems intended to be used in the normal office environment, relatively thin gage sheet metal, spot-welded together to form the rectangular structure, is sufficient. However, in military applications involving a shock qualification test, the subracks are usually fabricated from plate thick enough to withstand shock loading of 100 G without undergoing plastic distortion.

The subracks are assembled into a complete rack and housed in an enclosure. The assembly is rapid because the enclosure hardware accommodates the flange mounts on the subrack; see Fig. 7.22.

7.5.1 Card Guides and Retainers

Card guides are often divided into three groups depending on the mechanical and thermal requirements imposed on the design. In the first group, the devices act only to guide the card into the connector and to support the board. The board is maintained in its engaged position by a retainer bar attached to the subrack across one edge of the PCBs. Low cost guides molded from nylon, shown in Fig. 7.23*a*, are adequate in this application.

In the second group, the device serves to both guide and retain the card. The use of card edge retainers is essential in vibration environments because clamping one or more edges of the boards significantly increases the natural

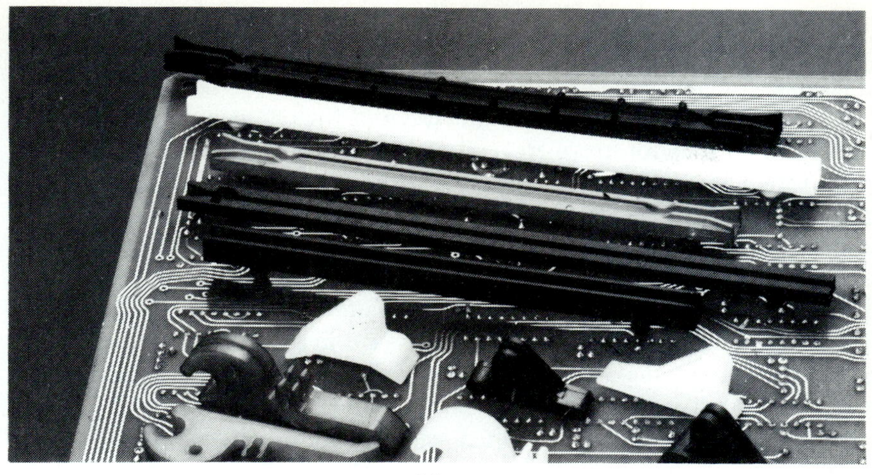

(a)

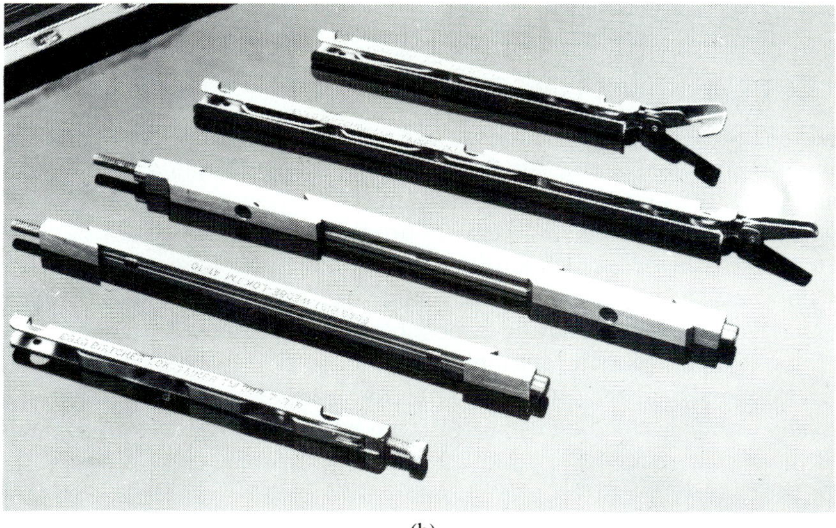

(b)

FIGURE 7.23
Card guides and retainers. (*a*) Nylon molded card guides. (*b*) Birtcher spring-type retainers and wedge lock retainers—slot is required for usage. (*Courtesy of EG & G Birtcher.*)

frequency of the circuit board. Also, in some installations conduction cooling is employed and it is necessary to clamp the heat frame, which supports the PCB, to a cold plate. Clamping the card is achieved by placing the edge of the PCB in a screw actuated cam device as shown in Fig. 7.23*b*. These retaining guides have two basic parts. The first is a channel-like housing that has a series of ramps stamped on its back side. The second part is a cam made of a strip of beryllium copper with a matching set of ramps. Turning a screw pulls the cam up onto the ramps on the channel and wedges the PCB between the cam and the housing. Retaining forces of 30–45 lb can be achieved with these spring-like retainers. The thermal contact resistance per unit length of the clamp varies from 4–17°C/W/in., depending on the detail design of the retainer.

The third group of retainers is for very heavy duty applications with clamping forces measured in hundreds of pounds. These devices use a wedge locking design as illustrated in Fig. 7.23*c*. The retainer is fabricated from aluminum bar stock with 45° angles cut on the end of each piece. A screw runs the length of each retainer. Tightening the screw causes the wedges to slide on one another and expand the width of the retainer. Because the retainer and circuit board are both contained in a slot that is cut in a cold plate, the expansion of the retainer forces the PCB against the wall of the slot. When conduction cooling is employed, a heat frame, which is bonded to the card, is clamped in place with a large uniformly distributed retaining force. Thermal contact resistance per unit length achieved with wedge lock devices range from 2.7–6.5 W/°C/in.

7.6 ELECTRONIC ENCLOSURES

Electronic enclosures serve to house the entire electronic assembly. The enclosures range in size from a case for a watch to large rugged military enclosures that weigh more than a ton. Whereas standards have been adapted for normal instrument cases and cabinets, the range of products that must be enclosed is so large as to make widespread standardization essentially impossible. Another fact that has impeded standardization is the volume of the product produced. When large annual volume is involved, it is possible to design unique cabinets with special features that facilitate assembly and reduce both weight and volume. These specially designed enclosures are often lower in cost when compared to a commercially available standard cabinet. A final but critical argument for unique design of the enclosures is the appeal of the package. In some office products the enclosure must serve as a piece of furniture and blend with the colors and function of the office, and while image is beyond the scope of this text, we should be well aware of its importance in marketing a product.

In this coverage of enclosures, we will first emphasis features common to the design of standard instrument cases and cabinets used in the office environment. As a second part, we will describe the common electronic equipment enclosure (CE^3) that was recently developed for the Navy.

7.6.1 Commercial Enclosures

Commercial enclosures, which include standardized instrument cases and cabinets, are intended for office and laboratory environments. These enclosures protect the circuits in a friendly environment. The environment is considered friendly if the temperature is maintained in a range comfortable for personnel, if

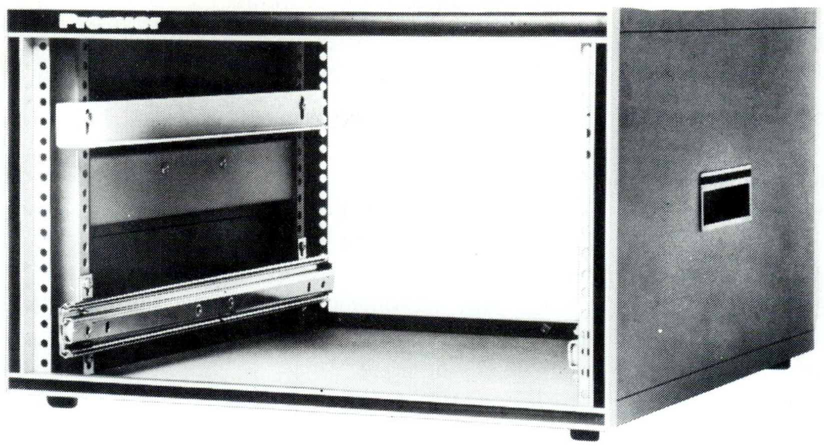

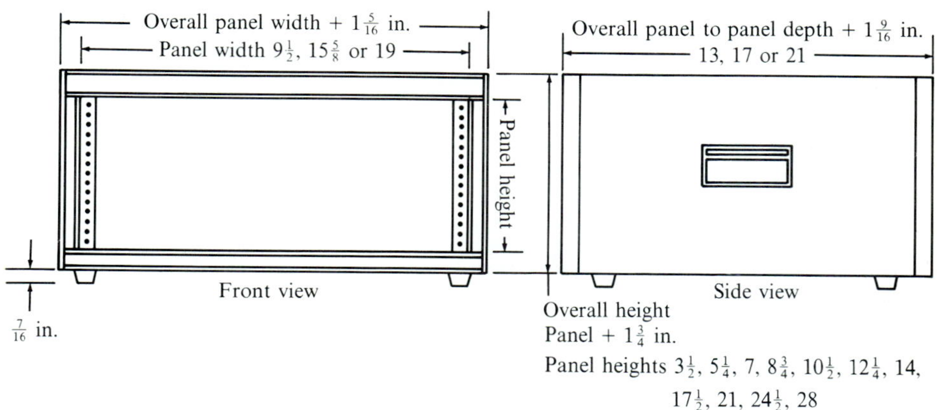

FIGURE 7.24
A commercially available series of instrument cases. (*Courtesy of Premier Metal Products.*)

the air is free of dirt and dust, the floors do not vibrate and the operating personnel are cooperative in that they respect ordinary locking devices.

A commercial instrument case, illustrated in Fig. 7.24, is intended to house an electronic system small enough to be placed on a table, desk or laboratory bench. A commercially available cabinet that houses larger systems, shown in Fig. 7.25, stands on the floor. Both enclosures are designed for the following features:

1. To be attractive.

2. To contain the circuitry on six sides.

3. To permit easy access for assembly and service personnel.
4. To be structurally sound.
5. To accept standard subracks, panels and accessories, such as slides, drawers and shelves.

The definition of attraction changes with time because we improve our understanding of features that appeal to the customer and those that improve the appearance of the product. Today, it is possible to provide cases that are painted with nonglare color coordinated finishes. Wood grain simulation in a variety of furniture woods are also available. Trim molding from anodized aluminum with vinyl inserts adds attractive markings and contrast at select locations. Clearly, the gray enamel box with rub on labels is long gone.

Containment of the circuitry is important in keeping dust and dirt from accumulating on the circuit board. It is also essential in protecting the operating personnel when high voltages and/or high currents exist within the enclosure. Closure is usually accomplished by using 16 gage steel sheets for the top, side and bottom panels. The back is often closed with a door and the front is closed with panels that contain circuitry and operating controls. When high voltages and currents are a concern, the door is fitted with a lock to prevent entry by

FIGURE 7.25
A commercially available cabinet-type enclosure. (*Courtesy of Premier Metal Products.*)

unauthorized personnel. In some cases it is necessary to permit air flow through the enclosure to improve component cooling. Cooling is enhanced by louvering the doors to allow the air to flow into the enclosure and then using a louvered top panel for the discharge flow.

Access during assembly and servicing is a necessary requirement in any enclosure. The enclosures are furnished with panels that can be removed to leave only the frame, which consists of the corner posts, headers and braces. In the stripped down condition, accessibility during initial assembly is insured. Accessibility during service is a different issue since the time and place of repair usually do not permit the service person to disassemble the cabinet. Instead, the circuits are made available to the service person by using a door to open the rear of the enclosure and by using either horizontal or vertical drawers that open to the front. Opening the drawer brings the circuit cards into view and allows the service person to use test probes and substitute cards as required for repair.

The structural strength and rigidity of an enclosure is determined by the corner posts and the interior headers and braces as illustrated in Fig. 7.25. Typically corner posts are fabricated from 14 gage steel (74 mil thick) and the headers and braces are fabricated from 16 gage steel (60 mil). Additional support is provided by the 12 gage (105 mil) panel mounting angles that are positioned along the vertical edges of the front and rear openings. Weight capacities are not usually specified by cabinet suppliers but interior loading in excess of 1000 lb is possible. Recall that additional support for the structure is provided during assembly when shelves and slides are bolted to the vertical rails. While the commercial enclosures are sufficiently rugged for the office or even factory environments, they will not perform well under most vibration and shock imposed on military electronics. The framing deforms plastically under the action of high shock loads, which increase the static loading by a factor of 10–100. In vibration, the light sheet metal panels and shelving resonate and fatigue failures often occur at the fasteners, the welded joints or the bracing.

The enclosures are fitted with vertical mounting angles along the edges of the front and rear openings. These angles are drilled and tapped with No. 10-32 threads at standard intervals. These tapped holes accommodate screws used to attach subracks, panels and control and display devices. The enclosures also accept a wide range of accessories, such as casters, leveling feet, writing surfaces, power strips, bus bars, shelves and drawers, which facilitate assembly, servicing and/or operation of the electronic system.

7.6.2 Military Enclosures

The design environment for most military electronics is not friendly and the enclosure must accommodate intense and enduring vibrations, high level shock loading, a large range of temperatures, fog, humidity, salt, spray, etc. Many of the enclosures for military applications are unique because they are designed to fit on a certain weapons platform with unusual space constraints. The annual volume of a particular product for the military is often small with production of 50–100 copies per year common. With unique enclosure designs, exacting durability requirements and small production runs, the cost for designing and manufacturing military enclosures is very high.

In some cases the weapons platforms are nearly the same and some commonality in the enclosure with significant cost savings is possible. The U.S. Navy recently developed the common electronic equipment enclosure CE^3 for use in shipboard applications. Aboard a ship the electronic enclosures are bolted to the deck in an electronics room. The enclosures are provided with an ample supply of chilled water for conduction cooling. Many of the electronic systems used by the Navy are advanced signal processors, which can be classified at the high end because they are large and very high performance machines. In some

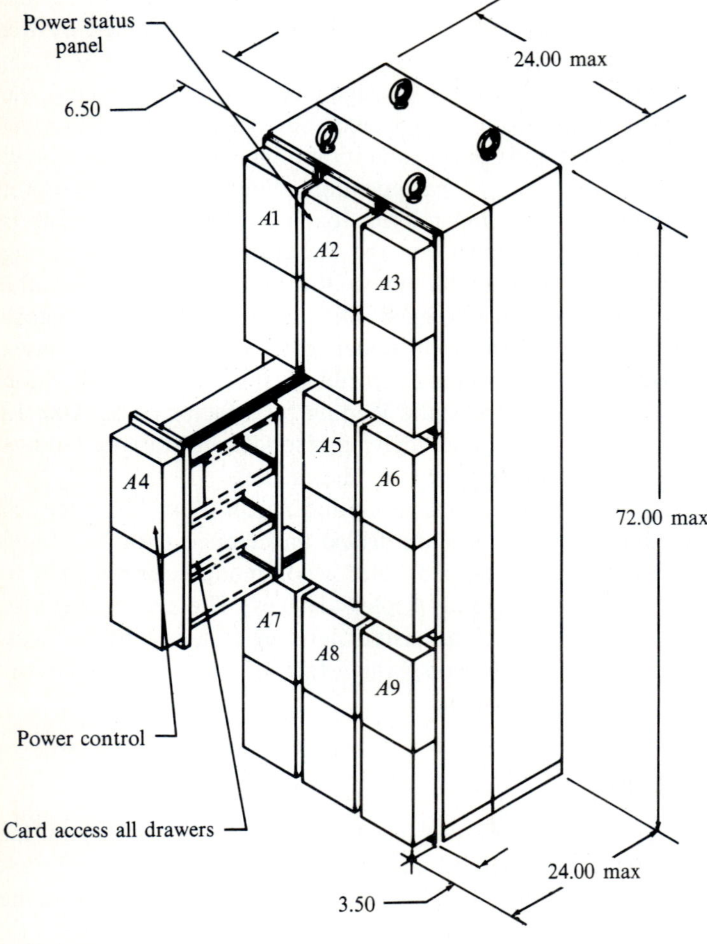

FIGURE 7.26
Common electronic equipment enclosure CE^3.

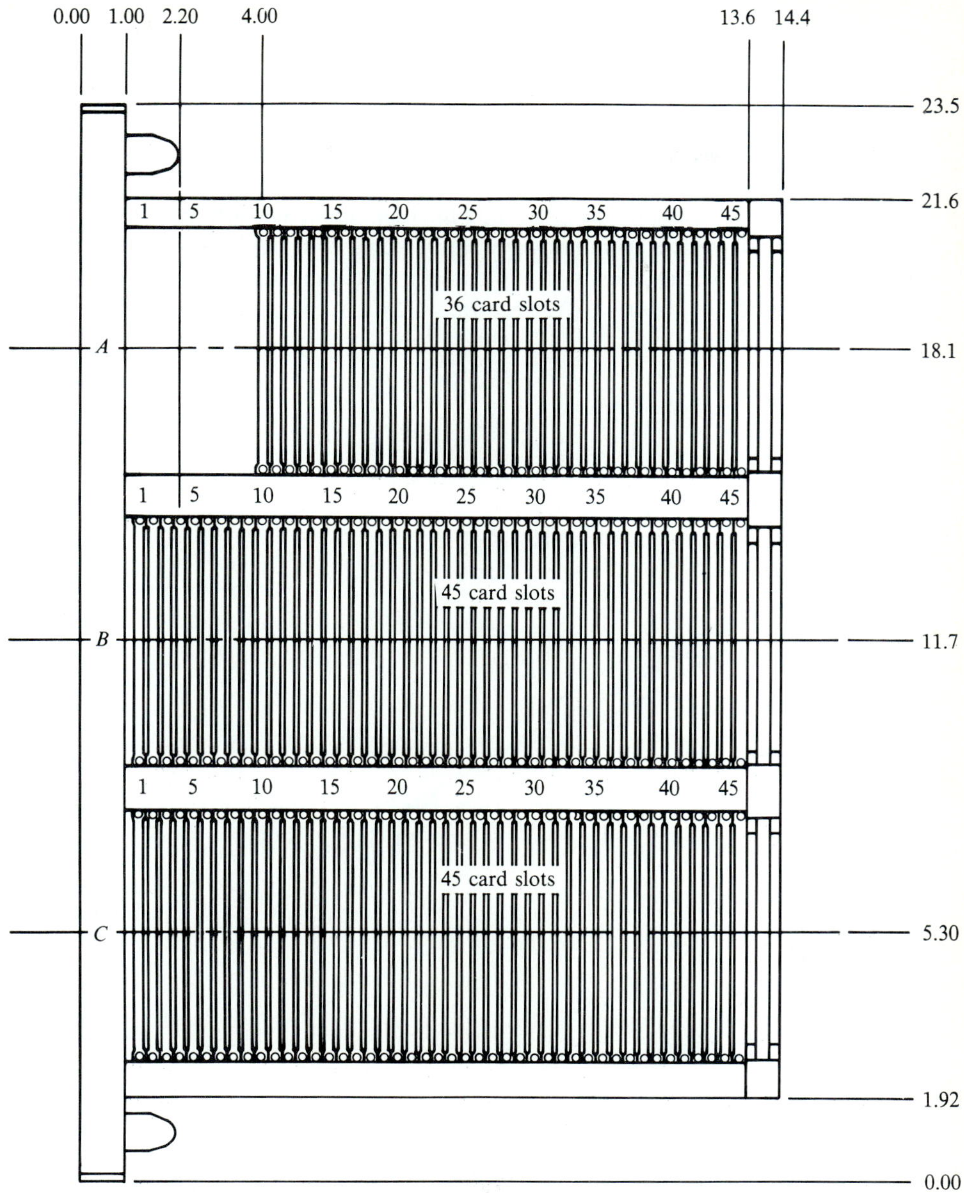

FIGURE 7.27
Water cooled drawers with circuit cards spaced on 0.3 in. centers.

integrated detection and control systems, the processors involve circuits containing 10^8–10^9 gates housed in several enclosures and connected together by massive signal busses.

With this background, let us consider the description of the common electronic equipment enclosure CE^3 that was developed for the U.S. Navy in the early 1980s. The enclosure is a free-standing unit (i.e., deck bolting but no top supports) with the dimensions given in Fig. 7.26. The enclosure has nine vertical drawers, each of which accommodates three rows of modules (circuit cards are often referred to as modules in Navy circles). The enclosure can be loaded with about 1100 modules having a 0.3 in. pitch. The modules are conduction cooled

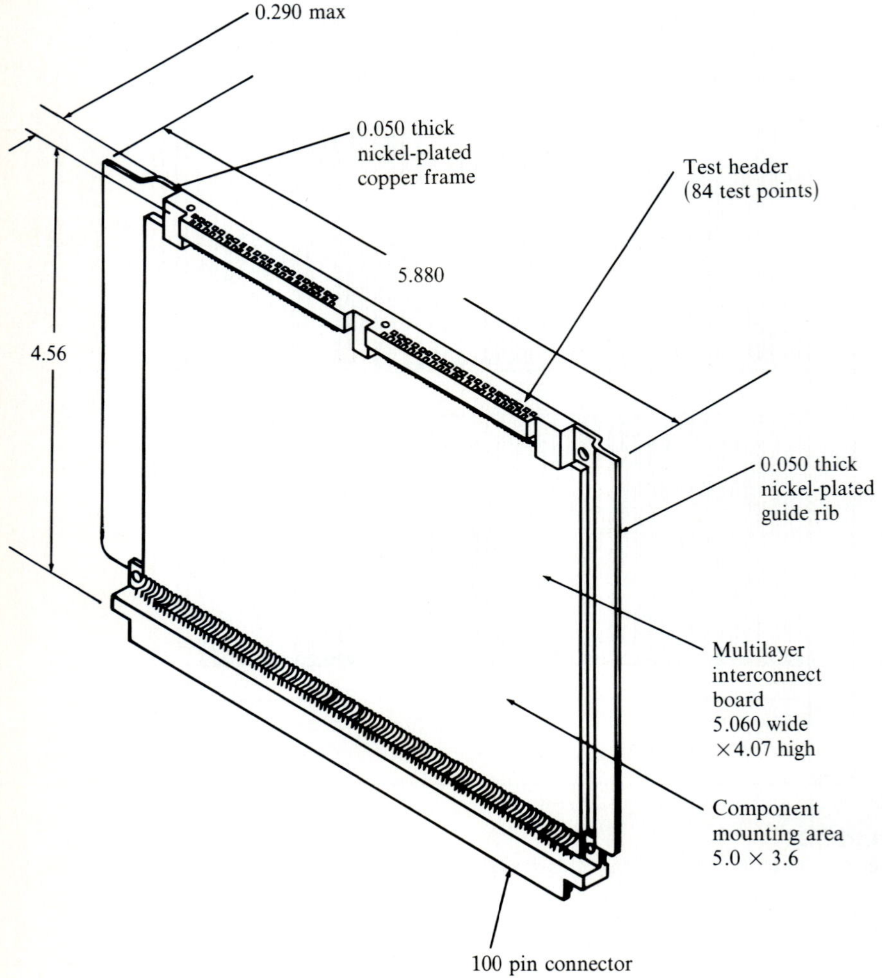

FIGURE 7.28
Standard electronic module (SEM)—format D module 0.3 in. pitch.

and the drawers are fabricated with cold plates that remove heat from two opposite edges of the module heat frames. Individual power supplies are mounted on the front of each drawer to provide the dc voltages required. The enclosure is capable of 18 kW dissipation with 25 gal/min of chilled water pumped through the drawers. The enclosure is built to withstand the severe vibration and shock requirements imposed by the Navy. It is possible to disassemble the enclosure, load it through a hatch and reassemble it in the equipment room of a ship. The

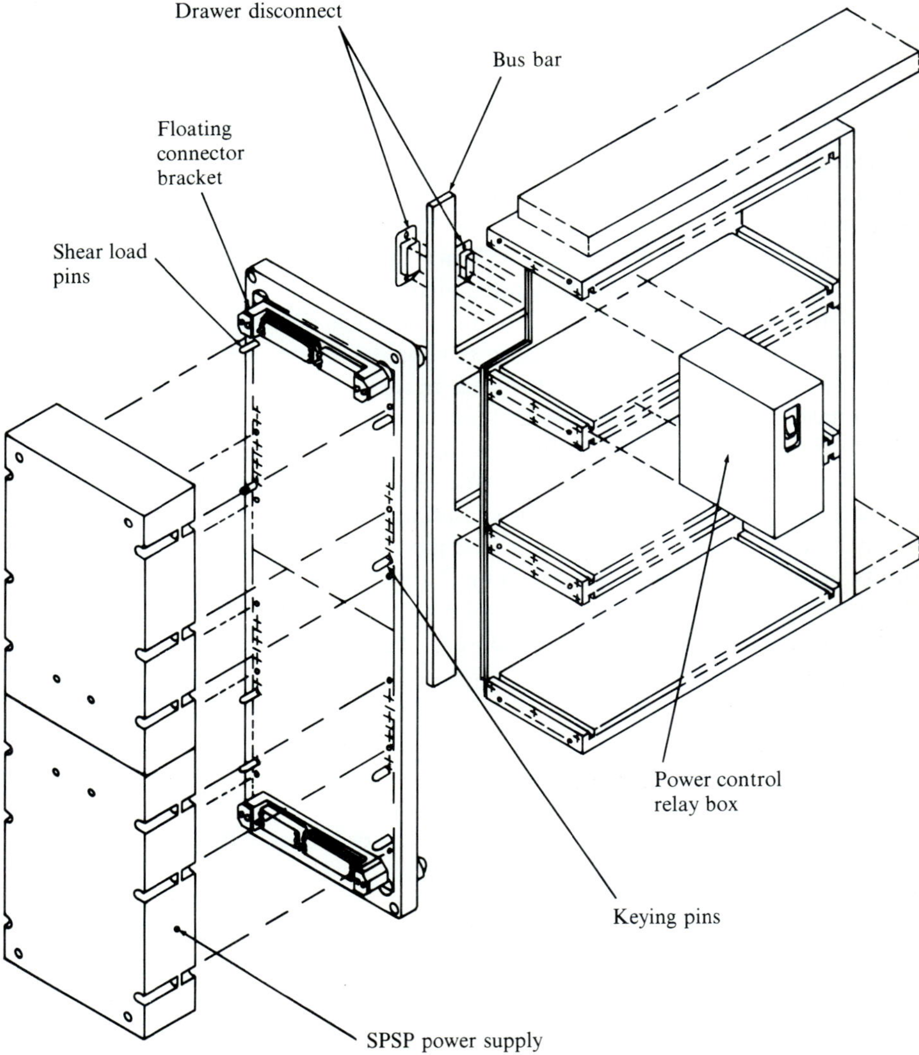

FIGURE 7.29
Expanded view showing power distribution to the back panels.

unit completely assembled with a full load of modules weighs slightly more than 1 ton.

Packaging the modules in a typical drawer is illustrated in Fig. 7.27. The drawer is divided into three compartments, labelled *A*, *B* and *C*, each accommodating a row of cards. The space for nine modules is reserved in compartment *A* for a power regulator used in distributing the power to the back panel. If all of the slots available are filled with modules, each drawer can house 126 modules. Each drawer is mounted on slides, permitting it to be extended its full length to provide accessibility for module testing and replacement.

The modules that are used with the CE^3 are standardized by the Navy and designated as format D modules. A format D module with a 0.3 in. thickness is shown in Fig. 7.28. This module includes a multilayer interconnect board with footprints for surface mounted devices, a copper heat frame, a 100 pin connector and a test header. The heat dissipated, 20 W, is conducted by the frame to the vertical edges. The card is clamped in the drawer slots with the spring-type retainers described in Section 7.5.1. The format D module is available in many combinations. A 0.6 in. thickness version has a multilayer circuit board bonded to both sides and a 250 pin connector. When this version of the module is used in the drawer every other card slot is populated.

With very dense packing of modules in each drawer, the power requirements are large. Indeed, each drawer is capable of supporting up to 2 kW of power. Two design tasks are involved in dealing with the power, namely, distribution and dissipation. The method of distribution is shown in Fig. 7.29. Power supplies are mounted on the front face of each drawer. These supplies are connected to a control box that contains a relay for switching the power on or off for the entire drawer. The control box interfaces with a bus bar that is formed in the shape of a π. Two arms of the π extended along a back panel and distribute supply voltages at select locations along the panel and at a single point ground. The current distributed is in the range of 300–500 A, depending on the particular voltage requirements of the circuits housed. The task of power dissipation will be described in Section 7.9.2.

7.7 WIRES AND CABLING

The printed circuit board, used as a circuit card or a back panel, was introduced to reduce the cost and lessen circuit degradation due to ordinary wiring. Ordinary wires were originally used, but the efficiency of the connection process was increased enormously and circuit degradation was minimized with PCBs. However, wiring and/or cables were not totally eliminated because it is necessary to connect units that are too far apart to effectively use either rigid or flexible printed circuit boards in a connection scheme. This type of wiring includes drawer-to-drawer connections in an enclosure, cabinet-to-cabinet connections in a large system and the transport of signals from room to room, building to building, city to city, country to country and continent to continent. We will concentrate on the design of the more local wiring systems with the enclosure and within a system.

7.7.1 Wire Conductors

The conductor used in almost all of the wiring is electrolytic tough pitch copper that is coated with tin, tin–lead or pure lead to a thickness of about 50 μin. The coating inhibits oxidation and other corrosive effects and assists in soldering. If the wire does not undergo flexing or vibration, solid conductors are used because they are lower in cost than stranded cable and can be handled and shaped more easily in assembly operations. The difficulty with solid conductors is their tendency to fail in fatigue due to cyclic bending. Recall that the strain produced by bending a long slender member is given by

$$\varepsilon = c/R \tag{7.14}$$

where $c = d/2$ is half the wire diameter
R is the radius of curvature of the deformed wire

The stress developed is

$$\sigma = E\varepsilon = Ec/R \tag{7.15}$$

For long life, in excess of 10^6 cycles, the bending stresses are compared to the endurance strength S_e of copper wire. The endurance strength of nonferrous alloys is often taken as

$$S_e = 0.35S_u \tag{7.16}$$

where S_u is the tensile strength. For copper the tensile strength depends strongly on the degree of cold-working in fabricating the wire as indicated in Table 7.6. The tensile strength ranges from 35,000 psi for annealed wire to 97,000 psi for hard cold-worked wire.

Stranded wire is nearly imperative when it is necessary to flex the cable assembly for a large number of cycles. Great increases in the fatigue life of stranded wire occur for two reasons. First, the diameter of the strand is an order of magnitude smaller than the wire diameter. Since c in Eq. (7.15) is related to

TABLE 7.6
Properties of copper wire

State	Coating	Strength (ksi)	Elongation (%)	Solder-ability	Oxidation resistance
Annealed	None	36	15	Fair	Good
Annealed	Tin	36	15	Good	Good
Annealed	Silver	36	15	Good	Good
Annealed	Nickel	36	15	Poor	Good
Med. Hard	None	60	0.88	Fair	Poor
Hard	None	97	0.86	Fair	Poor

TABLE 7.7
Parameters describing highly flexible wire for fatigue applications

AWG	Strand	Nominal Wall (in.)	Maximum voltage	Nominal O.D. (in.)
38	7/46	0.005	200	0.015
36	7/44	0.005	200	0.016
34	10/44	0.005	200	0.018
32	18/44	0.007	300	0.023
30	28/44	0.010	600	0.031
29	51/46	0.010	600	0.032
28	41/44	0.010	600	0.035
27	48/44	0.010	600	0.037
26	66/44	0.010	600	0.042
24	105/44	0.015	1000	0.058
22	168/44	0.015	1000	0.064
20	259/44	0.010	600	0.061

Source: Brim Electronics, Dept. GB, P.O. Box 336 G, Fairlawn, N.J. 07410.

the diameter of the strand, the stress imposed on the copper is reduced by the ratio of the conductor diameter D to the strand diameter d. The second reason is less obvious, but small nicks sometimes occur when removing the insulation from solid copper conductors. These nicks drastically reduce the fatigue life of the solid wire because they serve to concentrate the stresses. Nicks may also occur with stranded wire, but they only eliminate a strand or two of the many strands and result in small decreases in effective fatigue strength. For a given size conductor, an increase in the number of strands increases the flexibility of the conductor. A listing of stranded conductors, designed for use in applications involving significant cable flexing, is presented in Table 7.7. Note, that the number of strands used increases from 7 for AWG No. 38 wire to 259 for No. 20 wire. The diameter of the strand changes little with increasing wire size since only Nos. 44 and 46 wires are used in fabricating the conductor.

The lay of the individual strands, which is the axial length of one turn of a single strand, affects both the flexibility and the cost of the wire. Wires with short lays exhibit high flexibility because the strands are approaching the shape of a helix found in a coil spring. Of course, the cost increases since more time is involved in fabricating the conductor. In some applications, the strands are coiled around a high strength thread that serves as the core and carries any tensile loads that might be imposed on the conductor.

Finite fatigue life of cable assemblies is always difficult to predict and testing is essential to insure reliable performance when long cyclic life is involved. However, it is possible to adapt the Manson fatigue life equation to this application. This empirical equation permits one to relate strain range $\Delta\varepsilon$ and cyclic life N by

$$\Delta\varepsilon = 3.5(S_u/E)N^{-0.12} + D^{0.6}N^{-0.6} \tag{7.17}$$

where D is the ductility of the conductor material given by

$$D = \ln[1/(1 - \mathrm{RA})] \tag{7.18}$$

and RA is the reduction in area measured in a simple tension test. Combining the results of Eqs. (7.14) with (7.17) permits the fatigue life of either solid or stranded wire to be predicted.

7.7.2 Wire Insulation

The conductors are coated with a dielectric that serves to insulate it and to protect it from corrosive attack from the environment. A large number of different polymers are used as insulating materials for special purpose wiring. We will cover only the more commonly utilized polymers including polyvinyl chloride (PVC), polyethylene, rubber, polyurethane and Teflon. PVC is one of the most common insulating materials because it is low in cost and easy to process. PVC has a high dielectric strength (500 V/mil), as indicated in Table 7.8, and it is resistant to flame, water, oil and abrasion. Its primary limitation is a limited temperature range, −20–80°C. Recently, some of the PVCs have been irradiated, which increases the degree of cross linking in the molecular chains and raises the temperature rating to 105°C.

Polyethylene has a low dielectric constant, high dielectric strength, good resistance to solvents, low density and very good low temperature characteristics. Its primary disadvantages are its tendency to creep when subjected to stress and its poor flame retarding properties. Polyethylene is next to PVC in its low cost.

Natural and synthetic rubbers are often used as jackets for cables. They have excellent flexibility, very good resistance to abrasion and reasonable strength over a wide range of temperature. There are several different types of rubbers including natural, neoprene, hypalon, nitrile, butyl and silicone. The properties and costs vary widely and care should be exercised to select the most suitable of the large number of polymers and blends that are commercially available.

Polyurethane is an excellent insulation material with exceptional resistance to abrasion. It is frequently used as a thin coating insulation for magnet wire. Its main disadvantages are a low operating temperature and poor flame retarding properties.

Teflon is an outstanding insulating material with superb properties in almost every category. However, it is so expensive that it is used only in demanding applications. The advantage of the wide operating temperature range of Teflon, −70–250°C, is evident.

TABLE 7.8
Properties of select insulation materials used on wire

Material	Specific gravity (nominal)	Volume resistivity (nominal) (ASTM D257) (ohm-cm @ r.t.)	Voltage breakdown (nominal) (ASTM D149)[a] (volts / mil)	Res. to cold flow	Res. to abra-sion	Dielectric constant (nominal) (ASTM D150)
Rubber	0.93	10^{15}	150–500	Exc.	Exc.	2.3–3.0
Silicone rubber	0.97	10^{14}	100–600	Good	Poor	3.2
Neoprene	1.25	10^{11}	150–600	Exc.	Exc.	9.0
Hypalon[b]	1.15	10^{14}	500	Good	Exc.	7.0–10.0
Polyvinyl chloride (PVC)-standard	1.3	10^{11}	500	Fair	Fair	7.0
Polyvinyl chloride (PVC) premium	1.3	10^{12}	500	Fair	Good	7.0
Polyethylene—solid	0.95	10^{13}	600	Poor	Good	2.5
Polyethylene—foam	0.5	10^{13}	N.A.	Poor	Poor	1.5
Teflon[b] (TFE & FEP)	2.2	10^{13}	600	Fair	Exc.	2.1
Nylon	1.07	10^{14}	450	Good	Exc.	4.0
Polypropylene	0.91	10^{15}	650	Good	Exc.	2.2
Rulan[b]	1.3	10^{17}	420	Poor	Good	2.8
Irradiated polyvinyl chloride (PVC)	1.3	10^{12}	500	Fair	Good	5.0
Irradiated polyolefin[e] (polyalkene)	1.3	10^{15}	600	Fair	Good	2.5
Kynar[f]	1.76	2×10^{15}	250	Good	Exc.	5.0–6.0
Irradiated kynar	1.8	2×10^{15}	250	Good	Exc.	5.0–6.0
Polyurethane	1.1	10^{11-14}	500	Good	Exc.	5.0–8.0
Polysulfone	1.24	5×10^{15}	400	Good	Good	3.1
Kapton[b]	1.4	10^{13}	*g*	Good	Exc.	3.5
Ethylene-propylene copolymer (EPR)	0.86	10^{17}	900	Good	Good	3.3
Fluorosilicone	1.4	10^{14}	350	Good	Exc.	7.0
Tefzel 280[b]	1.70	10^{16}	400	Good	Exc.	2.6
Teflon PFA[b]	2.1	10^{18}	600	Good	Exc.	2.1
Halar[h]	1.68	10^{15}	490	Good	Exc.	2.6

[a] Values are for $\frac{1}{8}$ in. slab. A number of materials such as Kynar, irradiated polyolefin and Kapton have significantly improved.

[b] Trademark of DuPont.

[c] When properly pigmented to resist ultraviolet light.

[d] FEP Teflon 200°C max.

[e] Flame retarded.

[f] Trademark of Pennalt.

[g] Voltage breakdown for 1 mil thickness is 7 kV/mil for Kapton.

[h] Trademark of Allied Chemical Corp.

Flame retard. properties	Flexi- bility	Weather- ability	Temp. range (°C) (nominal)	Solvents—hydrocarbons: Aliphatic (alcohol–glycol)	Aromatic (gasoline–benzine)	Chlori- nated (trichloro- ethylene)
Poor	Exc.	Poor	−40–70°	Poor	Poor	Poor
Poor	Exc.	Exc.	−60–200°	Good	Fair	Good
Good	Exc.	Exc.	−30–90°	Good	Poor	Poor
Good	Good	Exc.[c]	−30–105°	Good	Poor	Poor
Exc.	Good	Exc.	−20–80°	Poor	Poor	Fair
Exc.	Good	Exc.	−55–105°	Poor	Poor	Fair
Poor	Fair	Exc.[c]	−60–80°	Good	Good	Good
Poor	Good	Exc.[c]	−60–80°	Poor	Poor	Poor
Exc.	Fair	Exc.	−70–250°[d]	Exc.	Exc.	Exc.
Poor	Poor	Exc.	−40–120°	Exc.	Good	Exc.
Poor	Poor	Exc.[c]	−40–105°	Good	Good	Good
Exc.	Good	Exc.	−50–80°	Good	Fair	Fair
Exc.	Good	Exc.	−55–115°	Good	Good	Good
Exc.	Good	Exc.	−50–125°	Exc.	Exc.	Exc.
Exc.	Good	Exc.	−40–150°	Exc.	Exc.	Exc.
Exc.	Good	Exc.	−55–175°	Exc.	Exc.	Exc.
Poor	Exc.	Exc.	−50–80°	Good	Good	Good
Exc.	Good	Good	−55–150°	Exc.	Good	Exc.
Exc.	Exc.	Exc.	−40–200°	Exc.	Exc.	Exc.
Poor	Exc.	Exc.	−40–80°	Good	Good	Good
Exc.	Exc.	Exc.	−60–200°	Exc.	Exc.	Exc.
Exc.	Fair	Exc.	−70–180°	Exc.	Exc.	Exc.
Exc.	Fair	Exc.	−70–250°	Exc.	Exc.	Exc.
Exc.	Fair	Exc.	−70–165°	Exc.	Exc.	Exc.

7.7.3 Wire to Cable

Wires of many different types are used in connecting components and circuit boards within an electronic enclosure. In this description we will begin with the simplest form of insulated wire—magnet wire—and add features to the wire or the insulation needed to develop cable assemblies with different degrees of complexity. These wires and cables, illustrated in Fig. 7.30, show important characteristics of several different design methods followed in this upper level of the connection process.

The simplest component is magnet wire, which consists of a solid copper conductor with a thin layer of enamel-like insulation. It is used primarily to wind coils on motors and inductors. However, it is often used as a soft wire on printed circuit boards when corrections to the circuit require some degree of rewiring. The thin wires conform to the surface of the board and effectively replace the circuit traces without detracting from the appearance of the PCB.

Single conductor hookup wire is used to make point-to-point connections when the points are distributed over a relatively large volume of the enclosure. The hookup wire is insulated and it may be shielded. To facilitate assembly and to minimize the space required in the enclosure, the hookup wires are tied together to form a wiring harness as illustrated in Fig. 7.30*a*.

In many applications the points to be connected are arranged at locations equally spaced along a line. Connections to a line of contacts are made most effectively with a flat ribbon cable. This cable, illustrated in Fig. 7.30*b*, consists of a sandwich of a number of parallel conductors with the insulation serving to space the wires and to form the ribbon. The flat ribbon cable has many advantages when compared to connections made with discrete wires. It requires only about half of the weight and volume as a harness. Also, the precise positioning of the conductors permits mass termination into connectors with a minimum of labor. Another advantage is the flexibility and handling characteristics. The flat ribbon can be rolled or flat folded like an accordion, allowing for compact storage in the enclosure.

Woven wire cable, illustrated in Fig. 7.30*c*, has many of the same characteristics as flat ribbon cable. It is fabricated from individually insulated round wires by weaving them with strong thread in a loom. The woven flat cable has the advantage of individual tailoring, that is, the individual wires in a cable can be different, depending on the requirements. The flexibility of the manufacturing process permits custom cable assemblies for specific packaging applications, like the cable with programmed breakouts presented in Fig. 7.30*d*.

The flat cables, either ribbon or woven, have electrical characteristics, distributed inductance and capacitance, that are important in circuit performance of the electronic system. These characteristics depend on insulation material, conductor lay and the proximity of a ground wire or plane. Teflon insulation, with its low dielectric constant, reduces capacitance and increases the propagation velocity of the signal pulse. The presence of a ground plane increases capacitance, decreases inductance and the net effect is to increase propagation

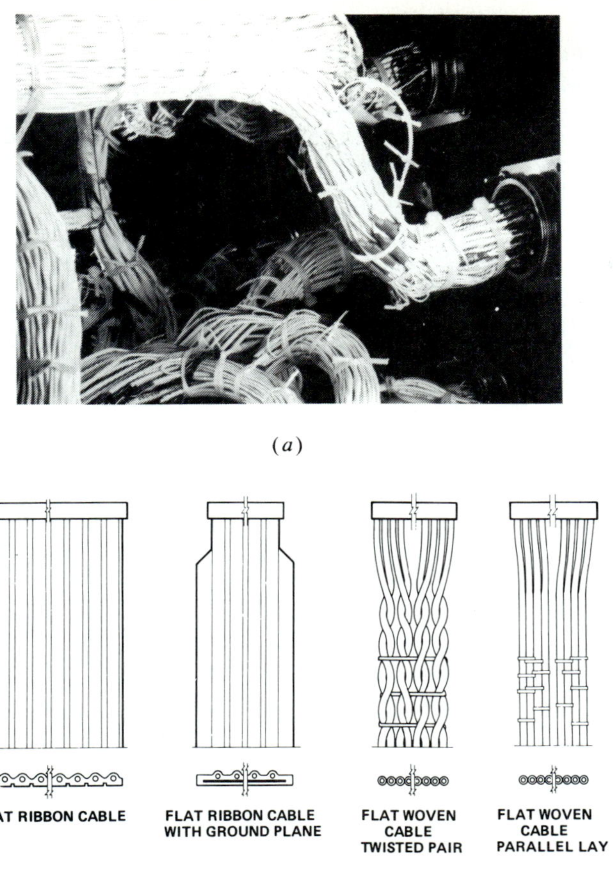

(*a*)

(*b*) (*c*)

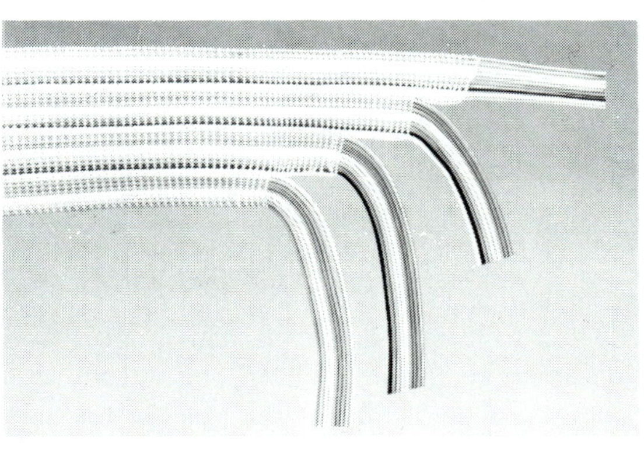

(*d*)

FIGURE 7.30
(*a*) Wiring harness utilizing round single conductor cable. (*b*) Flat ribbon cable with and without a ground plane. (*c*) Flat woven cable. (*d*) Flat woven cable with programmed breakouts. (*Figures* 7.30 (*a*) *and* 7.30 (*b*) *Courtesy of Woven Electronics.*)

TABLE 7.9
Electrical characteristics of flat ribbon and woven cable fabricated with stranded AWG No. 28 wire

Cable	Insulation material	Impedance (Ω)	Capacitance (pF / ft)	Inductance (μH / ft)	Propagation delay (ns / ft)
Flat ribbon parallel lay—ground plane	PVC	65	28	0.20	1.65
Flat ribbon parallel lay	PVC	142	8.0	0.30	1.40
Woven parallel lay	PVC	144	7.3	0.27	1.34
Woven parallel lay	FEP	150	6.2	0.23	1.26
Woven twisted pair	PVC	108	11	0.26	1.60
Woven twisted pair	FEP	128	8	0.26	1.45

delay. Twisting wire pairs to control the characteristic impedance of the cable results in an increase in the propagation delay and capacitance when compared to a flat lay ribbon cable. The electrical characteristics of six commonly employed flat cable assemblies are given in Table 7.9.

7.7.4 Wire Shielding

A shield is a conductor that is placed around a wire or group of wires to provide a barrier against external electromagnetic and electrostatic fields that could affect the signal being transmitted. The shield also serves to confine the transmitted signal to the central conductor, reducing the fields generated in the region external to the shield. In coaxial cables the shield also acts as the return wire necessary to complete the circuit.

Two different shielding techniques used with individual wires to control the effects of electrostatic fields are presented in Fig. 7.31. Flat tape shields are the most commonly used. The tape is fabricated by bonding a thin foil of copper or aluminum to a thin film of polyester to form a laminated, high strength conductive tape. The tape is wound in a spiral about the insulated central conductors as shown in Fig. 7.31*a*. A drain wire runs along the metal foil, which is connected to ground to terminate the shield.

Braided shields, shown in Fig. 7.31*b*, are woven conducting envelopes fabricated from small diameter (AWG No. 34 to 38) copper wire. These braided shields are used in most coaxial cables because they are effective as the return conductor and because they can be connected to the ground of the appropriate

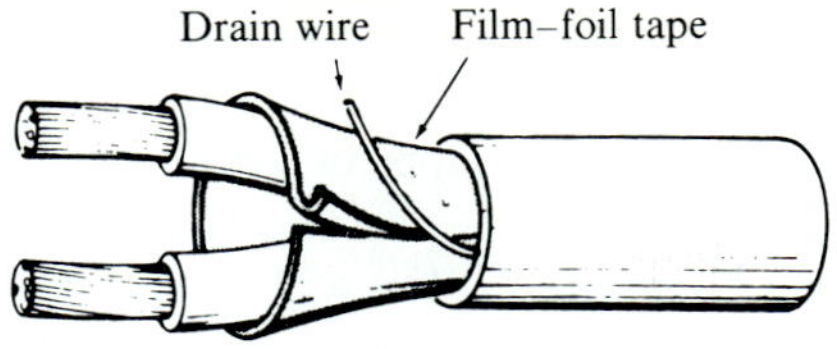

Polyester film–aluminum foil shield

(*a*) Spiral tape with drain wire

(*b*) Braided shield

FIGURE 7.31
Shielded wire. (*a*) Spiral tape with drain wire. (*b*) Braided shield.

connector. The braid is an effective shield over a wide range of frequencies from 1–140 kHz. However, the openings in the braid affect its performance. At the lower audiofrequencies a 75% area coverage is sufficient. However, as the frequencies increase, the area coverage is increased to 90–95% by using smaller diameter wires and a tighter weave.

Shielding against magnetic fields is much more difficult because the shielding materials must exhibit high permeability. Heavy layers of low carbon steel tape or permalloy are effective but they add significantly to the cost of the conductors. Availability of a wide range of magnetically shielded wire is another issue. A much lower cost approach, applicable in some applications, is to twist two conductors tightly together to achieve a balanced pair. When magnetic field lines intersect the twisted pair, they produce equal and opposite changes in the voltage that tend to cancel.

7.8 FANS AND COLD PLATES

In the design of an enclosure system, provisions must be made to dissipate the heat generated within the cabinet or case. For very low power systems, the heat is dissipated by natural convection and the primary concern is the placement of vents in the case, which will assist in the natural flow of the buoyant heated air. It is evident that vents placed on the bottom for air intake and at the top of the enclosure for exhaust are the most suitable. However, customer habits and preference often dictate the placement of vents at less desirable locations. In office environments, particularly with tabletop cases, the clearance allowed under the case is not sufficient for adequate flow and the vents are placed along the bottom of the side walls. Also, the top surface, which is horizontal, tends to become used as a table to accommodate anything from coffee service to stacks of papers. It is not advisable to place vents on this surface and, therefore, it is necessary to place the vents along the top side and rear walls to accommodate the air flow.

For forced air cooling fans are used to move larger volumes of air at higher velocities through the enclosure. With forced convection it is possible to accommodate significantly larger amounts of heat dissipation. We will cover the theoretical aspects of forced convection cooling in Chapter 9. The purpose here is to show fan placement in the enclosure and to discuss the problem of noise generation.

Conduction cooling is used for very high levels of heat dissipation in an enclosure. From an engineering viewpoint it is extremely effective because the coolant, usually water, can be taken very close to the source of heat and the thermal path and resistance is reduced to a minimum. The use of conduction cooling is usually limited to the very high performance systems where the need

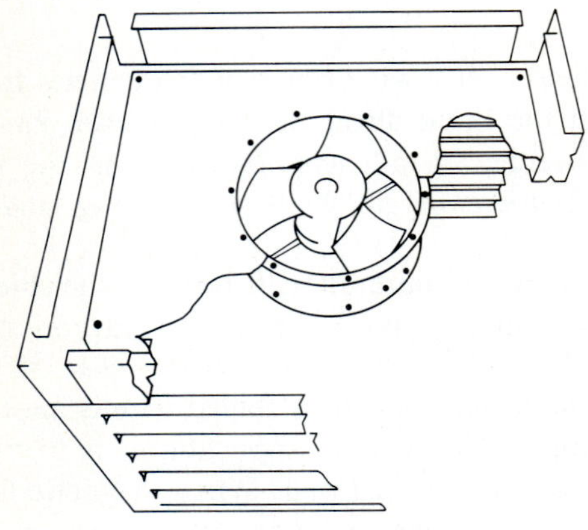

(*a*) Cabinet cooling base

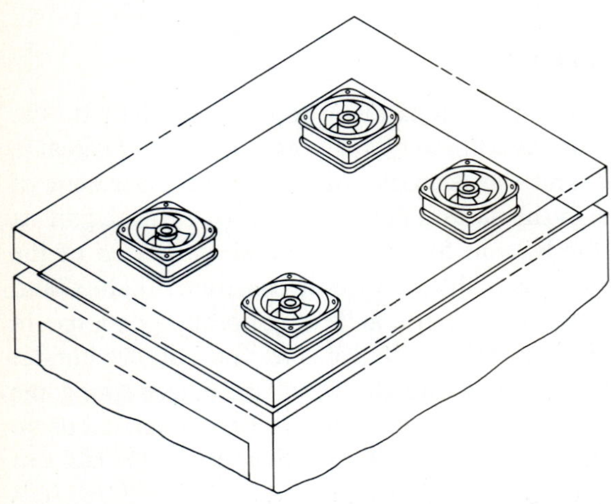

(*b*) Pagoda top incorporates 4 fans

FIGURE 7.32
Fan assemblies as enclosure accessories. (*Courtesy of Zero Corporation.*)

for extremely low junction temperatures with high power devices can only be achieved with extremely effective cooling procedures. For these systems, it is necessary to design cold plates and/or cold rails to provide suitable heat sinks at select locations within the enclosure. We will show an approach for the design of these heat sinks.

7.8.1 Fans and Fan Noise

Propeller fans and blowers are available in a wide range of sizes and capacities to force the air to flow through the enclosure. Convenient fan assemblies, as shown in Fig. 7.32, are available as accessory items from suppliers of commercial

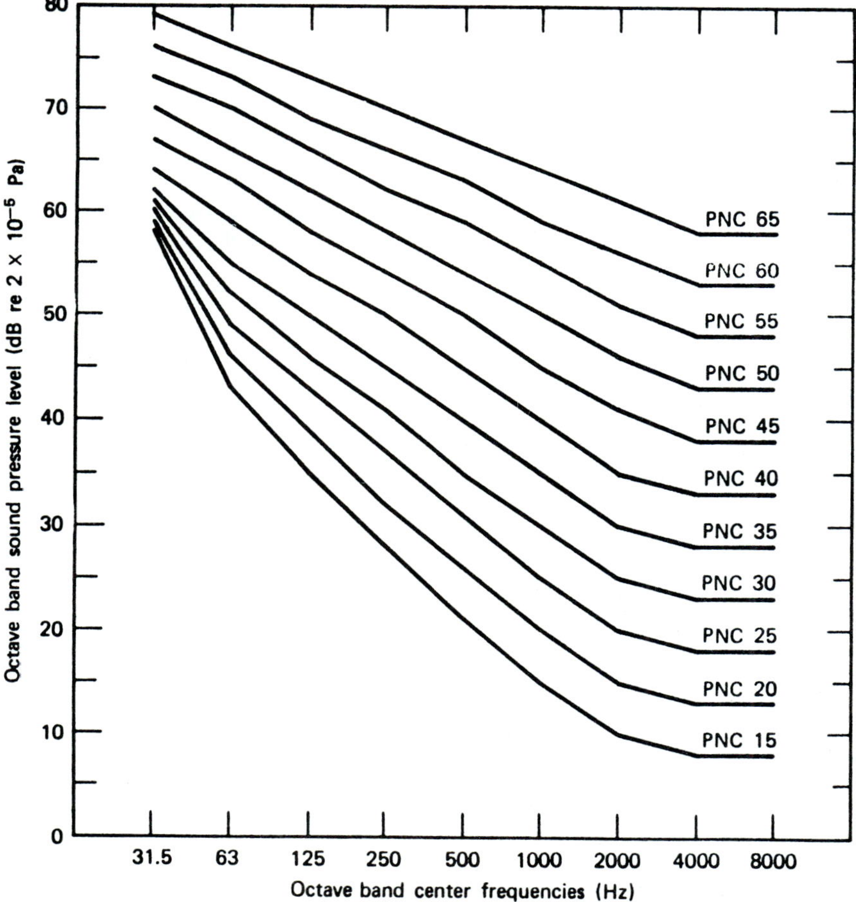

FIGURE 7.33
Recommended background noise levels at different octave bands for indoor rooms based on PNC curves. *(Source: L. L. Beranek, W. E. Blazier and J. J. Figawer, "Preferred Noise Criteria Curves and Their Applications to Rooms," J. Ac. Soc. Am., Vol. 50, No. 5, 1971.)*

cabinets and they permit mounting at the bottom, sides or top of the enclosures. It is relatively easy to select appropriate fans to deliver the required quantity of air. What is more difficult is to ensure that the fan noise will not pollute the environment. The rotating fan blades produce an oscillating pressure that generates a background noise. This noise can be measured as a sound pressure level given by

$$L_p = 10 \log_{10}\left(p^2/p_0^2 \right) \tag{7.19}$$

where L_p is the sound pressure level in decibels relative to p_0
p is the rms sound pressure in pascals
p_0 is the reference rms sound pressure equal to 2×10^{-5} Pa

Excessive levels of noise are detrimental in that noise interferes with speech, causes annoyance and interferes with task completion. Recommended background noise levels that avoid speech interference have been developed for

TABLE 7.10
Design objectives for indoor A-weighted sound pressure levels in rooms with various uses

Type or use of space	Approximate A-weighted sound level (dB A)
Concert halls, opera houses, recital halls	21–30
Large auditoriums, large drama theaters, churches (for excellent listening conditions)	Not above 30
Broadcast, television and recording studios	Not above 34
Small auditoriums, small theaters, small churches, music rehearsal rooms, large meeting and conference rooms (for good listening)	Not above 42
Bedrooms, sleeping quarters, hospitals, residences, apartments, hotels, motels (for sleeping, resting, relaxing)	34–47
Private or semiprivate offices, small conference rooms, classrooms, libraries, etc. (for good listening conditions)	38–47
Living rooms and similar spaces in dwellings (for conversing or listening to radio and television)	38–47
Large offices, reception areas, retail shops and stores, cafeterias, restaurants, etc. (moderately good listening)	42–52
Lobbies, laboratory work spaces, drafting and engineering rooms, general secretarial areas (for fair listening conditions)	47–56
Light maintenance shops, office and computer equipment rooms, kitchens, laundries (moderately fair listening conditions)	52–61
Shops, garages, power-plant control rooms, etc. (for just acceptable speech and telephone communication)	56–66

Source: L. L. Beranek, W. E. Blazier and J. J. Figawer, "Preferred Noise Criteria (PNC) Curves and their Application to Rooms," *J. Acoust. Soc. Am.*, vol. 50, pp. 1223–1228, 1971.

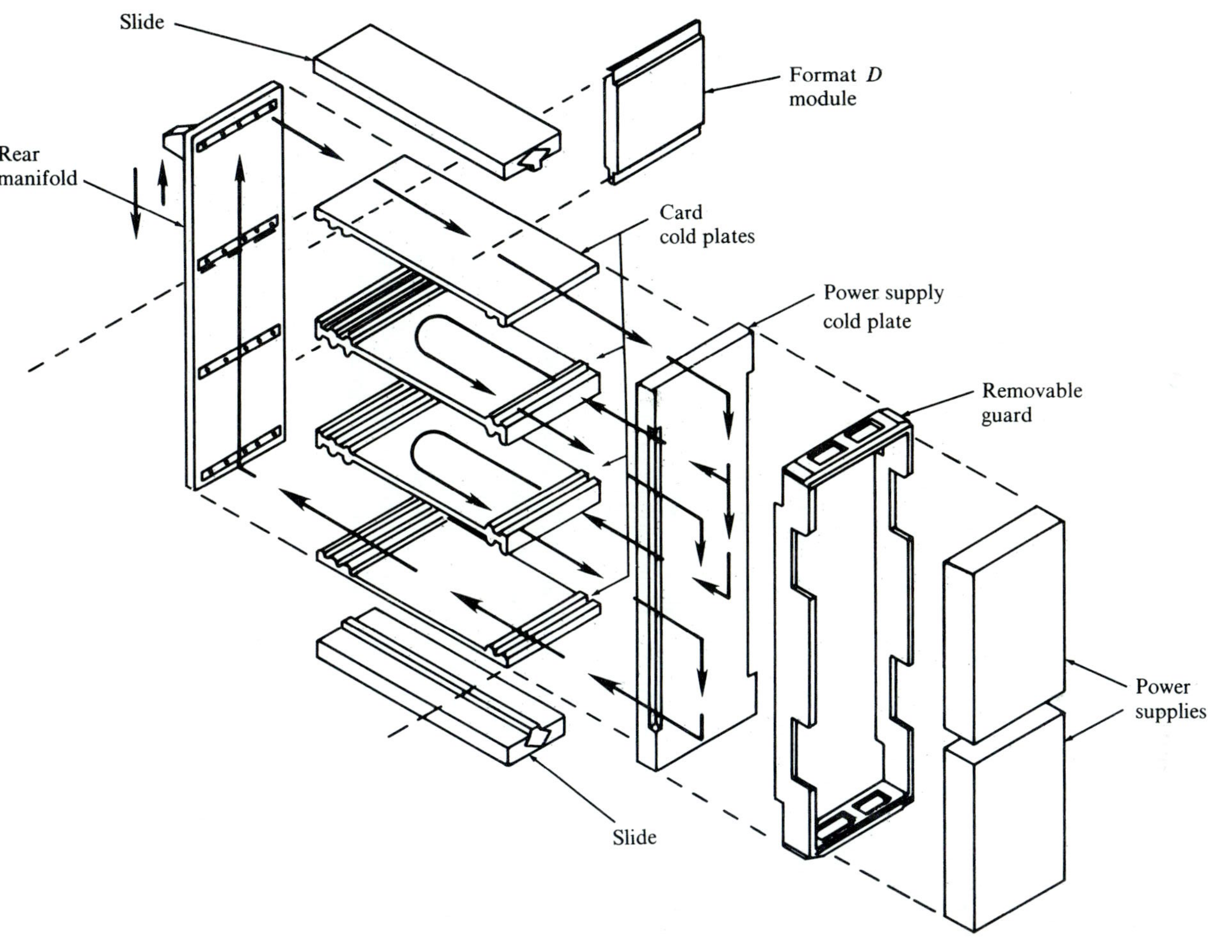

FIGURE 7.34
An exploded view showing design details of the drawer for the CE^3. The drawer provides for a cold plate attachment of the power supply and cold rail retention of the format D modules.

indoor applications. These results, known as the preferred noise criteria (PNC), are shown in Fig. 7.33. An examination of these curves indicates that the sound pressure level measured in each octave band should not exceed a specified decibel level, which varies with the frequency. Clearly, the higher frequencies interfere with speech more than the lower frequencies. The particular PNC curve that defines an acceptable level of noise pollution depends on the function of the room as indicated in Table 7.10. Note, that large offices should follow a PNC 45 curve, while small semiprivate offices should follow a PNC 40 curve. Annoyance and task interference are effects that depend on many parameters that affect the attitude of individuals and, as such, it is more difficult to specify an appropriate PNC curve. Additional information pertaining to these topics is given in reference 9.

7.8.2 Cold Plates and Cold Rails

Cold plates and cold rails are usually fabricated by drilling an interconnecting series of holes in an aluminum plate. Another method involves brazing copper tubing to a plate of steel or copper-based alloy to provide channels for fluid flow. With a cold plate, heat is conducted from the component (a power supply or a large module) to the surface of the plate and then through the plate to the cooling fluid flowing in the channels. A cold rail is usually incorporated in a drawer or a gate for cooling circuit cards with attached heat frames. Slots are cut into the cold rail to accommodate the edges of the heat frames and card retainers such as wedge locks. Heat is conducted across the heat frame–rail interface though the rail to the fluid flowing in the internal channels. Water is the most common fluid used although, in very high performance systems, a refrigerant may be employed to obtain very low cold rail temperatures. In some military systems where direct impingement of air on the components is not allowed, the air is used as the cooling fluid in heat exchangers, which serve as cold rails. An example of a water cooled plate and four cold rails employed in the drawer of the CE^3 is presented in Fig. 7.34.

Methods for determining the heat transferred to the cold plate are covered in Section 8.7.

REFERENCES

1. Ginsberg, G. L., ed.: *Connectors and Interconnections Handbook*, vol. 1, *Basic Technology*; vol. 2, *Connector Types*, Electronic Connector Study Group, Inc., Ft. Washington, Penn., 1979.
2. Bilotta, A. J.: *Connections in Electronic Assemblies*, Marcel Dekker Inc., New York, 1985.
3. Whitley, J. H.: "The Mechanics of Pressure Connections," AMP Inc., Harrisburg, Penn., December 1964.
4. Zimmerman, R. H.: "Engineering Considerations of Gold Electrodeposits in Connector Applications," AMP Inc., Harrisburg, Penn., May 1973.
5. Graham, A. H. and S. W. Updegraff: "Properties of Palladium Nickel Alloy and Pure Palladium for Connector Applications," *Proc. 1984 National Electronic Packaging and Production Conference*.

6. Bonner, R. F., J. A. Asselta and F. W. Haining: "Advanced Printed Circuit Design for High Performance Computer Applications," *IBM. J. Res. Develop.*, vol. 26, no. 3, pp. 297–305, May 1982.
7. Colomina, T.: "Selecting Circuit Card Guides and Retainers," *Machine Design*, vol. 60, June 9, 1988.
8. Manson, S. S.: "Fatigue: A Complex Subject—Some Simple Approximations," *Experimental Mechanics*, vol. 5, no. 7, pp. 193–226, 1965.
9. Cunniff, P. F.: *Environmental Noise Pollution*, Wiley, New York, 1977, pp. 101–114.
10. Beranek, L. L., W. E. Blazier and J. J. Figawer, "Preferred Noise Criteria (PNC) Curves and Their Application to Rooms," *J. Accoust. Soc. Am.*, vol. 50, pp. 1223–1228, 1971.

EXERCISES

7.1. List 10 hardware items involved in third level packaging. Can you list 20 or more items?

7.2. Consider a system housed in a single enclosure that contains 240 circuit boards. Sketch the design of the enclosure so that replacement of a malfunctioning board can be made within 5 min after arrival of the service person. List all of the factors that you have incorporated in your design to facilitate rapid access.

7.3. List the number of ways an accident can occur that will inflict injury to either an operator or a service person dealing with an electronic system.

7.4. In Fig. 7.4, we have shown connections from the circuit card to cables, back panels and chassis. Why do we encounter so many different types of receptacles.

7.5. Determine the force that can be exerted on a pin 25 mil in diameter and 0.75 in. long if it is fabricated from beryllium copper. Consider the pin prior to and after entry into the corresponding socket in the receptacle.

7.6. Prepare a graph showing buckling force on a circular pin for $d = 15–85$ mil and length $L = 0.25, 0.50$ and 0.75 in. The pins are fabricated from brass.

7.7. Describe the important mechanical and electrical properties you would expect in a material suitable for fabricating pin, blades and contacts. Write a materials specification that could be used by a purchasing department in procuring materials suitable for the design and manufacturing engineer.

7.8. Describe the difference between thermoplastics and thermosetting polymers. What are some of the differences in the properties of these two classes of polymers.

7.9. Describe the injection molding process and indicate why it is so effective in producing connector inserts.

7.10. Go to the library and find a complete description of the thermoplastic polyphenylene oxide. Write a short (two page) description suitable for briefing your supervisor on advantages and disadvantages of this polymer.

7.11. Why would anyone use a thermoset type polymer for fabricating inserts? Give at least three very good applications.

7.12. If a connector pin has a resistance of 50 mΩ and is conducting 20 A, determine the power dissipated at the connection. What would happen to this connector? How long would it take?

7.13. For the connector problem described in Exercise 7.13, determine the connection resistance requirement if the power loss is to be limited to (*a*) 0.1 W, (*b*) 0.2 W, (*c*) 0.5 W and (*d*) 1 W.

7.14. Using the results presented in Fig. 7.9, determine κ in Eq. (7.4) for copper, tin and nickel. Comment on the significance of these results in the selection of a material for a contact.

7.15. Describe the effect of the pores in a plated metal on the development of a film resistance in a contact. What procedures can be followed to reduce the effect of the pores in a plating? What is the most cost effective solution?

7.16. We plan to gold plate a very local region on the blade and contact where the two parts touch (selective plating). If the blade and contact are 25 mil wide and the plating thickness is 0.5 μm, determine the cost of the gold required to plate 1000 pairs of parts. Take the price of gold at $450 per troy ounce. Remember that there are 12 troy ounces in a pound.

7.17. What are the advantages of using a palladium–nickel alloy in place of hard gold as a plating material? What is hard gold?

7.18. Tin is used to plate contacts in low cost applications. Will tin always be effective as a plating material? Indicate the applications where tin will result in connectors that fail.

7.19. Why is the power through a connector an important factor to consider in the design of the pins and/or contacts? Describe a dry circuit condition for a connector. Should a connector be used in place of a switch for power on–power off applications? Why not?

7.20. Why is lubrication effective in extending the life of tin-plated contacts? Will lubrication extend the life of contacts plated with more precious metals? Explain your answer.

7.21. Prepare a family of curves showing the normalized insertion force F_i/N_i as a function of the lead angle θ with the coefficient of friction $\mu = 0.1$, 0.2, 0.3 and 0.5 as a parameter.

7.22. Verify Eqs. (7.5)–(7.10).

7.23. Derive an equation for the bending stress developed in the contact as a function of the interference Δh.

7.24. Determine the insertion force for a blade being inserted into the contact shown in Fig. 7.13*a*. The interference $\Delta h = 10$ mil, the contact is fabricated from beryllium–copper, the lead angle is 30°, the contact length $L = 600$ mil and the cross section of the contact arm is 30 × 30 mil. Let the friction be a variable ranging from $\mu = 0.2$–0.5.

7.25. For the single pin connection described in Exercise 7.24, find the force to insert and extract during the wiping phase of engagement.

7.26. How many pins of the type described in Exercise 7.24 could be placed in a multipin connector before it would be necessary to provide mechanical assistance for insertion and withdrawal of the plug portion of the connector.

7.27. Sketch the design of a simple pair of screws that could be used to both engage and disengage the two halves of a connector requiring a 100 lb insertion force. Indicate the minimum size screw that would be required. What size screw would you specify? Why?

7.28. Determine the interference necessary to develop a normal force of 200 g during the wiping phase for the contact described in Exercise 7.24. Find the maximum bending stress developed in the contact by this interference. Comment on the magnitude of this stress as it relates either to the selection of the material or the size of the contact.

7.29. Briefly describe a back panel and distinguish it from a circuit board.

7.30. What are the advantages of employing a very dense back panel with features like those described in Table 7.3?

7.31. Compare wire wrap panels with back panels, citing similarities and indicating differences. Also, give advantages and disadvantages of each.

7.32. What are the important differences between a pin plate and a wire wrap panel?

7.33. When can cable connected circuit boards be employed in place of wire wrapped panels and back panels.

7.34. Can you propose another method for connecting together circuit boards not covered in Section 7.3. Prepare an engineering sketch showing the key features of your design.

7.35. Describe three common types of batteries and indicate their important characteristics. Give examples of products that use each type of battery.

7.36. A large signal processor requires 12 kW of power to supply the ICc. Estimate the size and weight of the power supplies. Justify your result and comment on the significance of the size and weight.

7.37. Remove the cover from your PC and locate the power supply. Prepare a sketch showing the method used to connect the power and ground leads to the supply.

7.38. A power supply services eight back panels, each with a capacity of 300 A. The back panels are arranged in a 2 wide by 4 high vertical panel that is housed in a standard 19 in. wide by 70 in. high electronic enclosure. Prepare a sketch showing the locations of the back panels. Next, design a bus bar that will distribute the power from the supply to each back panel. Prepare a drawing of the bus bar. Be sure to include details pertaining to the terminations at the supply and at the back panels.

7.39. Perform an analysis for the bus bar designed in Exercise 7.38 showing the voltage drop from the power supply to the back panels. Also show the power losses.

7.40. Prepare a sketch of a design for a subrack intended to be used in an office environment. The rack should hold 24 circuit cards each 8×16 in. in size.

7.41. Prepare a three page description of the subrack designed in Exercise 7.40. Justify your design decisions in this engineering brief.

7.42. Describe the three groups of card guides and retainers used in subracks. Why does one encounter such a wide range of selection for something as simple as a card guide?

7.43. Prepare a sketch of a retainer–guide that will accommodate PCBs 0.060 in. thick. The retainer–guide should be a single piece part and clamp the edge of the card with a force of 2 lb/in. The goal of the design should be simplicity and ease of manufacture.

7.44. Prepare a design analysis and description intended for a critical engineering review of your project (Exercise 7.43).

7.45. A heat frame 6 in. long is clamped in a slot of a cold plate using a wedge lock retainer. Estimate the temperature difference across the heat frame cold plate interface if 20 W is transferred.

7.46. Obtain a copy of a catalog from a local electronics supply house and find the entry for an instrument case to accept a standard 5 × 19 in. panel. Determine the cost of the case and its weight. Determine the price per pound and note if the case is made of steel or aluminum. Is this price realistic?

7.47. Prepare an engineering brief that supports the cost of the instrument case described in Exercise 7.46.

7.48. Prepare an engineering brief that gives arguments indicating that the price of the case of Exercise 7.46 is too high.

7.49. How can you address both sides of the issue as required by Exercises 7.47 and 7.48.

7.50. Carefully remove the cover from your PC and examine its construction. Prepare a list of parts of those components that you would classify as part of the enclosure. Compare your list with others in the class and find the enclosure that is the simplest. Does this enclosure have the fewest number of parts? Can you make at least one suggestion that would reduce the cost of one of the components used. Can you make another suggestion that would modify a part to permit lower cost assembly. Can you make still another suggestion for a modification that would permit more rapid removal of the cover by service personnel. Do you think the team designing the enclosure for your PC considered all of these design issues?

7.51. Prepare a sketch showing the layout of a cabinet to hold 90 circuit cards. Include subracks in your layout and take the dimensions of the PCBs as 0.6 × 9 × 14 in. The average power to be dissipated is 20 W/card. Size the cabinet by selecting from the standard sizes in Fig. 7.25. Remember to leave space for the power supplies and the cables.

7.52. Using the dimensions of the CE^3 given in Figs. 7.26 and 7.27, prepare a drawing of the front portion of the enclosure. How would you design the fasteners to keep the drawers closed?

7.53. What material would you employ in fabricating the CE^3? Give reasons for this selection.

7.54. Estimate the temperature drop at the cold plate if a 0.3 format D module dissipating 20 W is clamped in place with Birtcher spring-type retainers. What would the result be if wedge locks were used in place of the spring-type retainers.

7.55. Why is it good practice to use a relay when switching large currents? What is a large current?

7.56. Determine the stress produced in bending a copper wire over a mandrel of variable radius. Show your results in a graph of stress as a function of R for wire sizes of AWG Nos. 20, 24, 28 and 32.

7.57. Superimpose on the graph of Exercise 7.56 a series of lines showing the fatigue strength S_e for copper wire. Consider that the wire may have tensile strengths that vary between 40 and 65 ksi depending on the state of cold work. Interpret the results shown on this new graph.

7.58. Determine the stress in an AWG No. 20 stranded wire 259/44 if it is flexed with a radius of 4 in. Compare this to the stress developed in a No. 20 solid conductor. Estimate the fatigue life for both wires.

7.59. A series of wires is terminated in a connector with soldered joints. The solder is not to be stressed by applications of load in service. The cable is attached to a drawer that is opened periodically by a service person and the cable may be stretched in this process. Design a strain relief device that will prevent the loads imposed on the cable from being transmitted to the solder connections.

7.60. Prepare a graph, using Manson's equation, that shows $\Delta\varepsilon$ as a function of cycles to failure N for copper wire.

7.61. For AWG No. 36 wire, used as a strand, find a relation showing fatigue life as a function of the radius of curvature.

7.62. Repeat Exercise 7.61 but use AWG No. 32 wire for the strand.

7.63. Describe the common polymeric materials used in insulating wires. What are the important properties of the polymers when used in wiring applications?

7.64. Teflon is an excellent insulating material for wiring applications. Why is its use limited to relatively few applications?

7.65. Consider a washing machine found in the ordinary laundromat. Lay out a sketch showing a wiring harness for this machine. Will the harness be two dimensional (i.e., all the wires lie in a plane) or will it be three dimensional (i.e., with wires in all three planes)? If the harness is two dimensional, indicate a procedure for manufacturing the harness. If the harness is three dimensional, modify the manufacturing procedure required to produce the cable assembly.

7.66. Compare woven and ribbon flat cable assemblies used in connecting subassemblies in electronic enclosure. Cite advantages and disadvantages for each approach.

7.67. Why is it necessary to shield wires in certain applications? Can you give two or three examples of problems you have experienced due to inadequate shielding?

7.68. A small instrument case 20 × 28 × 8 in. in size is to house circuits cooled by natural convection. Sketch a design of the case showing the vents you plan to use for air flow. Describe, in a short engineering brief, the key features of this design.

7.69. What is the pressure of the fluctuation required to produce a sound pressure level of 60 dB? Prepare a graph of sound pressure level as a function of pressure p over the range of 40–140 dB.

7.70. A field engineer provides you with the following measurements for the fan system you have recently designed:

L_p (dB)	Octave band center frequency (Hz)
65	63
60	125
55	250
52	500
48	1000
45	2000
44	4000
42	8000

Prepare a graph of these results and superimpose on this graph the appropriate PNC curve that can be satisfied. Prepare an engineering brief describing the types of rooms where the fan system can be used without interfering with speech.

PART III

ANALYSIS METHODS

CHAPTER 8

THERMAL ANALYSIS METHODS— CONDUCTION

8.1 INTRODUCTION

Thermal analysis of electronic systems involves predicting the junction temperatures of the integrated circuits used in both analog and digital circuits. It also involves predicting the temperature of other circuit components like resistors, capacitors and transformers. The analysis is extremely important because the system reliability and system availability is strongly dependent on component temperatures and relatively small temperature changes can have a marked affect on both component and system reliability.

There are several different methods used in designing cooling systems for dissipation of the heat generated during system operation to the environment. Natural convection, forced air convection, conduction to heat exchangers, radiation and boiling are all considered in the thermal management schemes. The particular design, which evolves to control the component temperatures, depends to a large extent on the price and performance characteristics of the end product. For price driven products, low cost cooling methods with natural and forced convection are the most common design approach. For high performance products, conduction is often used to efficiently transfer large quantities of heat with small temperature differences ΔT. Intermediate systems often use forced convection with specially designed duct systems and heat exchangers to reduce the temperature gradients required to transfer the heat load.

This treatment is divided into two parts, to facilitate the presentation of the material in a manner easily interpreted by the reader. First, we introduce the concept of failure rate, which is a function of both time and temperature, and then describe reliability of a component and finally the system reliability. With this background, the importance of temperature is established and basic heat transfer methods of analysis are presented. Because conduction is involved in every thermal design or analysis, it is treated in this chapter. Designs involving heat transfer by convection and radiation are covered later in Chapter 9. The presentation emphasizes a relatively simple analytical approach because this procedure best illustrates the key design techniques in developing an effective cooling system for electronic hardware. A more complete but complex approach is possible if special codes are employed for more precise estimates of junction temperatures. A detailed treatment of several of these codes is presented in reference 1.

8.2 RELIABILITY

Digital systems usually contain a large number (10^4–10^{10}) of components, such as ICs, transistors, diodes, resistors, etc. Often many thousands of these components are connected together in a series circuit. The successful operation of this circuit depends on the reliability of each component in the circuit and with so many components connected in series the reliability of each component must be extremely high. Even with high reliability components, system failures occur and it is necessary to provide for fault location in designing digital logic circuits and the operating software. With fault location, the digital system is periodically monitored and failures are identified and located within the system almost as soon as they occur. The location of the malfunctioning part is extremely important because identifying the faulty circuit board or a small set of suspect circuit boards greatly reduces the time necessary to replace the failed board and to return the system to an operating condition.

Analog systems usually have a much smaller number of components and adequate availability can usually be achieved without the need for fault location procedures. However, very high individual component reliability is still required to provide relatively trouble-free operation in even simple analog circuits.

8.2.1 Failure Rate

The failure rate λ is the measure of the number of failures that will occur in a specified number of hours of operation for a given component. Because we are concerned with extremely low failure rates, 10^6 hr of continuous operation is the usual time considered in specifying λ. The failure rate varies during the life of a given component and may increase or decrease with time. The failure rate is also a strong function of temperature. This functional relation with temperature is often described with a thermal acceleration factor defined as

$$A(T) = \lambda(T)/\lambda(T_0) \tag{8.1}$$

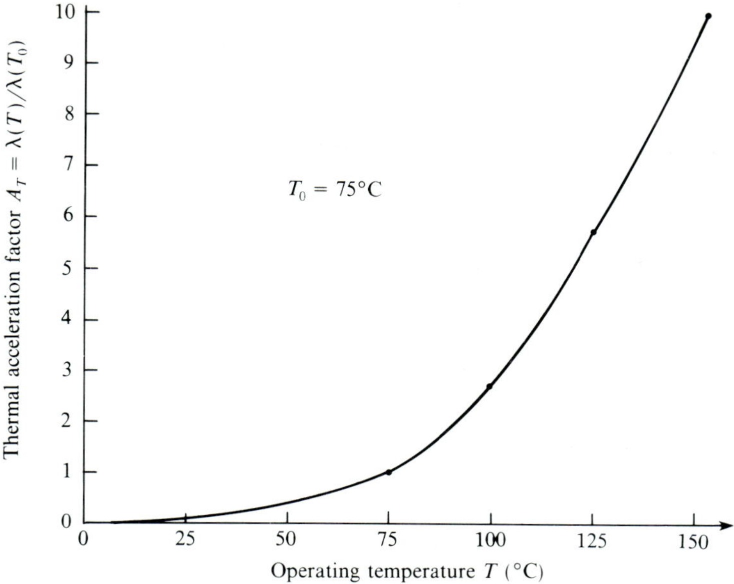

FIGURE 8.1
Increase in the failure rate of bipolar transistors as a function of operating temperature.

where T_0 is a reference temperature often taken as 25°C. The thermal acceleration factor $A(T)$, for a bipolar transistor, is presented as a function of temperature in Fig. 8.1. Failure rates for several types of components are given for two temperatures in Table 8.1. These data show the strong influence of temperature on failure and illustrate the importance of maintaining close control of the temperature of all of the components packaged in an electronic system.

TABLE 8.1
Number of failures after 1000 hr of operation per million units

	Test temperature (°C)	
Component	**25**	**75**
Thick film resistor	5	15
Chip capacitor	10	25
Power transistor	50	300
Diode	1	9
Logic ICs		
SSI	125	1125
MSI	250	2250
LSI	500	4500

Source: C. A. Harper, *Handbook of Thick Film Hybrid Microelectronics*, McGraw-Hill, New York, 1974.

Let us consider a simple test of components that is conducted to determine the failure rate λ of a given component with time. If a large number N_0 of components are placed in operation at time $t = 0$, then after some time a certain number N_f will have failed and the remainder N_s will have survived. Thus,

$$N_f(t) + N_s(t) = N_0 \tag{8.2}$$

The failure rate λ is determined from

$$\lambda(t) = N_f(t)/(N_0 t) \tag{8.3}$$

providing that each failed component is replaced immediately upon failure. Another measure of component failure rate is the mean time between failures (MTBF), which is

$$\text{MTBF} = 1/\lambda \tag{8.3a}$$

If the components that have failed during testing are not replaced during the test time t, then Eq. (8.3) is modified to account for the reduction in the number of component hours that have accumulated as shown by

$$\lambda(t) = N_f / \left[t_1 + t_2 + \cdots + (N_0 - N_f)t\right] \tag{8.4}$$

where t_1, t_2, etc., are the times associated with each failure.

The results of testing a large number of components and applying either Eq. (8.3) or (8.4) gives the well known bathtub mortality curve presented in Fig. 8.2. Examination of this curve shows three distinct regions of interest. During the initial period, $0 \leq t \leq t_1$, the failure rate is relatively high. Those components with small manufacturing flaws that escaped detection during inspection fail early in service. For $t_1 \leq t \leq t_2$, the failure rate λ_0 is relatively low and constant

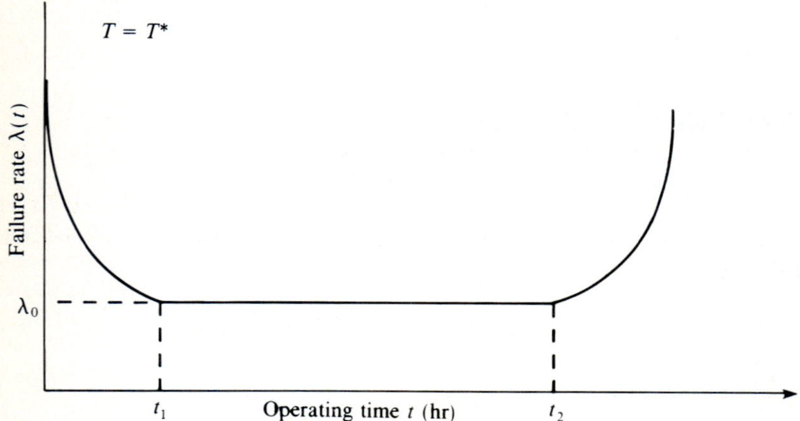

FIGURE 8.2
Bathtub-shaped mortality curve showing the change in failure rate for electronic components a function of time.

with time. This is the best period of operation from a reliability viewpoint. For $t > t_2$ the failure rate increases rapidly with time as the components fail due to the effects of aging.

For high reliability systems it is essential to eliminate the inherently defective components associated with the initial period of operation. This elimination is accomplished by subjecting the components to a burn-in period of simulated operation. Burn-in is the final phase of the manufacturing process prior to the real usage of the component. Assuming adequate burn-in procedures, $t_1 = 0$, the useful life of the component will be t_2. During this period of operation, $0 \leq t \leq t_2$, the failures will occur at a constant rate λ_0 at random locations in the system. The satisfactory operation of the system during this period depends on λ_0, the number of components in both series and in parallel circuits and the requirements for system availability. After t_2 hours of operation the failure rate λ increases rapidly, system reliability degrades and satisfactory performance cannot be anticipated.

8.2.2 Component Reliability

To illustrate component reliability consider a very simple circuit containing N_0 parallel components all exhibiting the same failure rate λ. In this case the component reliability is identical to the probability of survival of the individual components given by

$$P_s(t) = N_s(t)/N_0 \tag{8.5}$$

Note, that the probability of failure P_f is

$$P_f(t) = N_f(t)/N_0 \tag{8.6}$$

From Eq. (8.2) it is clear that

$$P_s(t) + P_f(t) = 1 \tag{8.7}$$

and

$$P_s(t) = 1 - [N_f(t)/N_0] \tag{8.8}$$

Differentiating Eq. (8.8) with respect to time gives

$$dP_s(t)/dt = (-1/N_0)[dN_f(t)/dt] \tag{8.9}$$

Equation (8.9) shows that the rate of change of the probability of component survival is related to the rate of component failure. Rearranging Eq. (8.9) and dividing both sides by $N_s(t)$ gives

$$[N_0/N_s(t)][dP_s(t)/dt] = [-1/N_s(t)][dN_f(t)/dt] \tag{8.10}$$

From Eq. (8.3) the component failure rate $\lambda(t)$ can be written as

$$\lambda(t) = [1/N_s(t)][dN_f(t)/dt] \tag{8.11}$$

and by Eqs. (8.5), (8.10) and (8.11) it is evident that

$$[dP_s(t)/P_s(t)] = -\lambda(t)\,dt \tag{8.12}$$

Integrating Eq. (8.12) yields

$$\ln[P_s(t)] = -\int \lambda(t)\,dt \tag{8.13}$$

or

$$P_s(t) = e^{-\int \lambda(t)\,dt} \tag{8.14}$$

For a constant failure rate with $\lambda(t) = \lambda_0$, consistent with Fig. 8.2, Eq. (8.14) reduces to

$$P_s(t) = e^{-\lambda_0 t} \tag{8.15}$$

Failure rates are usually quite low, of the order of 10^{-9} or 10^{-10}/hr. The rate depends on the scale of integration of the components considered as shown in Table 8.1. This fact indicates that component reliability is usually very high. For example, Eq. (8.15) predicts that a component with $\lambda_0 = 10^{-9}$ failures/hr can operate continuously for 10 yr with $P_s(87{,}600) = 0.9999124$.

8.2.3 System Reliability

The reliability of a system depends on the reliability of each component, the number of components and whether they are arranged in series or parallel. Consider first a series arrangement of four components as shown in Fig. 8.3, and note that the system fails in a series arrangement if any one component fails. It is clear that the reliability of the system is the same as the probability of all four components surviving for the specified time period. Thus,

$$P_s^s = P_{s1} P_{s2} P_{s3} P_{s4} \tag{8.16}$$

where the superscript s refers to the system of four components. Substitution of Eq. (8.15) into Eq. (8.16) and assuming λ constant with respect to time gives

$$P^s(t) = e^{-\lambda_1 t} e^{-\lambda_2 t} e^{-\lambda_3 t} e^{-\lambda_4 t} = e^{-(\lambda_1+\lambda_2+\lambda_3+\lambda_4)t} \tag{8.17}$$

If the number of components in the series arrangement is increased to N, then it is clear that

$$P_s^s(t) = e^{-\left[\sum_{i=1}^{N} \lambda_i\right] t} \tag{8.18}$$

The effect of a large series circuit is to degrade reliability. For example, a series circuit with N components all exhibiting the same probability of success is degraded in accordance with the number of components as indicated in Table 8.2. A system with 10^4 components in series requires component reliabilities in excess of 0.99999 to insure a system reliability of 90%.

Consider next a parallel arrangement of three redundant components, to define the system illustrated in Fig. 8.4. Failure of a parallel redundant system

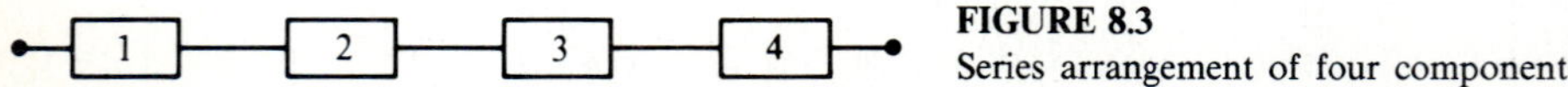

FIGURE 8.3
Series arrangement of four components.

TABLE 8.2
Probability of system success P_s^s as a function of component reliability and number of components in series. $P_s^s = (P_s)^N$

	P_s					
N	**0.900000**	**0.990000**	**0.999000**	**0.999900**	**0.999990**	**0.999999**
10^0	0.9000	0.9900	0.9990	0.9999	0.99999	0.999999
10^1	0.3487	0.9044	0.9900	0.9990	0.9999	0.99999
10^2	2.6×10^{-5}	0.3660	0.9048	0.9900	0.9990	0.9999
10^3	a	4.3×10^{-5}	0.3679	0.9048	0.9900	0.9990
10^4	a	a	4.5×10^{-5}	0.3679	0.9048	0.9900
10^5	a	a	a	4.5×10^{-5}	0.3679	0.9048
10^6	a	a	a	a	4.5×10^{-5}	0.3679

[a] Negligible.

requires failure of all of the components; thus,

$$P_f^s = P_{f1}P_{f2}P_{f3} \tag{8.19}$$

Substituting Eq. (8.7) into Eq. (8.19) gives

$$1 - P_s^s = (1 - P_{s1})(1 - P_{s2})(1 - P_{s3})$$

which reduces to

$$P_s^s = P_{s1} + P_{s2} + P_{s3} - P_{s1}P_{s2} - P_{s2}P_{s3} - P_{s1}P_{s3} + P_{s1}P_{s2}P_{s3} \tag{8.20}$$

Evaluation of Eq. (8.20) shows that the parallel arrangement of three components enhances system reliability. The extent of the improvement in system reliability P_s^s as a function of component or subsystem reliability P_s is shown in Fig. 8.5 for different redundancy $M = 1$, 2 and 3.

The enhanced reliability of a parallel circuit is a distinct advantage; however, the corresponding disadvantage is the added cost of the components and the added power, weight and size of the system. In the design of extremely reliable systems, parallel circuits that provide redundancy of two or three are often used. However, to minimize the added costs of the redundancy, parallel circuits are only used with subsystems that exhibit lower reliabilities. This practice leads to systems that incorporate combinations of both parallel and series circuits as shown in Fig. 8.6.

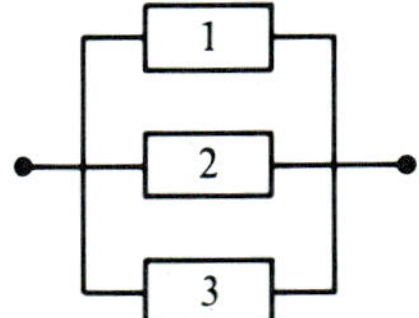

FIGURE 8.4
Parallel arrangement of three components.

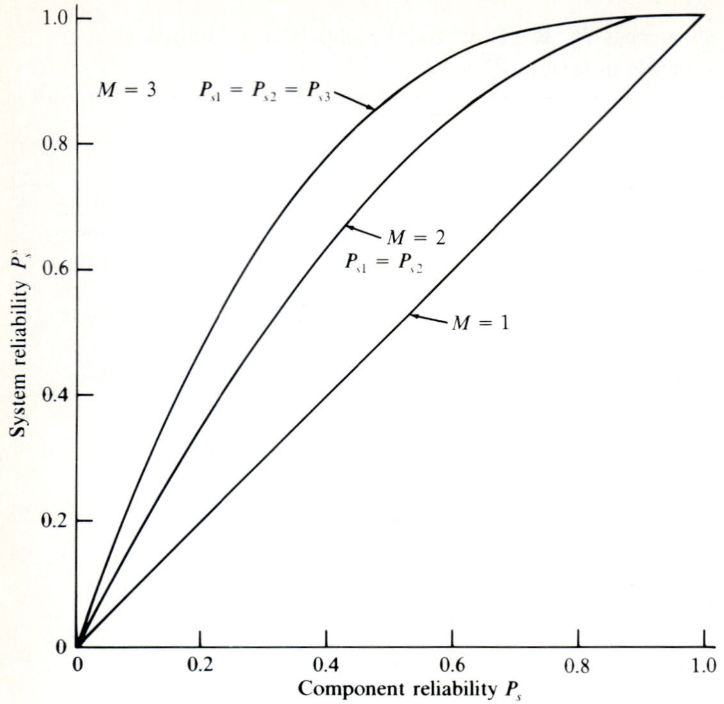

FIGURE 8.5
Improvement in system reliability by parallel connections of components.

Examine the series parallel arrangement shown in Fig. 8.6*a* and assume the reliability of each component or subsystem is the same value P_s. Then the probability of survival of the series line of N components may be written from Eq. (8.16) as

$$P_{s1} = P_s^N \tag{8.21a}$$

From Eq. (8.16) and (8.21a), it is clear that the probability of survival of M parallel lines is

$$1 - P_s^s = \left(1 - P_s^N\right)^M \tag{8.21b}$$

In this example (Fig. 8.6*a*), $N = 4$ and $M = 2$ and Eq. (8.21b) reduces to

$$P_s^s = P_s^4\left(2 - P_s^4\right) \tag{8.21c}$$

Taking $P_s = 0.9$, Eq. (8.21c) gives $P_s^s = 0.8817$. It is evident that the parallel portion of the circuit compensates for the series portion and the overall system reliability is about the same as the component reliability.

Next, consider the series parallel arrangement with a single cross strap as illustrated in Fig. 8.6*b*. In this case Eq. (8.21b) applies to both the left and the

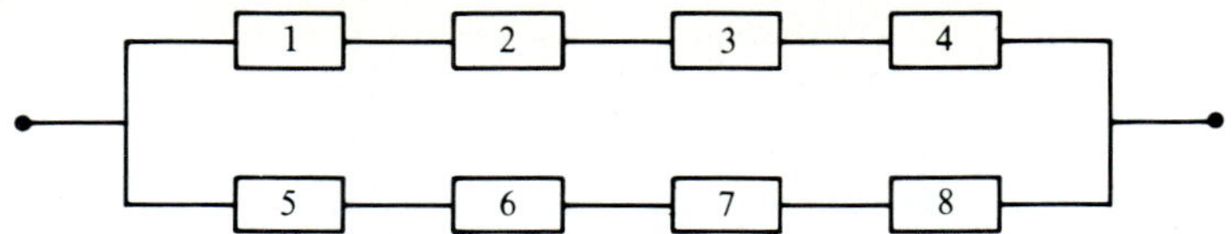

(*a*) Series parallel, no cross strapping

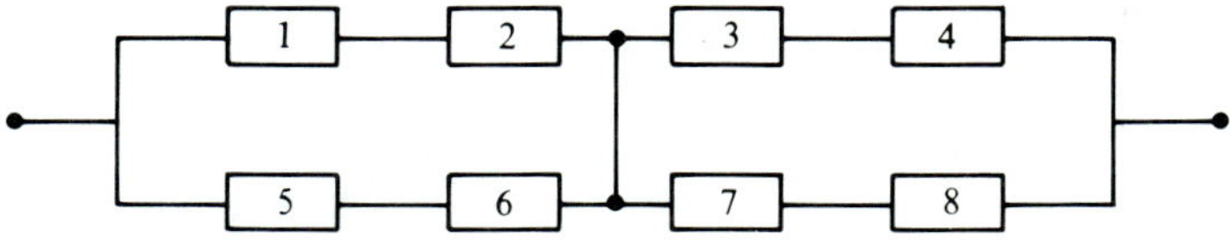

(*b*) Series parallel with one cross strap

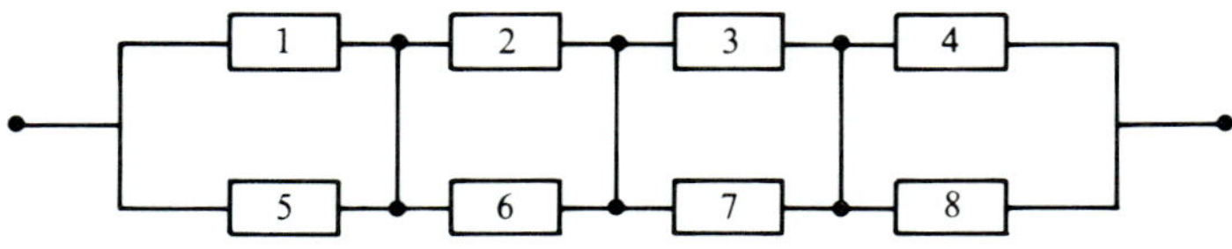

(*c*) Series parallel with maximum cross strapping

FIGURE 8.6
Series parallel circuits with different degrees of cross strapping.

right sides of the symmetric circuit and

$$P_{sL} = P_{sR} = 1 - \left(1 - P_s^N\right)^M \tag{8.22a}$$

Note, that the left and the right side of the circuit are series connected so that the system reliability is

$$P_s^s = P_{sL}P_{sR} = \left[1 - \left(1 - P_s^N\right)^M\right]^Q \tag{8.22b}$$

For the diagram shown in Fig. 8.6*b*, $N = M = Q = 2$ and Eq. (8.22b) reduces to

$$P_s^s = P_s^4\left(4 - 4P_s^2 + P_s^4\right) \tag{8.22c}$$

Again taking $P_s = 0.9$, Eq. (8.22c) gives $P_s^s = 0.9291$. This example indicates that the addition of a single cross strap has increased the reliability of the system from 0.8817 to 0.9291.

Finally, consider the case where the maximum number of cross straps are employed in the system configuration shown in Fig. 8.6*c*. With $N = 1$, $M = 2$ and $Q = 4$, Eq. (8.22b) reduces to

$$P_s^s = P_s^4(2 - P_s)^4 \tag{8.22d}$$

Continue the study of the effects of cross strapping by taking $P_s = 0.9$ and note that Eq. (8.22d) gives $P_s^s = 0.9606$. This example shows that maximum reliability is achieved when the cross strapping is used to the maximum extent possible.

These simple concepts in probability theory show the effects of series and parallel circuits on system reliability. It is evident that the series connections degrade reliability and that the parallel connections enhance reliability. Parallel or redundant circuits add to costs of the product, but in those cases where system availability is extremely critical, such as on-line banking transactions and airline reservations, the cost and inconvenience of a computer malfunction is so significant that redundancy in subsystems prone to failure is certainly warranted.

The importance of control of the component temperatures on system reliability cannot be overemphasized. The failure rate λ, which determines the component reliability P_s, is a function of temperature and a relatively small increase in temperature results in a large increase in failure rate and a subsequent loss in system availability.

8.3 STEADY STATE HEAT TRANSFER BY CONDUCTION

Heat is generated in electronic devices primarily by I^2R losses and must be dissipated to avoid excessive temperatures that can impair the reliability of the system. The amount of heat generated depends on the size and complexity of the system and can vary from a few milliwatts in a wristwatch to 10 kW or more in a large main frame computer. The design for dissipating the heat generated will depend on the product, the environment, the accessibility of a coolant, the reliability and the system availability requirements and the costs involved in providing an adequate cooling system. There is no single easy solution to the heat dissipation problem. Moreover, because the circuit densities increase with the scale of integration, the problem of reducing the thermal penalty ΔT, incurred in dissipating the heat, becomes more difficult.

Heat is transferred from the component (the source) to the environment (the sink) by four basic mechanisms, which include conduction, convection, radiation and boiling. Of these four mechanisms, conduction is the most important because it is always involved to some extent in transferring heat from the point of generation on the chip to some surface where another mechanism may be involved in the heat transfer process.

8.3.1 One-Dimensional Heat Transfer

Fourier's law for the rate of heat transfer by conduction is given by

$$q = -kA(dT/dx) \tag{8.23a}$$

where q is the rate of heat transfer (W)
A is the area normal to the path of heat flow (m^2)
dT/dx is the temperature gradient (°C/m)
k is the coefficient of thermal conductivity (W/m °C)

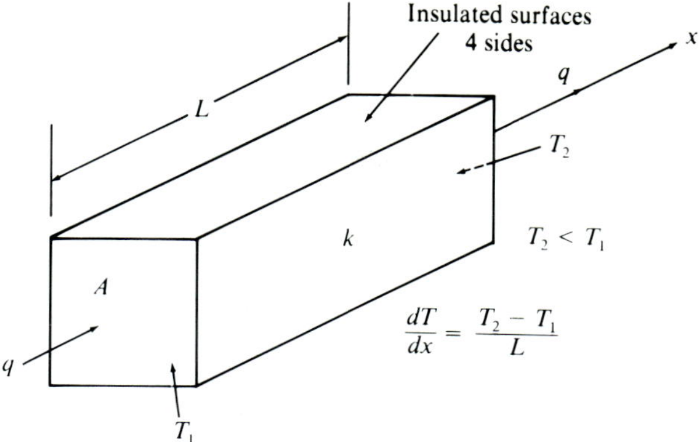

FIGURE 8.7
Schematic illustration of parameters affecting one-dimensional heat transfer.

Equation (8.23a) describes the transfer of heat at a constant rate in one dimension (along x) as illustrated in Fig. 8.7. The temperature gradient may be written as

$$dT/dx = (T_1 - T_2)/(0 - L) \tag{8.23b}$$

The gradient is a negative quantity since $T_1 > T_2$ and the heat flow is from the hot side to the cold side in the positive x direction. Substituting Eq. (8.23b) into Eq. (8.23a) gives

$$q = (kA/L)(T_1 - T_2) = K\,\Delta T \tag{8.24}$$

The term kA/L is the effective thermal conductance K of the one-dimensional conductor. The thermal conductance K is the reciprocal of the thermal resistance R_T given by

$$R_T = 1/K = L/kA \tag{8.25}$$

where R_T is expressed in terms of degrees Celsius per Watt.

It is possible to draw an analogy between heat flow and current flow by comparing Eq. (8.24) with Ohm's law Eq. (3.3) and equating corresponding terms:

$$I = E/R \qquad q = \Delta T/R_T \tag{8.26}$$

where I, the current, is the analog of q
E, the voltage, is the analog of $\Delta T = (T_1 - T_2)$
R, the resistance, is the analog of $R_T = L/kA$

The coefficient of thermal conductivity k is a material property. It is extremely important because of the very wide range of values that occur in materials commonly employed in packaging electronic systems. This fact is illustrated in Table 8.3, where the coefficient of thermal conductivity for these materials is shown. It is significant that k ranges over 5 orders of magnitude. To illustrate the importance of k, consider a conducting strip 2 mm thick, 10 mm wide and 100 mm long that is to conduct heat at the rate of 1 W. If the strip is

TABLE 8.3
Thermal conductivities of select materials

Material	k (W / m °C)
Aluminum	
Pure	126
2024 T4	121
6061 T6	156
7075 T6	121
Beryllium copper	82.7
Brass 70 Cu–30 Zn	100
Copper	
Pure	381
Drawn wire	287
Gold	296
Iron wrought	58.8
Kovar	15.6
Lead	32.7
Magnesium	157
Silicon	153
Steel 1020	55.4
Tin	62.3
Titanium	15.6
Zinc	102
Alumina	
95% pure	29.4
90%	12.1
Beryllia	
99.5%	242
95%	156
Glass	
Soft	0.98
Pyrex	1.26
Mica	0.59
Epoxy	
Unfilled	0.21
Filled	2.16
Fiberglass	0.26
Mylar	0.19
Nylon	0.24
Phenolic paper	0.28
Plexiglass	0.19
Polyvinyl chloride	0.16
Rubber	
Butyl	0.26
Silicone	0.19
Silicone grease	0.21
Teflon	0.19
Air	0.026
Water	0.658

fabricated from copper with $k = 381$ W/m °C and the sides of the conductor are insulated, then the temperature difference necessary to achieve the specified rate of heat transfer is given by Eq. (8.26) as

$$\Delta T = T_1 - T_2 = qR_T = (1 \times 0.1)/(381 \times 2 \times 10 \times 10^{-6}) = 13.1°\text{C}$$

However, if the strip is fabricated from alumina with $k = 29.4$ W/m °C, then the ΔT increases significantly since

$$\Delta T = 13.1(381/29.4) = 169.7°\text{C}$$

This simple example shows the marked influence of the coefficient of thermal conductivity on the temperature difference (a thermal penalty) necessary to transfer heat at a specified rate. The importance of the selection of materials in the design of packaging for electronic components and circuits cannot be overemphasized.

8.3.2 General Equations Governing Conduction

The previous section described a very important but specialized case of conduction dealing with steady state, one-dimensional heat flow with no heat generated in the body of the conductor. The more general case is transient, with temperatures changing relative to time, with heat flow in the three directions x, y and z and with internal heat generated at a rate of q_i throughout the volume of the conductor. A heat balance with a volume element of the conductor leads to the following classical differential equation that governs the most general solution:

$$\partial^2 T/\partial x^2 + \partial^2 T/\partial y^2 + \partial^2 T/\partial z^2 + q_i/k = (1/\alpha)\, \partial T/\partial t \qquad (8.27)$$

where q_i is the heating rate per unit volume (W/m^3).
t is time (s)
α is the thermal diffusivity (m^2/s) that is given by

$$\alpha = k/c\rho \qquad (8.28)$$

ρ is the density of the material (kg/m^3)
c is the specific heat of the material (J/kg °C)

In the absence of internal heating, with $q_i = 0$, Eq. (8.27) reduces to

$$\nabla^2 T = (1/\alpha)\, \partial T/\partial t \qquad (8.29)$$

where $\nabla = [\partial^2/\partial x^2 + \partial^2/\partial y^2 + \partial^2/\partial z^2]$ is the Laplace operator. Under steady state conditions, the temperatures have stabilized with respect to time ($\partial T/\partial t =$ 0) and Eq. (8.27) reduces to Poisson's equation

$$\nabla^2 T = -q_i/k \qquad (8.30)$$

Finally, with no internal heat generated in the conductor and steady state conditions, Eq. (8.27) simplifies to the classic Laplace equation

$$\nabla^2 T = 0 \qquad (8.31)$$

In most packaging applications, the geometry and boundary conditions for components, such as chip carriers, circuit boards and heat sinks, are complex. The complexities limit the availability of exact closed form solutions to the general equations. Because exact solutions are difficult, simplified models are developed that approximate the electronic assembly and permit reasonable estimates of the temperature of critical components. These simplified models also indicate design approaches to reduce ΔT for a specified rate of heating. Another method of analysis utilizes computer codes that give numerical solutions to problems with complex geometries. These codes based either on finite elements or finite differences are written to provide solutions for specific problems arising in electronic packaging. Reference 1 describes this computational approach and gives the source listing of two different programs.

The approach followed here is to emphasize the development of relatively simple models of complex electronic assemblies. The primary purpose of the analysis is to show design features and material selections that markedly affect temperatures and to give a first approximation of the temperature of the components.

8.3.3 Heat Flow in Cylindrical Packages

Consider the hollow cylinder shown in Fig. 8.8 and assume that the temperatures have achieved steady state and $q_i = 0$. In this case, the heat transfer is one dimensional, in the radial direction and from the inside to the outside wall. Modifying Eq. (8.23a) to account for the radial heat flow by letting $r = x$ and the area $A = 2\pi rL$ gives

$$q = -2\pi kLr(dT/dr)$$

Rearranging this equation gives

$$dT = -(q/2\pi kL)(dr/r) \tag{8.32}$$

Equation (8.32) can be integrated between the limits of r_1, r_2 and T_1, T_2 to give

$$\Delta T = T_1 - T_2 = q \ln(r_2/r_1)/(2\pi kL) \tag{8.33}$$

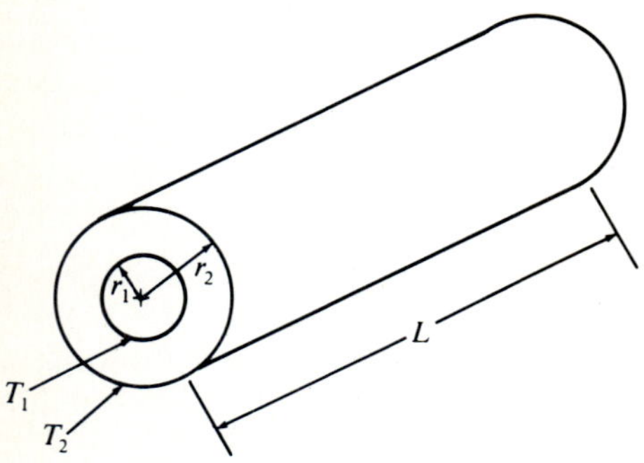

FIGURE 8.8
Conduction through a hollow cylinder.

For the hollow cylinder the thermal resistance is obtained from Eqs. (8.25) and (8.33) as

$$R_T = \ln(r_2/r_1)/(2\pi kL) \tag{8.34}$$

The results from Eqs. (8.33) and (8.34) are useful in determining the temperature difference ΔT required to conduct heat radially outward from cylindrical packages containing electronic circuits that act as heat sources.

8.3.4 Heat Flow through Layered Composites

Consider a composite made from four layers of different materials as shown in Fig. 8.9. Heat is transferred into the left side of the first layer where the surface temperature of the wall is T_i. The heat is transmitted in the x direction through the four layers with a ΔT developed so as to maintain a constant rate of heat transfer through the composite assembly. The problem is classified as one-dimensional steady state conduction with $q_i = 0$. Applying Eq. (8.24) to each of the layers gives

$$\begin{aligned} T_i - T_{1-2} &= qL_1/k_1A \\ T_{1-2} - T_{2-3} &= qL_2/k_2A \\ T_{2-3} - T_{3-4} &= qL_3/k_3A \\ T_{3-4} - T_0 &= qL_4/k_4A \end{aligned}$$

Adding these four equations yields

$$\Delta T = T_i - T_0 = q\sum_1^4 L_i/k_iA = q\sum_1^4 R_{Ti} \tag{8.35}$$

Examination of Eq. (8.35) shows that ΔT across a layered composite depends on

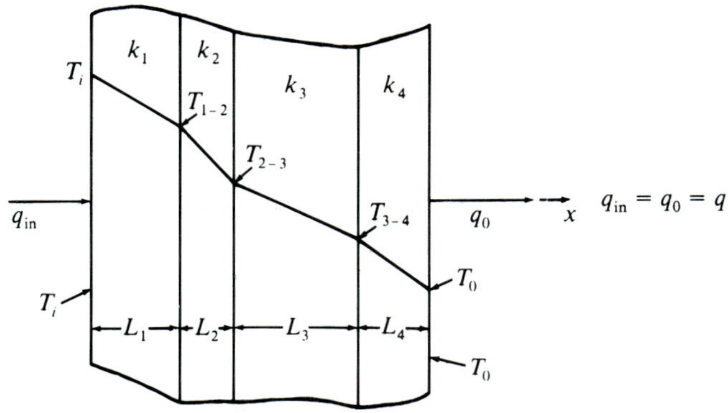

FIGURE 8.9
Conduction through a composite slab.

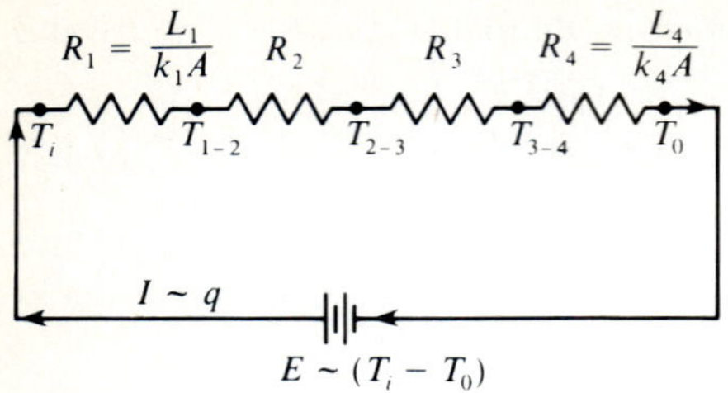

FIGURE 8.10
Electrical analogy for the composite slab.

the rate of heat transfer and the sum of the thermal resistance of each of the four layers. The electrical analogy corresponding to this result is presented in Fig. 8.10. The use of this approach to analyze heat conduction through composite layers will be described in Section 8.5.

8.3.5 Heat Transfer by Convection and Conduction

The examples illustrated in Figs. 8.7–8.9 indicated well defined temperatures at the input and output surfaces of the subject walls. This procedure is correct but not very realistic because the surface temperatures are not usually known. Instead, the temperature of the fluids adjacent to these surfaces is usually known. These fluid temperatures are different from the surface temperatures due to the development of a boundary layer at the surface. A more realistic representation of the temperature distribution, which accounts for the effect of the boundary layer between the fluid and the conducting surface, is presented in Fig. 8.11.

Heat transfer occurs between the fluid and the solid surfaces by convection, which is governed by the convection equation

$$q = hA(T_i - T_0) \tag{8.36}$$

where h is the convection coefficient (W/m^2 °C). The determination of h is

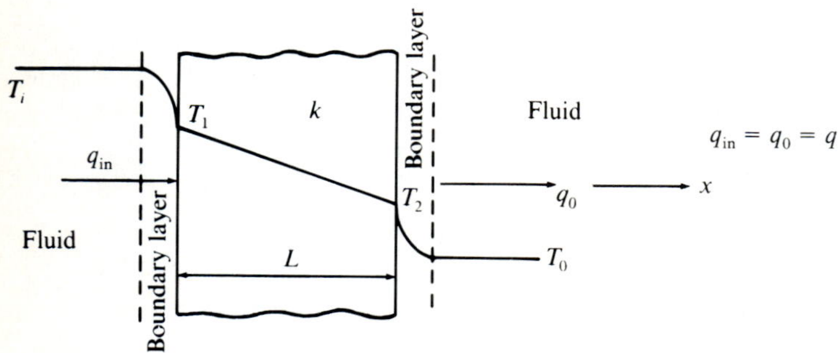

FIGURE 8.11
Heat transfer by convection and conduction.

involved and will be deferred until Chapter 9 where heat transfer by convection will be treated in considerable detail. At this point, consider h a known quantity and use Eqs. (8.24) and (8.36) to write the relations for the temperature differences across the boundary layers and the solid layer shown in Fig. 8.11 to obtain

$$T_i - T_1 = q/(h_i A)$$
$$T_1 - T_2 = qL/(kA)$$
$$T_2 - T_0 = q/(h_0 A)$$

Adding these equations gives

$$\Delta T = T_i - T_0 = (q/A)[(1/h_i) + (L/k) + (1/h_0)] \tag{8.37}$$

This result shows that the coefficients h_i, h_0 and k affect the rate of heat transfer for a specified ΔT. The coefficients can be combined by writing Eq. (8.24) in a modified form as

$$q = UA\,\Delta T \tag{8.38}$$

where U is the overall coefficient of heat transfer (W/m^2 °C):

$$U = 1/[(1/h_i) + (L/k) + (1/h_0)] \tag{8.39}$$

In terms of the electrical analog Eqs. (8.37), (8.38) and (8.39) indicate that

$$\Delta T = qR_T$$

where

$$R_T = (1/A)[(1/h_i) + (L/k) + (1/h_0)] \tag{8.40}$$

8.3.6 Contact Resistance

The previous discussions of heat conduction through composite slabs assumed that the layers were bonded together giving perfect thermal contact at the interfaces. The result of perfect thermal contact at the junction of two materials is to reduce the interface temperature difference to zero as indicated in Fig. 8.12*a*. However, when the two layers are not bonded together but are brought into simple contact, as illustrated in Fig. 8.12*b*, the fit is far from perfect. The surface roughness on both surfaces produces voids that inhibit the flow of heat across the interface and result in an addition thermal resistance R_c, which must be considered in the determination of R_T for a thermal path that includes an imperfect bond.

To show the effect of an imperfect bond, consider the heat flow through the composite body shown in Fig. 8.12*b*. Application of Eqs. (8.24) and (8.36) gives

$$T_i - T_1 = qL_1/k_1A$$
$$T_1 - T_2 = q/h_cA$$
$$T_2 - T_0 = qL_2/k_2A$$

Adding gives

$$\Delta T = T_i - T_0 = (q/A)[(L_1/k_1) + (1/h_c) + (L_2/k_2)] \tag{8.41}$$

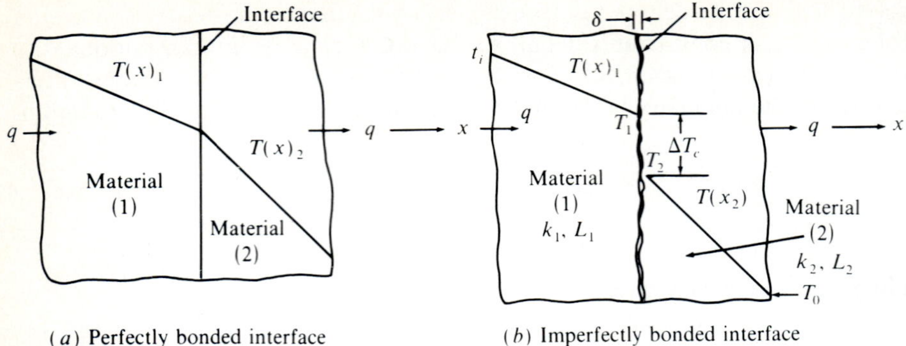

(*a*) Perfectly bonded interface (*b*) Imperfectly bonded interface

FIGURE 8.12
Examples of perfectly and imperfectly bonded interfaces showing temperature gradients for conduction.

where h_c is the contact coefficient (W/m² °C). The area A across the interface is not continuous and must be divided into the area A_c where contact occurs and the area A_v where voids exist. Thus

$$A = A_c + A_v \tag{8.42}$$

Heat flow across the interface gap δ is due to the conduction across the contact area A_c and conduction across the void area A_v. Take the interface gap δ and divide it between the two contacting surfaces as

$$\delta = \delta_1 + \delta_2 \tag{8.43}$$

where δ_1 and δ_2 are the surface roughness of materials 1 and 2. Considering parallel heat transfer paths through A_c and A_v gives

$$q = \frac{\Delta T_c}{\delta_1/(k_1 A_c) + \delta_2/(k_2 A_c)} + \frac{\Delta T_c}{\delta/(k_v A_v)} = h_c A \,\Delta T \tag{8.44}$$

where k_v is the thermal conductivity of the fluid that fills the voids. Solving Eq. (8.44) for h_c gives

$$h_c = (A_c/A)/[(\delta_1/k_1) + (\delta_2/k_2)] + (A_v/A)/(\delta/k_v) \tag{8.45}$$

The first term in Eq. (8.45) is due to conduction through the contact area and the second term is due to conduction through the fluid contained in the voids. While this equation is important in conceptual understanding of conduction across an interface, it is not usually employed to compute h_c because it is not possible to determine δ, δ_1, δ_2, A_c and A_v with sufficient accuracy to adequately describe the surfaces in contact.

A more pragmatic approach utilizes an empirical equation such as the one advanced by Tien:

$$h_c = 0.55m(k/\sigma)(p/H)^{0.85} \tag{8.46}$$

where

$$k = 2k_1k_2/(k_1 + k_2)$$

$$\sigma^2 = \sigma_1^2 + \sigma_2^2$$

σ is the rms value of the surface roughness (μm)
m is the rms value of the slope of the contacting asperities
p is the contact pressure (N/m^2)
H is the hardness of the softer of the two materials in contact, usually taken as three times the yield strength (N/m^2)

Experimental results showing h_c as a function of contact pressure are presented in Fig. 8.13 for aluminum-to-aluminum and steel-to-steel interfaces. The effect of surface finish is extremely important. Improving the surface finish from 120 to 10 μin. rms results in a threefold increase in h_c at $p = 100$ psi (0.69 MPa).

Several techniques can be used to improve the thermal resistance across an interface. The first is to plate a coating of a soft material, such as tin or zinc, on the contact area. This reduces the hardness H and it also concentrates the coating in the low portions of a surface to reduce the roughness σ and the slope m. Both of these factors improve h_c. The second technique is to use a thin layer of thermally conducting paste to fill the voids at the interface. The value of k for silicone grease is 0.208 W/m °C, which is much higher than that of air (0.0265 W/m °C). Reductions in thermal resistance of at least 2–3 can be achieved by filling the voids with paste. A third method involves increasing the pressure p by using wedge lock-type clamping.

The effect of interfaces that are not thermally bonded and that are intended for use in space applications is even more significant since hard vacuum is encountered in space. The air in the voids is lost and the value of h_c increases significantly. In these applications, the use of nonvolatile materials to fill the voids is even more important as indicated in Table 8.4.

8.3.7 Conduction from Small Area Heat Sources

The heat source in many electronic devices is often small in area as compared to the area of the channel used to conduct the heat from the source to the sink. The exact solution for problems of this type involves a three-dimensional analysis with a significant increase in complexity. Approximate solutions are possible, by retaining the one-dimensional form of the analysis, if an additional thermal resistance R_{Ts} is added to R_T to account for the effect of the relatively small area of the heat source. With this approach the temperature difference may be written as

$$\Delta T = q(R_{Ts} + R_T) \tag{8.47}$$

where R_{Ts} is the thermal resistance due to the constriction
R_T is the resistance of the conducting body

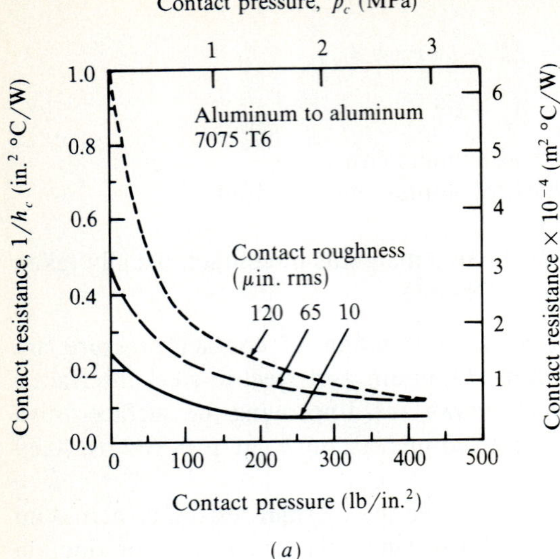

(a)

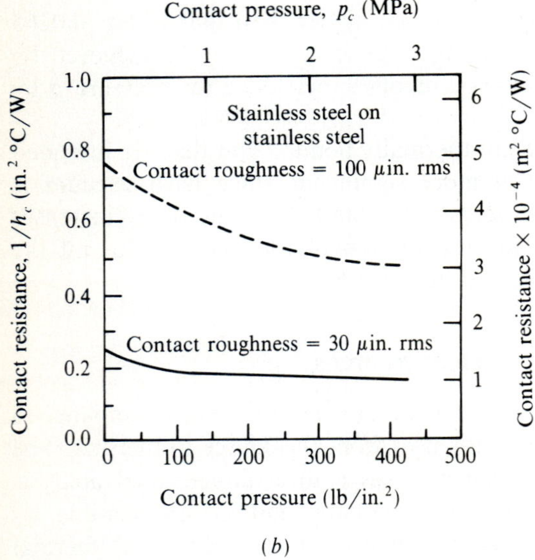

(b)

FIGURE 8.13
Heat transfer coefficient h_c as a function of pressure across the interface.

TABLE 8.4
Contact coefficient h_c for large aluminum plates[a] bolted at the corners

Interface material	h_c (W / m² °C) Air	Vacuum
0.05 mm aluminum foil	1703	568
0.5 mm rubber sheet	1135	227
0.12 mm beryllium copper foil	1135	227
No interface material inserted	1135	113

Source: Reference 13.
[a] Plate size 250 × 500 mm.

To illustrate this approach consider a heat source placed on the surface of a half-space as illustrated in Fig. 8.14. For the circular source, shown in Fig. 8.14*a*, with a uniform distribution of q, the thermal resistance R_{Ts} is

$$R_{Ts} = 16/(3\pi^2 Dk) = 0.54/Dk \tag{8.48}$$

For the source shown in Fig. 8.14*b*,

$$R_{Ts} = 0.55/Lk \tag{8.49}$$

If the conducting body is finite, the determination of the thermal resistance R_{Ts} is more difficult to estimate because the details of the geometry of the body must be taken into account. For a circular source on the cylindrical conductor, shown in Fig. 8.15, the thermal resistance due to the constrictive effect is

$$R_{Ts} = [1/(2\sqrt{\pi}\,ak)][1 - (a/b)]^{3/2} \tag{8.50}$$

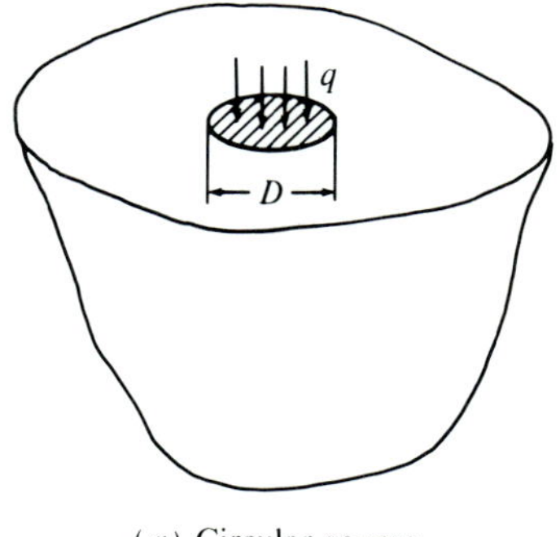

(*a*) Circular source

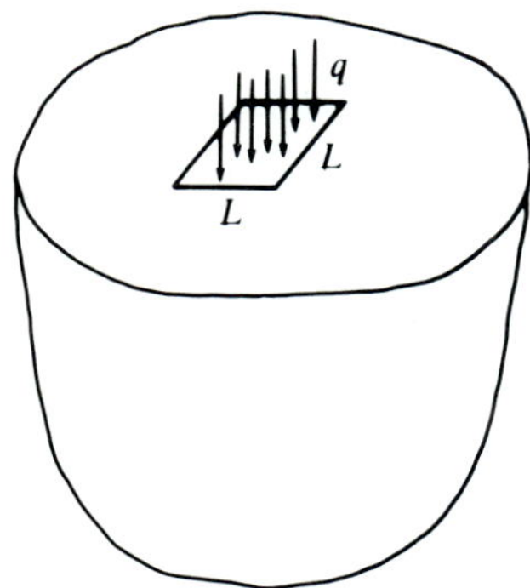

(*b*) Square source

FIGURE 8.14
Small area heat sources on a half-space.

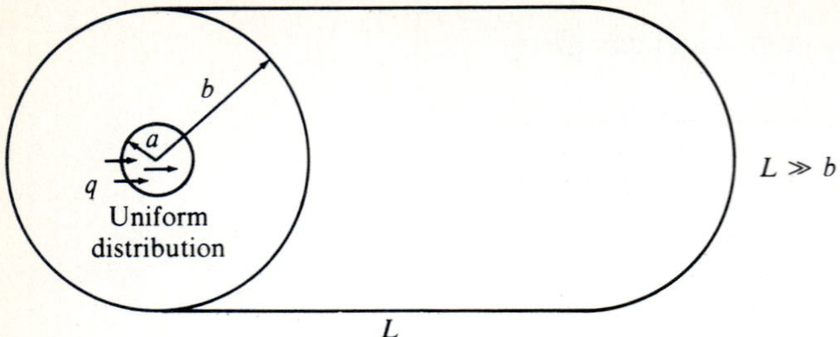

FIGURE 8.15
Uniform heat input over a small circular area with a circular conductor of heat.

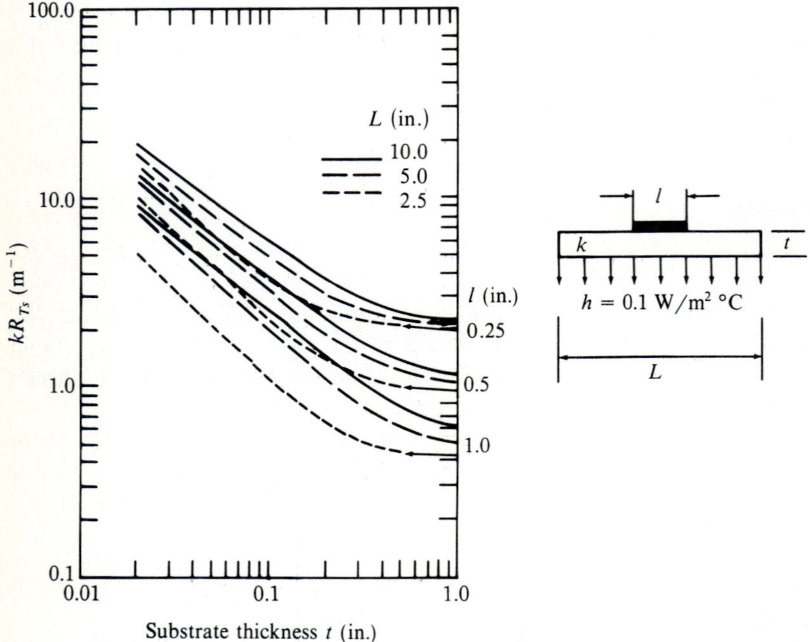

FIGURE 8.16
Thermal resistance for a square heat source $l \times l$ on a square conductor $L \times L$ with a finite thickness. (*After Ellison, reference 1.*)

The thermal resistance for a small square source on a square substrate with finite dimensions has been determined by Ellison as shown in Fig. 8.16.

8.4 TRANSIENT CONDUCTION

Transient thermal conditions occur during the start-up and shutdown of an electronic system. Transient behavior is important because it controls the time required for warm-up and thermal equilibrium of the system. After warm-up the temperature reaches a maximum and steady state conditions prevail. The maximum temperature controls the reliability of the various electronic devices as discussed in Section 8.2. Also, power-on and power-off cycling is very important in determining the reliability of mechanical components, such as connectors and solder joints. These joints may fail due to thermal strains induced by differential expansions and contractions that occur during warm-up and cool-down.

To illustrate the concept of warm-up time in a conduction cooled system, consider a device that generates heat at a rate q_g mounted on a totally insulated assembly with a weight w. The temperatures of the device and the component are at the ambient temperature T_0 at time $t = 0$ when the system power is turned on. Then Eq. (8.27) leads to

$$q_i = (k/\alpha)\, dT/dt$$

Note, that $q_i = q_g/V$ and that $w = \rho V$ where V is the volume of the assembly. Substituting the relations into the preceding expression gives

$$q_g = (wc)\, dT/dt \tag{8.51}$$

Integrating Eq. (8.51) gives

$$T - T_0 = (q_g/wc)t \tag{8.52}$$

Equation (8.52) predicts a linear increase in temperature with time until failure occurs due to excessively high temperature. This situation is obviously not satisfactory and occurs only when the system is insulated with no heat transfer from the component to the environment. The results show, in a very simple manner, the importance of providing a method of heat transfer from the system to the environment.

If a cooling mechanism is introduced into the system, the maximum temperature is bounded. We note that the thermal cooling mechanism exhibits a thermal resistance R_T, and it is necessary to modify Eq. (8.51) to account for the heat flow to the environment as indicated by

$$q_g - (T - T_0)/R_T = (wc)\, dT/dt$$

Rearranging this equation to show the differential equation in standard form yields

$$dT/dt + T/(wcR_T) = (1/wc)\left[q_g + (T_0/R_T)\right] \tag{8.53}$$

The homogeneous solution to Eq. (8.53) is

$$T_c = Ce^{-(t/wcR_T)} \tag{8.54}$$

and the particular solution that gives the steady state value of T is

$$T_p = q_g R_T + T_0 \tag{8.55}$$

Combining Eqs. (8.54) and (8.55) yields the general solution as

$$T = T_c + T_p = Ce^{-(t/wcR_T)} + q_g R_T + T_0 \tag{8.56}$$

The integration constant C is determined to satisfy the initial condition that

$T = T_0$ at $t = 0$, which leads to

$$C = -q_g R_T$$

Substituting this value of C into Eq. (8.56) gives

$$T = q_g R_T[1 - e^{-(t/wcR_T)}] + T_0 \tag{8.57}$$

Characteristic results for Eq. (8.57), illustrated in Fig. 8.17, show the temperature of the system as a function of time after the device power is turned on. The temperature increases exponentially with time and reaches a steady state value T_p given by Eq. (8.55) after a relatively long period of time. It is customary to take the warm-up time equivalent to three time constants (wcR_T is the time constant). At this time T has achieved 95% of the steady state value. The warm-up time t_w is then approximated by

$$t_w \simeq 3wcR_T \tag{8.58}$$

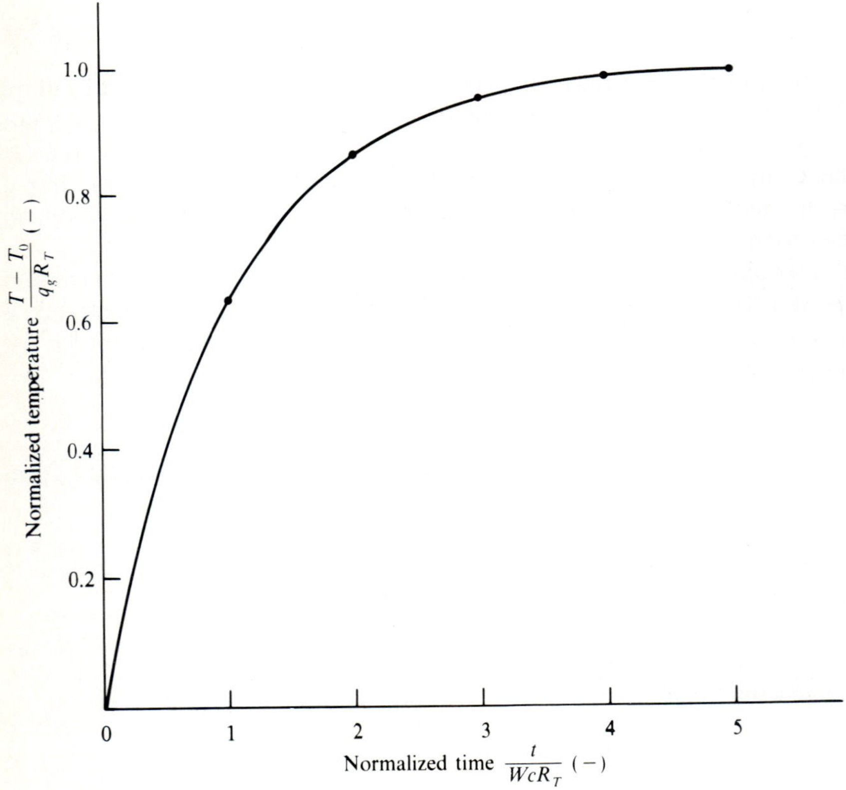

FIGURE 8.17
Normalized temperature as a function of normalized time after power is activated.

To determine the cool-down time after power off, let $q_g = 0$ in Eq. (8.53) to obtain

$$dT/dt + T/(wcR_T) = T_0/wcR_T \tag{8.59}$$

The homogeneous solution to Eq. (8.59) is the same as that given by Eq. (8.54) and the particular solution changes to $T_p = T_0$. The general solution for cool-down is then

$$T = Ce^{-(t/wcR_T)} + T_0 \tag{8.60}$$

The integration constant C is determined from the initial conditions for cool-down, which are $T = T_s$ at $t = 0$ where T_s is the steady state temperature given by Eq. (8.55). Substituting this initial condition into Eq. (8.60) gives $C = T_s - T_0$. It is now possible to recast Eq. (8.60) as

$$(T - T_0)/(T_s - T_0) = e^{-(t/wcR_T)} \tag{8.61}$$

This result indicates that the temperature of the system will decrease from T_s, when powered down, with an exponential decay to nearly the ambient temperature T_0, in three time constants. It is apparent that warm-up and cool-down times are identical. The time to achieve steady state conditions is dependent on the weight of the assembly, the effective specific heat of the materials employed and the thermal resistance from the source to the sink.

8.5 HEAT TRANSFER FROM CHIP TO CHIP CARRIER

An approximate solution for the temperature difference ΔT between the chip and the chip carrier is possible by using the simple methods of conduction described in Section 8.3. To illustrate this method of solution consider a power chip dissipating 0.5 W packaged in a dual in-line package (DIP) as illustrated in Fig. 8.18.

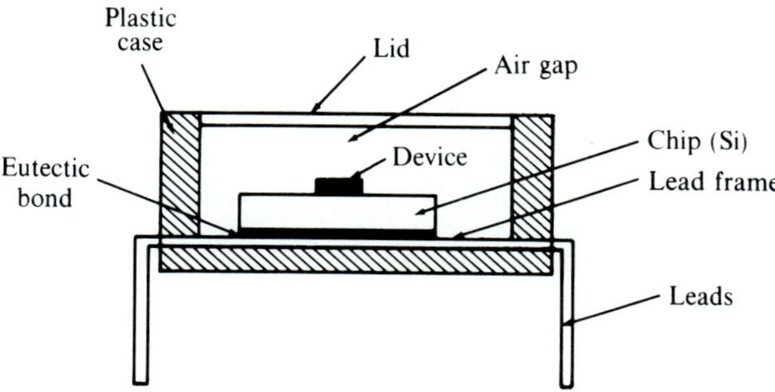

FIGURE 8.18
Conduction from a device packaged in a DIP.

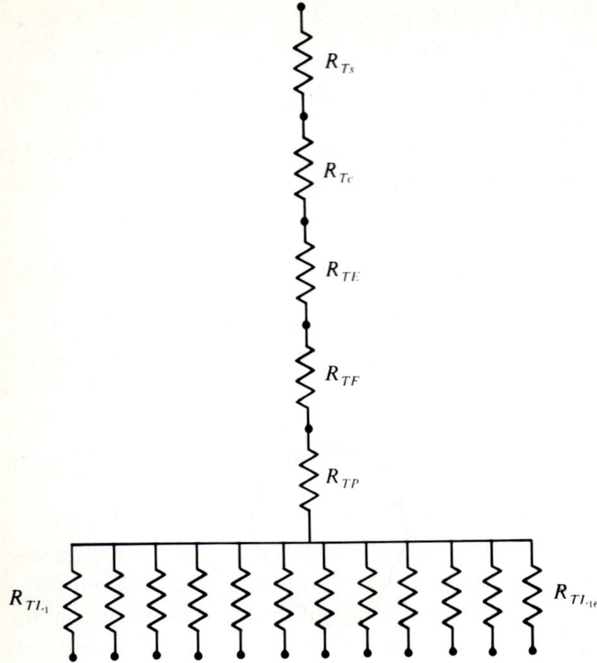

FIGURE 8.19
Resistance network representing the DIP packaged chip.

The first step is to model the package with a resistance network that represents the individual thermal constraints associated with each design feature involved in the heat flow path. The resistor network representing the DIP package is presented in Fig. 8.19. In constructing the resistor network assumptions are made concerning the predominate heat paths and the heat transfer mechanisms involved. In this case, it was assumed that the heat generated by the device flowed down through the chip, the eutectic bond, the lead frame and through the plastic insulation between the lead frame to the leads. The heat is then conducted outward along the leads to the outside of the package where it is dissipated by convection to air flowing along the surface of the PCB. In this example, convection from the leads to the air will not be considered because our treatment of convection as a heat transfer mechanism will be deferred until Chapter 9. Instead, the temperature of the leads will be taken as a constant T_L at the point where they exit the DIP body. It was also assumed that the heat transferred through the lid and the base of the DIP was negligible. The heat flow to and then through the lid is extremely small since the thermal resistance associated with heat flow across the evacuated space between the chip and the lid is very high. Heat transfer through the base is also small because of the low thermal conductivity of the plastic case and the presence of a layer of stagnant air between the base of the DIP and the printed circuit board.

The second step is to determine the individual resistances in the network by using the geometry of each design feature and the thermal properties of each material involved. Begin at the top of the resistance network with the first resistor R_{Ts}. This thermal resistance, due to the spreading of heat from the device to the chip, is estimated from Eq. (8.50) as

$$\begin{aligned} R_{Ts} &= \left[1/(2\sqrt{\pi}\,ak)\right][1 - (a/b)]^{3/2} \\ &= \left[1/(2\sqrt{\pi} \times 0.0005 \times 154)\right][1 - (0.5/2)]^{3/2} \\ &= 2.38°\text{C/W} \end{aligned}$$

where we have selected the following quantities:

$a = 0.5$ mm is the radius of the circular area of the device
$b = 2$ mm is the radius of an inscribed circle on a 4×4 mm chip
$k = 154$ W/m °C for silicon

The resistance of the chip, eutectic bond and the lead frame are determined from $R = L/(kA)$ as follows:

For the silicon chip:

$$R_{Tc} = 0.508 \times 10^{-3}/(154 \times 16 \times 10^{-6}) = 0.21°\text{C/W}$$

where $L = 0.508$ mm is the chip thickness
$A = 16\ \text{mm}^2$ is the chip area

For the eutectic bond:

$$R_{TE} = 0.05 \times 10^{-3}/(296 \times 16 \times 10^{-6}) = 0.01°\text{C/W}$$

where $L = 0.05$ mm is the bond thickness
$k = 296$ W/m °C is the coefficient of thermal conductivity for the AuGe solder

For the lead frame:

$$R_{TF} = 0.25 \times 10^{-3}/(381 \times 16 \times 10^{-6}) = 0.08°\text{C/W}$$

where $L = 0.25$ mm is the frame thickness
$k = 381$ W/m °C is the coefficient for copper

The thermal resistance from the base of the lead frame to the leads can only be approximated because of the complexity of the geometry as indicated in Fig. 8.20. Several simplifications are made in determining the thermal resistance R_{TP}:

$$R_{TP} = 0.2 \times 10^{-3}/(1 \times 4 \times 10^{-6}) = 50°\text{C/W}$$

where $L = 0.2$ mm the average space between the base and the leads
$k = 1$ W/m °C for the plastic insulation filling this space
$A = 4\ \text{mm}^2$ for all 16 leads each 0.25 mm thick and 1 mm wide

This thermal resistance is extremely high and, while the calculation is approximate, it does indicate the difficulty in dissipating heat in a molded plastic chip carrier.

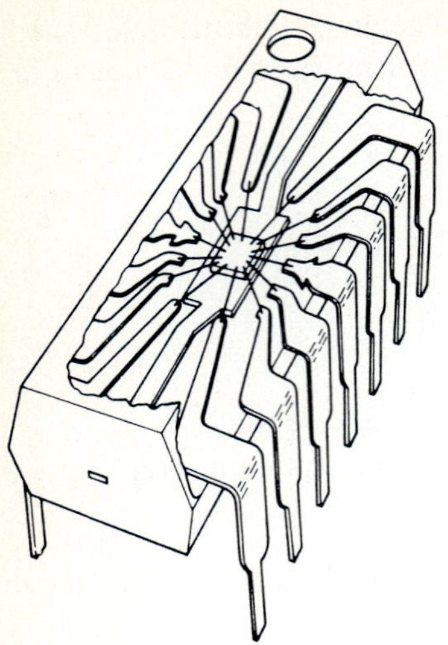

FIGURE 8.20
Internal geometry of a DIP showing spacing of the lead frame and the leads.

The final thermal resistance to be determined is for the individual leads from the heat frame to the exterior of the DIP:

$$R_{TL} = 6 \times 10^{-3}/(381 \times 0.25 \times 10^{-6}) = 63°\text{C/W}$$

where $L = 6$ mm is the average lead length. The 16 individual leads are equivalent to 16 resistances in parallel that can be combined to give

$$R_{TLe} = R_{TL}/N = 63/16 = 3.94°\text{C/W}$$

Summing the resistances in the network to obtain the total resistance $R_T = 56.62°\text{C/W}$ is shown in Table 8.5. For a chip dissipating 0.5 W housed in this plastic DIP-type chip carrier, the ΔT between the junction and the case is determined from Eq. (8.25) as

$$\Delta T = qR_T = 0.5 \times 56.62 = 28.31°\text{C}$$

TABLE 8.5
Listing of resistances for the various components in the heat path for a plastic DIP

Component	Symbol	Resistance (°C / W)
Spreading	R_{Ts}	2.38
Chip	R_{Tc}	0.21
Eutectic bond	R_{TE}	0.01
Lead frame	R_{TF}	0.08
Plastic	R_{Tp}	50.0
Leads (effective)	R_{TLe}	3.94
Total	R_T	56.62

These results show the dominance of the frame to lead resistance R_{TP} in the thermal resistance of the DIP. The separation between the frame and the leads is necessary to electrically isolate the individual leads; however, filling this separation with a material having a very low coefficient of thermal conductivity significantly degrades the ability of this chip carrier to house higher powered chips. A much better approach to the design of a chip carrier is to separate the electrical and thermal requirements by using one side of the chip for the electrical connections and one side of the chip for a heat transfer path. In the design of the DIP, the placement of the lead frame on the bottom side of the chip and the bonding pads on the top of the chip effectively committed both surfaces of the chip to electrical requirements. This placement eliminated the possibility of developing a low resistance path for heat flow by surrounding the heat source with thermally insulating materials.

8.5.1 Commercial Description of Heat Transfer in Chip Carriers

Most manufacturers of ICs package the chip in appropriate carriers and specify the heat transfer characteristics of their packages in terms of a thermal resistance termed θ_{j-a}. A typical curve showing θ_{j-a} as a function of air velocity over a single package mounted on a circuit board is shown in Fig. 8.21. The term θ_{j-a} is the total thermal resistance, which refers to the temperature difference between the junction temperature T_j of the electronic device and the ambient temperature

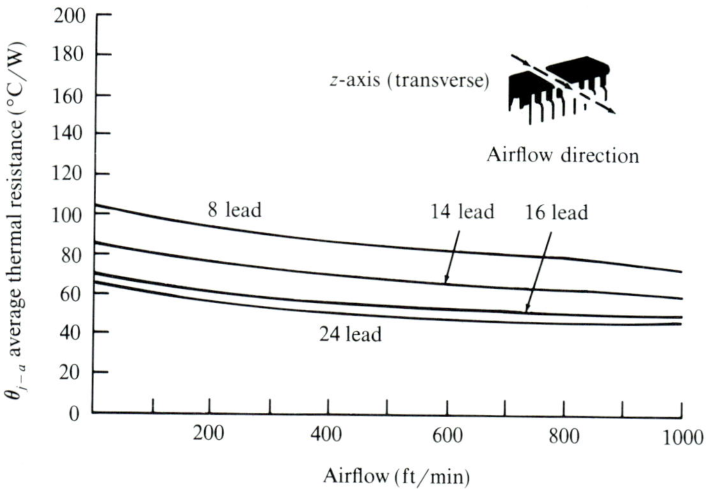

FIGURE 8.21
Thermal resistance θ_{j-a} (junction to air) for a plastic DIP as a function of air flow. (*Courtesy of Motorola Semiconductor Products, Inc.*)

T_a of the air flowing over the chip carrier. This temperature difference is

$$T_j - T_a = q\theta_{j-a} \tag{8.62}$$

which is similar to Eq. (8.25). The thermal resistance θ_{j-a} incorporates two different thermal resistances. The first resistance is due to conduction from the chip to the package as described in Section 8.5. This part of θ_{j-a} involves the detailed design of the chip carrier. The second part of the resistance is due to convection and the formation of a boundary layer between the case and its leads and the air. Thus

$$\theta_{j-a} = \theta_{j-c} + \theta_{c-a} \tag{8.63}$$

where θ_{j-c} is the thermal resistance of the junction to the case
θ_{c-a} is the resistance of the case/leads to the ambient air

The term θ_{j-c} is equivalent to the thermal resistance R_T that was determined in

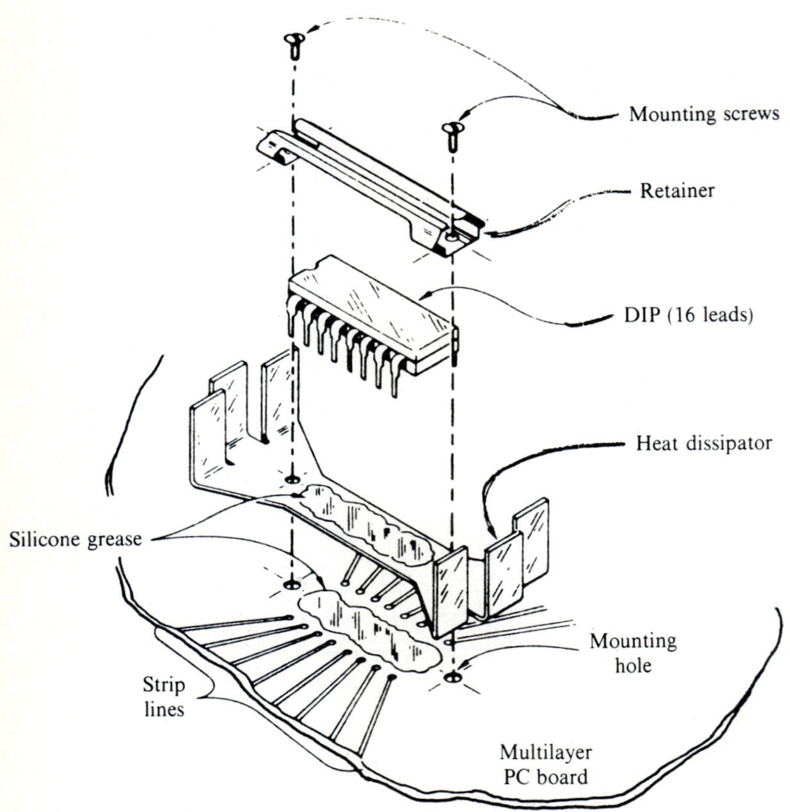

FIGURE 8.22
Method of enhancing heat dissipation from a DIP by utilizing conduction through the base and convection from an added heat dissipator. (*Courtesy of Motorola Semiconductor Products, Inc.*)

the example presented in Section 8.5. The term θ_{c-a} will be treate Chapter 9.

In some products cooling by passing the air directly over the chip not possible. Indeed, in some military systems direct air flow over the devices is prohibited. In these instances it is necessary to transfer the heat by conduction from the package, through the circuit board and along a heat frame to the air cooled sink. Contact with the circuit board enhances the heat transfer from the chip carrier to the circuit board. The DIP, which is designed to stand off the circuit board, may require thermal enhancement. The enhancement, illustrated in Fig. 8.22, which reduces the thermal resistance from the package to the board, is often used with higher powered chips. Another method for improving the thermal path is to fill any air gap with some type of heat conducting paste. Note in Fig. 8.22 that silicone grease is used at the two interfaces to reduce the contact resistance between the parts in the path of heat flow.

8.6 CONDUCTION IN CIRCUIT CARDS

In most conduction cooled systems the heat flows from the chip through the chip carrier to the circuit board, where it must be conducted to the next element in the heat dissipation system. Circuit boards are manufactured from electrically insulating materials such as glass–epoxy laminates and are not good conductors of heat. The thermal penalties involved in including the circuit card in the path of heat flow are considerable as will be demonstrated in the following subsection.

8.6.1 Conduction in the Plane of the Circuit Board

To show the poor performance of a conducting system containing a section of circuit board, take the example illustrated in Fig. 8.23. In this case we attempt to

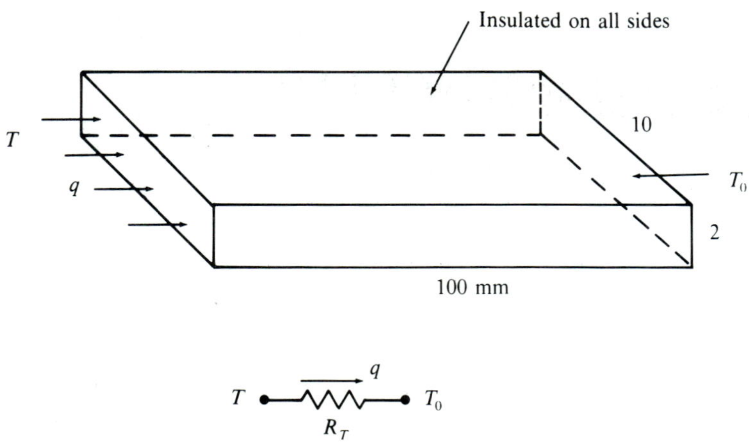

FIGURE 8.23
Heat conduction along the length of a circuit board.

conduct heat in the plane of the circuit board. The resistance of the conducting element of the circuit board is

$$R_T = L/kA = 100 \times 10^{-3}/(0.26 \times 2 \times 10 \times 10^{-6}) = 19{,}230°\text{C/W}$$

This is clearly an extremely high value of thermal resistance and the example illustrates that it is not possible to conduct heat in the plane of the PCB for any significant distance without incurring an excessively high thermal penalty.

The design of a multilayer circuit board (MLB) incorporates several ground and power planes embedded through the thickness as illustrated in Fig. 8.24. To determine if these copper planes improve the thermal resistance, consider the same conducting geometry as shown in Fig. 8.23 except for the inclusion of four ground and power planes, each fabricated from 2 oz copper cladding. The resistance network that provides the thermal model for this conductor is given in Fig. 8.24*b*. The first resistance network contains nine resistors, four for the copper planes and five for the intermediate glass–epoxy layers. This network can be simplified by neglecting the small amount of heat that will flow through the glass–epoxy layers. The simplified resistance network consists of four parallel resistances R_{cu} given by

$$\begin{aligned} R_{cu} &= \left[100 \times 10^{-3}/(381 \times 0.0027 \times 25.4 \times 10 \times 10^{-6})\right]\left[1/0.717\right] \\ &= 533.8°\text{C/W} \end{aligned}$$

Note that the term [1/0.717] is a correction to account for the loss of 28.3% of the area due to perforations in the ground and power planes, illustrated in Fig. 8.24*c*. These perforations are 60 mil in diameter on 100 mi centers and require removal of 28.3% of the copper. They serve to electrically isolate the power planes from the plated through holes.

Using the rule for parallel resistance gives the total thermal resistance for the MLB containing four copper planes as

$$R_T = 533.8/4 = 133.5°\text{C/W}$$

This result shows that the copper planes markedly improved conduction in the plane of the circuit board, but the thermal resistance is still much too high for planar conduction to be seriously considered in the design of a heat dissipation system.

8.6.2 Conduction through the Thickness of a Circuit Board

Consider the segment of circuit board shown in Fig. 8.25 where the heat q is transmitted through the thickness of the board. In this case the thermal resistance is

$$R_T = L/kA = 1 \times 10^{-3}/(0.26 \times 25 \times 25 \times 10^{-6}) = 6.15°\text{C/W}$$

This is a much lower thermal resistance than those described previously. It is clear that if the circuit board must be in the path of heat flow, it should be orientated so that the flow is through the thickness. Also, the thermal resistance

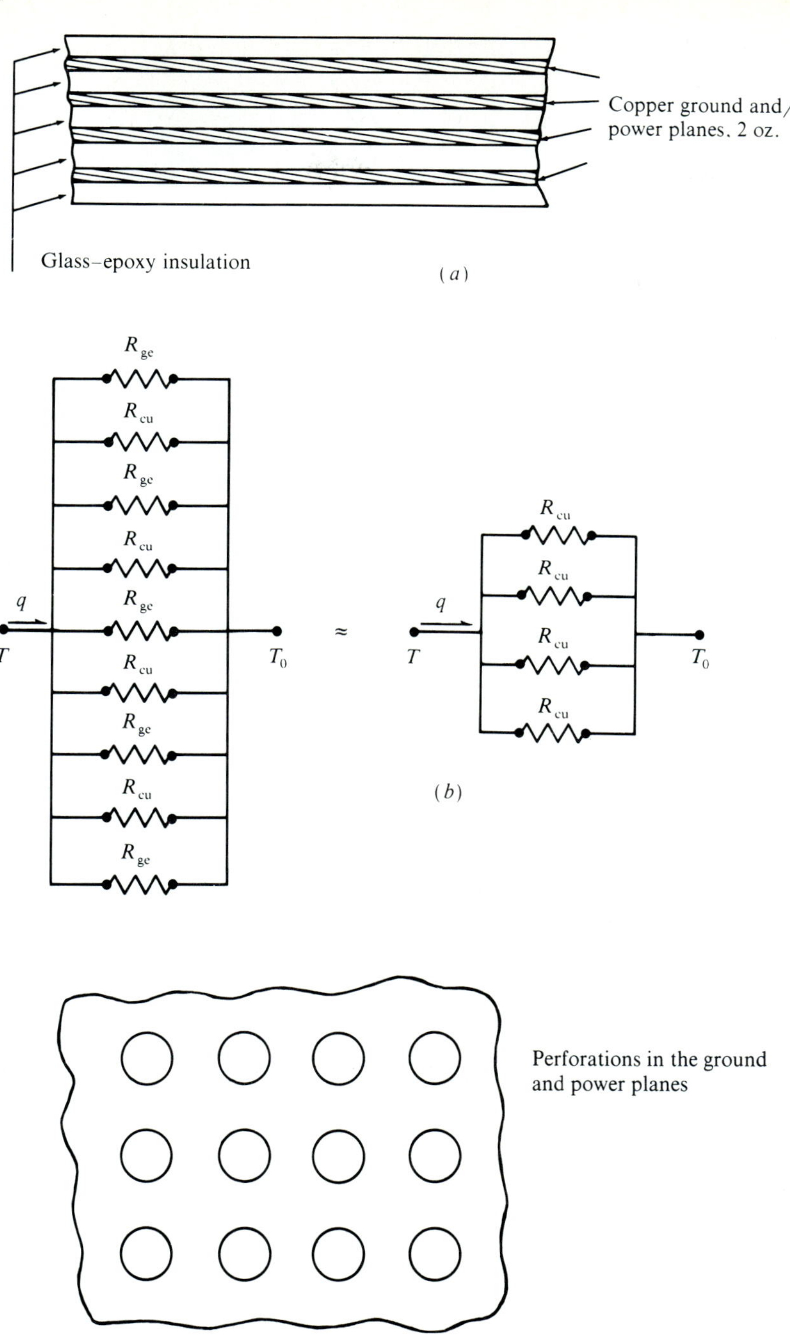

FIGURE 8.24
Geometry and thermal modeling of a MLB.

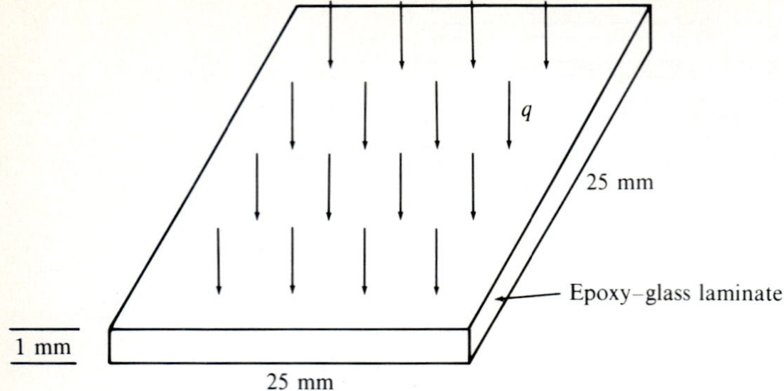

FIGURE 8.25
Heat transmission through the thickness of the board.

of the board in the through thickness direction can be reduced substantially by incorporating thermal vias into the design.

Thermal vias are typically plated through holes (PTHs) that have been filled with solder. The PTHs are usually 1 mm in diameter and are on 2.5 mm centers over the entire area of the board as shown in Fig. 8.26. The fraction of the area covered by these solder-filled vias is

$$A_v/A = \pi(1)^2(100)/(4 \times 25 \times 25) = 0.126$$

The heat is conducted in parallel paths through the vias and the circuit board. The resistance due to the vias is

$$R_{Tv} = 1 \times 10^{-3}/(50 \times 25 \times 25 \times 10^{-6} \times 0.126) = 0.254°\text{C/W}$$

and the resistance of the circuit board surrounding the vias is

$$R_{Tb} = 6.15[1/(1 - 0.126)] = 7.04°\text{C/W}$$

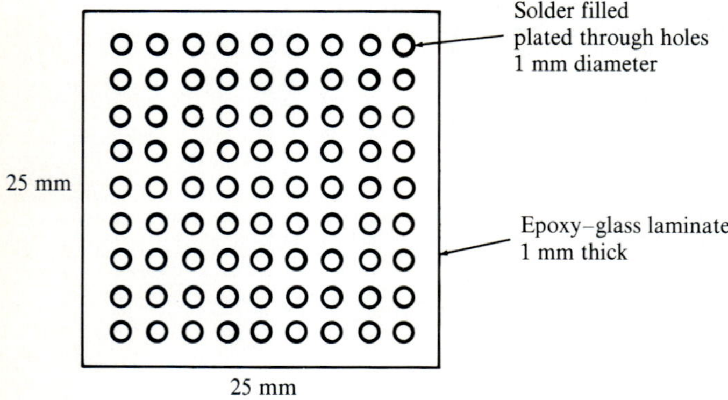

FIGURE 8.26
Thermal vias enhanced through thickness heat transfer.

Using the parallel resistance rule,

$$R_{Te} = R_{Tv}R_{Tb}/(R_{Tv} + R_{Tb}) \tag{8.64}$$

to find the effective thermal resistance gives

$$R_{Te} = (0.254 \times 7.04)/(0.254 + 7.04) = 0.245°\text{C/W}$$

The effect of the thermal vias is dramatic because the through thickness thermal resistance has been reduced by a factor of 6.15/0.245 = 25 and the thermal penalty imposed on the heat dissipation system by the circuit board has been reduced to less than $\frac{1}{4}$°C/W.

8.6.3 Conduction Cooling with a Heat Frame

In the design of a product where it is not possible to pass air directly over the chip carriers to dissipate the heat, the conduction method illustrated in Fig. 8.27 is employed. In this system the heat is transmitted from the chip carriers to the circuit board and then to a heat frame. The heat frame provides a low resistance path to a heat sink. We will consider a typical circuit card consisting of a multilayer board bonded to a heat frame with the dimensions given in Fig. 8.28. The board is divided at the center line, taking into account symmetry of the construction and the heat input. The right side of the board is divided into eight segments with a heat input q_1 to q_8 imposed on each segment. Since the PCB is a multilayered board, it is assumed that sufficient copper ground and power planes exist to spread the heat input uniformly over each area segment. With these assumptions the resistance network, shown in Fig. 8.29, represents a thermal model that can be used to predict temperatures at different locations on the board. The network contains resistances R_a, R_f, R_{f_1} and R_c, which may be determined by considering the detailed features of the electronic module. Consider first the resistance R_a, which represents the combined resistance to the heat flow through the thickness of the board, through the bonding adhesive and to the center of the heat frame. We compute R_a as

$$R_a = R_b + R_{\text{ep}} + R_{\text{cu}\perp} \tag{a}$$

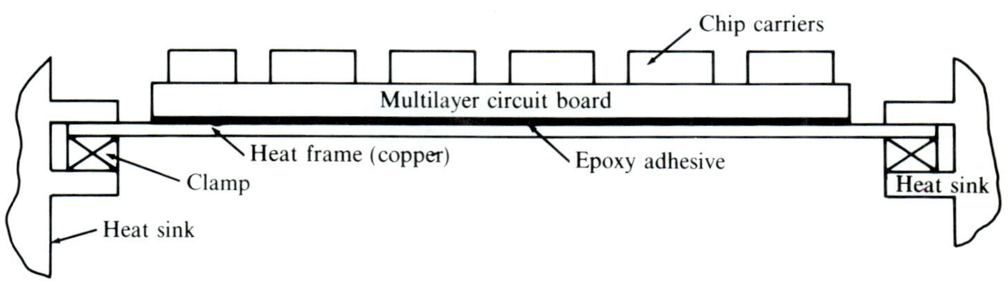

FIGURE 8.27
Enhancement of conduction in an electronic module by using an attached heat frame.

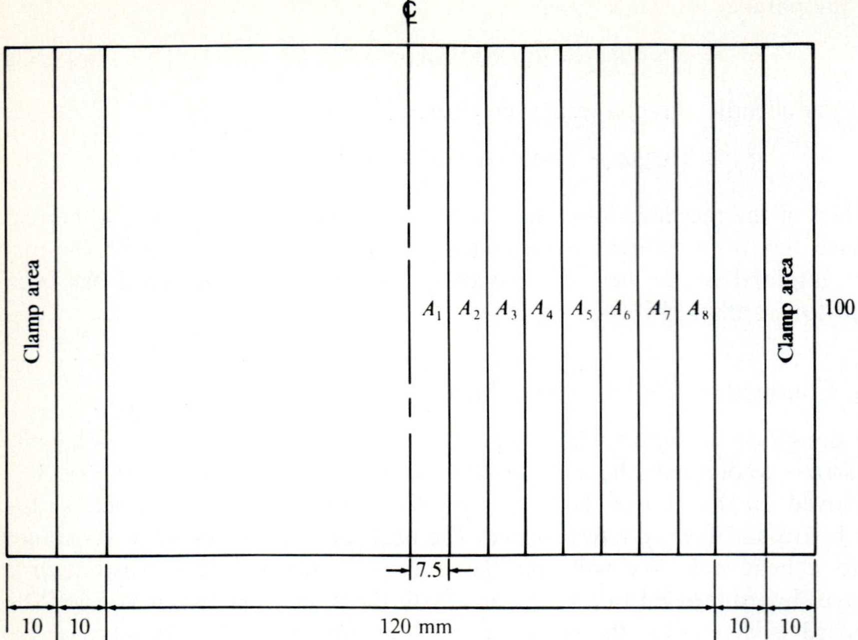

FIGURE 8.28
Front view of an electronic module showing the right side of the circuit card divided into eight equal sized areas A_1 to A_8.

To determine R_b, note that the board thickness is 2 mm and that the area of a segment is 750 mm². Since the thermal vias cover the entire area of the board, as illustrated in Fig. 8.26, we may use the results for the effective thermal resistance given by Eq. (8.64) to write

$$R_b = 0.245(2 \times 25 \times 25/750) = 0.409°\text{C/W}$$

For the thermal resistance for the epoxy adhesive R_{ep}, we take the thickness

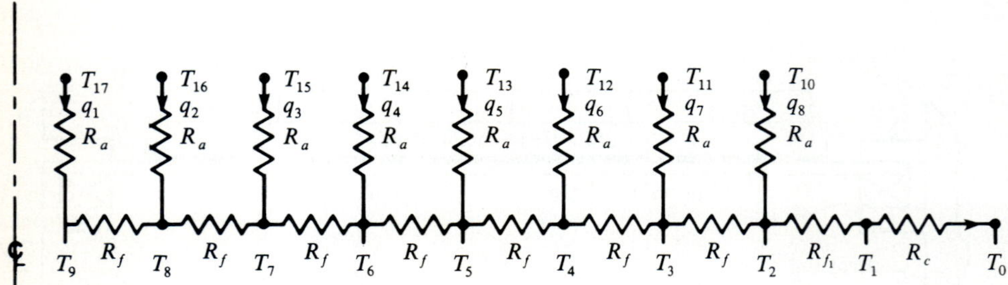

FIGURE 8.29
Resistance network representing a thermal model of the module shown in Fig. 8.27.

$L = 0.125$ mm and $k = 2$ W/m^2 °C and write

$$R_{ep} = 0.125 \times 10^{-3}/(2 \times 750 \times 10^{-6}) = 0.083°\text{C/W}$$

For $R_{cu\perp}$ we take half the frame thickness to be 0.5 mm and write

$$R_{cu\perp} = 0.5 \times 10^{-3}/(381 \times 750 \times 10^{-6}) = 0.002°\text{C/W}$$

Adding these three resistances gives

$$R_a = 0.494°\text{C/W} \tag{b}$$

Next we determine the resistance R_f due to the heat frame as

$$R_f = 7.5 \times 10^{-3}/(381 \times 1 \times 100 \times 10^{-6}) = 0.197°\text{C/W} \tag{c}$$

The resistance R_{f_1} is larger than R_f because the distance from the center of A_8 to the center of the clamp area ($L = 18.75$ mm) is larger than the center-to-center distance between the segment areas (7.5 mm). R_{f_1} scales from R_f according to distance as

$$R_{f_1} = (18.75/7.5) \times 0.197 = 0.493°\text{C/W} \tag{d}$$

The last resistance to be determined is the contact resistance R_c, which is due to the interface between the heat frame and the heat sink. Taking $h_c = 10^4$ W/m^2 °C and noting that the clamp area is 10×100 mm^2, the value of R_c is

$$R_c = 1/(10^4 \times 10^3 \times 10^{-6}) = 0.1°\text{C/W} \tag{e}$$

It is now possible to compute the temperature at any location on the board if the heat inputs q_1 to q_8 are specified. In this example, take each of the qs to be equal to 1 W. This assignment implies that the total heat to be dissipated is 16 W with 8 W being conducted to each heat sink.

To determine the temperature T_1, note from Eq. (8.25) that

$$T_1 = \left[\sum q_i\right] R_c + T_0 = 8 \times 0.1 + 20 = 20.8°\text{C} \tag{f}$$

where the heat sink temperature T_0 was taken as 20°C. The temperatures at several other locations are determine as

$$T_2 = \left[\sum q_i\right] R_{f_1} + T_1 = 8 \times 0.493 + 20.8 = 3.94 + 20.8 = 24.74°\text{C} \tag{g}$$

$$T_3 = \sum_1^7 q_i R_f + T_2 = 7 \times 0.197 + 24.74 = 26.12°\text{C} \tag{h}$$

$$T_4 = \sum_1^6 q_i R_f + T_3 = 6 \times 0.197 + 26.12 = 27.31°\text{C} \tag{i}$$

$$T_{10} = q_8 R_a + T_2 = 1 \times 0.494 + 24.74 = 25.24°\text{C} \tag{j}$$

A complete listing of the temperatures is presented in Table 8.6. Note, that the maximum temperature is in the center of the board with $T_{max} = T_{17} = 30.75°$C. The temperature difference necessary to transfer the heat from this central

TABLE 8.6
A complete listing of the temperatures in degrees Celsius at locations defined in Fig. 8.29

$T_0 = 20$
$T_1 = 8R_c + T_0 = 20.8$
$T_2 = 8R_{f1} + T_1 = 24.74$
$T_3 = 7R_f + T_2 = 26.12$
$T_4 = 6R_f + T_3 = 27.30$
$T_5 = 5R_f + T_4 = 28.29$
$T_6 = 4R_f + T_5 = 29.08$
$T_7 = 3R_f + T_6 = 29.67$
$T_8 = 2R_f + T_7 = 30.06$
$T_9 = R_f + T_8 = 30.26$
$T_{10} = q_8 R_a + T_2 = 25.24$
$T_{11} = q_7 R_a + T_3 = 26.62$
$T_{12} = q_6 R_a + T_4 = 27.80$
$T_{13} = q_5 R_a + T_5 = 28.78$
$T_{14} = q_4 R_a + T_6 = 29.57$
$T_{15} = q_3 R_a + T_7 = 30.16$
$T_{16} = q_2 R_a + T_8 = 30.55$
$T_{17} = q_1 R_a + T_9 = 30.75$

location to the heat sink was 10.75°C. The temperature at the edge of the circuit card was much lower, $T_{10} = 25.24$°C with $\Delta T = 5.24$°C.

This example illustrates that a circuit board can be incorporated into a heat dissipation system while maintaining reasonable ΔTs to conduct a heat flux exceeding 1000 W/m^2. The detrimental effect of the poor conductivity of the epoxy–glass was mitigated by reducing the length of travel (through thickness), by incorporating an array of thermal vias on 2.5 mm centers and by employing a heat frame to conduct the heat from all locations on the board to the heat sink.

8.7 COLD PLATES AND COLD RAILS

Reference to Fig. 8.27 shows that the heat frame is clamped to a heat sink, called a cold rail, which is at a temperature $T_0 = 20$°C. The temperature is maintained at T_0 by passing water through channels bored in the rail or through tubing welded to a plate as illustrated in Fig. 8.30.

The heat transferred through the cold rail can be approximated by

$$q = kS(T_2 - T_1) \tag{8.65}$$

where S is a shape factor dependent on the geometry of the cross section of the cold rail. For the shape illustrated in Fig. 8.30, S is given by

$$S = 2\pi L/\ln\{[e/(\pi r)]\sinh(2\pi H/e)\} \tag{8.66}$$

The symbols used in Eq. (8.66) are defined in Fig. 8.30. The temperature T_1 of the wall of the hole in the cold rail will be slightly higher than the water temperature due to the ΔT across the boundary layer and will depend on the convection coefficient associated with the flow conditions.

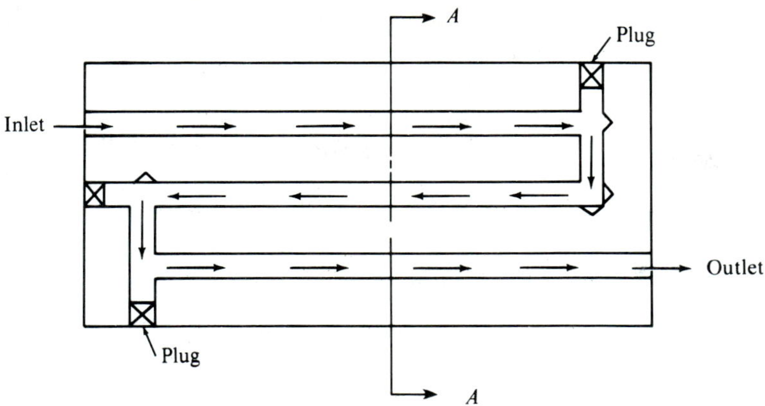

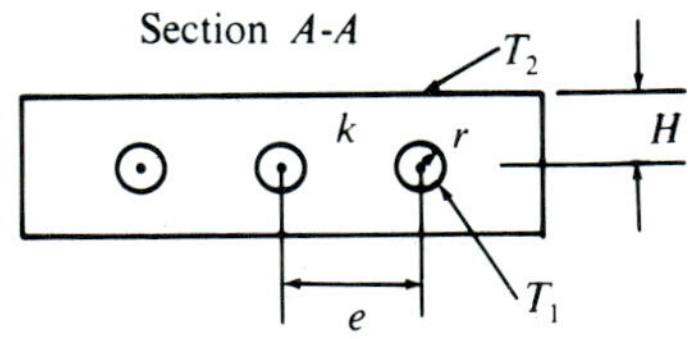

FIGURE 8.30
Construction details of a cold rail.

As the coolant flows through the cold rail, heat is transferred through the rail and across the boundary layer to the fluid. The temperature increase in the fluid is given by

$$\Delta T = T_o - T_i = q/\left[(dm/dt)c_p\right] \tag{8.67}$$

where dm/dt is the mass flow rate (kg/s), c_p is the specific heat (J/kg) and the subscripts i and o refer to the inlet and outlet temperatures. For water as a coolant at an average temperature of 27°C, $c_p = 4177$ J/kg °C or 9.78 W-min/lbm °C. Also, the mass flow rate dm/dt, written in units of pounds mass per minute is

$$dm/dt = 8.31G$$

where G is the volume flow rate in gallons per minute. Using these conversion factors in Eq. (8.67) gives a relation that can be used to determine ΔT in commonly employed units of Watts for q and gallons per minute for the flow rate

$$\Delta T = 0.0123q/G \tag{8.68}$$

Cold rails are very effective heat sinks because they offer low thermal resistance when properly designed and operate at relatively low flow rates. For example, a heat rate of 1 kW can be dissipated with a flow rate of 2 gal/min with an increase in the water temperature from inlet to outlet of only about 6°C. For very high reliability and high performance products, a refrigerant, say freon 12, can be employed as the coolant. Refrigerated coolant reduces the temperature of the heat sink to about $-30°C$ and lowers the junction temperatures by about 50°C. Refrigerated cold rails were used to great advantage in the heat management system employed in the Cray I computer in the early 1970s.

8.8 ADVANCED COOLING METHODS

For dense logic chips, the heat flux is usually quite high and as VLSI becomes more common, power dissipated on many chips will exceed 10 W. As ULSI is introduced in the mid 1990s, one may anticipate chip power approaching 50 W. It will be extremely difficult to maintain low junction temperatures in these high powered devices. The common thermal management techniques involving traditional chip carriers and circuit boards introduce large thermal resistance that results in high junction temperatures. To successfully accommodate high powered ICs will require a new approach to thermal management that abandons most of the traditional packaging concepts. The new approaches will be based on two fundamental premises. First, one side of the chip will be reserved for wiring to accommodate the signal I/O and the power requirements. It is implied that the other side of the chip is reserved for heat dissipation. Second, the new thermal management methods will take the coolant to the chip and eliminate the thermal resistance between the sink (i.e., the coolant) and the chip.

The IBM Corporation has implemented the first premise, on a commercial scale, in developing the thermal conduction module (TCM), which utilizes one side of the chip for electrical wiring and the other side of the chip for a cooling scheme. Pease and Tuckerman, in an academic laboratory, have implemented both of these premises by bringing coolant directly to the back side of the chip while reserving the front for the electrical functions.

8.8.1 The Thermal Conduction Module

In the early 1980s the IBM corporation introduced the thermal conduction module (TCM) in the 3081 main frame computer system. This method of packaging chips was markedly different than previous designs. With the TCM, the first and second level packages were combined into a single unit. Also, the electrical and mechanical functions of the chip were divided, with the circuitry confined to one side of the chip and the heat transfer system confined to the other side of the chip. Construction details of the TCM are shown in Fig. 8.31.

The module is capable of handling a minimum of 100 logic chips on a multilayer ceramic substrate 90 × 90 mm in size. The chips are bonded to the substrate with the C_4 process where an area array of solder balls between the chip and the substrate pads are reflowed to make the required electrical connections. The chips and the substrate are both contained in a sealed metallic housing. To enhance heat transfer, helium is used to fill the housing cavity.

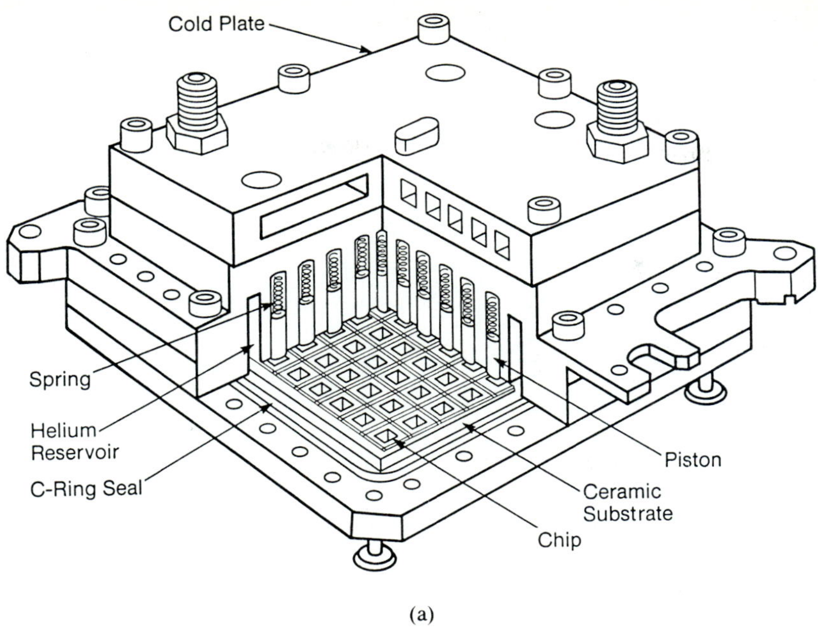

(a)

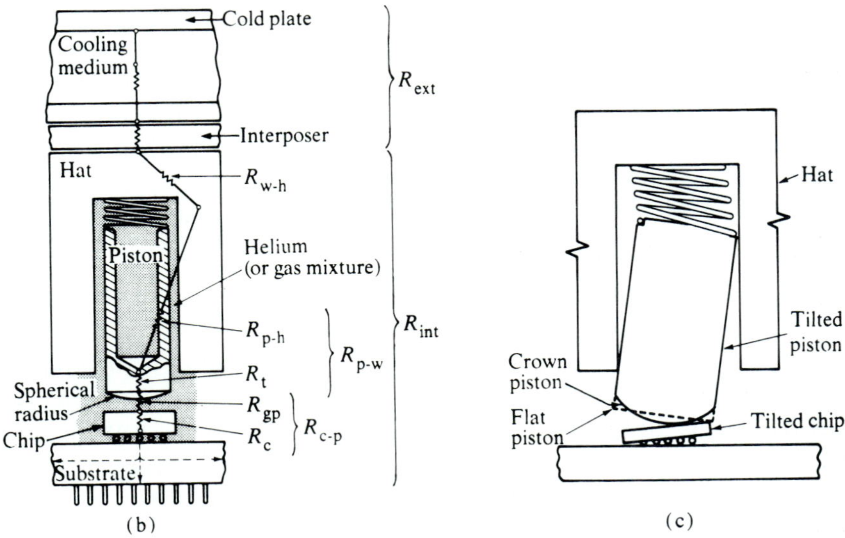

FIGURE 8.31
Details of the thermal conduction module. (*a*) Cutaway view showing the complete system. (*b*) Resistance network corresponding to the design features. (*c*) Crowned pistons to insure contact with tilted chips. (*Copyright 1982 by International Business Machines Corporation; reprinted with permission.*)

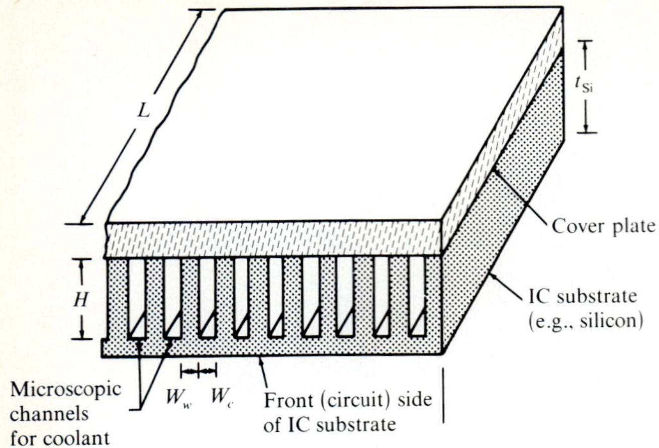

FIGURE 8.32
Microchannels to accommodate fluid flow for cooling in the IC substrate. (*After Tuckerman.*)

Spring-loaded pistons with spherical ends are pressed against the back of each chip. These pistons serve to carry the heat from the chip to the hat. A cold plate positioned on the hat is supplied with water that serves as the coolant and the ultimate heat sink.

The thermal performance of this unit is excellent, with a total thermal resistance of about 7°C/W. This resistance permits junction temperatures of about 50–60°C to be maintained on relatively high powered logic chips.

A second advantage of this type of packaging is the high density of the chip placement. The compactness permits extremely short lead lengths that markedly reduce signal propagation delays in the circuitry. The resulting improvement in system cycling time, which was achieved with the TCM, is illustrated in Fig. 4.24.

8.8.2 Pease and Tuckerman Experiments

David Tuckerman, in an outstanding Ph.D. thesis (1984), showed the feasibility of entirely new concepts in cooling chips with extremely low thermal resistance. Pease was the faculty advisor. The new concept, illustrated in Fig. 8.32, utilizes microchannels etched into the silicon substrate. With short channels about 50 μm wide the flow of the coolant is laminar and the rate of heat transfer is excellent. Fluids such as water, mercury, Fluorinert and liquid nitrogen all show merit as coolants for these high performance heat exchangers.

Typical interfacial thermal resistances of 0.02–0.08°C/W were measured in a series of carefully controlled experiments incorporating the microchanneling process in both the back of the chip and in a substrate in contact with a normal chip. These results indicate that thermal loading of 10^7 W/m^2 can be removed from wafer devices at normal operating temperatures (100°C).

REFERENCES

1. Ellison, G. N.: *Thermal Computations for Electronic Equipment*, Van Nostrand Reinhold Co., New York, 1984, pp. 183–368.
2. Thornell, J. W., W. A. Fahley and W. L. Alexander: "Hybrid Microcircuit Design and Procurement Guide," DOC AD 705974, National Technical Information Service, Springfield, Va., 1972.
3. Fourier, J. B.: *Theorie Analytique de la Chaleur*, Paris, 1822 (trans., Dover, New York, 1955).
4. White, F. M.: *Heat Transfer*, Addison-Wesley Publishing Co., Reading, Mass., 1984.
5. Krauss, A. D. and A. Bar-Cohen: *Thermal Analysis and Control of Electronic Equipment*, Hemisphere Publishing Corp., Washington, D.C., 1983.
6. Cooper, M.G., B. B. Mivkic and M. M. Yovanovich: "Thermal Contact Conductances," *International Journal of Heat and Mass Transfer*, vol. 12, pp. 279–300, 1969.
7. Tien, C. L.: "A Correlation for Thermal Contact Conductance of Nominally Flat Surfaces in a Vacuum," Special Publication 302, National Bureau of Standards, September 1968.
8. Barzelay, M. E., K. N. Tong and G. Holloway: "Effect of Pressure on Thermal Conductance of Contact Joints," NACA Technical Note 3295, May 1955.
9. Fried, E. and F. Costello: "Interface Thermal Contact Resistance Problem in Space Vehicles," *ARS Journal*, February 1962.
10. Tuckerman, D. B.: "Heat Transfer Microstructures for Integrated Circuits," Ph.D. Dissertation, Stanford University, February 1984. Available from University Microfilms International, Ann Arbor, Mich.
11. Sloan, J. L.: *Design and Packaging of Electronic Equipment*, Van Nostrand Reinhold Company, New York, 1985.
12. Blodgett, A. J. and D. R. Barbour: "Thermal Conduction Module: A High Performance Package," *IBM J. Res. Develop.*, vol. 26, no. 1, pp. 30–36, 1982.
13. Steinberg, D. S.: *Cooling Techniques for Electronic Equipment*, John Wiley, New York, 1980, pp. 66–69.

EXERCISES

8.1. Describe why fault location methods are important in system availability in large digital processors. Define system availability.

8.2. A sample of 500 components is tested under continuous operation for a year. The three components that fail at 32, 60 and 290 days are replaced during the test. Find $\lambda(t)$ and MTBF. Define MTBF. Comment on the results.

8.3. Determine $\lambda(t)$ if the components that failed in Exercise 8.2 were not replaced during the test.

8.4. Explain the reasons for using components that have been subjected to a burn-in operation. Why not perform the burn-in after the system is assembled?

8.5. Determine the probability of success $P_s(t)$ for the component tested in Exercise 8.2.

8.6. Prepare a data presentation showing $P_s(t)$ for a component with $\lambda_0 = 10^{-3}, 10^{-5}, 10^{-7}, 10^{-9}, 10^{-12}$ and $t = 1, 2, 5, 10, 20$ and 50 yr. It may be helpful to use a spread sheet in performing these computations and displaying the results.

8.7. Prepare an engineering brief that interprets the results of Exercise 8.6.

8.8. A system consists of four series connected subsystems. The failure rates of the subsystems are 10^{-4}, 10^{-5}, 10^{-6} and 10^{-9}/hr. Prepare a graph showing the system reliability as a function of time for a period of 20 yr.

8.9. Using one redundant subsystem with the system described in Exercise 8.8, improve the system reliability. Indicate where the subsystem would be placed, give reasons for the choice of the subsystem and prepare a revised reliability time graph that illustrates the improvement.

8.10. Derive the system reliability for the series parallel arrangement with the single cross strap that is shown in Fig. E8.10. Note that all of the devices shown exhibit the same probability of failure $P_f = 0.001$.

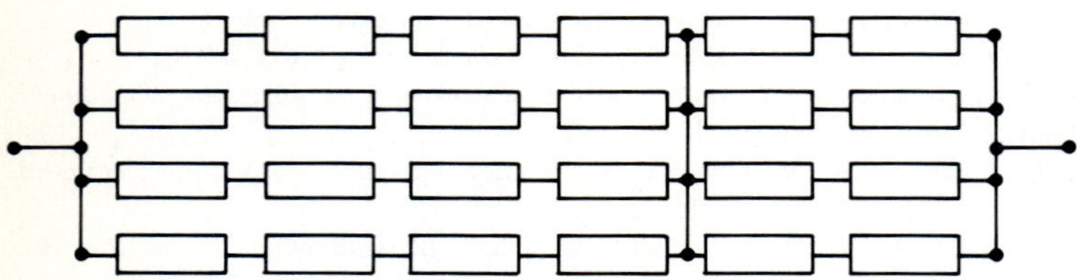

FIGURE E8.10

8.11. Progressively add cross straps to the block arrangement described in Exercise 8.10 and determine the enhanced system reliability. Discuss the merits of adding additional cross straps. Also, describe any disadvantages associated with cross strapping.

8.12. Consider a conductor $0.5 \times 0.5 \times 4$ in. with 10 W to be conducted in the 4 in. direction. The sides are insulated. The conductor is fabricated from aluminum 6061T6, Kovar, fiberglass composite, alumina 96% and beryllia 99.5%. The cold end of the conductor is maintained at 25°C. Find the temperature of the hot end. Prepare an outline showing conclusions that one may draw from this simple exercise.

8.13. Determine the thermal resistance for the conductors described in Exercise 8.12.

8.14. Examine the thermal coefficient of conductivity given in Table 8.3 and select from this list those materials that are good electrical insulators. Tabulate k for these materials and draw a conclusion on the merits of the design of a thermal path that must also provide for insulating circuits.

8.15. An alumina cylinder is used to contain a thyratron that dissipates 800 W. The cylinder has an inside and outside diameter of 2.6 and 2.8 in., respectively, and a length of 4 in. Determine the thermal resistance of the cylinder and the temperature difference ΔT. Discuss the validity of the cylinder as a thermal model for this tube.

8.16. Determine the heat transferred across the plate defined in Fig. 8.11 if the dimensions are 8 mm $\times$ 0.4 m $\times$ 0.6 m. The plate is made from aluminum 2024. The convection coefficient h is 6 W/m^2 °C on one side of the plate and 3 W/m^2 °C on the other. The temperature difference available is 20°C. Examine the three components of the thermal resistance and discuss the factor limiting the amount of heat being transferred across the plate. How could you change the amount transferred?

8.17. Determine the contact resistance at the interface between two aluminum plates. Each plate is 0.25 in. thick, the surface finish is 125 μin. rms, the contact area is 1.5 in.^{2} and the contact pressure is 100 psi. If we were to change from aluminum to stainless steel for the plate material, give the new contact resistance.

8.18. If you were to increase the pressure on the contact area described in Exercise 8.17 from 100 to 300 psi, what would be the contact resistance. Now decrease the pressure to 20 psi and again find the contact resistance. Prepare an engineering brief describing the effect of contact pressure on contact resistance for aluminum–aluminum and stainless steel–stainless steel interfaces.

8.19. A small heat source with an area A is to be placed on a large conductor, which we will treat as a half space. The heat source can be distributed over a circle with a diameter D or a square with a side L. Determine the difference in the thermal resistance due to constriction between these two options. Plot your results on Fig. 8.16 and comment on the position of your points relative to the curves for a body with finite dimensions.

8.20. The thermal resistance of a cooling system is 2°C/W. The system employs conduction and dissipates 10 W through an aluminum assembly that weighs 0.75 lb. Determine the RC constant for this system and prepare a graph of the temperature as a function of time if the ambient temperature is 25°C.

8.21. Write Eq. (8.57) in a nondimensional form and discuss the significance of each side of the new equation. Compare this result with Eq. (8.61) and comment.

8.22. Suppose the case shown in Fig. 8.18 was filled with a thermally conductive and electrically insulating fluid. Modify the thermal model represented by the resistive network, shown in Fig. 8.19, to take this fluid into account.

8.23. A silicon chip 4 × 4 mm × 0.3 mm thick is soldered to a ceramic circuit card that is bonded to a cold plate as shown in Fig. E.8.23. The surface temperature of the cold plate is maintained at 20°C. Determine the temperature of the junctions on the chip as a function of the heat dissipated. Consider the range from 1–10 W.

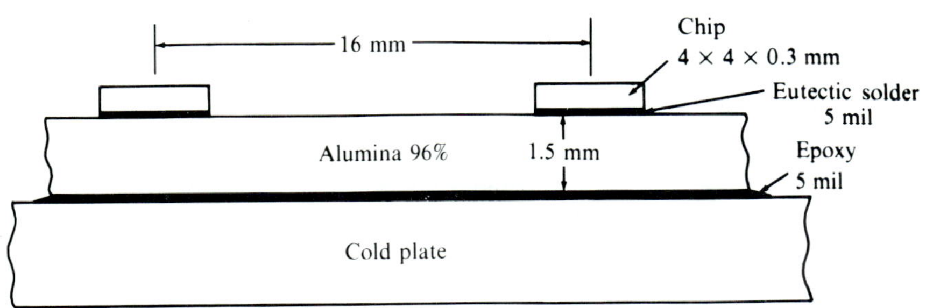

FIGURE E8.23

8.24. Discuss the heat dissipation in a DIP-type chip carrier. Would it improve the thermal resistance if alumina were used in place of the plastic for the carrier material? Quantify this answer.

8.25. Sketch the essential features of a design modification for a DIP that improves the thermal resistance by a factor of 10. The basic outline of the DIP and the pin spacing cannot be changed in the new design. Write an engineering brief that describes your approach and includes an analysis of the new design.

8.26. A 24 lead plastic cased DIP that dissipates 0.6 W is cooled by direct exposure to an air stream. If the flow is 600 ft/min with the air at 32°C, find the junction temperature. Comment on the result.

8.27. Determine the thermal resistance of a thermal conductor 10 × 10 mm in cross section by 100 mm long. The following circuit board materials are to be considered: FR-4, XXXP, Teflon, polyimide–glass, alumina, silicon and SiC. Write a paragraph with the conclusion you have drawn from this exercise.

8.28. A long unit strip of multilayer board, shown in Fig. E8.28, is made from five planes of 2 oz copper and four layers of glass–epoxy. A source of heat uniformly distributed over 1 mm by unity is dissipated through the thickness of the board. The bottom plane of copper is held at a constant temperature $T_s = 20°C$. Find the temperature in the x-y plane for x and y ranging from 0–4 mm. Are the copper planes useful in distributing the heat in the y direction. Prepare a sketch of the isothermal lines that demonstrates this fact.

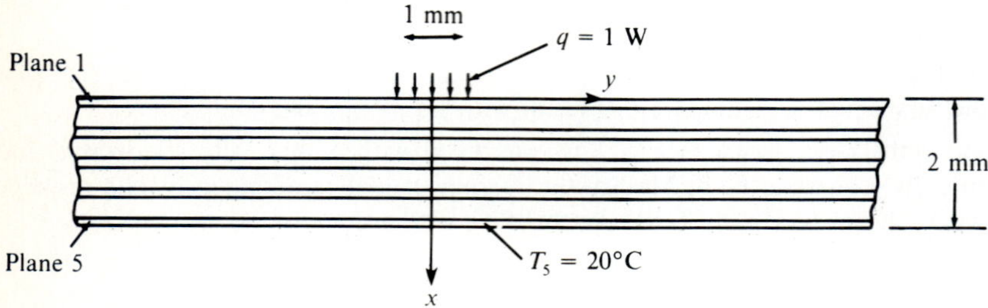

FIGURE E8.28

8.29. A mistake was made and the vias shown in Fig. 8.26 were not filled with solder. Only those vias supporting pins from the chip carriers were filled (14%). Determine the thermal resistance for this segment of circuit board. Who made the mistake—design or manufacturing—and how could you tell?

8.30. Explain the purpose of a heat frame. Why would one use a heat frame in air cooled systems? Would you ever use a heat frame in space applications? Why?

8.31. Weight is a premium in space applications. What material would you use for a heat frame design if you were told it cost $100,000 to put a pound in orbit. Prepare an engineering brief showing the procedure used to select the material and include a cost analysis to support your position.

8.32. If the heat input to each segment of the thermal model shown in Fig. 8.29 is $q_1 = 1$ W, $q_2 = 1.2$ W, $q_3 = 1.4$ W, $q_4 = 1.6$ W, $q_5 = 1.8$ W, $q_6 = 2.0$ W, $q_7 = 2.2$ W and $q_8 = 2.4$ W, determine the temperatures T_1 to T_{17}.

8.33. The heat input to the board and heat frame given in Exercise 8.32 is not uniform. Those devices with the highest rate of dissipation were placed near the heat sink. Comment on the advisability of the component placement resulting in the heat distribution shown in the preceding exercise. Be sure to include reliability in your discussion.

8.34. A cold plate is made by drilling 0.5 in. diameter holes on 1 in. centers in an aluminum plate 0.75 in. thick and 24 in. long. Determine the shape factor S for this plate. What is the thermal resistance of the cold rail?

8.35. If the cold plate in Exercise 8.34 is 8 in. wide, determine the amount of heat that can be transferred as a function of ΔT.

8.36. Water flowing at 1.5 gal/min is used with a cold plate for cooling a wide variety of power supplies. Prepare a graph showing the heat that can be dissipated as a function of ΔT for this cold plate.

8.37. Examine Eqs. (8.65) and (8.66) and explain the difference between T_1, T_2, T_o and T_i. Draw a sketch of the cold plate and locate on it the four temperatures. Are you entirely satisfied that these four temperatures completely describe the temperature distributions. Explain?

8.38. Explain why the ends of the pistons used in the TCM are crowned. Would a flat end provide better thermal contact?

8.39. Your task is to modify the TCM to reduce the thermal resistance without making major design changes. Make at least three suggestions for design changes. Prepare an engineering brief that supports your new design. If possible show an analysis that predicts the reduction in the thermal resistance.

8.40. Design a microgroove heat exchanger that is to be fabricated by etching a wafer of silicon. The chips are to be bonded to one side of the wafer heat exchanger with eutectic solder. Show an arrangement to accommodate the manifolds with attention to detail to reduce the pressure required to drive water through the microchannels. Prepare an analysis that shows design trade-offs on size, heat capacity, flow rates, pressure drop and thermal resistance.

CHAPTER

9

THERMAL ANALYSIS METHODS—RADIATION AND CONVECTION

9.1 INTRODUCTION

In addition to conduction, heat transfer by radiation and convection are important methods for dissipating heat to the environment. Indeed, in space applications of electronic systems radiation is the final mechanism used to dissipate the heat through the vacuum of outer space. Radiation is the electromagnetic transmission of energy from one opaque surface to another. The rate of heat transfer is dependent on the fourth power of the temperature difference between the transmitting and the receiving bodies. This fourth power relationship makes radiation very important in high temperature applications like combustion and incandescent lighting; however, in electronics where the aim is to keep the temperatures low, radiation is not a major factor in heat dissipation schemes. Only in space where it is the only mechanism available is it a dominant factor. In earth-bound applications radiation is usually a minor factor that is considered, in a detailed and accurate analysis, together with convection and conduction to determine the heat transfer rates and the temperatures of the components.

Convection is extremely important as a means of heat dissipation in electronic systems since air cooling is used in most low to medium performance systems. With convection the heat transfer is from a solid to a fluid and then the fluid is moved from the system to the atmosphere or sink. When possible, free or natural convection is used and the buoyancy of the air serves to transport the heat away from the system. For higher heat transfer rates buoyancy does not provide sufficient air velocity and forced convection is employed. In these cases, fans are used to drive the air at higher velocities, which increases the volume flow rate of the air and the heat transfer coefficient.

We will describe both radiation and convection as heat transfer mechanisms and then show methods utilizing the basic heat transfer relations, with applications to the cooling of electronic systems packaged in enclosures.

9.2 LAWS GOVERNING HEAT TRANSFER BY RADIATION

Heat transfer by radiation is very important in high temperature applications where the flux density E_λ^b is relatively high. Planck's radiation law, shown graphically in Fig. 9.1*a*, clearly indicates the sharp drop in E_λ^b with the lower operating temperatures that are typical for electronic devices and enclosures. Also evident is the shift of the peak flux to a higher wavelength λ as the temperature decreases. These features are shown more clearly in Fig. 9.1*b*, where the data for the temperature range from 25–100°C are presented. Integrating the relation for E_λ^b over λ from 0 to ∞ leads to the Stefan–Boltzmann relation given by

$$E^b = \sigma T^{*4} \tag{9.1}$$

where $\sigma = 5.670 \times 10^{-8}\ \mathrm{W/m^2\ K^4}$ is the Stefan–Boltzmann constant
E^b is the total energy radiated from a black body ($\mathrm{W/m^2}$)
$T^* = T + 273°$ the absolute temperature (K) of the surface
T is the temperature (°C)

From Eqs. (8.36) and (9.1) it is evident that the equivalent radiation heat transfer coefficient is

$$h_r = (q^b/A)/\Delta T = E^b/\Delta T \tag{9.2}$$

The value of h_r depends strongly on the temperature of the surface as E^b increases as the fourth power of this temperature.

9.2.1 Shape Factors and Energy Exchange

Heat transfer by radiation is more complex than indicated by Eqs. (9.1) and (9.2) since there is radiation exchange between two bodies and the geometries of both of these bodies affect this exchange of heat. The exchange concept is illustrated in Fig. 9.2 where surfaces S_1 and S_2 are shown and heat is exchanged between incremental areas ΔA_1 and ΔA_2. The incremental heat transferred from body 1 to 2 is

$$\Delta q_{1\text{-}2} = E_1^b \Delta A_1 [\cos\theta_1 \cos\theta_2 \Delta A_2/\pi r^2] \tag{a}$$

and from body 2 to 1 is

$$\Delta q_{2\text{-}1} = E_2^b \Delta A_2 [\cos\theta_2 \cos\theta_1 \Delta A_1/\pi r^2] \tag{b}$$

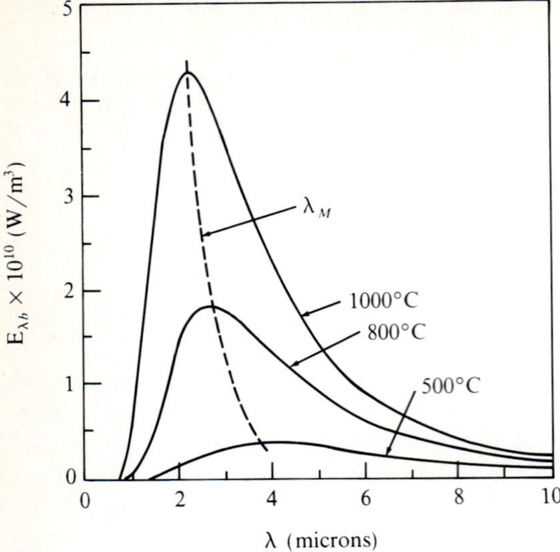

(*a*) Moderate temperatures

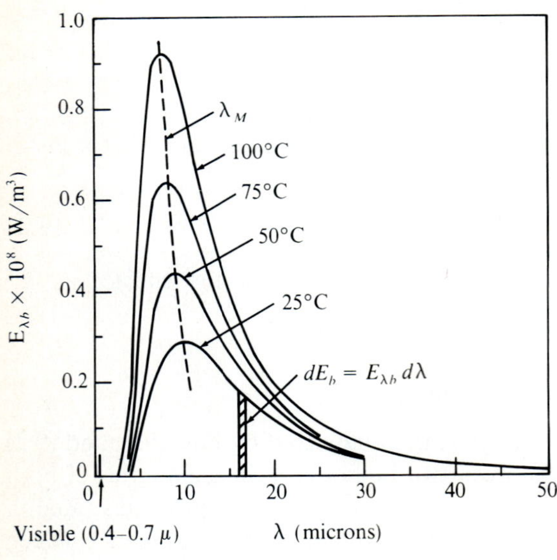

(*b*) Low temperatures

FIGURE 9.1
Distribution of emissive power from a black body as a function of wave length λ of the radiation source.

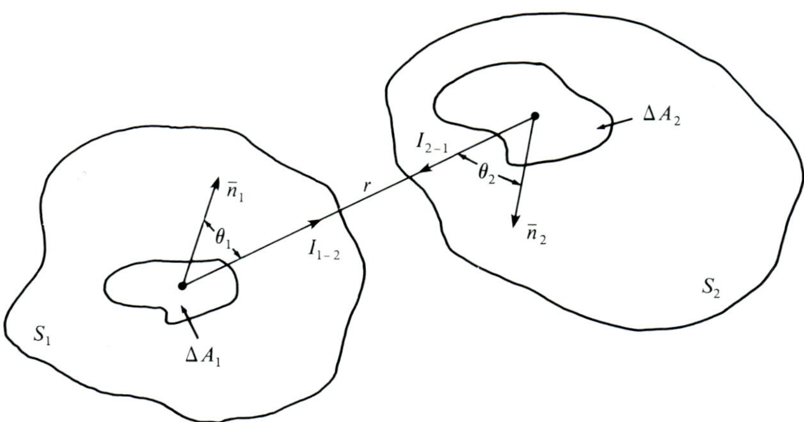

FIGURE 9.2
Radiant energy exchange between bodies.

The net exchange of heat between the two incremental areas is

$$\Delta q = \Delta q_{1-2} - \Delta q_{2-1} \tag{c}$$

Substituting Eqs. (a) and (b) into (c) gives

$$\Delta q = \left(E_1^b - E_2^b\right)\left[\cos\theta_1 \cos\theta_2 \,\Delta A_1 \,\Delta A_2/(\pi r^2)\right] \tag{9.3}$$

Equation (9.3) is integrated over the areas of S_1 and S_2 to obtain

$$q^b = \left(E_1^b - E_2^b\right) A_1 \mathbb{F}_{1-2} \tag{9.4}$$

where q^b is the black body heat transfer
$\mathbb{F}_{1-2}$ is the shape factor

The shape factor indicates the portion of the total radiation emitted from surface 1 that is intercepted by surface 2. Because the product of the shape factor and the area exhibits reciprocity, we may write

$$A_1 \mathbb{F}_{1-2} = A_2 \mathbb{F}_{2-1} \tag{9.5}$$

Combining Eqs. (9.1) and (9.4) gives

$$q^b = \sigma \mathbb{F}_{1-2} A_1 \left(T_1^{*4} - T_2^{*4}\right) \tag{9.6}$$

The shape factors for common geometries have been determined by integration and are represented in Figs. 9.3 to 9.5. These results can be used with Eq. (9.6) to evaluate the black body heat transferred q^b by radiation between two bodies. For more complex applications involving multiple bodies, the reader is referred to reference 2 where methods are described to obtain the required shape factors.

Equation (9.6) is valid for black surfaces where either reflection does not occur or it is so small that it can be neglected. When the surfaces are "gray,"

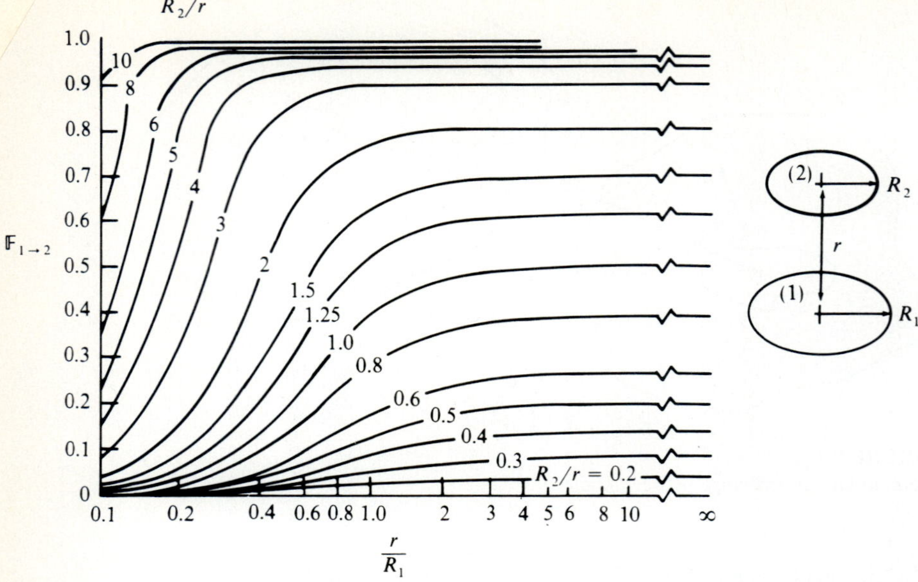

FIGURE 9.3
Shape factor for parallel coaxial disks [1].

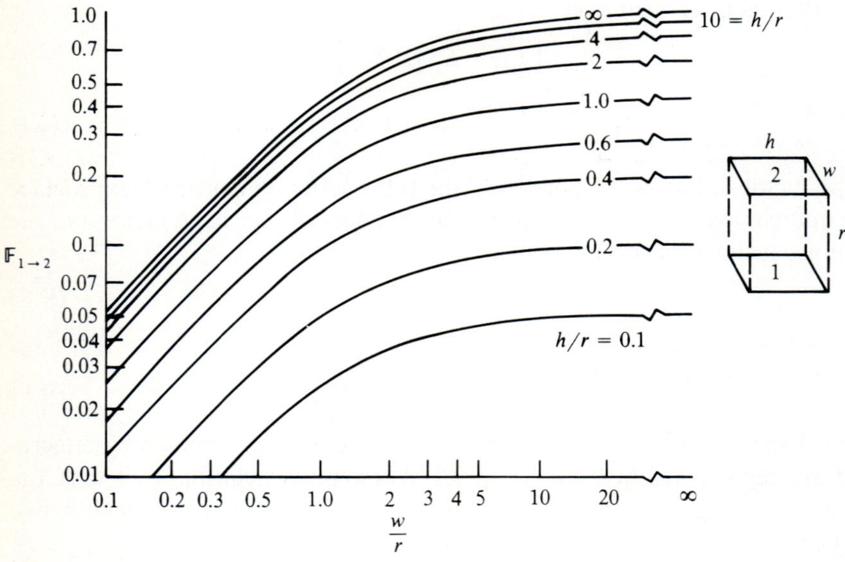

FIGURE 9.4
Shape factor for equal size opposing rectangles [1].

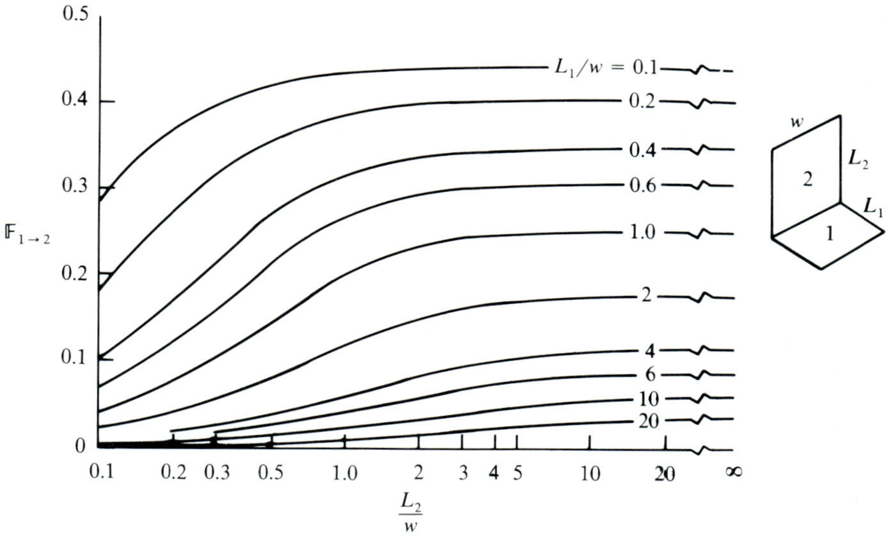

FIGURE 9.5
Shape factor for perpendicular rectangles [1].

reflection takes place and it is often necessary to account for this effect to improve the accuracy of the thermal analysis.

9.2.2 Influence of Surface Emissivity on Radiation Heat Transfer

In practice surfaces are never totally black and it is necessary to modify Eq. (9.6) to account for surface emittance and reflection. Noting from Fig. 9.6*a* that irradiation impinging on a surface can be absorbed, transmitted or reflected leads to three coefficients that sum to 1 as indicated by

$$\alpha + \rho + \tau = 1 \tag{a}$$

where α is the ratio of the irradiation absorbed
ρ is the ratio reflected
τ is the ratio transmitted

For the typical gray body with an opaque and diffusing surface,

$$\tau = 0 \qquad \alpha = \varepsilon \quad \text{and} \quad \rho = 1 - \varepsilon \tag{b}$$

where ε is the emissivity.

Consider a gray body with a diffusing surface receiving energy and emitting its own energy as indicated in Fig. 9.6*b*. The total energy propagating away from the surface is termed the radiosity J, which is given by

$$J = E + \rho H \tag{9.7}$$

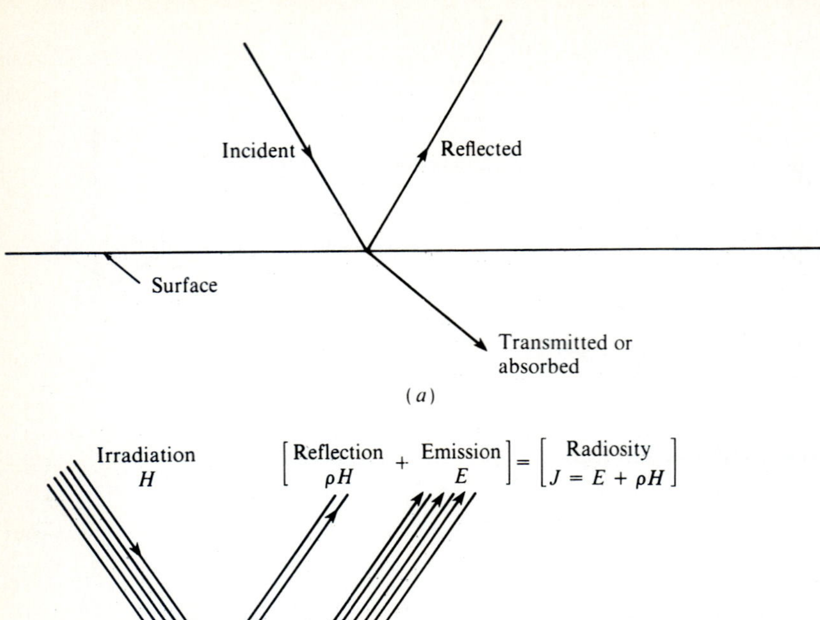

FIGURE 9.6
(*a*) Irradiation impinging on a surface. (*b*) Conceptual illustration showing radiosity.

Note in Eq. (9.7) that E is the energy radiated from the gray surface, which is related to the black body energy E^b by

$$E = \varepsilon E^b \tag{9.8}$$

Combining these two equations yields

$$J = \varepsilon E^b + \rho H \tag{9.9}$$

Next observe that the irradiation H_1 falling on surface A_1 is due to only one other surface of area A_2. Then it follows that

$$H_1 = \mathbb{F}_{2\text{-}1} A_2 J_2 \tag{9.10}$$

where J_2 is the radiation flux leaving A_2. Combining Eqs. (9.9) and (9.10) gives, for surface 1,

$$J_1 A_1 = \varepsilon_1 E_1^b A_1 + \rho_1 \mathbb{F}_{2\text{-}1} A_2 J_2 \tag{9.11a}$$

Using reciprocity of the shape factor in Eq. (9.11a) leads to

$$J_1 = \varepsilon_1 E_1^b + \rho_1 \mathbb{F}_{1\text{-}2} J_2 \tag{9.11b}$$

Solving for $\mathbb{F}_{1-2}J_2$ gives

$$\mathbb{F}_{1-2}J_2 = \left(J_1 - \varepsilon_1 E_1^b\right)/\rho_1 \tag{9.12}$$

The net heat transferred from surface 1 is

$$q_1 = J_1 A_1 - H_1 = J_1 A_1 - \mathbb{F}_{2-1} J_2 A_2 \tag{9.13}$$

Noting reciprocity of the shape factor gives

$$q_1 = A_1\left(J_1 - \mathbb{F}_{1-2}J_2\right) \tag{9.14a}$$

Substituting Eq. (9.12) into Eq. (9.14a) results in

$$q_1 = \left[\varepsilon_1 A_1/(1-\varepsilon_1)\right]\left(E_1^b - J_1\right) \tag{9.14b}$$

These equations may be used to determine the heat transferred due to radiation; however, it is usually easier to use an electrical analogy developed by Oppenheim [3], which follows the concepts developed previously in Chapter 8. We again consider the case of only two surfaces, as shown in Fig. 9.7*a*. The analogy indicates that q is similar to the current and that the radiation fluxes E_1

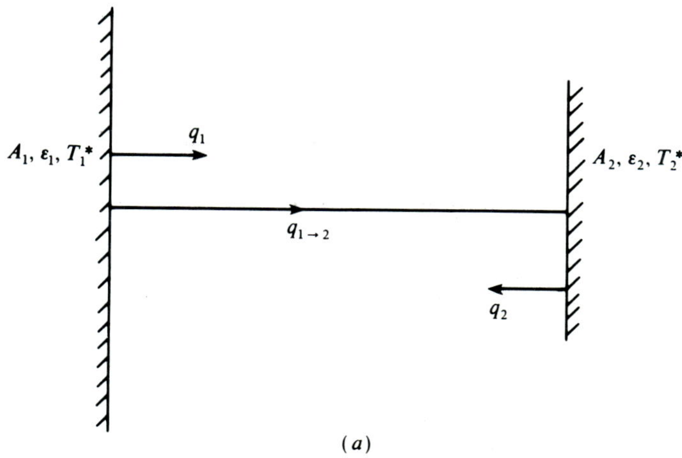

FIGURE 9.7
Radiation heat transfer between two surfaces. (*a*) Radiation exchange between two surfaces. (*b*) Electrical network for the analogy.

and E_2 are similar to the voltage. In the two surface problem there are three resistances to be considered, namely, R_1, R_{sf} and R_2 as indicated in Fig. 9.7*b*. Note that the intermediate voltages between the resistors are analogous to the energy densities J_1 and J_2.

The resistance R_1 is determined by noting from the electrical analogy that

$$q_1 = \left(E_1^b - J_1\right)/R_1 \tag{9.15}$$

Direct comparison of Eqs. (9.14b) and (9.15) shows that

$$R_1 = (1 - \varepsilon_1)/\varepsilon_1 A_1 \tag{9.16}$$

Because R_1 is the resistance due to emissivity ε_1 associated with surface 1, it is evident that the resistance for surface 2 may be written as

$$R_2 = (1 - \varepsilon_2)/\varepsilon_2 A_2 \tag{9.17}$$

Finally, R_{sf} is related to q_1 by

$$q_1 = (J_1 - J_2)/R_{sf} \tag{9.18a}$$

where

$$R_{sf} = 1/(A_1 \mathbb{F}_{1-2}) \tag{9.18b}$$

The rate of heat transfer by radiation between the two gray surfaces can be calculated from

$$q_{1-2} = \left(E_1^b - E_2^b\right)/(R_1 + R_{sf} + R_2) \tag{9.19}$$

From Eq. (9.1) it is clear that

$$E_1^b - E_2^b = \sigma\left(T_1^{*4} - T_2^{*4}\right) \tag{9.20}$$

The electrical analogy provides a simple and direct approach for calculating the heat transferred by relating it to the black body surface energies while minimizing the algebra involved. Values of the emissivity ε that are employed in Eqs. (9.16) and (9.17) are given in Table 9.1.

9.2.3 The Radiation Heat Transfer Coefficient

It is interesting to consider the rate of heat transfer by radiation between two gray bodies in terms of a radiation heat transfer coefficient h_r. If we rewrite the equation for q_{1-2} as

$$q_{1-2} = h_r A_1 (T_1 - T_2) \tag{9.21}$$

then the relation for h_r can be derived from Eq. (9.19) and (9.21) as

$$h_r = \sigma\left(T_1^{*4} - T_2^{*4}\right)/\left[A_1(T_1 - T_2)(R_1 + R_{sf} + R_2)\right] \tag{9.22a}$$

It is easy to show the following identity, which is useful in developing Eq. (9.22a):

$$a^4 - b^4 = (a - b)\left(a^3 + a^2 b + ab^2 + b^3\right) \tag{9.22b}$$

TABLE 9.1
Emissivity ε of different materials $T = 100°C$

Material	Condition	
Alleghany metal	No. 4 polish	.13
Alleghany alloy No. 66	Polished	.11
Aluminum	Commercial sheet	.09
Aluminum	Polished	.095
Aluminum	Rough polish	.18
Brass	Polished	.059
Carbon	Rough plate	.77
Carbon, graphitized	Rough plate	.76
Chromium	Polished	.075
Copper	Polished	.052
Copper–nickel	Polished	.059
Iron	Dark gray surface	.31
Iron	Roughly polished	.27
Lampblack	Rough deposit	.84
Molybdenum	Polished	.071
Nickel	Polished	.072
Nickel–silver	Polished	.135
Paint, white	Clean	.79
Paint, cream	Clean	.77
Paint, black	Clean	.84
Paint, bronze	Clean	.51
Silver	Polished	.052
Stainless steel	Polished	.074
Steel	Polished	.066
Tin	Polished	.069
Tin	Commercial coat	.084
Tungsten	Polished coat	.066
Zinc	Commercial coat	.21
Fuzed quartz	1.96 mm thick	.775
Covex D (glass)	3.40 mm thick	.83
Nonex (glass)	1.57 mm thick	.835
Aluminum paint		.29

Substituting Eq. (9.22b) into Eq. (9.22a) gives

$$h_r = \sigma\left(T_1{}^{*3} + T_1{}^{*2}T_2{}^{*} + T_1{}^{*}T_2{}^{*2} + T_2{}^{*3}\right)/\left[A_1(R_1 + R_{sf} + R_2)\right] \tag{9.23}$$

Next, if we substitute Eqs. (9.16), (9.17) and (9.18b) into Eq. (9.23), we obtain

$$h_r = \frac{\sigma\left(T_1{}^{*3} + T_1{}^{*2}T_2{}^{*} + T_1{}^{*}T_2{}^{*2} + T_2{}^{*3}\right)}{(1-\varepsilon_1)/\varepsilon_1 + (1/\mathbb{F}_{1-2}) + (A_1/A_2)(1-\varepsilon_2)/\varepsilon_2} \tag{9.24}$$

Note also that the thermal resistance associated with the radiation heat transfer rate is given by

$$R_r = 1/(h_r A_1) \tag{9.25}$$

The application of the radiation heat transfer coefficient is illustrated in the exercises following this chapter.

9.3 CONVECTION HEAT TRANSFER

In the preceding section, we considered heat transfer by radiation and noted that a fluid was not necessary as a transport media for heat to be dissipated by radiation. In Chapter 8 we considered conduction, where heat transfer took place in a solid or in a composite that consisted of a combination of solids in thermal contact. In this section, we introduce convection as a means for heat transfer. With convection, heat is transferred from a solid to a fluid, where a boundary layer forms at the interface. As the boundary layer impedes the amount of heat transferred, it is an important parameter that must be incorporated into the determination of the convective heat transfer coefficient.

To illustrate the importance of convection, consider the boundary layer shown conceptually in Fig. 9.8 where a wall of area A is maintained at a constant temperature T_w. The fluid (air) in the boundary layer is moving and heat is transferred by this motion across the layer to the external sink, which is at a temperature T_a. The heat transferred across the boundary layer is given by

$$q_{cv} = hA(T_w - T_a) \tag{9.26}$$

where h is the convective heat transfer coefficient.

Consider again the same boundary layer but without fluid motion. In this case, convection cannot occur without fluid motion and the heat is transferred across the layer entirely by conduction through the fluid. From Eq. (8.24) it is evident that

$$q_{cd} = kA(T_w - T_a)/\mathscr{L} \tag{9.27}$$

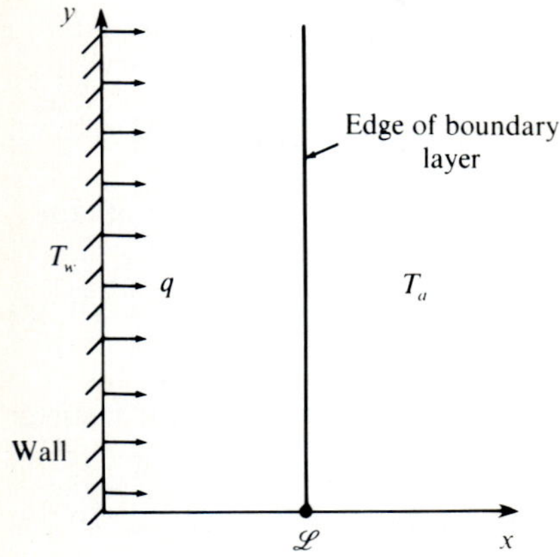

FIGURE 9.8
Heat transfer across a boundary layer that separates the wall from the external sink.

The ratio q_{cv}/q_{cd} gives the Nusselt number Nu, defined by

$$\mathrm{Nu}_{\mathscr{L}} = q_{cv}/q_{cd} = h\mathscr{L}/k \tag{9.28}$$

where $\mathscr{L}$ is a characteristic length parameter.

A relatively low value of the Nusselt number, about 1, indicates a small contribution due to convection in the heat transfer process because of low velocity in the boundary layer. On the other hand, if the Nusselt number is large, about 100, then convection is dominant and the boundary layer exhibits turbulent flow.

Convective heat transfer is divided into two different categories, namely, free or natural convection, where buoyancy produces the movement of the fluid, and forced convection, where fans or blowers drive the fluid. In electronic systems the fluid is usually air, which serves as the transport media to carry the heat to the environment. Air can be blown over the electronic devices directly to dissipate the heat from the chip carriers, or the air can be passed over plate- or duct-type heat exchangers to remove the heat from one or more subsystems. In either case it is necessary to determine the convective heat transfer coefficient h associated with the dissipation of heat from the circuit boards or heat exchangers.

A mathematical theory for boundary layers and convection across these layers has been developed and is presented in standard textbooks on heat transfer [4]. However, the theory is very complex and the few meaningful solutions that exist are not in a form that can be easily employed. To obtain more useful solutions, experiments are conducted and empirical relations are developed in terms of dimensionless parameters. The dimensionless parameters that are used in describing the convective heat transfer coefficient include the Nusselt, Prandtl, Grashof and Reynolds numbers.

Another factor that markedly affects the rate of heat transfer is the type of flow. Is the flow laminar with a velocity profile that varies smoothly with distance across the boundary layer or is the flow turbulent with strong mixing occurring in the boundary layer? We classify the type of flow as laminar or turbulent, and then use the most suitable empirical relation to obtain a solution that is sufficiently accurate for engineering design. This approach covers an analysis of air cooled electronic systems. It should be recognized that convective heat transfer with a much wider range of fluids (say liquid metals to oils) is a much more difficult topic, but we avoid the added complexity due to different cooling fluids in this treatment.

9.4 FREE OR NATURAL CONVECTION

An example of free convection is presented in Fig. 9.9 where the boundary layer for a heated vertical plate is shown. Note that the free stream velocity U is zero and that the velocity u of the flow in the boundary layer is achieved by the local elevation in the temperature of the air. The heated air has a lower density ρ and buoyant forces develop, which produce velocities u in the boundary layer. The

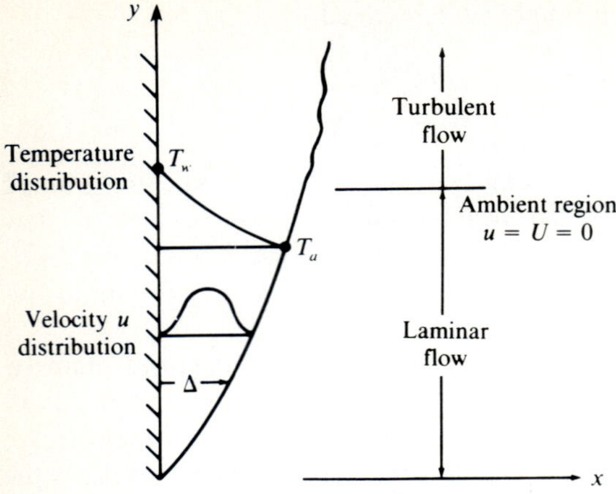

FIGURE 9.9
Boundary layer, velocity distribution and temperature distribution for natural convection associated with a heated vertical plate.

convection velocity u increases with the position parameter y according to

$$u_{\text{ave}} \simeq \left[(1 - \rho_w/\rho_a) gy\right]^{1/2} \tag{9.29}$$

where ρ is the density of air
g is the gravitational constant

and subscripts w and a refer to the wall and ambient locations. If the wall is high the velocities become large and turbulent flow occurs. Clearly, as the boundary layer develops, its thickness Δ and its mass flux both increase with y. The added mass of fluid is due to entrainment from the ambient region into the boundary layer.

The local Nusselt number $\text{Nu}(y)$ associated with free convection depends on the Grashof number, the Prandtl number and the shape of the surface under consideration. The local Grashof number is defined as

$$\text{Gr}_y = (g\beta/\nu^2)(T_w - T_a)y^3 \tag{9.30}$$

where β is the coefficient of thermal expansion
ν is the kinematic viscosity

The term $g\beta/\nu^2$ depends only on temperature for any fluid. For air, which is our primary concern, this term together with other useful properties is given in Table 9.2.

Next, recall the definition of the Prandtl number as

$$\text{Pr} = \mu c_p/k \tag{9.31}$$

Substituting Eq. (8.28) into Eq. (9.31) gives

$$\text{Pr} = \nu/\alpha \tag{9.32}$$

Reference to Table 9.2 shows that the Prandtl number is nearly constant at $\text{Pr} = 0.72$ over the temperature range normally encountered in air cooling of

TABLE 9.2
Properties of dry air at atmospheric pressure

T		ρ	c_p	μ	ν	k		β	$g\rho^2/\mu^2$	$g\beta\rho^2/\mu^2$
°C	°F	(g / in.^{3})	(J / g °C)	(10^{-4}g / in. s)	(in.^{2} / s)	(10^{-4} W / m °C)	Pr	(10^{-3} / °C)	(10^6 / in.^{3})	(10^3 / in.^{3} °C)
− 18	0	0.0227	1.000	4.195	0.0187	5.803	0.73	3.916	1.12	4.38
0	32	0.0212	1.003	4.403	0.0209	6.168	0.72	3.661	0.899	3.29
19	50	0.0204	1.004	4.519	0.0220	6.374	0.72	3.528	0.812	2.90
38	100	0.0186	1.004	4.856	0.0259	6.945	0.72	3.216	0.569	1.83
66	150	0.0171	1.008	5.150	0.0301	7.473	0.72	2.952	0.446	1.36
93	200	0.0158	1.008	5.442	0.0344	8.000	0.72	2.729	0.324	0.885
121	250	0.0147	1.012	5.780	0.0392	8.440	0.71	2.538	0.260	0.674
149	300	0.0137	1.012	6.085	0.0441	8.968	0.71	2.370	0.195	0.462
177	350	0.0129	1.016	6.305	0.0491	9.495	0.70	2.214	0.161	0.366
204	400	0.0121	1.024	6.614	0.0544	9.979	0.69	2.094	0.128	0.269

electronic components. Recognizing this simplification permits Pr to be treated as a constant (0.72) in all subsequent developments for both free and forced convection.

9.4.1 Free Convection on a Vertical Plane with Laminar Flow

If the buoyant flow along the vertical plate is laminar, experimental results lead to an expression for the local Nusselt number:

$$\mathrm{Nu}_y = f(\mathrm{Pr})\mathrm{Gr}_y^{1/4} \tag{9.33}$$

This equation is valid in the laminar flow region $10^4 \le \mathrm{Gr}_y \le 10^9$. The functional relation for $f(\mathrm{Pr})$ varies to some degree depending on the reference used. Here we will use the results of Ostrach [5] and LeFevre [6] to obtain

$$f(\mathrm{Pr}) = \tfrac{3}{4}\mathrm{Pr}^{1/2}/\left[2.435 + 4.884\mathrm{Pr}^{1/2} + 4.953\mathrm{Pr}\right]^{1/4} \tag{9.34}$$

If we set Pr = 0.72, then $f(\mathrm{Pr}) = 0.357$ and Eq. (9.33) can be written as

$$\mathrm{Nu}_y = 0.357\mathrm{Gr}_y^{1/4} \tag{9.35}$$

The local value of h_y is

$$h_y = k\mathrm{Nu}_y/y = 0.357k\mathrm{Gr}^{1/4}/y \tag{9.36}$$

Substituting Eq. (9.30) into Eq. (9.36) we obtain

$$h_y = 0.357k\left(g\beta/\nu^2\right)^{1/4}\left(T_w - T_a\right)^{1/4}y^{-1/4} \tag{9.37}$$

where $g\beta/\nu^2$ is determined for $T_{\mathrm{ave}} = (T_w + T_a)/2$.

We note from the form of Eq. (9.37) that the convective heat transfer coefficient decreases with increasing y as shown in Fig. 9.10. This reduction is due to the increase in the thickness of the boundary layer with y, which causes an increase in the thermal resistance. The average convection coefficient h_{ave} over the height H of the vertical plate is

$$h_{\mathrm{ave}} = (1/H)\int_0^H h_y\,dy \tag{9.38}$$

Integrating Eq. (9.38) gives

$$h_{\mathrm{ave}} = \tfrac{4}{3}\left[h_y\right]_{y=H} \tag{9.39}$$

This result indicates that the average heat transfer coefficient is larger by a factor of $\frac{4}{3}$ than the value of h at $y = H$.

The results of Eq. (9.35) are for a constant temperature T_w at the wall. However, if the wall temperature varies while the heat flux transferred remains constant with position y, the relation for the local Nusselt number changes to

$$\mathrm{Nu}_y = f_1(\mathrm{Pr})\mathrm{Gr}_y^{*1/5} \tag{9.40}$$

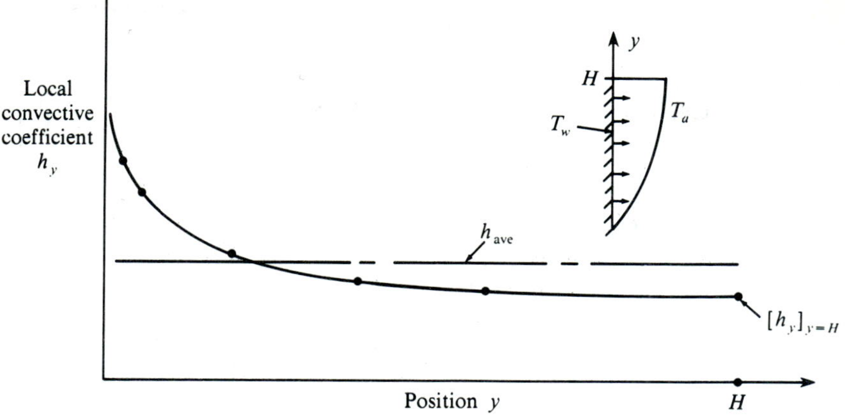

FIGURE 9.10
Variation in h_y with position y on a vertical plate with laminar flow due to free convection.

where $f_1(0.72) = 0.487$
$\quad\quad Gr_y^* = (q/A)g\beta y^4/(k\nu^2)$

The results from Eq. (9.40) indicate slightly higher values of Nu_y and h_y than the results for constant wall temperature given in Eq. (9.35). Exercises are provided to show the change in wall temperature for this case and to indicate the difference in h.

9.4.2 Free Convection on a Vertical Plate with Turbulent Flow

When the length of the vertical plate is extended, a transition from laminar flow to turbulent flow occurs that begins when $Gr = 10^7$. The turbulence becomes fully developed at $Gr = 10^9$. The Rayleigh number Ra, which is given by

$$Ra = GrPr \tag{9.41}$$

is often used to indicate the onset of turbulence in free convective flow. The conversion from laminar to turbulent flow depends on local disturbances and it is common to account for these local effects by considering transition at a relatively low value of $Ra = 10^9$. For values of $Ra \geq 10^9$, where the flow is turbulent, the local Nusselt number is

$$Nu_y = f_2(Pr)Gr^{2/5} \tag{9.42}$$

where $f_2(0.72) = 0.0221$. Substituting Eqs. (9.28) and (9.30) into Eq. (9.42) indicates that $h_y \propto y^{1/5}$; hence, integration with respect to y gives

$$Nu_{ave} = (5/6)[Nu_y]_{y=H} \tag{9.43}$$

Exercise 9.24 is provided to show the difference between heat transfer under laminar and turbulent flow conditions.

9.4.3 An Approximate Relation for Free Convection for the Vertical Plate

It is possible to avoid determining if the convective flow is laminar or turbulent by using an approximate relationship for the average Nusselt number, developed by Churchill and Chu [7], that is given by

$$[\mathrm{Nu}_H]_{\mathrm{ave}}^{1/2} = 0.825 + 0.325\mathrm{Ra}_H^{1/6} \tag{9.44}$$

where Ra_H is the Rayleigh number given by $\mathrm{Ra}_H = 0.72\mathrm{Gr}_H$. The subscript H indicates the Rayleigh and Grashof numbers are evaluated at $y = H$. Equation (9.44) is for a constant temperature wall and is valid for $0.1 \leq \mathrm{Ra}_H \leq 10^{12}$. Some accuracy is lost in employing this approximate relation, but it should give predictions accurate to $\pm 30\%$, which is within the scatter in the data reported for convection coefficients.

9.4.4 Free Convection Heat Transfer with Other Shapes

We noted in the introduction to Section 9.3 that the shape of the body was important in determining the coefficient of heat transfer. In this section we will present in summary form the empirical results for shapes of general interest in cooling electronic systems. In all of the relations listed in the following text, we have used air as the fluid and set $\mathrm{Pr} = 0.72$. Consider first a vertical cylinder of diameter D and height H. The average Nusselt number is given [8] as

$$[\mathrm{Nu}_H]_{\mathrm{ave}}^{\mathrm{cyl}} = [\mathrm{Nu}_H]_{\mathrm{ave}}^{\mathrm{plate}}[1 + \zeta^{0.9}] \tag{9.45}$$

where $\zeta = (H/D)\mathrm{Gr}_H^{-1/4}$. The average Nusselt number for the plate is determined from Eq. (9.44).

For a horizontal cylinder of diameter D [9], $\mathrm{Nu}_{\mathrm{ave}}$ is given as

$$[\mathrm{Nu}_D]_{\mathrm{ave}}^{1/2} = 0.60 + 0.322\mathrm{Ra}_D^{1/6} \tag{9.46}$$

This relation is valid for $10^{-5} \leq \mathrm{Ra}_D \leq 10^{12}$.

For a plate with a characteristic length L oriented horizontally, [10] gives the following empirical relations:

With a hot upper surface and a cool lower surface:

$$[\mathrm{Nu}_L]_{\mathrm{ave}} = 0.54\mathrm{Ra}_L^{1/4} \quad \text{for } 10^4 \leq \mathrm{Ra}_L \leq 10^7 \tag{9.47}$$

$$[\mathrm{Nu}_L]_{\mathrm{ave}} = 0.15\mathrm{Ra}_L^{1/4} \quad \text{for } 10^7 \leq \mathrm{Ra}_L \leq 10^{11} \tag{9.48}$$

With a cool upper surface and a hot lower surface:

$$[\mathrm{Nu}_L]_{\mathrm{ave}}^{1/4} = 0.27\mathrm{Ra}_L \quad \text{for } 10^5 \leq \mathrm{Ra}_L \leq 10^{11} \tag{9.49}$$

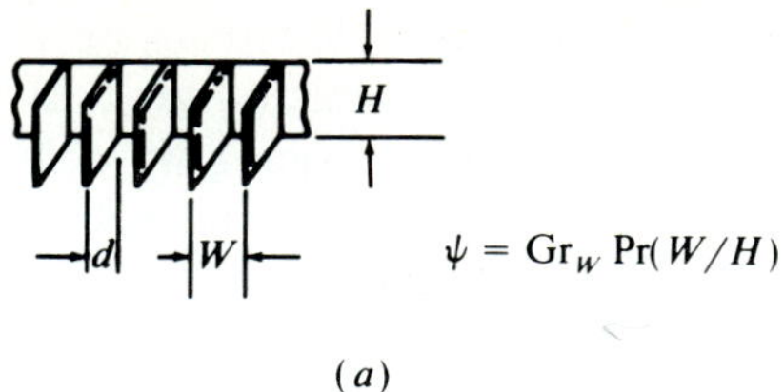

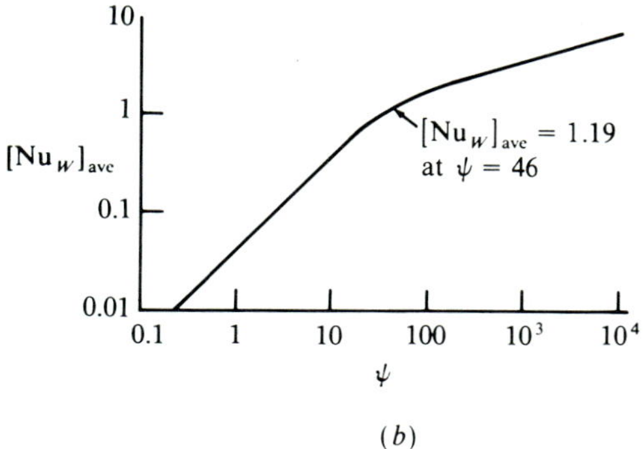

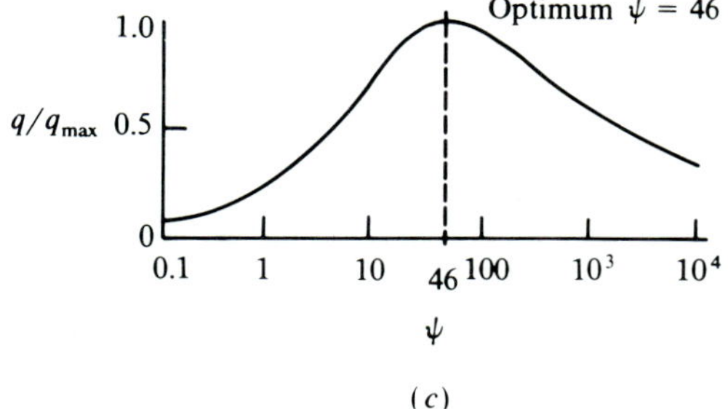

FIGURE 9.11
$[\mathrm{Nu}_W]_{\mathrm{ave}}$ and q/q_{max} as a function of ψ showing optimum spacing of the fins. (*a*) Geometry of a fin-type heat exchanger. (*b*) $[\mathrm{Nu}_W]_{\mathrm{ave}}$ as a function of ψ. (*c*) q/q_{max} as a function of ψ.

The characteristic length L is taken as the average of the length and width of a rectangular plate. It is equivalent to $0.9D$ for a circular plate. For a plate with an irregular shape, L is determined by dividing the plate area by its perimeter.

Finally, consider an array of equally spaced fins (vertical plates) as illustrated in Fig. 9.11*a*. Elenbaas [11] developed an empirical solution to this problem, which is given by

$$[\mathrm{Nu}_W]_{\mathrm{ave}} = (\psi/24)(1 - e^{-35/\psi})^{3/4} \tag{9.50}$$

where $\psi = \mathrm{Gr}_W \mathrm{Pr}(W/H)$
H is the height of each fin
W is the distance between fins

Results for Nu_W as a function of ψ are presented in Fig. 9.11*b*.

The heat transfer coefficient is $h = k\text{Nu}_W/W$ and the total heat transferred is $q = hA\,\Delta T$. The value of q/q_{max}, presented as a function of ψ in Fig. 9.11*c*, shows a maximum at $\psi = 46$, indicating an optimum spacing W for the fins. The heat transferred increases initially by decreasing the spacing between fins and adding more fins for a specified length L of heat exchanger. However, a point is reached where the fins are so dense that they inhibit flow and the capacity for heat transfer begins to decrease. The optimum spacing W for the fins is obtained from

$$W = 63.9\,H/\text{Gr}_W \tag{9.51}$$

for air with $\text{Pr} = 0.72$. At this optimum value for W, $[\text{Nu}_W]_{ave} = 1.19$.

9.5 FORCED AIR CONVECTION COEFFICIENTS

While heat dissipation from electronic systems by natural convection has the advantage of simplicity and economy, the amount of heat that can be transferred at a specified ΔT is not always sufficient to accommodate the heat generated by the system. In these cases the convective heat transfer coefficient is enhanced by increasing the velocity of the air flow over the heat exchangers or, in some cases, over the chip carriers directly. The increased air velocity is achieved by the use of one or more fans placed in the electronic enclosure with ducts formed to carry the proper mass flow of air to the various subsystems housed in the enclosure.

The approach in determining the heat transferred by forced convection is to calculate the Nusselt number and then the convection coefficient h. Again the complexities of the problem require approximate solutions based on empirical equations developed from experimental results. In this treatment we will cover two different geometries, flat plates and rectangular ducts, because they are important in the design of air cooled electronic systems.

9.5.1 Laminar Flow over a Flat Plate

Consider a flat plate inserted in a flow field as shown in Fig. 9.12. The flow field is external to one side of the plate with a velocity U in the x direction and a temperature T_a. The heat is transferred from the plate to the flow field through the boundary layers that exhibit thickness $\delta(x)$ and $\Delta(x)$. The local velocity u is a function of y in the velocity boundary layer and is represented by the polynomial

$$u/U \simeq 2\eta - 2\eta^3 + \eta^4 \tag{9.52}$$

where $\eta = y/\delta(x)$. By considering the momentum integral equation for a flat plate boundary layer, we may write

$$\tau_W = (d/dx)\int_0^\infty \rho u(U - u)\,dy = \mu(\partial u/\partial y)_{y=0} \tag{9.53}$$

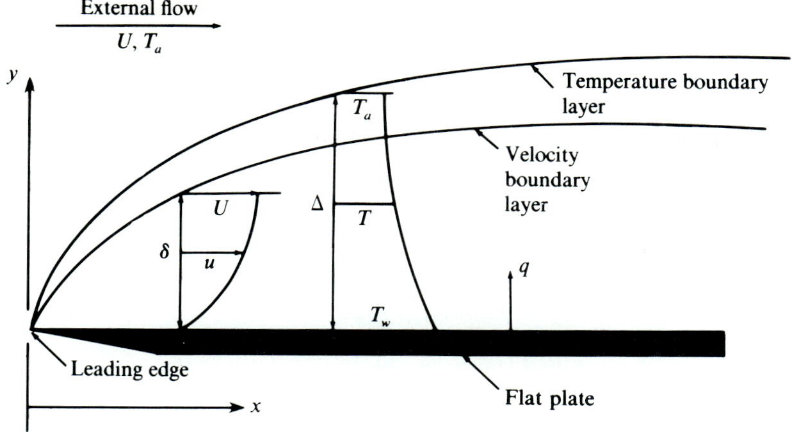

FIGURE 9.12
Boundary layers for velocity u and temperature T shown with profiles for u and T.

where τ_W is the shear stress at the surface of the plate. Substituting Eq. (9.52) into Eq. (9.53) leads to

$$\delta = 5.84x/\mathrm{Re}_x^{1/2} \tag{9.54}$$

$$\tau_W = 2\mu U/\delta \tag{9.55}$$

where

$$\mathrm{Re}_x = \rho U x/\mu = Ux/\nu \tag{9.56}$$

Inspection of these results shows that the thickness δ of the boundary layer increases with $x^{1/2}$. However, experiments indicate that the constant 5.84 is too high. In practice the constant in Eq. (9.54) is replaced with 5.0 to give improved accuracy. Hence

$$\delta = 5.0x/\mathrm{Re}_x^{1/2} \tag{9.57}$$

Noting that the friction factor f for the plate is

$$f = 2\tau_W/\rho U^2 \tag{9.58}$$

and combining Eqs. (9.55) and (9.58) gives

$$f = 0.685/\mathrm{Re}_x^{1/2} \tag{9.59}$$

The heat transferred from the flat plate is based on a polynomial approximation of the temperature profile given by

$$(T - T_W)/(T_a - T_W) = 2\zeta - 2\zeta^3 + \zeta^4 \tag{9.60}$$

where $\zeta = y/\Delta$. Note that the energy integral equation for a boundary layer is

$$q_W/A_x = (d/dx)\int \rho c_p u(T - T_a)\, dy = [-k(\partial T/\partial y)]_{y=0} \tag{9.61}$$

Substituting T from Eq. (9.59) and u from Eq. (9.52) into Eq. (9.61) leads to

$$q/A_W = 2k(T_W - T_a)/\Delta \tag{9.62}$$

The local convection coefficient h_x is

$$h_x = q_W/(T_W - T_a) = 2k/\Delta \tag{9.63}$$

This equation is not commonly used because of the presence of the unknown Δ in the denominator. Instead, h_x is determined from the local Nusselt number, which was derived by Pohlhausen [12] as

$$\mathrm{Nu}_x = 0.332\mathrm{Re}_x^{1/2} f(\mathrm{Pr}) \tag{9.64}$$

where $f(\mathrm{Pr}) = \mathrm{Pr}^{1/3}$. For air with $\mathrm{Pr} \simeq 0.72$, Eq. (9.65) reduces to

$$\mathrm{Nu}_x = 0.298\mathrm{Re}_x^{1/2} \tag{9.65}$$

If we convert Eq. (9.65) into a dimensional form and solve for h_x, we obtain

$$h_x = 0.298\rho^{1/2}U^{1/2}k\mu^{-1/2}x^{-1/2} \tag{9.66}$$

It is evident from this relation that the convection coefficient increases with velocity as $U^{1/2}$ and decreases with distance x along the length of the plate as $1/x^{1/2}$.

The average value of the convection coefficient for a plate of length L is obtained from

$$h_{\mathrm{ave}} = (1/L)\int_0^L h_x\,dx = f(T)(1/L)\int_0^L x^{-1/2}\,dx \tag{9.67}$$

Integrating gives

$$h_{\mathrm{ave}} = 2f(T)/L^{1/2} \tag{9.68}$$

Next, evaluate h at the position $x = L$ from Eq. (9.66) as

$$h_L = f(T)/L^{1/2} \tag{9.69}$$

It is clear from these two equations that

$$h_{\mathrm{ave}} = 2h_L \tag{9.70}$$

The procedure is to calculate $\mathrm{Re}_L = UL/\nu$ and to substitute this value in Eq. (9.65) to obtain Nu_L. Then $h_L = k\mathrm{Nu}_L/L$ is substituted into Eq. (9.70) to determine h_{ave}. Finally, the heat transferred from the plate to the flow field is

$$q = h_{\mathrm{ave}}A(T_W - T_a) \tag{9.71}$$

The properties of the air μ, ν, k and ρ are functions of temperature that change with y across the boundary layer. This complication is avoided by using the average temperature

$$T_{\mathrm{ave}} = (T_W + T_a)/2 \tag{9.72}$$

across the boundary layer in determining α, ν, k and ρ from Table 9.2.

9.5.2 Turbulent Flow over a Flat Plate

If the velocity U is sufficiently large, the boundary layer is more involved than that shown in Fig. 9.12. At the higher velocities, the flow in the boundary layer undergoes a transition from laminar to turbulent flow. The form of the boundary layer, as it is developed from the leading edge of the plate, includes the laminar, transition and turbulent regions as presented in Fig. 9.13. The position of the transition x_c is a function of the Reynolds number and the surface roughness. Usually the transition position for a flat plate is determined from the critical Reynolds number, which is usually taken as

$$\mathrm{Re}_{xc} = 500{,}000 \tag{9.73}$$

where the free field velocity U is used in calculating Re. Using Eq. (9.56) together with Eq. (9.73) gives the location of the transition region as

$$x_c = (5\nu/U) \times 10^5 \tag{9.74}$$

For flat plates with $L > x_c$, the flow is turbulent and different relations exist for the friction factor and the Nusselt number. The friction factor is

$$\begin{aligned} f_x &= 0.0592\mathrm{Re}_x^{-1/5} \\ f_{\mathrm{ave}} &= 0.074\mathrm{Re}_L^{-1/5} \end{aligned} \tag{9.75}$$

providing $5 \times 10^5 < \mathrm{Re}_x < 10^7$. The Nusselt number is

$$\mathrm{Nu}_x = 0.0296\mathrm{Re}_x^{4/5}\mathrm{Pr}^{1/3} \tag{9.76}$$

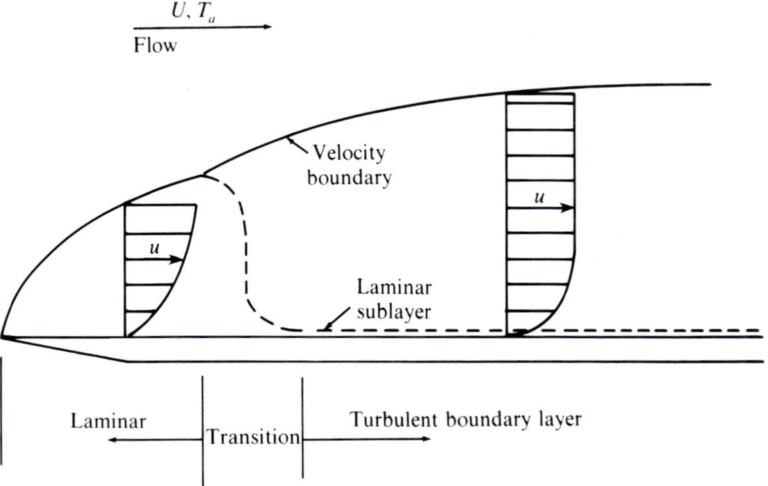

FIGURE 9.13
Development of a laminar and turbulent boundary layers on a flat plate.

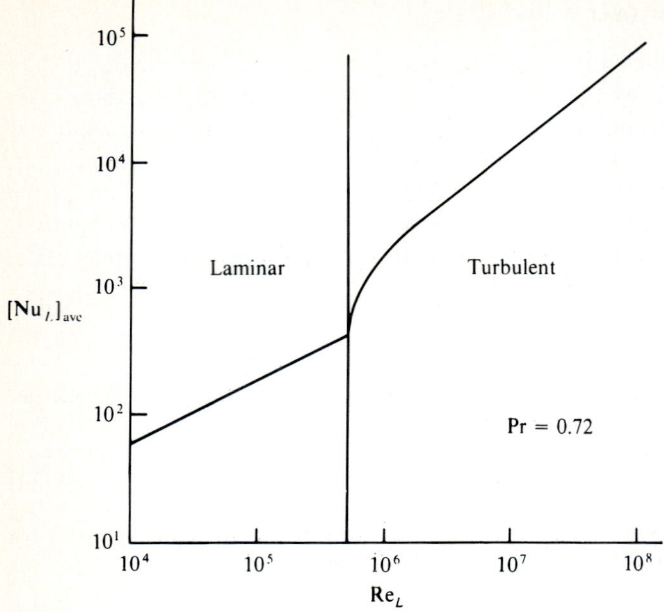

FIGURE 9.14
Change in the average Nusselt number with Reynolds number for a flat plate $\mathrm{Re}_c = 500{,}000$.

Noting Pr = 0.72 for air reduces Eq. (9.76) to

$$\mathrm{Nu}_x = 0.0265\mathrm{Re}_x^{4/5} \tag{9.77}$$

Finally, the average value for Nu over a plate of length L is

$$[\mathrm{Nu}_L]_{\mathrm{ave}} = 0.0332\mathrm{Re}_L^{4/5} \tag{9.78}$$

A comparison of $[\mathrm{Nu}_L]_{\mathrm{ave}}$ for turbulent and laminar flow is presented in Fig. 9.14. It is clear that $[\mathrm{Nu}_L]_{\mathrm{ave}}$ increases markedly with Reynolds number and that higher velocity flow that produces turbulence markedly increases the convection heat transfer coefficient $[h_L]_{\mathrm{ave}}$.

9.5.3 Combined Laminar and Turbulent Flow over a Flat Plate

We have treated laminar and turbulent flow as two separate cases and introduced turbulence when $\mathrm{Re}_L > 500{,}000$. When the flow velocity produces Re_L larger than 500,000, both laminar and turbulent flow occur together and the heat transferred in both the laminar and turbulent regions of the plate must be considered. The convection coefficient averaged over the length of the plate is the sum of

$$h_{\text{ave}} = (k/L)\left[\int_0^{xc} (\text{Nu}_x^L/x)\,dx + \int_{xc}^{L} (\text{Nu}_x^T/x)\,dx\right] \tag{9.79}$$

where the superscripts L and T refer to laminar and turbulent flow conditions. It is possible to write Eq. (9.79) as

$$[\text{Nu}_L]_{\text{ave}} = (h_{\text{ave}}L/k) = [\text{Nu}_{xc}^L]_{\text{ave}} + [\text{Nu}_L^T]_{\text{ave}} - [\text{Nu}_{xc}^T]_{\text{ave}} \tag{9.80}$$

Substituting Eqs. (9.65), (9.70) and (9.78) into Eq. (9.80) gives

$$(h_{\text{ave}}L/k) = 0.596\text{Re}_{xc}^{1/2} + 0.0332(\text{Re}_L^{4/5} - \text{Re}_{xc}^{4/5}) \tag{9.81}$$

Noting that $\text{Re}_{xc} = 5 \times 10^5$ in Eq. (9.81) gives

$$(h_{\text{ave}}L/k) = 0.0332\text{Re}_L^{4/5} - 781.7 \tag{9.82}$$

These results clearly indicate that higher heat transfer coefficients can be achieved by increasing the Reynolds number since h is proportional to $U^{4/5}$. However, higher velocities require more power from the fans since the shear stress produced on the surface of the plate τ_W is proportional to $U^{9/5}$. More importantly, the noise level is increased with higher velocity fans and the noise must not become objectionable when compared to the background sound levels associated with the operating environment.

9.5.4 Convection from a Flat Plate with a Uniform Heat Flux

The previous developments for heat transfer from the flat plate to an external flow field considered the plate at a constant temperature. However, in some applications the heat flux to the plate is held constant and the plate temperatures change with x. The relations describing the heat transfer coefficients are obtained in a similar manner and only the final results will be summarized here:

$$\text{Nu}_x = 0.406\text{Re}_x^{1/2} \quad \text{for laminar flow} \tag{9.83a}$$

$$\text{Nu}_x = 0.0276\text{Re}_x^{4/5} \quad \text{for turbulent flow} \tag{9.83b}$$

The temperature of the plate varies with position x according to

$$T_W = T_a + (q_* x)/(k\text{Nu}_x) \tag{9.84}$$

where $q_* = q/A$ is the heat flux transferred by the plate.

9.6 FORCED CONVECTION WITH INTERNAL FLOW—DUCTS

In most design applications it is necessary to transport the air to electronic subsystems that are arranged within the enclosure. The air passes through a heat exchanger for the subsystem, undergoes an increase in temperature and is then transported to another subsystem or exhausted into the atmosphere. Ducts are the passages used in transporting the air within the enclosures. In some cases portions of the duct also serve as the heat exchangers. Components are mounted on the duct walls and heat is conducted through the wall to be dissipated into the air stream. Analysis of the head required to drive the air and the heat transferred in a duct is affected by the flow field. It is necessary to first consider the duct entrance where the flow is developing and then to classify the flow as laminar or turbulent. The equations used in the analysis depend on the characteristics of the flow field.

First, examine the entrance of a duct and observe that the requirements for continuous temperature distribution and zero velocity at the wall produce boundary layers similar to those described in Section 9.5. As we move down the duct, the boundary layers increase in thickness and at some location x_{fd} the layer boundaries meet, as illustrated in Fig. 9.15. For $x > x_{fd}$, the flow field is fully developed and the duct is filled with boundary layer flow. In the fully developed region, the velocity profile is independent of position x. Also, the shape of the temperature profile is independent of x but the magnitude of the temperature

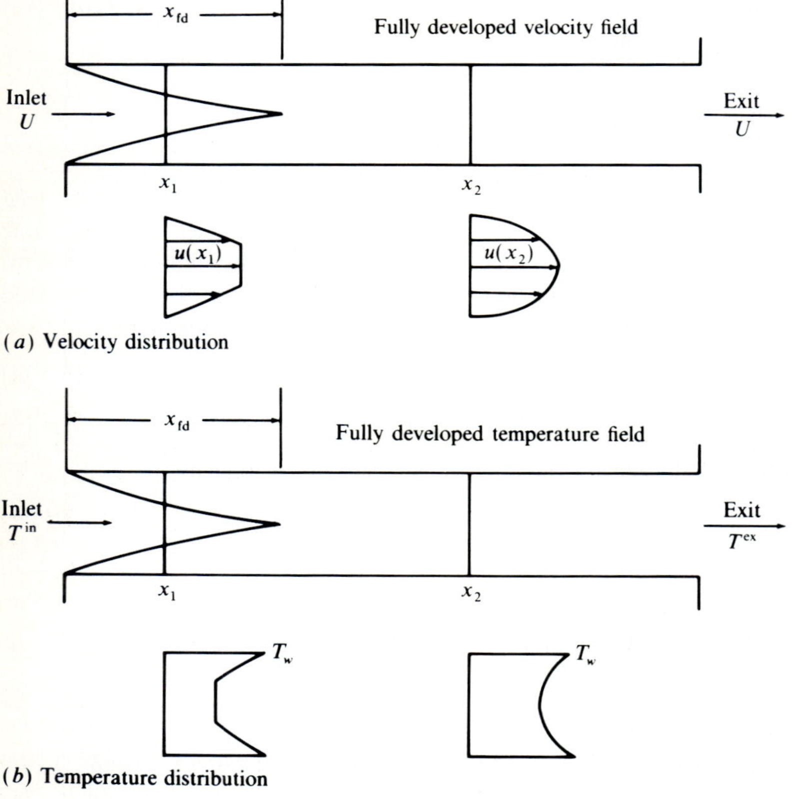

FIGURE 9.15
Temperature and velocity distributions for air flow in a duct with $\text{Pr} \simeq 1$.

usually varies due to the heat transferred from the walls of the duct to the air stream.

The entrance region $x < x_{fd}$ exhibits velocity and temperature profiles that change with location x. The heat transfer coefficient and the friction factor in the entrance regions are high because the boundary layer is thin. For general flow conditions, $\mathrm{Pr} \neq 1$ and the value of x_{fd} is different for velocity and temperature. However, for air, in the temperature range of interest, $\mathrm{Pr} \simeq 0.72$, which is close enough to 1 to permit the assumption that the fully developed conditions for temperature and velocity occur at the same x_{fd} given by

$$x_{fd} = 0.04\mathrm{Re}_D D \tag{9.85}$$

9.6.1 Average Velocity and Temperature

We noted in Fig. 9.15 that both the temperature and the velocity varied with distance from the walls in both developing flow and in fully developed flow. These profiles affect the head requirements and the heat transferred, but we normally do not seek detailed information on the flow field. Instead we work with average values of the velocity and temperature and find suitable relations for the unknowns Nu and f in terms of Re_D, position x and duct shape.

For steady flow in a straight duct with a cross-sectional area A, the mass flow rate $\dot{m}$ is constant and is given by

$$\dot{m} = \rho U A = G A \tag{9.86a}$$

where the density ρ is considered as a constant with respect to position x along the length of the duct. The velocity U is an average over the cross section defined by

$$U = Q/A = (1/A)\int u\, dA \tag{9.87a}$$

where Q is the volume flow rate through the duct. The mass velocity G is defined as

$$G = \dot{m}/A = \rho U \tag{9.86b}$$

Conservation of energy indicates that the heat transferred from the duct walls q_W increases the enthalpy of the air according to

$$q_W = \dot{m} c_p (T^{\mathrm{ex}} - T^{\mathrm{in}}) \tag{9.88}$$

where the superscripts refer to the exit and inlet temperature. The average temperature at some location x along the length of the duct is determined from the integration of the enthalpy flux over a specified cross section. Hence

$$T_{\mathrm{ave}} = (\rho/\dot{m})\int uT\, dA \tag{9.89}$$

The relations for $u(y)$ and $T(y)$, which are substituted into Eq. (9.89), depend on whether the flow is laminar or turbulent and these two flow conditions are covered later in Section 9.7. After integration as indicated by Eq. (9.89), the temperature T_{ave} depends only on x. We can develop the relation for $T_{\mathrm{ave}}(x)$ if

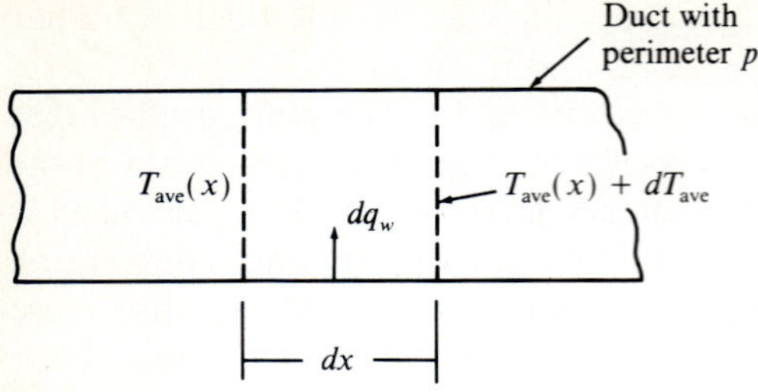

FIGURE 9.16
Control volume for heat exchange between the walls and the air stream.

the steady state energy equation for the control volume shown in Fig. 9.16 is written as

$$dq_W = h(T_W - T_{ave})p\,dx = \dot{m}c_p\,dT_{ave} \tag{9.90}$$

where p is the perimeter of the duct. If we take the case of a constant wall temperature T_W, the integration of Eq. (9.90) leads to

$$\int_{T^{in}}^{T_{ave}} dT/(T - T_W) = -(p/\dot{m}c_p)\int_0^x h(x)\,dx \tag{9.91}$$

Equation (9.91) reduces to

$$T_{ave} = T_W + (T^{in} - T_W)e^{-\beta(x)} \tag{9.92}$$

where

$$\beta = -(p/\dot{m}c_p)\int_0^x h\,dx \tag{9.93}$$

The distribution of T_{ave} with x is presented in Fig. 9.17. In the fully developed region h is a constant with respect to x and Eq. (9.93) indicates that β increases as a linear function of x. For this reason T_{ave} approaches T_W as an exponential for large x.

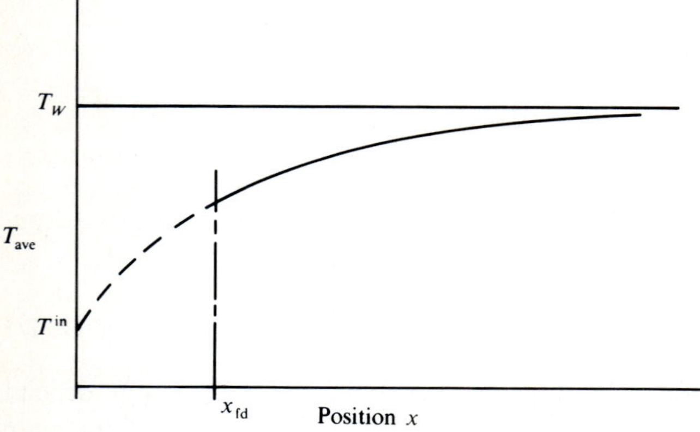

FIGURE 9.17
Temperature T_{ave} as a function of position x for steady state air flow in a duct with T_W constant.

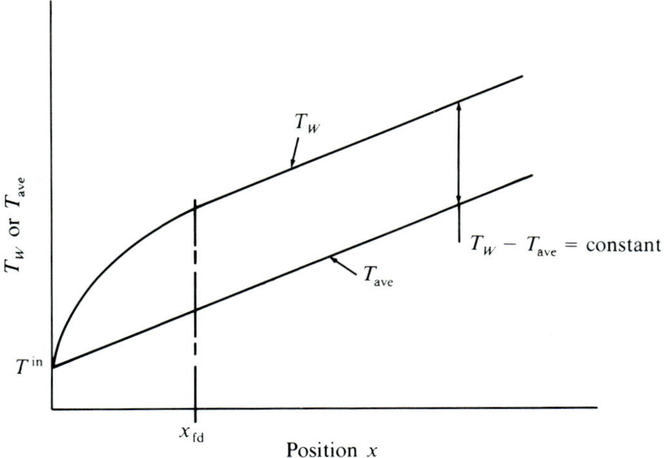

FIGURE 9.18
Temperature distributions along the duct for q_{W*} equal a constant.

As a second case, consider that the heat flux q_{W*} is a constant with x along the wall of the duct. Noting that

$$dq_W = q_{W*} p \, dx \tag{9.94}$$

then integration of Eq. (9.90) leads to

$$\left(pq_{W*}/\dot{m}c_p \right) \int_0^x dx = \int_{T^{in}}^{T_{ave}} dT \tag{9.95}$$

which become

$$T_{ave} = T^{in} + \left(q_{W*} p/\dot{m}c_p \right) x \tag{9.96}$$

This result indicates that the average temperature of the air increases linearly with position as it flows down the duct as shown in Fig. 9.18. With q_{W*} constant, the wall temperature of the duct must increase as a function of position. The relation for $T_W(x)$ is determined from Eq. (9.90) as

$$T_W = q_{W*}/h(x) + T_{ave} \tag{9.97}$$

The distribution of the wall temperature is illustrated in Fig. 9.18. In the region of fully developed flow, the difference between T_{ave} and T_W is constant for $x_{fd} < x < L$.

9.6.2 Temperature Difference

When designing ducts, we are interested in determining the total heat transferred over the length L of the duct. We determine q_W by applying Eq. (9.26), which is recast as

$$q_W = h_x pL(T_W - T_{ave})_x \tag{9.98}$$

to show the relation with subscripts previously introduced in this section. For the case of a duct with a constant temperature wall, the application of Eq. (9.98) is difficult because T_{ave} varies with x as shown in Fig. 9.17. To simplify the approach, Eq. (9.98) is rewritten in terms of average values for h_x and $(T_W - T_{ave})_{ave}$ as

$$q_W = h_{ave} pL(T_W - T_{ave})_{ave} \tag{9.99}$$

Note, in this expression that we are considering two averages. The first across the cross section to get T_{ave} and the second over the length L to obtain $(T_W - T_{ave})_{ave}$. The total heat transferred to the air is

$$q_a = \dot{m}c_p(T^{ex} - T^{in}) = \dot{m}c_p\left[(T^{ex} - T_W) - (T^{in} - T_W)\right] \tag{9.100}$$

For $x = L$, Eq. (9.92) gives

$$T_{ave} = T^{ex} = T_W + (T^{in} - T_W)e^{-\beta} \tag{9.101}$$

where

$$\beta = pLh_{ave}/\dot{m}c_p \tag{9.102}$$

From Eqs. (9.101) and (9.102), it is evident that

$$pLh_{ave}/\dot{m}c_p = \ln\left[(T^{ex} - T_W)/(T^{in} - T_W)\right] \tag{9.103}$$

Now, if we set $q_a = q_W$ and use Eq. (9.103), it is clear that

$$(T_W - T_{ave})_{ave} = \frac{\left[(T^{ex} - T_W) - (T^{in} - T_W)\right]}{\ln\left[(T^{ex} - T_W)/(T^{in} - T_W)\right]} \tag{9.104}$$

This expression is often called the log-mean temperature difference. If the inlet and exit temperatures do not differ by more than 50% Eq. (9.104) can be replaced without introducing significant error by the arithmetic average

$$(T_W - T_{ave})_{ave} = \tfrac{1}{2}\left[(T^{ex} - T_W) + (T^{in} - T_W)\right] \tag{9.105}$$

The temperatures T^{in}, T^{ex} and T_W are either known or easily measured and are used with Eqs. (9.105) and (9.99) to determine the heat transferred as the air passes through the duct.

9.7 HEAT TRANSFER AND FRICTION COEFFICIENTS FOR DUCT FLOW

The coefficients of interest for duct flow are determined from the Nusselt number and the Reynolds number with both of these numbers referenced to the diameter D of the duct by

$$\mathrm{Nu}_D = hD/k \tag{9.106}$$

$$\mathrm{Re}_D = UD/\nu \tag{9.107}$$

For ducts with cross sections that are not circular, D is the hydraulic diameter given by

$$D = 4A/p \tag{9.108}$$

We will consider first the coefficients associated with fully developed flow and summarize the design equations for laminar and then turbulent flow.

For fully developed laminar flow in circular ducts:

$$f = 16/\mathrm{Re}_D \tag{9.109}$$

$$\mathrm{Nu}_D^q = 4.36 \quad \text{with constant heat flux} \tag{9.110a}$$

$$\mathrm{Nu}_D^T = 3.66 \quad \text{with constant wall temperature} \tag{9.110b}$$

$$\mathrm{Re}_D < 2100 \quad \text{for laminar duct flow} \tag{9.111}$$

For fully developed laminar flow in rectangular ducts [13]:

w/h	Nu_D^q	Nu_D^T	$f\,\mathrm{Re}_D$
1	3.61	2.98	14.23
2	4.12	3.39	15.55
3	4.79	3.96	17.09
4	5.33	4.44	18.23
6	6.05	5.14	19.70
8	6.49	5.60	20.58
∞	8.24	7.54	24.00

(9.112)

The symbols w and h refer to the width and height of the rectangular cross section.

For fully developed turbulent flow [14]:

$$(4f)^{-1/2} = -1.8\log\left[(\varepsilon/3.7D)^{1.11} + 6.9/\mathrm{Re}_D\right] \tag{9.113}$$

$$\mathrm{Nu}_D = 0.023\mathrm{Re}_D^{0.8} \quad \text{and} \quad \mathrm{Nu}_D^q = \mathrm{Nu}_D^T \tag{9.114}$$

$$\mathrm{Re}_D > 10{,}000 \quad \text{for turbulent duct flow} \tag{9.115}$$

where ε is the surface roughness.

Consider next the entrance region of the duct where the flow is developing. The boundary layers are thin in this region and the friction factor f and the convection coefficient are elevated. The theory and the experiments are also much more complex and the analysis of the developing flow region is beyond the scope of this text. However, we will indicate some empirical results that may be useful for circular ducts.

For laminar developing flow in circular ducts: The friction coefficient is shown as a function of a normalized position parameter in Fig. 9.19. The average Nusselt numbers for a constant temperature wall and for a uniform heat flux are presented in Fig. 9.20. These graphs are both based on Pr = 0.7, which is a close

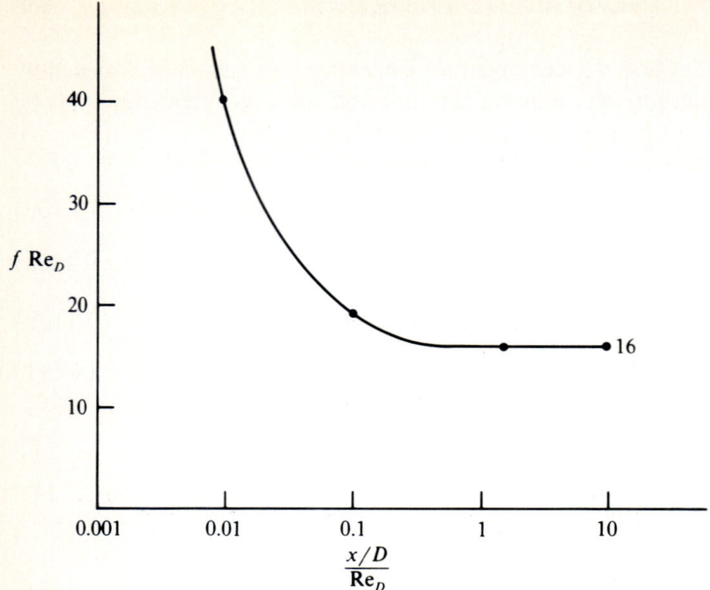

FIGURE 9.19
Friction factor for developing laminar flow in a circular duct.

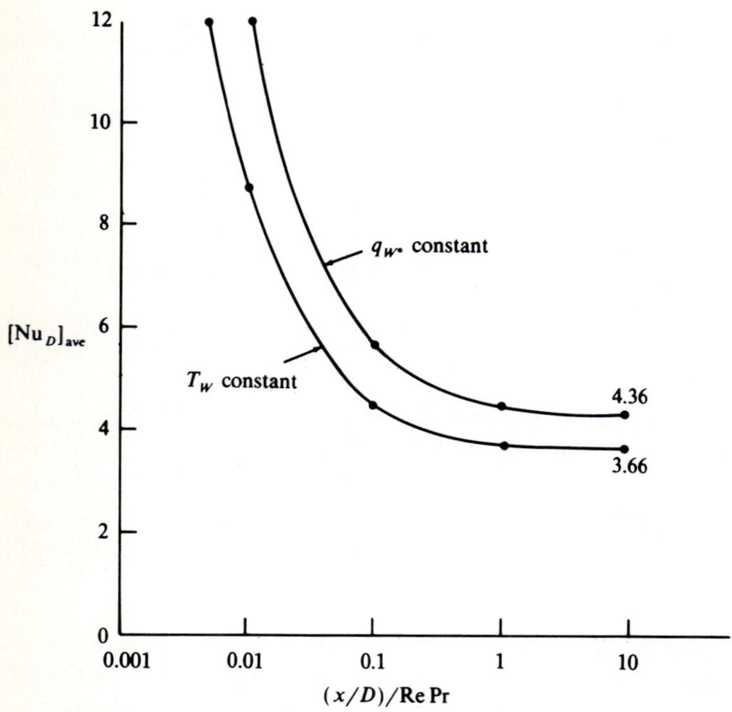

FIGURE 9.20
Average Nusselt number for developing laminar flow in a circular duct.

approximation to Pr for air-cooled electronic systems. The graphical results can be used to adjust results for the effects of the entrance. Recall that the extent of the developing region is given by x_{fd} and is determined from Eq. (9.85).

For developing turbulent flow:

$$[\mathrm{Nu}_x]_{ave}/[\mathrm{Nu}]_{fd} = 1 + 2/(x/D) \tag{9.116}$$

where $[\mathrm{Nu}]_{fd}$ is given in Eq. (9.114).

9.7.1 Summary

We have presented results for both free and forced convection that can be used to design cooling systems for electronic equipment. In both cases we showed that the convection coefficient was related to the Nusselt number. For free convection empirical equations were given that showed Nu $= f(\mathrm{Gr, shape})$. The influence of the Prandtl number was eliminated by setting Pr $= 0.72$ for air. For forced convection Nu $= f(\mathrm{Re, shape})$ and the distinction between laminar and turbulent flow becomes more important.

It should be made clear to the reader that the equations given here are approximate. Errors of $\pm 30\%$ can be anticipated. This fact implies that the designer either provides for comfortable margins of capacity in the layout of the air cooling system or that models be constructed and tested to determine actual heat transfer rates and temperatures in designing cooling systems with low margins of capacity.

9.8 AIR FLOW IN ELECTRONIC ENCLOSURES

The flow of air in an electronic enclosure is extremely important in reducing the ΔT associated with cooling the components and subsystems. We noted in the previous section that the velocity U strongly influenced the Re and in turn Re affects Nu and the convection coefficient. The mass flow rate $\dot{m}$ controls the temperature of the cooling air and must be sufficiently large to maintain $\Delta T = T^{ex} - T^{in}$ at a reasonably small value. The analysis of air flow in electronic enclosures is usually based on the volume flow rate Q, defined in Eq. (9.87a). For a duct system that distributes the flow to the various parts of the electronic enclosure we determine Q at each heat exchanging surface. Because the pressures in the duct system are relatively low, we treat the fluid (air) as incompressible and simplify the analysis.

9.8.1 Fluid Statics

A typical air delivery system in an electronic enclosure, illustrated in Fig. 9.21, includes a duct that turns, divides, changes in cross-sectional area and changes in

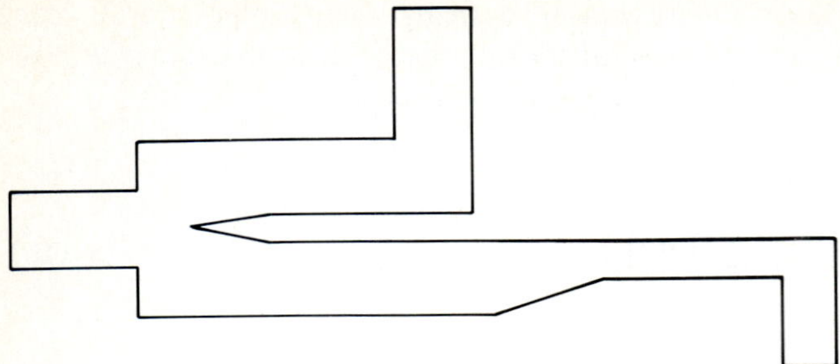

FIGURE 9.21
Duct work with turns, divisions and changing elevation.

elevation. Some basic relations in fluid statics are necessary to determine flow parameters in this type of duct. First, consider the pressure–density–height relation, which is based on equilibrium:

$$z_2 - z_1 = h = (p_1 - p_2)/w \tag{9.117}$$

where w, the density of the fluid, is treated as a constant
p is the pressure
h and z are defined in Fig. 9.22

Continuity of the mass flow through a duct with changing cross-sectional areas leads to

$$A_1U_1 = A_2U_2 \tag{9.118}$$

The volume flow rate is

$$Q = UA \tag{9.87b}$$

Conservation of energy leads to Bernoulli's equation

$$p_1/w + U_1^2/2g + z_1 = p_2/w + U_2^2/2g + z_2 \tag{9.119}$$

This relation implies that no energy is added or lost as the fluid flows between station 1 and station 2. In practice, energy can be added by inserting a fan in the

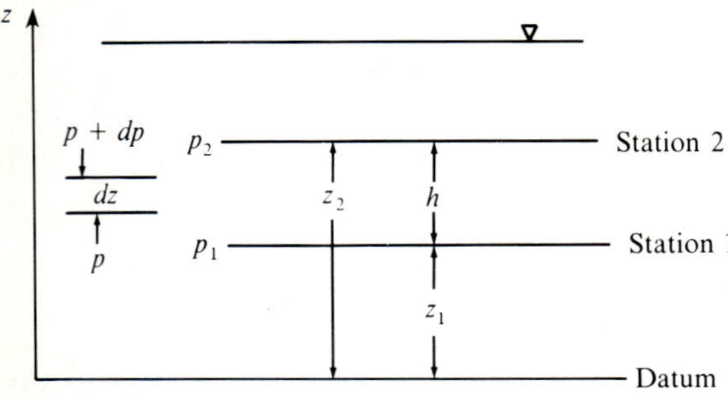

FIGURE 9.22
Definition of pressures in terms of heights z_1 and z_2.

system and energy can be lost due to wall friction and changes in the duct configuration. To account for these gains from fans h_f and friction losses h_L, we modify Eq. (9.119), when appropriate, by

$$p_1/w + U_1^2/2g + z_1 + h_f = p_2/w + U_2^2/2g + z_2 + h_L \qquad (9.120)$$

Each term in this relation represents a vertical linear distance, called a head (h) as shown in Fig. 9.22. We rewrite Eq. (9.120) in terms of these heads as

$$h_{s1} + h_{d1} + h_{P1} + h_f = h_{s2} + h_{d2} + h_{P2} + h_L \qquad (9.121)$$

where h_s is the static or pressure head
h_d is the dynamic or velocity head
h_P is the potential head
h_f is the head due to a fan between stations 1 and 2
h_L is the total head loss between stations 1 and 2

As an example of the head loss due to friction recall from fluid mechanics that

$$\Delta p = 4(L/D)(\rho U^2/2)f \qquad (9.122)$$

where Δp is the pressure change
ρ is the mass density equal to w/g
D is the hydraulic diameter
f is the friction factor
L is the length between stations 1 and 2

This relation reduces to

$$h_L = \Delta p/w = 4(L/D)(U^2/2g)f \qquad (9.123)$$

We will use these relations later in the analysis of forced air flow in enclosures.

9.8.2 Enclosure Impedance and Fan Characteristics

A typical electronic enclosure with a system for heat removal by forced air cooling is presented in Fig. 9.23. The air is brought into the enclosure by fans mounted at the bottom of the rear panel. Duct work, not shown in the illustration, distributes the air to the necessary locations within the enclosure. Several design rules are provided in the illustration, giving practical details regarding the layout of forced air cooling systems. After cooling the electronic components distributed throughout the enclosure, the air is exhausted through vents, which are usually distributed near the top of the side panels, into the atmosphere. Head losses occur as the air flows through the enclosure, which depend on the volume flow rate Q as indicated in Fig. 9.24. It is clear from Eq. (9.123) that the head losses increase in proportion to Q^2.

To provide the pressure necessary to offset the head losses, fans are employed. The fan manufacturers [15] provide performance curves for fans that give the static head h_s developed, the power required and the efficiency as a

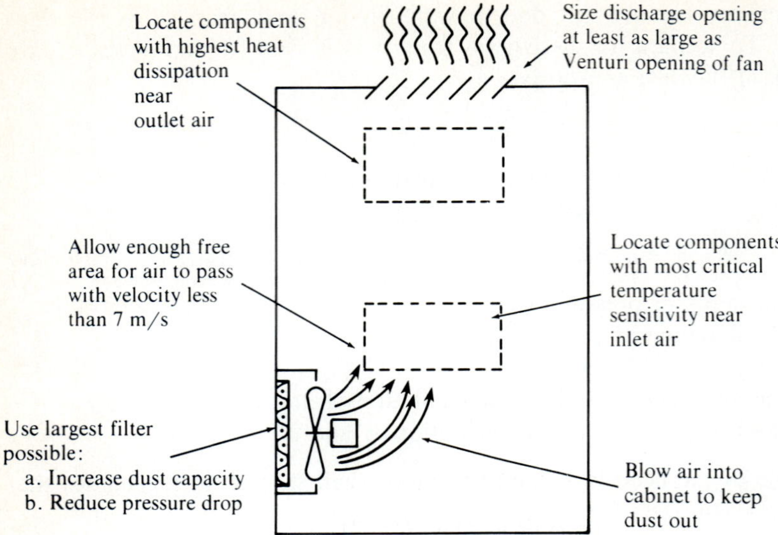

FIGURE 9.23
Typical electronic enclosure with forced air cooling. Some design rules for layout of the cooling system are illustrated.

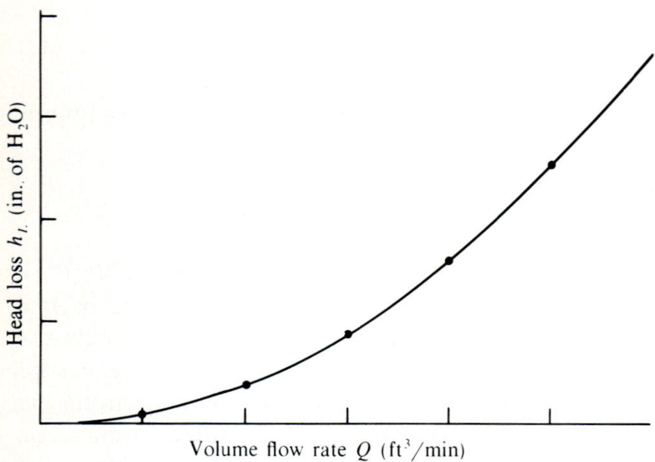

FIGURE 9.24
Head loss h_L increases as a function of Q^2.

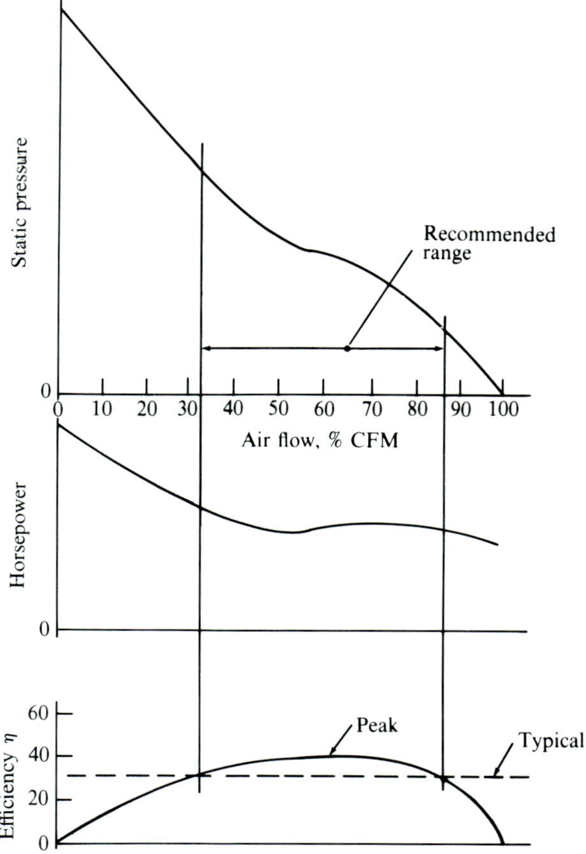

FIGURE 9.25
Performance characteristics of a propeller fan.

function of Q. A typical set of performance curves is shown in Fig. 9.25 for a propeller-type fan. The system volume flow rate Q is determined by superimposing the h_L-Q curve for the enclosure with the h_s-Q curve for the fan. The intersection of the two curves, as illustrated in Fig. 9.26, defines the operating point with the corresponding volume flow rate Q_0. Fans are selected so that Q_0 is within the recommended range of operation for the specified fan, which results in efficiencies between 30 and 40%.

In some instances sufficient flow cannot be achieved with a single fan and it is necessary to consider adding a second fan to the system to enhance the flow rate. Two possibilities exist for the placement of the second fan. It can be placed in parallel, like the arrangement shown in Fig. 9.23, or it can be placed in series, where the first fan provides the input for the second fan. The correct arrangement for placement depends on the relative head loss of the enclosure. If the head loss is low, as illustrated in Fig. 9.27*a*, then the parallel arrangement provides a significant increase in air flow with a modest increase in head loss. However, if the system impedance is high, a parallel arrangement is not effective and a series

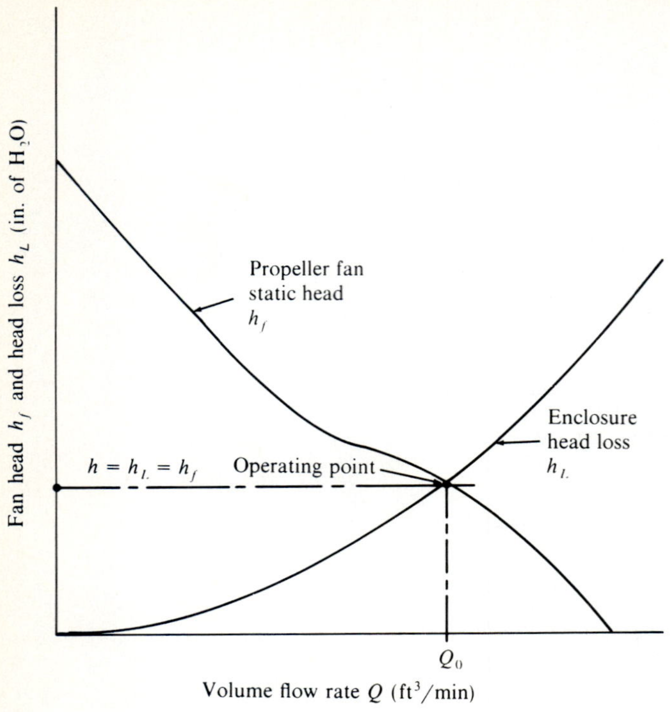

FIGURE 9.26
Intersection of fan head and enclosure loss curves determines the operating point for a forced air cooling system.

arrangement is necessary. Reference to Fig. 9.27*b* shows that flow enhancement is possible in high impedance enclosures but the increase, even with the preferred series fan arrangement, is small and is accomplished with a relatively large increase in head loss. In some cases it may be necessary to redesign the ducts to decrease the system impedance to obtain a sufficient flow rate.

9.8.3 Fan Placement in the Enclosure

There are three possible positions for placing a single fan in the duct system of an enclosure, namely, at the inlet, at the exit or at an intermediate location within the ducts. When the fan is placed at the exit, we have the advantage that the heat due to the power to drive the fan is dissipated into the atmosphere and is not transported through the system. Another advantage is the flexibility of the air distribution in the enclosure. Inlet vents can be placed at most locations on the vertical panels to achieve cooling at each of the required positions in the enclosure. The main disadvantage with fans mounted at the exhaust side is the

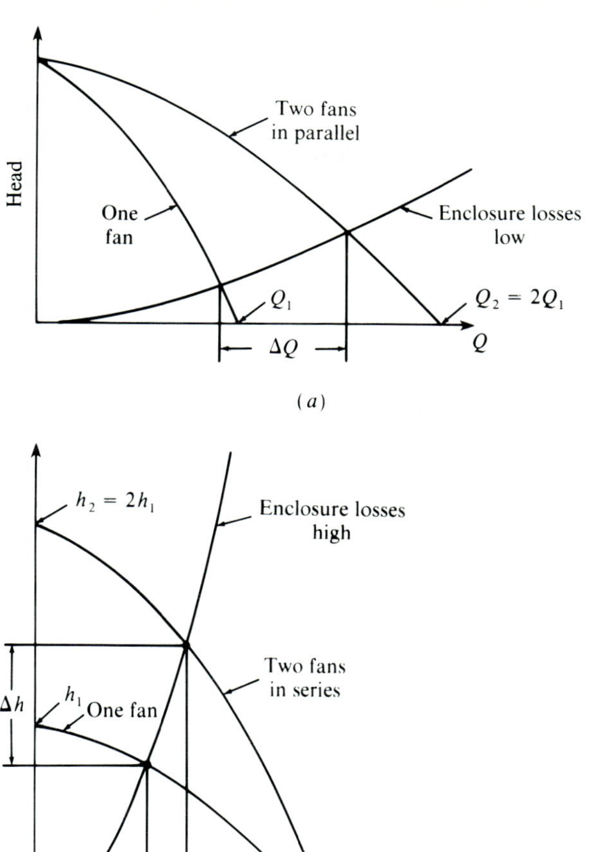

FIGURE 9.27
Use of two fans to enhance air flow. (*a*) Fans in parallel. (*b*) Fans in series.

difficulty in filtering the air. Inlet air enters the enclosure from many inlet vents and from cracks or other openings not intended for air flow. In hostile environments unfiltered air carries dirt and dust particles that collect on circuit components and in time cause malfunctions.

When the fan is placed at the inlet, the enclosure is pressurized and the air can be filtered. This enables the electronic systems to be operated in a cleaner environment and enhances life and reduces maintenance. Another advantage is that the fan handles cooler and more dense air, which increases its capacity. The primary disadvantage is that the power to operate the fan is converted to heat, which is transported through the enclosure. Of course, this heat load adds to the heat that must be dissipated from the electronics and increases the size of the cooling system that is required.

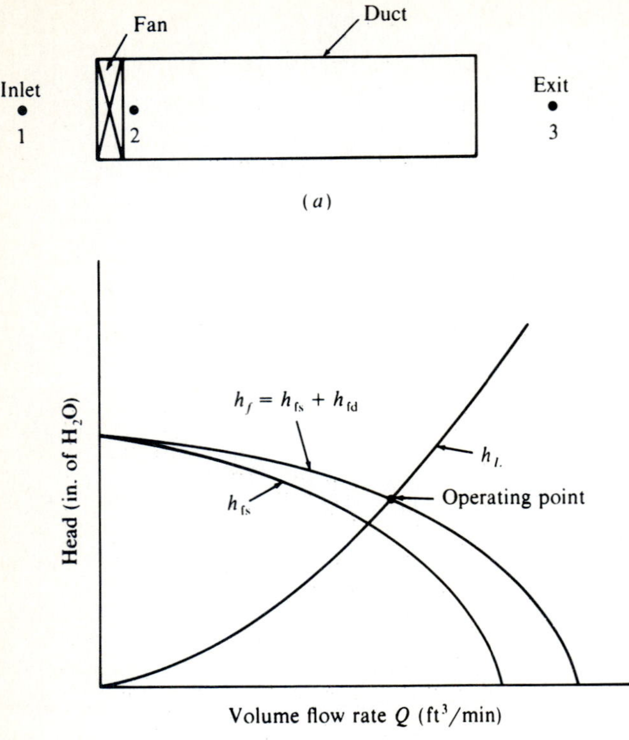

FIGURE 9.28
Influence of fan position on volume flow rate. (*a*) Station locations for the fan positioned at the inlet. (*b*) Change in operating point due to the placement of the fan at the inlet.

The intermediate fan has the disadvantages of both the exhaust and inlet fan positions. Its only advantage is the safety provided by the inaccessibility afforded by the location of the fan within the enclosure. To insure the safety of fans positioned at inlet and exit locations, grilles are employed, which prevent the accidental insertion of fingers into the fan blades.

The operating point of the system may be affected by the placement location. Consider first the placement of the fan at the inlet position as shown in Fig. 9.28*a*. Applying Eq. (9.121) across the fan from stations 1 to 2, we obtain

$$h_f = h_{s2} + h_{d2} \tag{9.124}$$

Clearly, the head produced by the fan is divided into two parts. The static head as given in the rating curve of Fig. 9.25 and the dynamic head, which depends on the volume flow rate Q. Next, write Eq. (9.121) again between stations 2 and 3 to show

$$h_L = h_f = h_{s2} + h_{d2} \tag{9.125}$$

This result indicates that a fan positioned at the inlet produces a head loss h_L equal to the sum of both the static and dynamic head developed by the fan. This fact causes the operating point of the system to change as shown in Fig. 9.28*b* with an increase in the flow rate Q_0. A similar analysis, for the fan placed at the exit, shows that the duct losses are equal to the static head developed by the fan and the operating point shown in Fig. 9.26 is correct.

9.8.4 An Electrical Analogy for Head Loss Determination

We noted in Eq. (9.123) that the head loss due to flow in a duct is proportional to Q^2 or U^2. This fact leads to

$$h_L = R_A Q^2 \tag{9.126}$$

where R_A is the constant of proportionality.

Examination of Eq. (9.126) shows that it is analogous to Ohms law $E = IR$ with $h_L \Rightarrow E$, $Q^2 \Rightarrow I$ and $R_A \Rightarrow R$. The analogy is very useful since it gives a relation for the head loss for any volume flow rate if the resistance R_A can be determined for the duct system.

In a typical duct system with turns and changes in cross-sectional area we encounter head losses due to each discontinuity in the duct as well as the losses due to wall friction. The total resistance R_A is given by

$$R_A = \sum R_i \tag{9.127}$$

Relations for R_i associated with typical duct discontinuities are presented in Fig. 9.29. The effect of wall friction is determined from the friction factor f that was covered previously in Section 9.7. The relation between R_i and f can be determined from Eqs. (9.87a) and (9.123), which gives

$$h_L = 2(L/D)\left[Q^2/(A^2 g)\right] f \tag{9.128}$$

Combining Eqs. (9.126) and (9.128) gives

$$R_i = 2(L/D)\left[1/(A^2 g)\right] f \tag{9.129}$$

This relation can be used to determine the resistance to air flow due to wall friction along the length L of the duct.

The resistance due to circuit cards with electronic components exposed to the air stream is difficult to characterize because the pattern of components is so variable. A first approximation to the resistance R_i is

$$R_i = 2L \times 10^{-3}/A^2 \tag{9.130}$$

where L is the length of the circuit card
A is the cross-sectional area of the channel over the card

Dimensions used in these relations for R_i are in inches for the duct sizes and cubic foot per minute for the volume flow rate. While these are mixed units, they

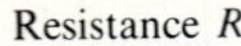

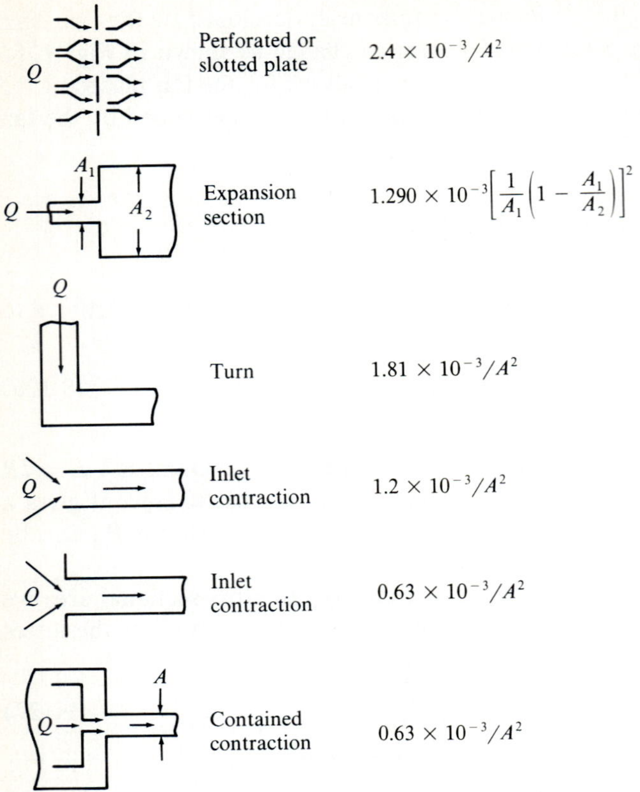

FIGURE 9.29
Empirical relations for determining R_i for geometric discontinuities in the duct system.

are consistent with standard practice in determining the head loss in inches of water.

In the analysis of head loss in a duct, it is clear that the combined resistance R_T of segments of the duct that are arranged in a series combination is given by

$$R_T = R_1 + R_2 + \cdots + R_n \tag{9.131}$$

On the other hand, if the ducts are in a parallel arrangement as shown in Fig. 9.30, then the head loss over each element is the same:

$$h_L = h_{L1} = h_{L2} \tag{9.132}$$

However, the volume flow rate Q_T is divided into Q_1 and Q_2 with

$$Q_T = Q_1 + Q_2 \tag{9.133}$$

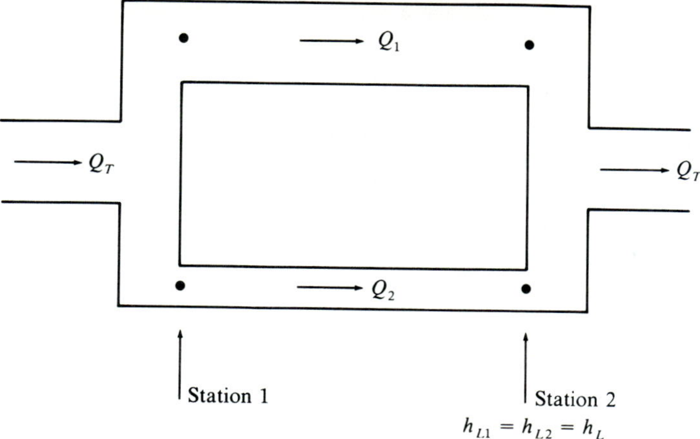

FIGURE 9.30
Parallel arrangement of ducts.

Combining Eqs. (9.126), (9.132) and (9.133) leads to

$$(1/R_T)^{1/2} = (1/R_1)^{1/2} + (1/R_2)^{1/2} \tag{9.134}$$

Finally, from Eqs. (9.133) and (9.134) we can show that the flow divides according to

$$Q_1 = (R_T/R_1)^{1/2} Q_T \tag{9.135a}$$

$$Q_2 = (R_T/R_2)^{1/2} Q_T \tag{9.135b}$$

These relations permit the determination of the volume flow rate within an enclosure and the sizing and specification of the fans required to remove the heat from the electronics. The thermal penalty ΔT associated with the cooling system can be established by using the results from Section 9.7 to determine the heat transfer coefficients and the temperature of the air as it moves from the inlet to the exit of the duct system. Exercises are provided to illustrate the application of these relations.

REFERENCES

1. Hamilton, D. C. and W. R. Morgan: "Radiant Interchange Configuration Factors," NACA Technical Note 2836, 1952.
2. White, F. M.: *Heat Transfer*, Addison Wesley, Reading, Mass., 1984.
3. Oppenheim, A. K.: "Radiation Analysis by the Network Method," *ASME Transactions*, vol. 78, pp. 725–735, 1956.
4. Jaluria, Y.: *Natural Convection Heat and Mass Transfer*, Pergamon Press, New York, 1980.
5. Ostrach, S.: "An Analysis of Laminar Free Convection Flow and Heat Transfer about a Flat Plate Parallel to the Direction of the Generating Body Force," NACA Report 1111, 1953.

6. LeFevre, E. J.: "Laminar Free Convection from a Vertical Plane Surface," in *Proc. 9th Int. Congress of Applied Mech.*, vol. 4, p. 168, 1956.
7. Churchill, S. W. and H. H. S. Chu: "Correlating Equations for Laminar and Turbulent Free Convection from a Vertical Plate," *J. Heat Transfer*, vol. 18, pp. 1323–1329, 1975.
8. Minkowycz, W. J. and E. M. Sparrow: "Local Nonsimilar Solutions for Natural Convection on a Vertical Cylinder," *J. Heat Transfer*, vol. 96, pp. 178–183, 1974.
9. Churchill, S. W. and H. H. S. Chu: "Correlating Equations for Laminar and Turbulent Free Convection from a Horizontal Cylinder," *Int. J. Heat Mass Transfer*, vol. 18, pp. 1049–1053, 1975.
10. Goldstein, R. J., E. M. Sparrow and D. C. Jones: "Natural Convection Mass Transfer Adjacent to Horizontal Plates," *Int. J. Heat Mass Transfer*, vol. 16, pp. 1025–1035, 1973.
11. Elenbaas, W.: "Heat Dissipation of Parallel Plates by Free Convection," *Physica*, vol. 9, no. 1, pp. 1–28, 1942.
12. Pohlhausen, E.: Der Wärmeaustausch zwischen fasten Körpern und Flüssigkeiten mit kleiner Reibung und kleiner Warmeleitung. *Z. Angew Math. Mech.*, vol. 1, pp. 115–121, 1921.
13. Shah, R. K. and A. L. London: *Laminar Flow Forced Convection in Ducts*, Academic Press, New York, 1979.
14. Haaland, S. E.: "Simple and Explicit Formulas for the Friction Factor in Turbulent Pipe Flow," *J. Fluids Engineering*, vol. 105, pp. 89–90, 1983.
15. "Technical Considerations in the Application of Rotron Fans and Blowers," EG & G Rotron Custom Division.
16. "Laboratory Methods of Testing Fans for Rating Purposes," ASHRAE Standard 51-75, 1975.

EXERCISES

9.1. A flat plate 6 × 8 in. in size serves as a radiation heat exchanger for an electronic assembly. The heat dissipated if the plate is considered to be a black body is $q^b = 50$ W. The heat is radiated to another flat, parallel and black plate of the same size positioned 60 in. away.

(*a*) Determine the shape factor $\mathbb{F}_{1-2}$.

(*b*) If the temperature of the receiving plate is maintained at $-100°C$, find the temperature of the transmitting plate.

9.2. Consider the heat exchanger described in Exercise 9.1. If a temperature limit $T_1 = 80°C$ is imposed on the heat exchanger, find the maximum q^b that can be dissipated.

9.3. A flat plate 4 × 8 in. radiates heat to another parallel flat plate of equal size positioned 18 in. away. Both plates are considered as black bodies. Construct a family of curves showing q^b as a function of T_1 for $0 < T_1 < 150°C$ if:

(*a*) $T_2 = -100°C$

(*b*) $T_2 = -50°C$

(*c*) $T_2 = 0°C$

(*d*) $T_2 = 50°C$

9.4. A long rectangular duct is designed with a 4 in. width and 2 in. high sides. The top is maintained at 100°C and the other three walls are held at 0°C. Treat the duct as a black body and determine the heat transferred from the top surface per foot of duct to:

(*a*) One vertical side

(*b*) The bottom

(*c*) Both sides and the bottom

9.5. A pair of circular disks 6 in. in diameter are concentric and parallel. If one disk is maintained at 50°C and the other at −50°C, determine their separation distance if q^b is: 1, 5, 10 and 20 W.

9.6. A 1 × 2 × 10 in. angle section illustrated in Fig. E9.6 is used as a radiation heat exchanger. The horizontal side is the heat sink, which is maintained at 20°C. Determine the temperature of the vertical side if the black body heat transferred is: 0.5, 1, 2 and 5 W.

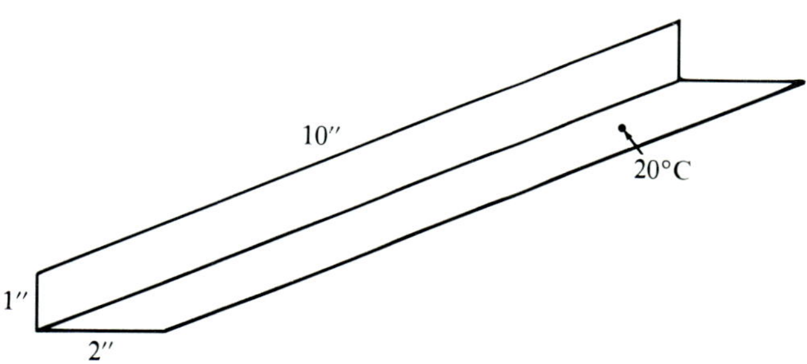

FIGURE E9.6

9.7. Repeat Exercise 9.1 if the two plates have gray surfaces with an emissivity $\varepsilon = 0.4$.

9.8. Repeat Exercise 9.2 using gray surfaces with emissivity $\varepsilon = 0.4$.

9.9. Repeat Exercise 9.3 but consider that the plates are fabricated from polished copper.

9.10. Repeat Exercise 9.4 with the duct constructed from commercial sheet aluminum.

9.11. Repeat Exercise 9.5 with emissivities $\varepsilon_1 = 0.8$ and $\varepsilon_2 = 0.4$.

9.12. Repeat Exercise 9.6 with $\varepsilon = 0.9$ for the vertical side and $\varepsilon = 0.3$ for the horizontal side.

9.13. Determine the radiation heat transfer coefficient for the heat exchanger described in Exercise 9.1 if the gray surfaces have emissivities of 0.4. Take the heat transfer as $q = 15$ W and the temperature of the receiving plate as $T = 10$°C.

9.14. Determine the radiation heat transfer coefficient for the heat exchanger described in Exercise 9.3 if the emissivities of the plates are 0.5 and 0.3. Consider temperatures T_1 and T_2 in degrees Celsius as listed in the following table

	T_1	T_2
(*a*)	50	20
(*b*)	100	20
(*c*)	150	20
(*d*)	200	20

9.15. Using the results from Exercise 9.13 determine the thermal resistance R_r due to the radiation heat exchanger. Compare this value of R_r with the thermal resistance offered by a copper bar 1 × 1 × 25 mm conducting heat with the same temperature differences.

9.16. A vertical plate 4 in. wide and 6 in. high is maintained at a temperature of 80°C. The temperature of the ambient air on one side of the plate is 20°C. Consider that the other side of the plate is insulated. Determine:

(*a*) The convection coefficient as a function of the position y and show a graph of $h(y)$.

(*b*) The average convection coefficient for the plate.

(*c*) The heat transferred from the plate to the ambient air.

9.17. A vertical plate 25 mm high and 200 mm wide is held at a constant temperature of 90°C. Both sides of the plate are exposed to ambient air at 15°C. Determine:

(*a*) The convection coefficient as a function of the position y and show a graph of $h(y)$.

(*b*) The average convection coefficient for the plate.

(*c*) The heat transferred from the plate to the ambient air.

9.18. Repeat Exercise 9.17 with the plate turned through 90° so that the 200 mm edge is vertical. Describe why there is a difference in the results.

9.19. Find the Grashof number as a function of position y for the plate described in Exercise 9.16. Determine the average and the maximum value of Gr. Is the flow over the plate laminar or turbulent?

9.20. Consider a plate of height H with $T_w = 100°\text{C}$ and $T_a = 20°\text{C}$. Find the position y_t where the flow undergoes a transition from laminar to turbulent flow assuming $y_t < H$.

9.21. Determine the heat flux q/A for a single-sided vertical plate heat exchanger with its height H equivalent to the position y_t where the laminar flow undergoes a transition to turbulent flow.

9.22. A vertical plate 4 in. wide and 6 in. high transfers heat from one side. The plate has a constant heat flux q/A with a wall temperature that varies with position y. Determine the value of $(T_w - T_a)$ as a function of the heat flux q/A. Truncate your set of solutions when the flow becomes turbulent at the top of the plate.

9.23. Compare the results of Exercises 9.22 and 9.1 and show the differences between constant temperature and constant flux heat transfer for free convection.

9.24. Consider a vertical plate heat exchanger that is sufficiently high to have both laminar and turbulent flow regions. For the case where $(T_w - T_a) = 100°\text{C}$:

(*a*) Construct a graph showing Nu(y).

(*b*) Determine the average value of Nu in the laminar flow region.

(*c*) Determine the average value of Nu in the turbulent flow region.

(*d*) Select the height of the plate so that the portions of the plate subject to laminar and turbulent flow are equal and then find the relative amount of heat transfer from each part.

9.25. Repeat Exercise 9.24 but use instead Eq. (9.44) to determine the average value of Nu_H.

9.26. A cylindrical electronic component 20 mm in diameter and 60 mm long is mounted with its axis in the vertical direction. If the component dissipates 5 W to the ambient air, which is at a temperature of 25°C, find the surface temperature of the component.

9.27. Repeat Exercise 9.27 but consider the cylindrical component to be oriented in the horizontal direction.

9.28. A flat plate that serves as the top panel for an electronic enclosure also acts as a heat exchanger. The panel is 10 × 19 in. in size. If the surface temperature is limited to 60°C, find the capacity of the heat exchanger. Why would we limit the temperature of the top panel of an enclosure to the value specified here.

9.29. Repeat Exercise 9.28 but consider the panel to be the bottom panel of an electronic enclosure.

9.30. A fin-type heat exchanger as illustrated in Fig. 9.11*a* is constructed with 20 fins having $W = 5$ mm, $H = 40$ mm and $d = 15$ mm. Determine the average value of Nu for the heat exchanger and the heat capacity as a function of $(T_w - T_a)$.

9.31. Are the fins used in the heat exchanger of Exercise 9.30 optimized? If not, determine W necessary to optimize the configuration. With this new spacing find the new capacity of the 20 fin exchanger. What is its new length? If we increase the number of fins and hold the original length the same, give the new capacity. Use $(T_w - T_a) = 60°C$ in this analysis.

9.32. Consider a flat plate 24 in. long placed in an air stream with a temperature $T_a = 20°C$. Prepare a graph showing the Reynolds number as a function of position x along the plate. The graph should include separate curves for free stream velocities of 1, 2, 5, 10, 20 and 50 m/s. Identify on each curve the position of the transition from laminar to turbulent flow.

9.33. For a plate with single-sided heat transfer in an air stream with a velocity $U = 5$ m/s, find the average value of h. The temperatures are $T_W = 90°C$ and $T_a = 25°C$. The plate is 6 in. long and 4 in. wide. Also find the heat transferred from the plate to the air stream.

9.34. Repeat Exercise 9.33 with $U = 1$ m/s and $U = 10$ m/s. Comment on the effect of air stream velocity on the capacity of the plate as a heat exchanger.

9.35. Repeat Exercise 9.33 but consider the plate to have lengths of 2, 4 and 8 in. Prepare a graph showing the heat flux q/A as a function of the length of the plates.

9.36. For the plate in Exercise 9.33 find the maximum velocity U before turbulent flow occurs. Is there anything wrong with permitting turbulent flow to occur in a flat plate heat exchanger for use in cooling electronic components?

9.37. A flat plate, one-sided heat exchanger 6 in. wide by 12 in. long is in an air stream with a velocity of 50 m/s. If the air stream is at $T_a = 300$ K and the plate is at $T_W = 400$ K determine:
(*a*) The location of the transition region.
(*b*) The average Nu over the laminar region.
(*c*) The average Nu over the turbulent region.
(*d*) The average Nu over the entire plate.
(*e*) The heat transferred from the plate to the air stream.

9.38. Repeat Exercise 9.37 but consider that the plate dissipates a uniform heat flux $q/A = 1$ W/in.^{2}. In addition to finding solutions to parts (*a*) to (*e*), find the temperature of the plate T_W as a function of the position x.

9.39. For a rectangular duct with dimensions of 2 × 4 in., determine the distance that is required for fully developed flow in the duct. The temperature of the air is 20°C and its velocity is 10 m/s.

9.40. For the duct described in Exercise 9.39, determine the volume flow rate Q, the mass flow rate $\dot{m}$ and the mass velocity G. Express the results in both English and SI units.

9.41. For a duct with a 20 × 90 mm cross section with the two 90 mm walls held at a temperature $T_W = 75°C$, find the temperature of the air as it moves from the inlet to the exit of the duct. The velocity of the air stream $U = 10$ m/s and the inlet temperature is 20°C. The duct is 0.6 m long. Construct a graph showing ΔT as a function of position x.

9.42. Repeat Exercise 9.41 but consider a constant heat flux $q/A = 1600$ W/m² on the two 90 mm duct walls rather than a constant temperature.

9.43. Determine $(T_W - T_{\text{ave}})_{\text{ave}}$ from Eqs. (9.104) and (9.105) with $T^{\text{in}} = 20°C$ and $T_W = 75°C$. Let T^{ex} vary in 10° increments from 30 to 70°C and show the differences between the results from these two equations as a function of T^{ex}.

9.44. For the duct described in Exercise 9.41, find the heat transferred to the air stream. Also determine the friction factor f.

9.45. For the duct described in Exercise 9.42, find the heat transferred to the air stream.

9.46. Repeat Exercise 9.41 with $U = 4$ m/s. Explain the difference in the results when the velocity of the air stream is lowered.

9.47. Repeat Exercise 9.41 with $U = 50$ m/s. Explain the difference in the results when the velocity of the air stream is increased.

9.48. Find the head in inches of air corresponding to a pressure of 1 psi. Find the head in inches of water for a pressure of 1 psi.

9.49. Find the head in inches of water corresponding to an air stream velocity of 10 m/s. Determine the head if the velocity is change to 100 ft/s.

9.50. For the duct system illustrated in Fig. E9.50 show the equivalent resistance network for the electrical analogy pertaining to head loss.

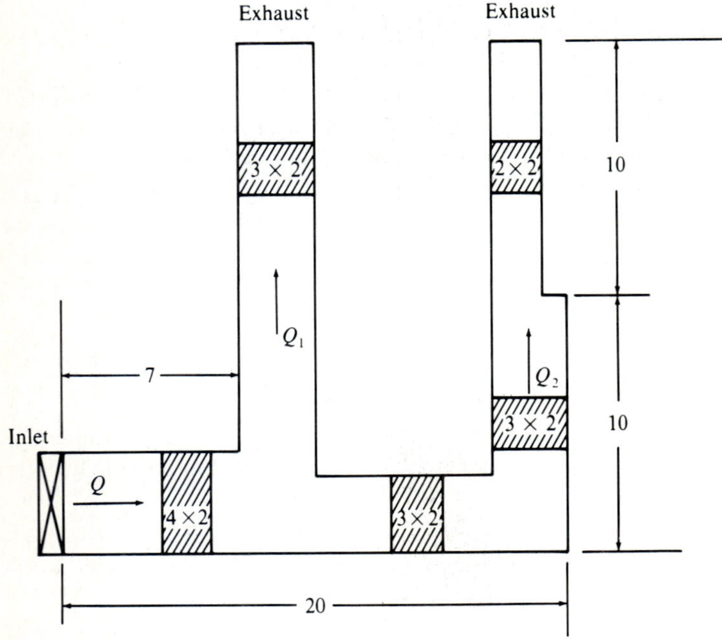

FIGURE E9.50
A duct system. Dimensions in inches.

9.51. Determine the value of the resistances R_i for the duct system illustrated in Fig. E9.50. Note that adjustable baffles on the exhausts permit control of the volume flow rate so that Q_1 and Q_2 are 30 and 20 ft^3/min, respectively.

9.52. Determine the head loss for the duct system in Fig. E9.50 for the volume flow rates given in Exercise 9.52.

9.53. Prepare a graph showing the impedance of the duct system for total volume flow rates, which range from 10–200 ft^3/min. Keep the ratio $Q_1/Q_2 = 1.5$.

9.53. Show the resistance network for the electrical analogy to determine head loss in the duct shown in Fig. E9.53.

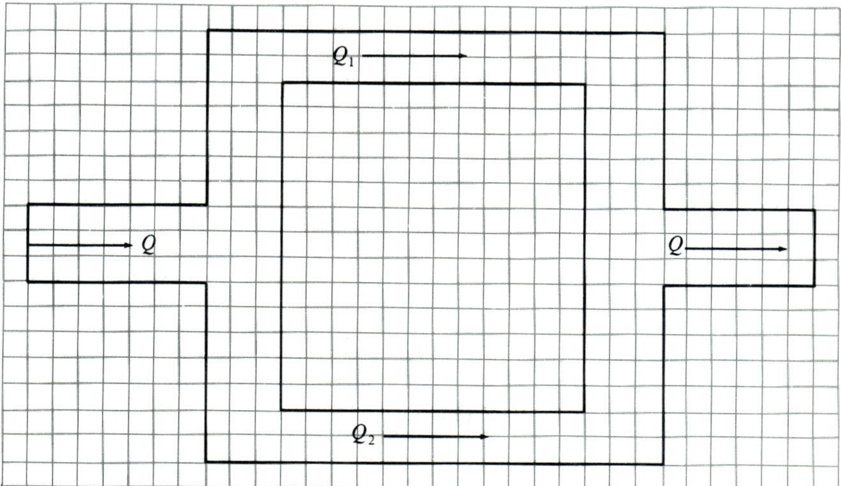

FIGURE E9.53
Duct system with parallel flow. The duct is 1 in. deep. Lateral dimensions are to scale with each grid line = 1 in.

9.54. Determine the resistances R_i for this network. Combine these resistances to obtain the equivalent resistance for the entire system.

9.55. Prepare a graph of the system impedance as the flow rates vary from 5–50 ft^3/min showing data points at increments of 5 ft^3/min.

CHAPTER 10

ANALYSIS OF VIBRATION OF ELECTRONIC EQUIPMENT

10.1 INTRODUCTION

An electronic product is subjected to shock and/or vibration during three different phases of its life. The first exposure often occurs at the end of the manufacturing process when the product is subjected to a combination of temperature cycling and random vibration as a screening environment to eliminate production flaws. The second exposure is during shipping from the assembly plant through the distribution chain to the customer where the product in its shipping container may be loaded and off-loaded on several occasions. The final exposure occurs during the operation of the product in the environment provided by the customer.

The reliability of even well designed products is usually degraded to some extent in the manufacturing process. A small but finite number of defective components are used in the construction and assembly errors occur. To maintain a high product reliability in field operation, these defects must be eliminated in the factory before the product is shipped. Manufacturing screening tests are often used in both commercial and military systems to eliminate the majority of these defects [1, 2]. The screening tests are adapted to a specific product. Usually, the product is subjected to a random vibration with the power spectral density of the exposure varied with the frequency as shown in Fig. 10.1. The product under test is positioned on the vibration table so that the axis of motion is perpendicular to

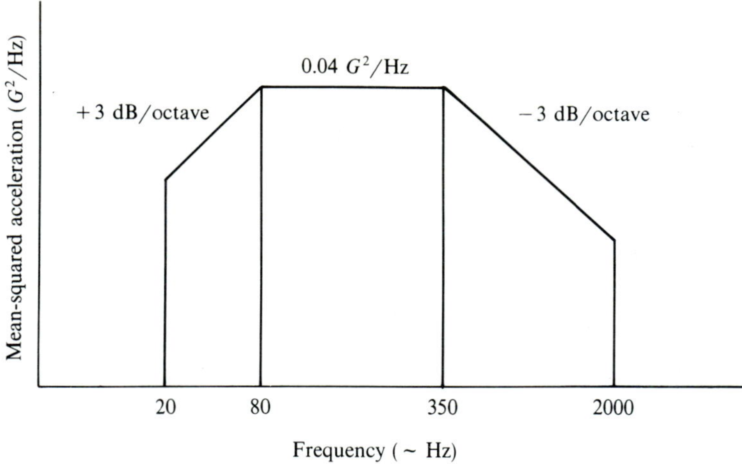

FIGURE 10.1
Power spectral density for random vibration recommended for a manufacturing screening test [1].

the printed circuit boards. If the critical components in the product are oriented in more than one plane, the vibration test is repeated sequentially with the vibration applied along the three orthogonal axes. The duration of the random vibration is usually 10 min if a single axis of vibration is sufficient. When it is necessary to repeat the vibration along the orthogonal axes, 5 min applied to each axis is considered sufficient exposure. The equipment under evaluation is operational during the test. Malfunctions in the operation indicate that a latent defect has been exposed by the random vibration. Inspection of the product reveals the defect, which can be corrected in the plant prior to shipping.

In addition to vibration, the manufacturing screen may also incorporate temperature cycling. The range of temperature in the cycling is usually from −55 to +55°C. Again, the equipment is functional during the temperature cycling and malfunctions indicate failure due to a latent defect that can be identified, located and the faulty component replaced. The number of temperature cycles used in the screen depends on the complexity of the product. One cycle is sufficient for simple products with 100 electronic components, but moderately complex equipment with 500 parts employ three cycles in testing. Up to 10 cycles are specified for very complex systems with 4000 or more components. An example of the reduction in failures per unit as a function of the number of temperature cycles is presented in Fig. 10.2.

The shipping environment imposed on the product, as it moves from the factory through the distribution chain to the customer, differs from product to product and even from shipment to shipment. The differences are due to placement in the transport vehicle, vehicle operator, drops and/or tosses during handling, routing, road conditions, season, etc. The American Society for Testing

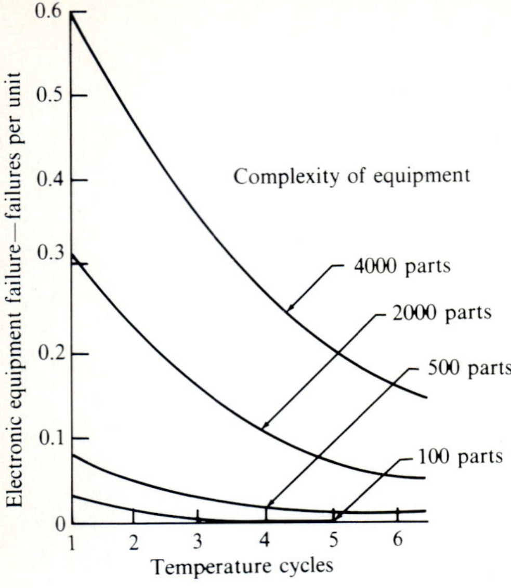

FIGURE 10.2
Improvement in failure rate with the number of temperature cycles for electronic equipment with different degrees of complexity.

Materials (ASTM) has characterized the shock and vibration exposures that may be encountered [3]. The shock environment, presented in Table 10.1, results from handling when the product is transferred from one location to another. The drop height is related to the product weight, with the lighter product likely to be dropped from a higher position. The assurance level is related to the number of units to be shipped and their value. Assurance level II is for average size shipments of moderately priced units.

The vibration environment is often specified in terms of a random vibration spectrum. While the spectrum to be incorporated in the ASTM standard has yet to be finalized, the one shown in Fig. 10.3 may be considered representative for the random vibration environment produced by a truck [4].

TABLE 10.1
Shock environments

Shipping weight [lb (kg)]	Drop height [in. (mm)] assurance level		
	I	II	III
0–20 (0–9.1)	24 (610)	15 (381)	9 (229)
20–40 (9.1–18.1)	21 (533)	13 (330)	8 (203)
40–60 (18.1–27.2)	18 (457)	12 (305)	7 (178)
60–80 (27.2–36.3)	15 (381)	10 (254)	6 (152)
80–100 (36.3–45.4)	12 (305)	9 (229)	5 (127)
100–200 (45.4–90.7)	10 (254)	7 (178)	4 (102)

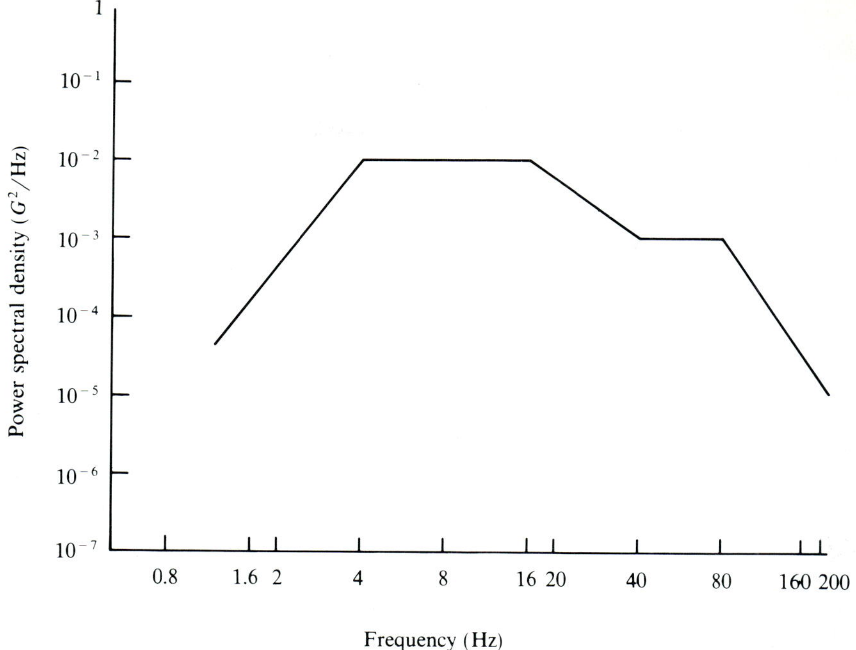

FIGURE 10.3
Random vibration spectrum representing vibration produced by a truck.

The shock and vibration environment encountered in operation at the customer location clearly depends on the application. Often the equipment is installed in an office or a laboratory where the environment is innocuous. However, in some cases military customers place the equipment on a weapons platform. In this instance, the environment is certainly not benign and the equipment must be designed to resist strong shocks and prolonged vibration at high G levels. To facilitate the initial design, the military customer will define the shock and vibration environment in the procurement documentation. Examples of military specifications for shock and vibration are given in references 5–8. Another application, with severe vibration and shock loading, involves systems installed on machine tools and on vehicles. Rugged electronic equipment is becoming more common as processors of all types are moving from the office to the workplace. Indeed, electronics on automobiles for engine control include tens of millions of systems subject to severe vibration, shock and temperature fluctuations.

In this chapter we will introduce the basic equations describing motion due to both shock and vibration. Applications of these equations to the design of select electronic components will then be treated. Finally, exercises will be presented to demonstrate the approach used in designing systems for shock and vibration environments.

10.2 VIBRATING SYSTEMS WITH A SINGLE DEGREE OF FREEDOM

Mechanical systems respond to a dynamic disturbance, imposed either by forces or displacements, by vibrating. The vibratory response is often quite complex with several different time dependent motions occurring simultaneously. We are fortunate in that of these many motions, one is usually dominant and essentially independent of the other much smaller motions. In these instances, the mechanical system can be treated as a single degree of freedom system without introducing significant error.

10.2.1 Free Vibrations

To begin the discussion of vibrating systems, consider the simple single degree of freedom system represented by the spring and weight shown in Fig. 10.4. After the weight W is attached to the spring, the origin O is defined as the equilibrium position and y, the time dependent position of the weight, is measured positive upward from this point. If the weight is deflected downward by a distance δ and released from rest, the system will vibrate at its natural frequency. We analyze this vibration by writing the equation of motion of the weight as

$$m\ddot{y} + ky = 0 \tag{10.1}$$

where $\ddot{y}$ is the second derivative of y with respect to t
k is the spring rate

Solution of this ordinary second-order differential equation gives

$$y = A \sin \omega_n t + B \cos \omega_n t \tag{10.2}$$

where ω_n is the natural frequency of the system given by

$$\omega_n^2 = k/m \tag{10.3}$$

The coefficients A and B are constants of integration that are determined from the initial conditions: $t = 0$, $y = -\delta$ and $\dot{y} = 0$. Substituting these initial conditions into Eq. (10.2) gives

$$y = -\delta \cos \omega_n t \tag{10.4}$$

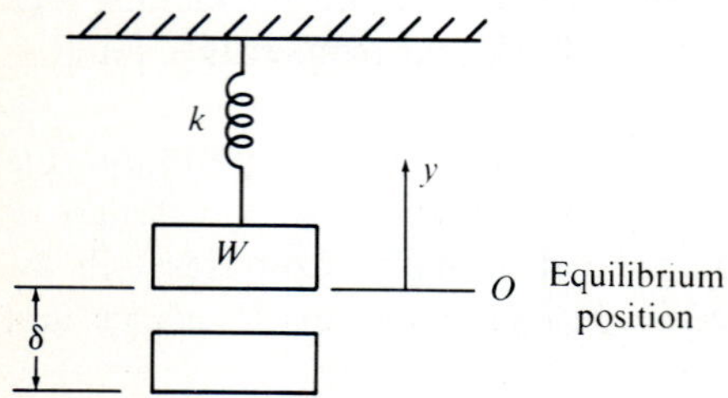

FIGURE 10.4
Spring and weight assembly representing a single degree of freedom system without damping.

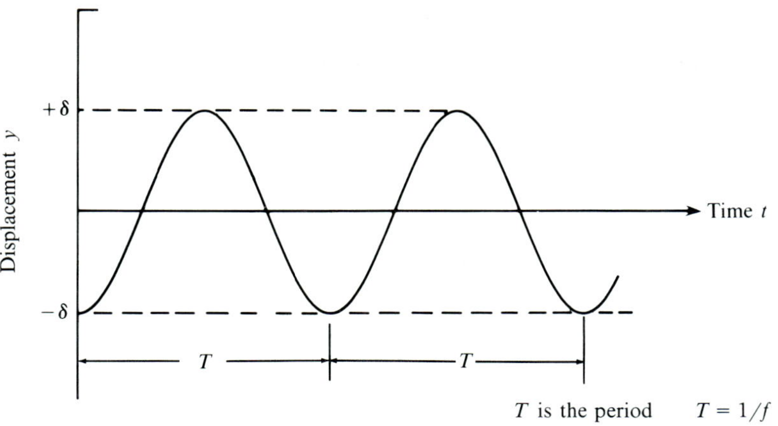

FIGURE 10.5
Displacement y as a function of time for a system with a single degree of freedom and no damping.

A graph of the initial response of the motion of the system, presented in Fig. 10.5, shows that the weight oscillates at a constant amplitude δ with a period T. The relation between the period and the natural frequency is given by

$$T = 1/f_n \tag{10.5a}$$

$$f_n = (1/2\pi)\omega_n = (1/2\pi)\sqrt{k/m} \tag{10.5b}$$

Note that combining Eqs. (10.3) and (10.5b) leads to:

$$f_n = (1/2\pi)\sqrt{g/\delta_s} \tag{10.6}$$

where g is the gravitational constant.

These results indicate that the vibratory motion given by the displacement y will continue indefinitely. This is not realistic because some damping is always present, which will cause the amplitude of the oscillation to decay with time.

When viscous damping is introduced in the vibrating system, a dash pot is added in parallel with the spring as shown in Fig. 10.6. The weight is again deflected downward by an amount δ_s and released from rest. The differential equation of motion for the weight is given by

$$m\ddot{y} + \eta\dot{y} + ky = 0 \tag{10.7}$$

where η is the damping coefficient. Substituting $y = e^{\lambda t}$ into Eq. (10.7) leads to a quadratic auxiliary equation. The roots of this equation are

$$\lambda = -\eta/2m \pm \left[(\eta/2m)^2 - (k/m)^2\right]^{1/2} \tag{10.8}$$

The solution for the displacement y depends on whether λ is a real or a complex number. Three different solutions are described that characterize the type of

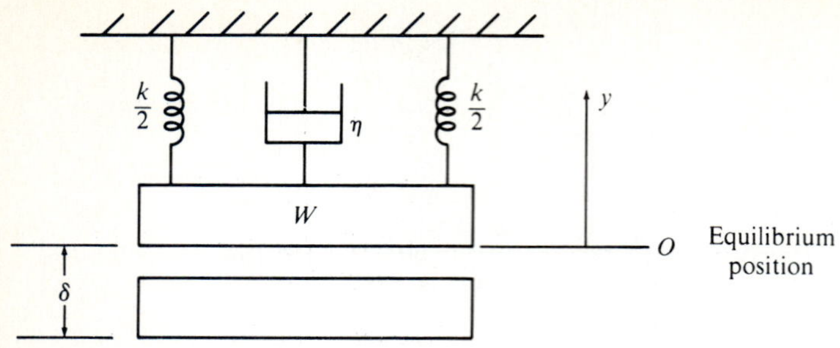

FIGURE 10.6
Spring and weight assembly representing a single degree of freedom system with damping.

damped vibration that can occur, depending on the values of k/m and $\eta/2m$.

Case 1. Critical damping occurs if $k/m = \eta/2m$ and

$$y/\delta_s = 1 - e^{-\alpha t(1+\alpha t)} \tag{10.9}$$

Case 2. Overdamped motion occurs if $\eta/2m > k/m$ and

$$y/\delta_s = 1 - e^{-\alpha y}[(\alpha/\beta)\sinh \beta t + \cosh \beta t] \tag{10.10}$$

Case 3. Underdamped motion occurs if $\eta/2m < k/m$ and

$$y/\delta_s = 1 - e^{-\alpha t}[(\alpha/\beta)\sin \beta t + \cos \beta t] \tag{10.11}$$

where $\alpha = \eta/2m$ and $\beta^2 = (\eta/2m)^2 - k/m$.

The response of the system to the initial displacement δ_s is illustrated in Fig. 10.7. It is clear, that the system recovers the initial equilibrium position without oscillation for the critical and the overdamped cases. Vibration, with its oscillatory motion, occurs only with the underdamped case.

Since the amount of damping associated with mechanical systems is usually small, critical and overdamped motion are rare events and are not usually considered in a study of vibration of mechanical systems. We are concerned with the underdamped case where vibrations that can damage electronic components occur because of either excessive displacements or forces generated during the oscillatory motion.

Returning to Eq. (10.11), which describes the motion for the underdamped case, we modify this relation by introducing the natural frequency and the damping ratio defined by

$$d = \eta/\eta_c \tag{10.12}$$

and

$$\eta_c = 2m\omega_n \tag{10.13}$$

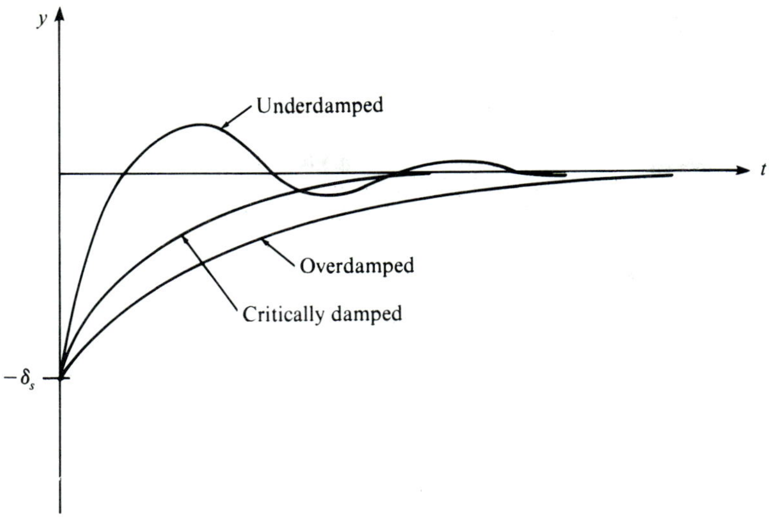

FIGURE 10.7
Displacement y as a function of time for a system with a single degree of freedom and different amounts of damping.

Substituting Eqs. (10.3), (10.12) and (10.13) into Eq. (10.11) leads to

$$y = e^{-d\omega_n t}\left[A \sin \omega_n\sqrt{1-d^2}\,t + B\cos \omega_n\sqrt{1-d^2}\,t\right] \tag{10.14}$$

The constants of integration A and B in this relation are determined by using the initial conditions that $y = -\delta_s$ and $\dot{y} = 0$ at time $t = 0$. With $A = 0$ and $B = -\delta_s$, we find that Eq. (10.14) reduces to

$$y = -\delta_s e^{-d\omega_n t}\cos \omega_n\sqrt{1-d^2}\,t \tag{10.15}$$

Examination of this relation indicates that the motion is expressed in terms of a product. The first term is the damping multiplier $e^{-d\omega_n t}$, which reduces the amplitude of the oscillation exponentially with time. The second term, a cosine function, gives the time variance of the oscillatory motion. Note that the damped natural frequency,

$$\omega_{\text{nd}} = \left(1 - d^2\right)^{1/2}\omega_n \tag{10.16}$$

is always smaller than the natural frequency ω_n.

The examination of a simple single degree of freedom system with damping in free vibration permitted us to define important terms like the natural frequency, the period and the damping ratio, and to distinguish between critically damped, overdamped and underdamped systems. However, free vibration is not a major concern in the design of electronic systems. Instead, forced vibration, where a periodic external force is imposed on the system for some time interval, is of major importance. With forced vibrations we encounter resonance where the forces constraining the system and the rigid body displacements of the system are amplified.

10.2.2 Forced Vibrations

To treat the problem of forced vibrations, consider a periodic external force applied to the weight W as illustrated in Fig. 10.8. We apply the external force as a cosine function, but the results would be essentially the same if we were to choose a sine function. The equation of motion for the weight can be written as

$$m\ddot{y} + \eta\dot{y} + ky = P\cos\omega t \tag{10.17}$$

where P is the amplitude of the external force.

Since the external force is applied for an extended interval of time, the system will achieve steady state motion. For this reason we seek the particular solution to Eq. (10.17) that describes this steady state motion. The particular solution is written as

$$y_p = A\cos\omega t + B\sin\omega t \tag{10.18}$$

The unknown coefficients A and B are determined by substituting Eq. (10.18) into Eq. (10.17) to obtain

$$A = P[k - m\omega^2]\Big/\left[(\eta\omega)^2 + (k - m\omega^2)^2\right] \tag{10.19a}$$

$$B = P\eta\omega\Big/\left[(\eta\omega)^2 + (k - m\omega^2)^2\right] \tag{10.19b}$$

Next, substituting Eqs. (10.19) into Eq. (10.18) gives

$$y = \frac{P\cos(\omega t - \phi)}{\left[(k - m\omega^2)^2 + (\eta\omega)^2\right]^{1/2}} \tag{10.20}$$

where ϕ is the phase angle given by

$$\tan\phi = \eta\omega/(k - m\omega^2) \tag{10.21}$$

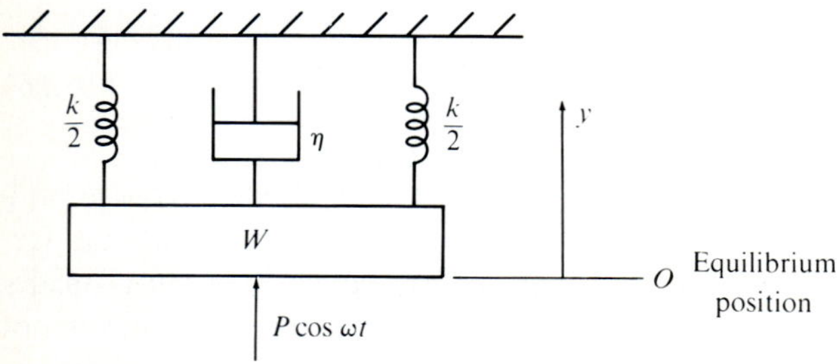

FIGURE 10.8
Spring and weight assembly subjected to a periodic external force.

The form of Eq. (10.20) is not convenient for analysis, so we modify it by introducing Eqs. (10.3), (10.12) and (10.13) to obtain

$$y_p = A_\omega \cos(\omega t - \phi) \tag{10.22}$$

where the amplitude of the forced vibration A_ω is a function of the driving frequency ω given by

$$A_\omega = (P/k)\Big/\left[(1 - r^2)^2 + (2\,dr)^2\right]^{1/2} \tag{10.23}$$

where $r = \omega/\omega_n$.

We will defer the discussion of A_ω because it is an essential element in the amplitude transmission coefficient that is introduced in the next subsection.

10.2.3 Transmission Coefficients

When an electronic system is subjected to forced vibrations we are concerned with two different features of the system response. The first is the amplitude of the vibration, which is important since an excessive amplitude can result in impact where an electronic subassembly can strike an adjacent structure. The second is the maximum force that is developed at the constraints. In resonance this force may become large enough to cause plastic deformation or fracture of the support structure or the attachments of the electronic components. We will determine the transmission coefficients for both the amplitude and the forces involved as a function of the driving frequency ω and the damping ratio d.

Consider the amplitude of the forced vibration A_ω given by Eq. (10.23) and normalize this quantity by

$$\mathbb{A} = A_\omega/\delta_s \tag{10.24}$$

Note that $\mathbb{A}$ is the amplitude ratio and $\delta_s = P/k$ is the equivalent static deflection. Combining Eqs. (10.23) and (10.24) yields the transmission coefficient $\mathbb{A}$ for the amplitude as

$$\mathbb{A} = 1\Big/\left[(1 - r^2)^2 + (2\,dr)^2\right]^{1/2} \tag{10.25}$$

The amplitude ratio depends only on the frequency ratio r and the degree of damping d. The $\mathbb{A}$-r relationship, shown in Fig. 10.9, indicates that a resonance condition occurs when $r = \omega/\omega_n = 1$. At resonance, $\omega = \omega_n$ and Eq. (10.25) reduces to

$$\mathbb{A} = 1/(2d) \tag{10.26}$$

It is evident that the response of the system can become very large as the damping ratio d goes toward zero. Unfortunately, the damping ratio is most electronic assemblies is small and values of $\mathbb{A} > 10$ are common.

With very low damping, it is advisable to design the electronic assembly with a natural frequency that is higher than the driving frequency. To explore the

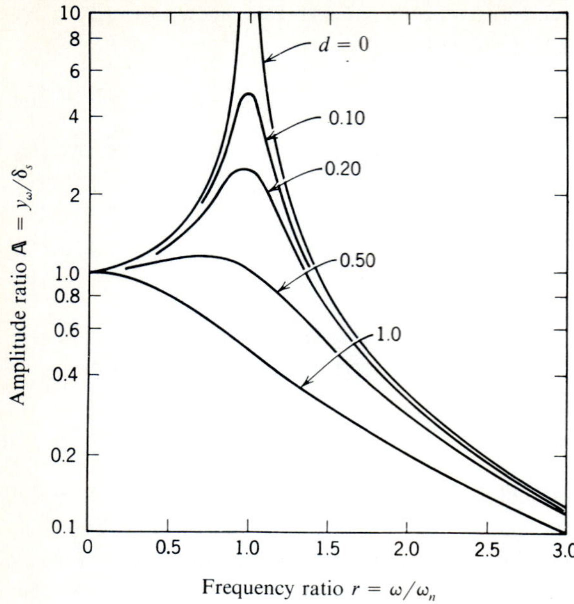

FIGURE 10.9
Amplitude ratio $\mathbb{A}$ as a function of frequency ratio ω/ω_n for different degrees of damping.

influence of the frequency ratio for those cases where the damping ratio is low, let $d = 0$ and note that Eq. (10.25) reduces to

$$\mathbb{A} = 1/(1 - r^2) \tag{10.27}$$

The results of Eq. (10.27), presented in Fig. 10.10, indicate that $\mathbb{A} < 2$ can be achieved if $r < 0.7$. If it is not possible to avoid the resonance region $0.9 < r < 1.1$, the use of a vibration isolation system to protect the system from the detrimental influence of the excessive amplitudes may be required.

Now that we have established the transmission coefficient $\mathbb{A}$ for the amplitude, consider next the constraining forces that are required to support the spring and dash pot shown in Fig. 10.8. The total force F transmitted to the support system is given by

$$F = ky + \eta\dot{y} \tag{10.28}$$

Substituting Eqs. (10.22) and (10.23) into Eq. (10.28) yields

$$F = (P/k)\left[k^2 + (\eta\omega)^2\right]^{1/2} \Big/ \left[(1 - r^2)^2 + (2\,dr)^2\right]^{1/2} \tag{10.29}$$

Next, define the force transmission coefficient $\mathbb{F}$ as

$$\mathbb{F} = F/P \tag{10.30}$$

Then by substitution of Eq. (10.29) into Eq. (10.3), we obtain

$$\mathbb{F} = (1 + 2\,dr^2)^{1/2} \Big/ \left[(1 - r^2)^2 + (2\,dr)^2\right]^{1/2} \tag{10.31}$$

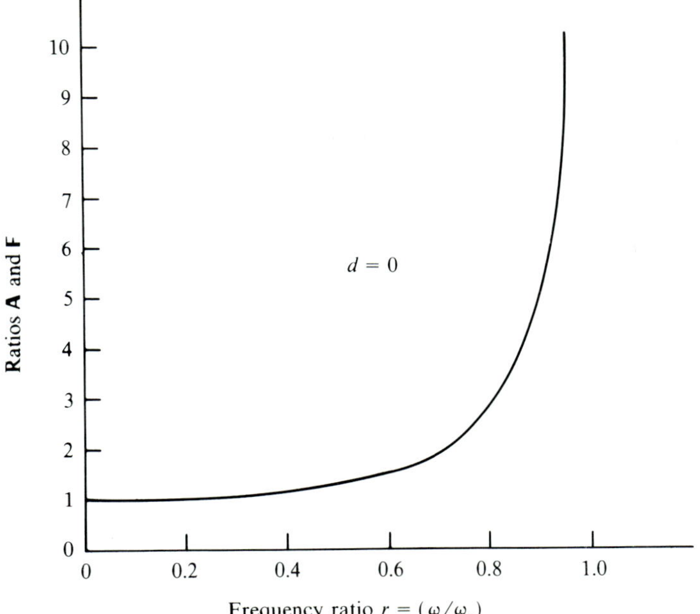

FIGURE 10.10
Ratios $\mathbb{A}$ and $\mathbb{F}$ as a function of the frequency ratio for systems with very low damping.

A graph showing $\mathbb{F}$ as a function of the frequency ratio r, presented in Fig. 10.11, indicates resonance at $r = 1$ with very high values of $\mathbb{F}$. The amplification of the forces depends strongly on the damping ratio d. However, since the damping ratio is small (i.e., less than 0.1) for most electronic assemblies and/or components, extremely high forces will be generated if the system is exposed to frequencies near ω_n.

It is of interest to consider the resonance condition where $r = 1$. At resonance, Eq. (10.31) reduces to

$$\mathbb{F} = \left[1 + (2d)^2\right]^{1/2} \Big/ 2d \simeq 1/2d \tag{10.32}$$

Also, observe that for small damping $d \Rightarrow 0$ and the force transmission coefficient can be approximated by

$$\mathbb{F} = 1/(1 - r^2) \tag{10.33}$$

Comparison of Eqs. (10.32) and (10.33) with Eqs. (10.26) and (10.27) shows that $\mathbb{A} = \mathbb{F}$ either at resonance or if $d \Rightarrow 0$. A significant difference in $\mathbb{A}$ and $\mathbb{F}$ occurs only when the damping ratio is large and the system is not at resonance.

Since damping in electronic assemblies is usually very low, one can anticipate large amplification of the vibrating amplitude and the forces transmitted to the support structure if the system is driven at a frequency that approaches ω_n.

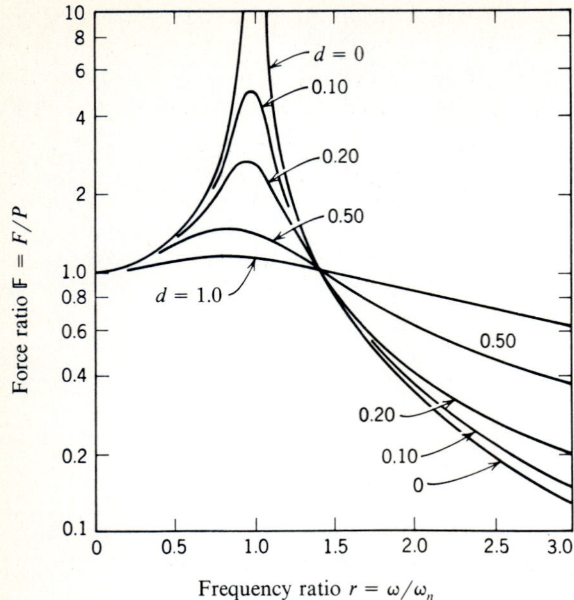

FIGURE 10.11
Force ratio $\mathbb{F} = F/P$ as a function of frequency ratio ω/ω_n for different degrees of damping.

Clearly, extended exposure of the system to resonance conditions must be avoided to prevent damage due to fatigue failure of solder joints, lead wires and fasteners. If it is not possible to design with $\omega_n > 2\omega$, then isolation of the electronic system from the disturbing force may be required.

10.3 ISOLATION OF SYSTEMS FROM EXCITING FORCES

The concept of vibration isolation is presented in Fig. 10.12 where the electronic system, with a mass m, is suspended with springs and a damper in an enclosure. We seek to reduce the amplitude of vibration of the mass m. The motion of the

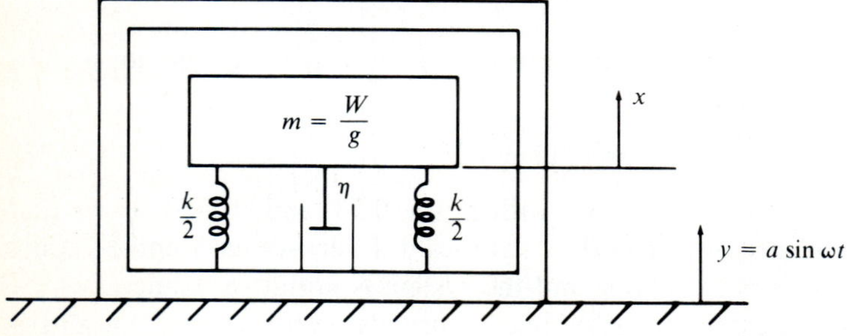

FIGURE 10.12
System defined by x isolated from motion y of the supporting enclosure.

electronic assembly is identified with the displacement coordinate x. We also seek to minimize the forces imposed on the mass, the suspension and the enclosure. The enclosure is subjected to a vibratory motion described in terms of the coordinate y. The suspension is entirely separate from the electronics. The independence of the spring and the dash pot enables the designer to select these components to isolate the mass m from the imposed vibrations. To analyze this isolator, we write the equation of motion for the mass m as

$$m\ddot{x} = -k(x - y) - \eta(\dot{x} - \dot{y}) \tag{10.34}$$

Next, we rearrange the terms to obtain

$$m\ddot{x} + \eta\dot{x} + kx = ky + \eta\dot{y} \tag{10.35}$$

Now consider the input $y(\omega)$ to the housing to be periodic with

$$y = y_0 e^{j\omega t} \tag{10.36}$$

The response of the mass will also be periodic of the form

$$x = x_\omega e^{j(\omega t - \phi)} \tag{10.37}$$

where ϕ is the phase angle between the input and output motions. Substituting Eqs. (10.36) and (10.37) into Eq. (10.35) leads to

$$x_\omega/y_0 = (k + j\omega\eta)e^{j\phi}/[(k - m\omega^2) + j\omega\eta] \tag{10.38}$$

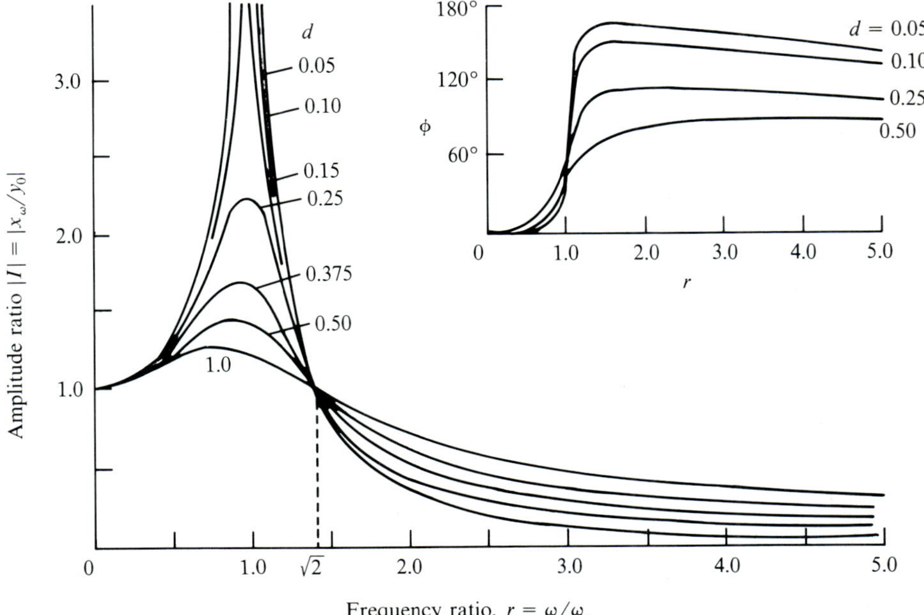

FIGURE 10.13
Amplitude ratio and phase angle with frequency ratio for an isolated system.

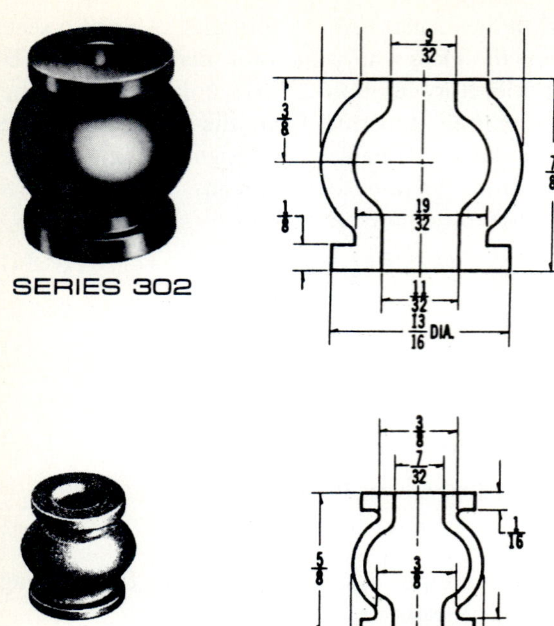

(*a*) Molded elastomer mounts

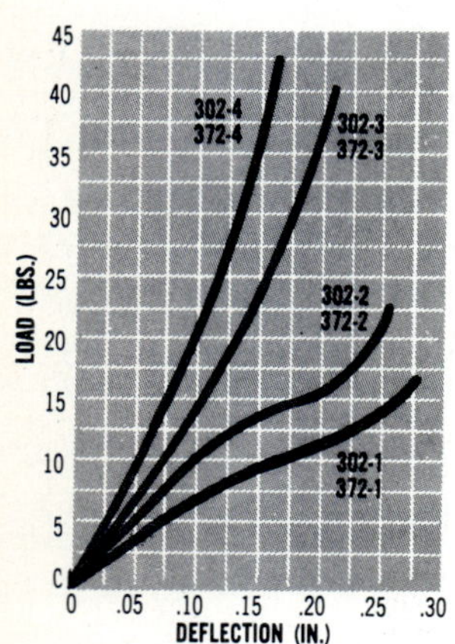

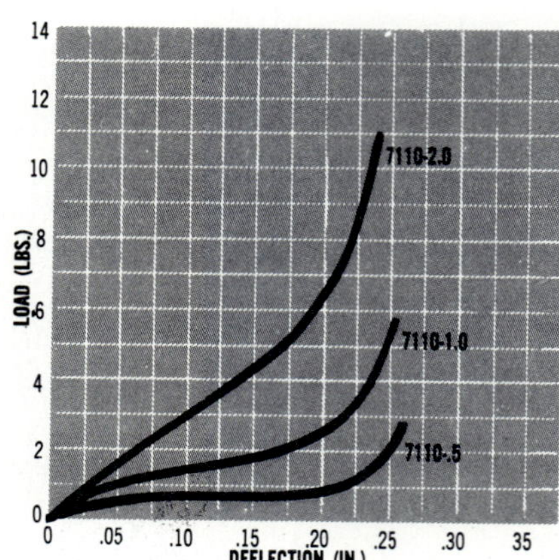

(*b*) Load-deflection relations for different elastomer mounts

FIGURE 10.14
A molded elastomer isolator used for vibration and noise isolation of small loads. (*Courtesy of Barry Controls.*)

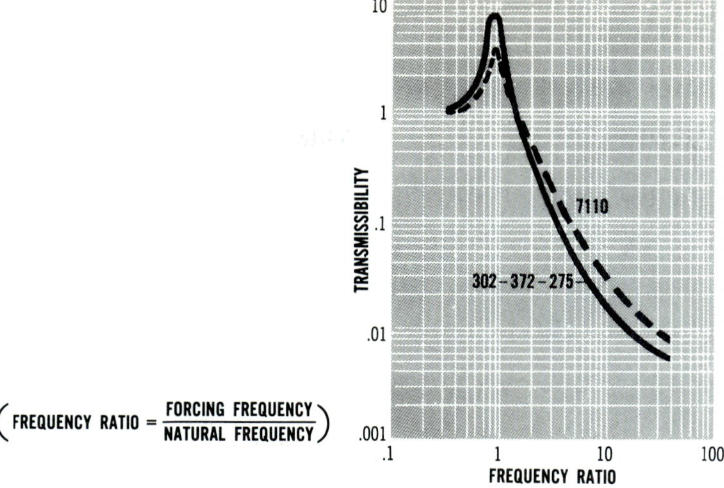

(c) Transmissibility as a function of frequency ratio

FIGURE 10.14
(cont.)

Using methods from complex variables, it is easy to show that the absolute value of the amplitude ratio $|I|$ is

$$|I| = |x_{\omega}/y_0| = \left\{\left[1 + (2\,dr)^2\right]\Big/\left[(1 - r^2)^2 + (2\,dr)^2\right]\right\}^{1/2} \tag{10.39}$$

where the phase angle m is given by

$$\tan\phi = 2\,dr^3\Big/\left[1 - r^2 + (2\,dr)^2\right] \tag{10.40}$$

A graph of the isolated amplitude ratio $|I|$ is presented in Fig. 10.13. Again the effects of a resonance condition are obvious with the large amplification of the input displacement. However, the isolation system as described by k and η is selected especially to avoid a resonance condition and to reduce both the displacements and the forces imposed on the electronics package. This isolation is accomplished by designing a suspension system with a frequency ratio $r > 3$, which gives an isolation ratio $|I| < 0.1$ for systems with small damping.

There are many spring dash pot assemblies called isolation mounts that are commercially available [9]. An example of a molded elastomer mount, which is effective for many lower weight electronic applications, is presented in Fig. 10.14a. The natural frequency of these mounts at rated loads is about 15 Hz, so that isolation of frequencies above 22 Hz will occur. Of course, the transmissibility of the isolation system improves at higher frequencies as indicated by the response curves shown in Fig. 10.14b. Methods used to install the isolation mounts are illustrated in Fig. 10.15.

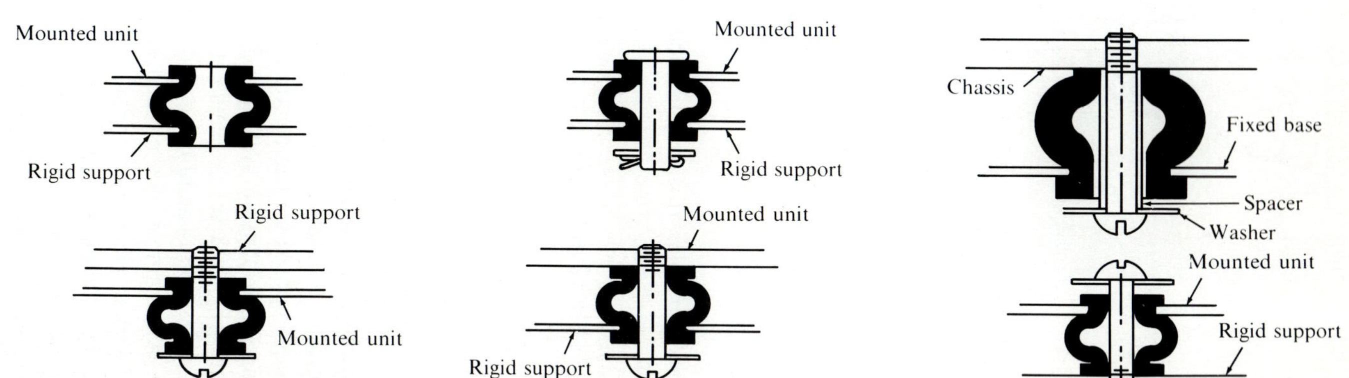

FIGURE 10.15
Several methods for assembly of isolators between the electronic subsystems and the support structure. (*Courtesy of Barry Controls.*)

10.4 VIBRATION OF AXIAL LEADED COMPONENTS

Many electronic components are packaged with axial leads and are soldered into a circuit board with the geometries indicated in Fig. 10.16. We are interested in determining the natural frequency of the axial leaded component and in finding the stresses that develop in the lead wires and the solder joints.

As an example, consider the wire wound resistor shown in Fig. 10.17*a* as the axial leaded component. We note that the body of the resistor is much larger than the diameter of the wire, which leads us to model the resistor leads as assembled on a circuit board as a beam with built-in ends. The leads are represented by the beam and the body of the resistor provides the end constraints and the applied load W as indicated in Fig. 10.17*b*. The static deflection at the load point is obtained from beam theory as

$$\delta = WL^3/192EI \tag{10.41}$$

Substituting Eq. (10.41) into Eq. (10.6) yields

$$f_n = (1/2\pi)\sqrt{192EIg/WL^3} \tag{10.42}$$

where $I = \pi d^4/64$

This is a simple and direct approach for determining the natural frequency of the resistor as mounted on a circuit card or some other structure. Examination of Eq. (10.42) shows that f_n can be increased significantly by decreasing the lead length L between the body and the solder joint. However, a minimum L is required to prevent the heat from the soldering operation from damaging the component. The magnitude of this minimum depends on the soldering process used in the assembly operation.

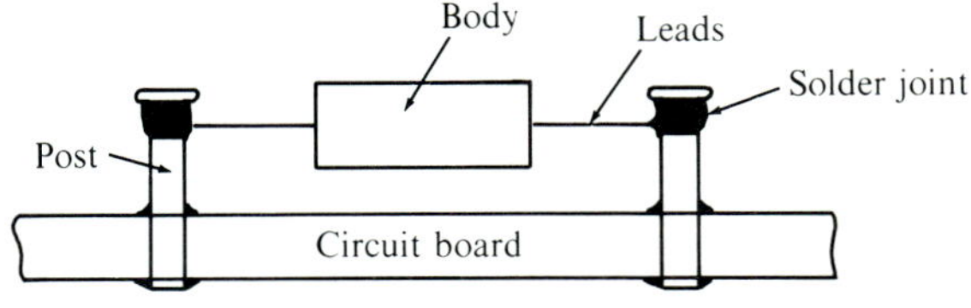

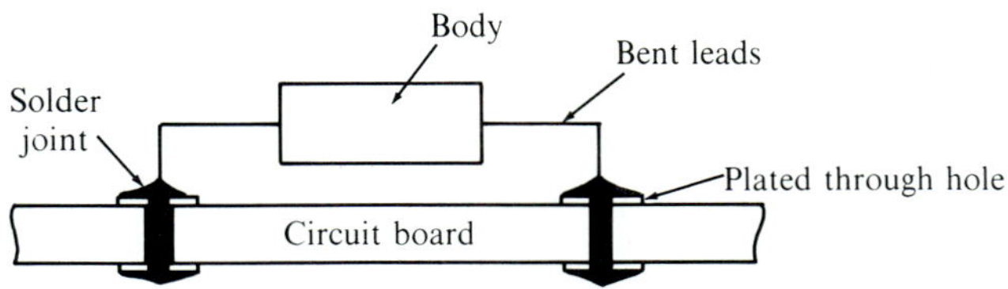

FIGURE 10.16
Axial lead components attached to a circuit board.

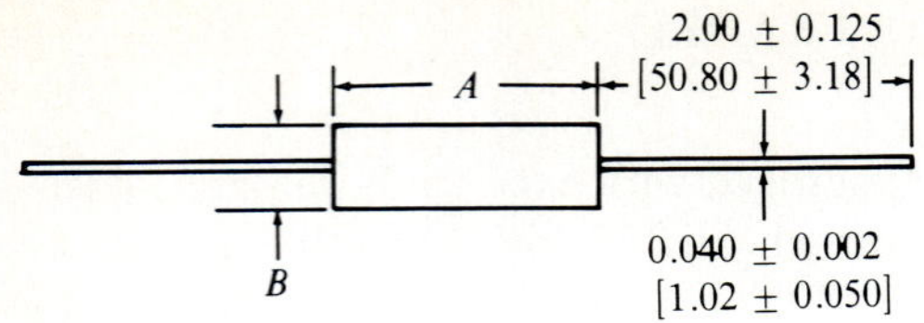

Power rating (W)	Resistance range (Ω)	A Length	B Diameter	Max. weight (g)
2	0.008–0.332	0.675 ± 0.010 [17.15 ± 0.254]	0.250 ± 0.010 [6.35 ± 0.254]	3.0
5	0.010–0.659	0.927 ± 0.010 [23.55 ± 0.254]	0.343 ± 0.010 [8.71 ± 0.254]	5.0
10	0.010–0.800	1.828 ± 0.010 [46.43 ± 0.254]	0.392 ± 0.010 [9.96 ± 0.254]	11.0

(a)

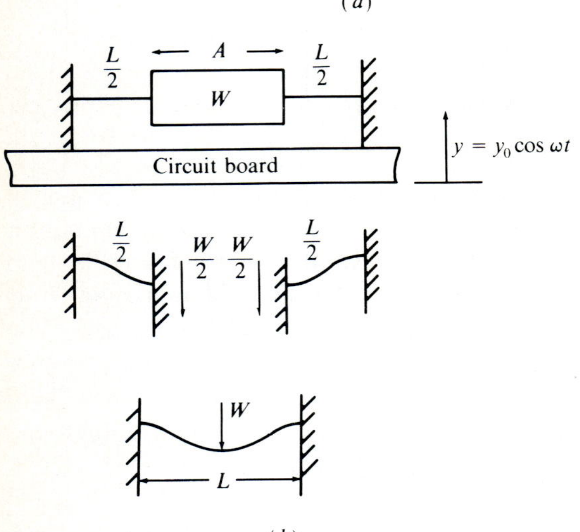

(b)

FIGURE 10.17
(a) Typical specifications for wire wound resistors. (b) Modeling the axial leaded component as a beam with built-in ends.

To find the stresses in the lead wires, we make use of elementary solutions from beam theory. Note that the lead wires are subjected to a maximum moment at $x = 0$ and $x = L/2$ given by

$$M_{\max} = FL/8 \tag{10.43}$$

Recall the flexure formula

$$\sigma = Mc/I \tag{10.44}$$

Observe that the dynamic force F is related to the weight of the component by

$$F = \mathbb{F} W \tag{10.45}$$

where $\mathbb{F}$, the force transmission coefficient, is obtained from Fig. 10.11. Combining Eqs. (10.43), (10.44) and (10.45) gives

$$\sigma = 4\mathbb{F} WL/(\pi d^3) \tag{10.46}$$

The bending stress induced in the lead wire is strongly influenced by $\mathbb{F}$ and since the damping is so small in this assembly, $\mathbb{F}$ varies from 20 to 50 at resonance. For small diameter wires the stresses can become excessive and fatigue failures may result. We will consider fatigue of lead wires in the next section.

10.5 FATIGUE ANALYSIS OF COMPONENT LEADS

Lead wire materials in common usage include tin-plated copper, nickel A and a low coefficient of expansion alloy known as Kovar. All of these materials are ductile and we are not concerned with the stresses and deformations imposed by a single application of load on the lead wires. Indeed, in the assembly of the circuit boards the leads are clipped and then bent so that they can be inserted into the plated through holes. Lead failure does not occur during the bending operation because the wire materials are ductile and will accommodate, by plastic flow, the large deformations that are applied only once.

We are concerned with repeated application of loading, due to vibrations, that will cause the leads to fail in fatigue. If a component is subjected to a vibratory input at the resonance frequency, high stresses are applied repeatedly to the leads, a large number of stress cycles are accumulated and a fatigue crack is initiated that will eventually fracture the lead, causing failure of the component.

The resistance of a material to cyclic loading is measured in terms of its fatigue strength S_f as described previously in Section 7.7.1. The fatigue strength S_f is a function of the number of cycles N of reversed bending that have been applied to the lead wires. An empirical relationship between the fatigue strength S_f and N is given by

$$\log_{10} S_f = a - b \log_{10} N \tag{10.47}$$

which is valid for $N > 10^3$. Clearly, the fatigue strength decreases as a log function of the cycle count. The strength-cycle relation for some common lead wire materials is shown graphically by *S-N* diagrams presented in Fig. 10.18.

A fatigue analysis compares the fatigue strength with the bending stresses imposed on the lead wires. This comparison is the basis for determining the safety factor $\mathbb{S}$ according to

$$\mathbb{S} = S_f/(K\sigma) \tag{10.48}$$

where σ is the bending stress given by Eq. (10.46) and K is the stress concentration associated with the geometric discontinuity at the lead wire/solder joint or

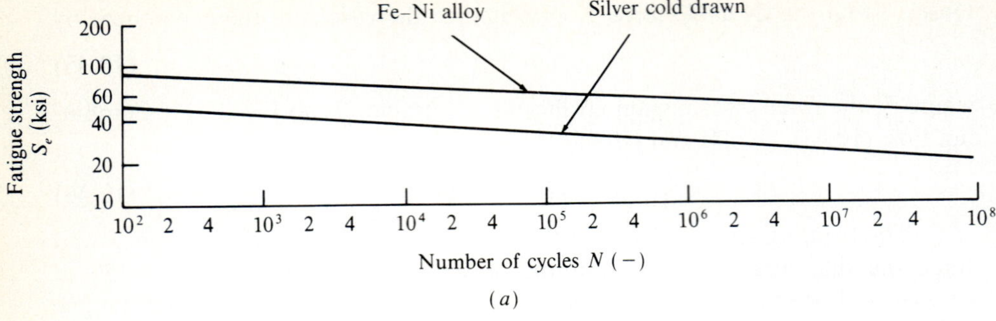

(*a*)

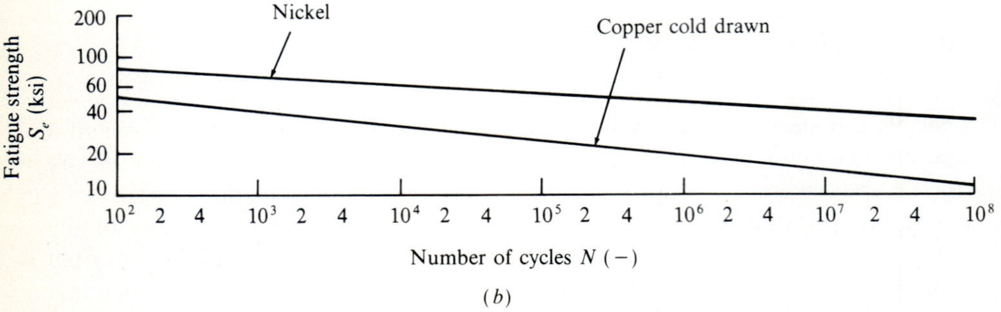

(*b*)

FIGURE 10.18
Fatigue strength S_e as a function of number of cycles for lead wires subjected to reverse bending.

lead wire/component body interface. We usually take K between 1.5 and 2, but little data are available to justify these choices.

The value of S_f used in Eq. (10.48) is dependent on the number of cycles anticipated in service. For severe exposures, where resonance conditions can be anticipated for extended intervals, infinite life design is required. In this case, the value of S_f corresponding to a cycle life of $N = 10^8$ is used. In other situations where exposure to resonance conditions is limited, finite life design may be appropriate. To employ finite life design, it is necessary to estimate the number of cycles of imposed loading and then to use Fig. 10.18 to select the finite life fatigue strength.

As an example of the finite life design approach, consider an electronic system that is required to pass a qualification test that requires a resonance dwell of 30 min. The number of cycles anticipated during the dwell period t_d is

$$N = f_n t_d \tag{10.49}$$

If $f_n = 500$ Hz and $t_d = 30 \times 60 = 1800$ s, then $N = 9 \times 10^5$ cycles and S_f can be determined from the appropriate *S-N* diagram for this cycle count.

In some qualification tests the electronic equipment is subjected to a number of sweeps on the vibration table where the frequency of vibration is

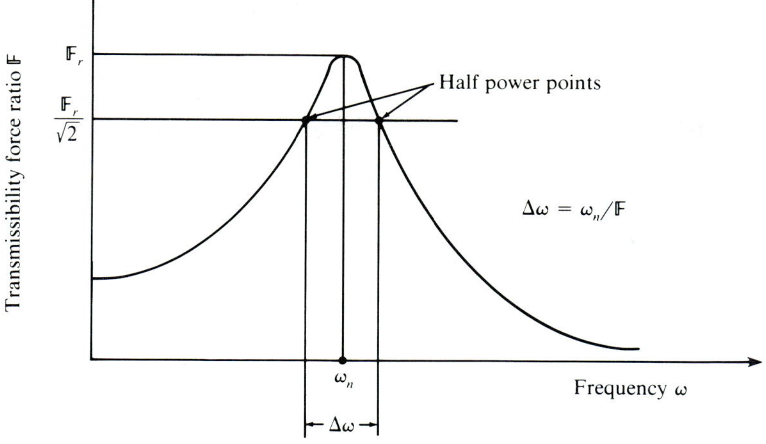

FIGURE 10.19
Location of half-power points on the force transmissibility curve.

varied between 10 and 2000 Hz. In this type of vibration testing, fatigue damage occurs at frequencies near resonance where the transmissibility factor $\mathbb{F}$ is high. An approach for approximating the cycle count with sweeping frequencies involves determining the time to sweep through the half-power points. The half-power points are illustrated in Fig. 10.19. The time to sweep from the first to the second half-power point is given by

$$t_s = \ln[(2\mathbb{F} + 1)/(2\mathbb{F} - 1)]/R_s \ln(2) \tag{10.50}$$

where R_s is the sweep rate in octaves per minute. By substituting the results from Eq. (10.50) into Eq. (10.49), we obtain the cycle count as

$$N = nf_nt_s \tag{10.51}$$

where n is the number of single sweeps required in the qualification test.

The approach described here permits us to perform finite life fatigue analysis. However, we should be aware that fatigue strength exhibits relatively large statistical variation and the use of a safety factor $\mathbb{S}$ of 2 or more in Eq. (10.48) is advisable. Another factor that is significant is the care exercised in preparing the lead wires. If any wire is nicked in bending or in clipping, a large concentration of stress, occurs at the notch, with $K = 10$ or more, and the lead wire may fail prematurely. One purpose of the random vibrations imposed on the electronics in the manufacturing screening tests is to detect lead wire nicks and to replace the affected components prior to shipping the product.

In addition to infinite and finite life fatigue, we will consider low life fatigue. Low life fatigue involves lead wires, cables and other parts that will be subjected to a relatively low number of cycles of stress (less than 10^4) during the anticipated life of the product. With low life fatigue, the stress levels imposed on the material are greater than the yield strength. The life of the part is determined to a large degree by the ductility of the material from which it is fabricated. Manson [10] has developed a useful relation for predicting cyclic life N that is based on the applied strain range $\Delta\varepsilon$:

$$\Delta\varepsilon = (3.5S_u/E)N^{-0.12} + (D/N)^{-0.6} \tag{10.52}$$

where the strain range $\Delta\varepsilon = \varepsilon_{max} - \varepsilon_{min}$
S_u is the tensile strength
$D = \ln[1/(1 - RA)]$ is the ductility
RA is the reduction of area

While Eq. (10.52) is approximate, it does provide a simplified method for determining fatigue life N based on results from a simple uniaxial tension test of the lead wire material to determine RA and S_u.

10.6 VIBRATION OF CIRCUIT BOARDS

Printed circuit boards, which were described in Chapter 5, are also subject to vibration that can cause damage to the components mounted on the board or to the connectors used to connect the PCB to the back panel. Since most PCBs are rectangular, they can be treated as rectangular plates. We are concerned with the out-of-plane vibration of the plate as indicated in Fig. 10.20. The natural frequency for a rectangular plate depends on the boundary conditions, which may be clamped, simply supported or free. Altogether, there are 21 possible combinations of these simple boundary conditions. Fortunately, solutions to these boundary value problems exist in a form that can be readily used in design.

Leissa [11], in an excellent summary of available results for frequency and mode shapes, presents the solutions of Janich [12]. The solutions for the natural frequency ω_n are given in the form

$$\omega_n^2 = (\pi^4 D C_1)/(a^4 \rho C_2) \tag{10.53}$$

where C_1 and C_2 are given in Table 10.2
a and b are the plate dimensions
ρ is the mass density per unit area of the plate
D is the flexural rigidity defined by

$$D = Eh^3/[12(1 - \nu^2)] \tag{10.54}$$

where h is the thickness, ν is Poisson's ratio and E is the modulus of elasticity.

The out-of-plane deflection w is given in the form

$$w = w_\omega f[(x/a),(y/b)]\cos\omega t \tag{10.55}$$

where the deflection function $f[(x/a),(y/b)]$ is listed in Table 10.2. From Eq. (10.55), it is evident that the deflection of the circuit board in vibration is a function of the size of the board given by a and b, the exciting frequency ω and the amplitude w_ω, which is controlled by the input.

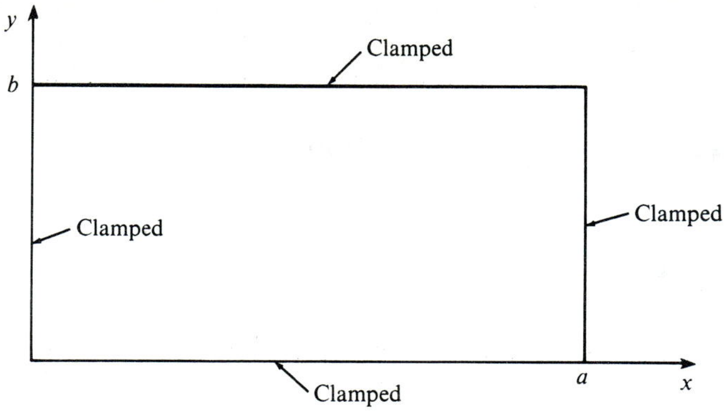

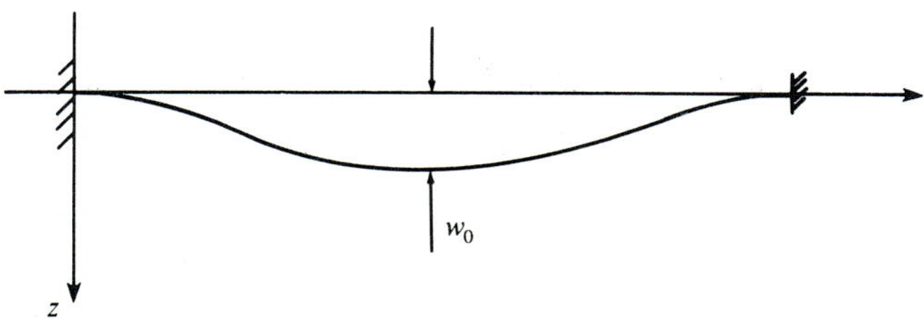

FIGURE 10.20
Definition of the plate dimensions and the coordinate system.

Let us consider the simplest of the cases listed in Table 10.2 as an example. This last listing in Table 10.2 (case no. 18) describes a rectangular plate with two opposing edges simply supported and the other two opposing edges free. The deflection for this plate may be written as

$$w = w_{\omega} \sin(\pi x/a)\cos \omega t \tag{10.56}$$

The maximum amplitude will occur along the center line $x = a/2$ at the resonance frequency when $\omega = \omega_n$. With these conditions, Eq. (10.56) reduces to

$$w_{\max} = w_{\omega n} \cos \omega_n t \tag{10.57}$$

Differentiating Eq. (10.57) twice with respect to time gives the maximum out-of-plane acceleration $\ddot{w}$ as

$$\ddot{w}_{\max} = -w_{\omega n}\omega_n^2 \cos \omega_n t \tag{10.58}$$

Let us examine the amplitude of the acceleration and define a^*_{out} as

$$a^*_{\text{out}} = \ddot{w}_{\max}/g = -w_{\omega n}\omega_n^2/g \tag{10.59}$$

In this form, the acceleration is given in terms of Gs. Solving this relation for $w_{\omega n}$

TABLE 10.2
Parameters c_1 and c_2 for rectangular plates with different boundary conditions used to compute the natural frequency $\omega_n^2 = (\pi^4 D c_1)/(a^4 \rho c_2)$

Boundary conditions	Deflection function or mode shape	c_2	c_1
1. a, b	$\left(\cos\frac{2\pi x}{a} - 1\right)\left(\cos\frac{2\pi y}{b} - 1\right)$	2.25	$12 + 8\left(\frac{a}{b}\right)^2 + 12\left(\frac{a}{b}\right)^4$
2. a, b	$\left(\cos\frac{3\pi x}{2a} - \cos\frac{\pi x}{2a}\right)\left(\cos\frac{2\pi y}{b} - 1\right)$	1.50	$3.85 + 5\left(\frac{a}{b}\right)^2 + 8\left(\frac{a}{b}\right)^4$
3.	$\left(1 - \cos\frac{\pi x}{2a}\right)\left(\cos\frac{2\pi y}{b} - 1\right)$	0.340	$0.0468 + 0.340\left(\frac{a}{b}\right)^2 + 1.814\left(\frac{a}{b}\right)^4$
4.	$\left(\cos\frac{2\pi x}{a} - 1\right)\sin\frac{\pi y}{b}$	0.75	$4 + 2\left(\frac{a}{b}\right)^2 + 0.75\left(\frac{a}{b}\right)^4$
5.	$\left(\cos\frac{2\pi x}{a} - 1\right)\frac{y}{b}$	0.50	$2.67 + 0.304\left(\frac{a}{b}\right)^2$
6.	$\cos\frac{2\pi x}{a} - 1$	1.50	8
7.	$\left(\cos\frac{3\pi x}{2a} - \cos\frac{\pi x}{2a}\right)\left(\cos\frac{3\pi y}{2b} - \cos\frac{\pi y}{2b}\right)$	1.00	$2.56 + 3.12\left(\frac{a}{b}\right)^2 + 2.56\left(\frac{a}{b}\right)^4$
8.	$\left(\cos\frac{3\pi x}{2a} - \cos\frac{\pi x}{2a}\right)\left(1 - \cos\frac{\pi y}{2b}\right)$	0.227	$0.581 + 0.213\left(\frac{a}{b}\right)^2 + 0.031\left(\frac{a}{b}\right)^4$
9.	$\left(1 - \cos\frac{\pi x}{2a}\right)\left(1 - \cos\frac{\pi y}{2b}\right)$	0.0514	$0.0071 + 0.024\left(\frac{a}{b}\right)^2 + 0.0071\left(\frac{a}{b}\right)^4$
10.	$\left(\cos\frac{3\pi x}{2a} - \cos\frac{\pi x}{2a}\right)\sin\frac{\pi y}{b}$	0.50	$1.28 + 1.25\left(\frac{a}{b}\right)^2 + 0.50\left(\frac{a}{b}\right)^4$
11.	$\left(\cos\frac{3\pi x}{2a} - \cos\frac{\pi x}{2a}\right)\frac{y}{b}$	0.333	$0.853 + 0.190\left(\frac{a}{b}\right)^2$
12.	$\cos\frac{3\pi x}{2a} - \cos\frac{\pi x}{2a}$	1.00	2.56
13.	$\left(1 - \cos\frac{\pi x}{2a}\right)\frac{\pi^2}{b^2}\sin\frac{\pi y}{b}$	0.1134	$0.0156 + 0.0852\left(\frac{a}{b}\right)^2 + 0.1134\left(\frac{a}{b}\right)^4$
14.	$\left(1 - \cos\frac{\pi x}{2a}\right)\frac{y}{b}$	0.0756	$0.0104 + 0.0190\left(\frac{a}{b}\right)^2$
15.	$1 - \cos\frac{\pi x}{2a}$	0.2268	0.0313
16.	$\sin\frac{\pi x}{a}\sin\frac{\pi y}{b}$	0.25	$0.25 + 0.50\left(\frac{a}{b}\right)^2 + 0.25\left(\frac{a}{b}\right)^4$
17.	$\left(\sin\frac{\pi x}{a}\right)\frac{y}{b}$	0.1667	$0.1667 + 0.0760\left(\frac{a}{b}\right)^2$
18.	$\sin\frac{\pi x}{a}$	0.50	0.50

gives the amplitude of the maximum displacement of the PCB as

$$w_{\omega n} = -ga^*_{\text{out}}/\omega_n^2 \tag{10.60}$$

From Eq. (10.24) it is evident that

$$a^*_{\text{out}} = \mathbb{A}\, a^*_{\text{in}} \tag{10.61}$$

Substitution Eq. (10.61) into Eq. (10.60) yields

$$w_{\omega n} = \mathbb{A}\, ga^*_{\text{in}}/\omega_n^2 \tag{10.62}$$

This relation permits the maximum displacement of the PCB to be determined for a given input a^*_{in} that is specified in the qualification test procedure. Since ω_n^2 is in the denominator, it is clear that the circuit boards should be designed with a high natural frequency to reduce the amplitude of the displacement.

One difficulty occurs in using Eq. (10.62), namely, the transmission coefficient is often unknown. There are three different approaches for determining $\mathbb{A}$. The first is to construct a model of the PCB and to perform vibration sweeps on a shaker table. By measuring the ratio of the acceleration at the center of the board and the shaker table, as illustrated in Fig. 10.21, the natural frequency of the board and the transmission coefficient $\mathbb{A}$ can be established. While vibration testing is the most accurate method of determining the displacement, natural frequency and the transmission coefficient, it is expensive and time-consuming.

Another approach, due to Steinberg [13], is to estimate $\mathbb{A}$ at the resonance frequency as

$$\mathbb{A} = [f_n]^{1/2} \tag{10.63}$$

In using this technique, we determine the natural frequency from Eq. (10.53),

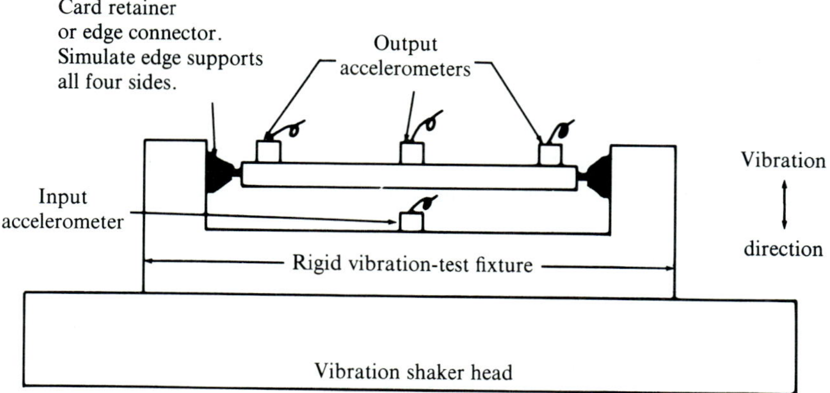

FIGURE 10.21
Vibration test arrangement for printed circuit boards.

estimate $\mathbb{A}$ from Eq. (10.63) and then compute the maximum displacement for a given input a_{in}^* from Eq. (10.62).

The remaining technique for estimating $\mathbb{A}$ is based on the knowledge of the degree of damping for a particular type of construction and size of a PCB. If the degree of damping can be estimated, the transmission coefficient $\mathbb{A}$ is determined from Eq. (10.26). Exercises are provided to demonstrate the application of these methods to vibration analysis of circuit boards.

10.6.1 Improving the Vibration Behavior of Circuit Boards

There are two characteristics of a PCB used to judge its merit to withstand a vibratory environment, namely, the natural frequency and the degree of damping, which markedly affects the transmission coefficient. We seek to make the natural frequency as high as possible to limit the amplitude of the oscillation as shown by Eq. (10.62). Also, we attempt to increase the degree of damping to reduce the transmission coefficient at resonance. Reference to Eq. (10.53) clearly indicates that the natural frequency decreases as the square of the size of the circuit board. Large circuit cards are advantageous from an electrical viewpoint in that they can support many components. However, large circuit boards have relatively low natural frequencies. To illustrate this fact, consider a square card with all four edges simply supported. From the results listed in Table 10.2 and Eq. (10.53), it is clear that

$$f_n = (\pi/a^2)[D/\rho]^{1/2} \tag{10.64}$$

In this example, f_n decreases with the square of the edge dimension a of the card. While large card areas have advantages for efficient packaging of a large number of components, the size must be limited if the system is to operate in a harsh vibratory environment.

The board thickness also affects ω_n. From Eqs. (10.64) and (10.54), with $\nu = \frac{1}{4}$ and $\rho = \gamma h/g$, we obtain

$$f_n = (\pi/a^2)[4Eg/45\gamma]^{1/2}h \tag{10.65}$$

where γ is the weight density of the board material. Clearly Eq. (10.65) shows that the natural frequency increases linearly with thickness. This equation also shows the effect of size with f_n decreasing with a^2, which corresponds with the results given in Eq. (10.64).

Another approach that markedly stiffens the card and increases f_n is to bond a heat frame to the circuit card. This approach, often used in conductive cooling arrangements, provides a composite cross section with a flexural rigidity D that is very large.

The edge supports of the circuit card are clearly important in determining the natural frequency. Reference to Table 10.2 indicates that clamped edges are

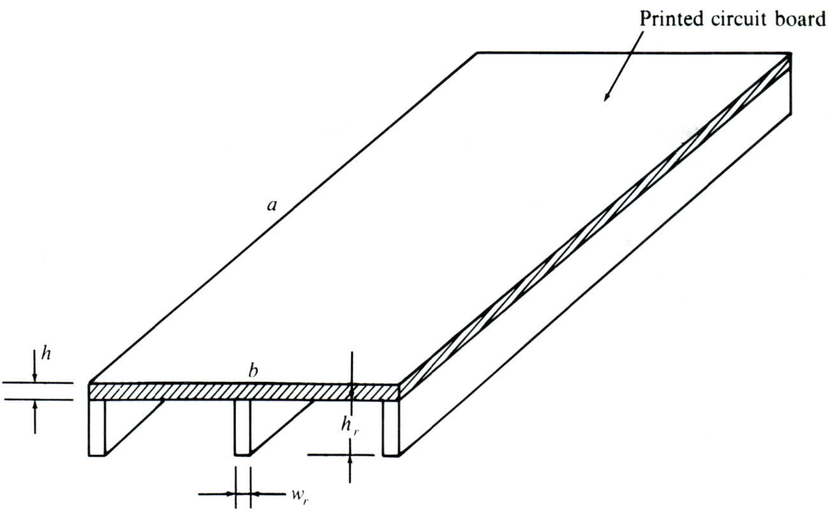

FIGURE 10.22
Ribs attached to a circuit board to increase its stiffness.

preferred in enhancing ω_n. However, clamped edges are difficult to achieve in many installations because they require wedge lock-type retainers (see Chapter 7) that are costly. More common are edge guides that essentially provide simple support to two opposing sides of the card. Edge card connectors are frequently employed on one side of the card and they provide a simple support edge condition. In many instances one side of the card is free.

Ribs may be attached to circuit boards to increase their stiffness as shown in Fig. 10.22. The ribs are usually strips of aluminum or stainless steel that are adhesively bonded to the circuit board. Composite beam theory is used to determine the effective inertia I_e of the cross section. The flexural rigidity is then determined from

$$D = EI_e/b \tag{10.66}$$

Damping enhancement is another approach for mitigating vibration effects. Clearly, increasing the degree of damping d decreases both $\mathbb{A}$ and $\mathbb{F}$; however, space is usually too limited to place mechanical damping devices on the boards. Application of a viscous coating will improve the degree of damping. Currently conformal coatings are applied to some circuit boards to protect the traces on the surface from effects of humidity. These coatings are usually too thin and too elastic to absorb enough energy to provide significant damping enhancement. A thicker more viscous coating that could also serve as a moisture barrier would be helpful in providing protection against vibration as well as humidity.

10.6.2 Stresses in Circuit Boards Due to Vibration

When a circuit board is deflected during exposure to vibration, stresses are produced in the plane of the board that may produce failure by fatigue damage. The magnitude of the stresses produced depends on the deflected shape of the board, which is given by $w(x, y)$. The deflection function $w(x, y)$ is controlled by the boundary conditions imposed in constraining the board as indicated in Table 10.2.

We will consider a simple example to, first, illustrate the method to follow in determining stresses induced by vibration and, second, to generalize on the relative importance of this topic. The example selected is a board with simple supports as shown in Fig. 10.23. This is an example illustrating severe vibration because the board is completely free on two edges and free to rotate about the supported edges. This freedom leads to large deflections and relatively high stress at the center of the circuit board. For the simply supported PCB the deflection function given in Table 10.2 is

$$w = w_0 \sin(\pi x/a) \tag{10.67}$$

where w_0 is the displacement at $x = a/2$. The moment per unit width of the board is related to $w(x)$ by

$$M = -D(d^2w/dx^2) \tag{10.68}$$

The distribution of the moment along the length of the board is obtained by substituting Eq. (10.67) into Eq. (10.68) to give

$$M = Dw_0(\pi/a)^2 \sin(\pi x/a) \tag{10.69}$$

which is a maximum at $x = a/2$ given by

$$M_{max} = Dw_0(\pi/a)^2 \tag{a}$$

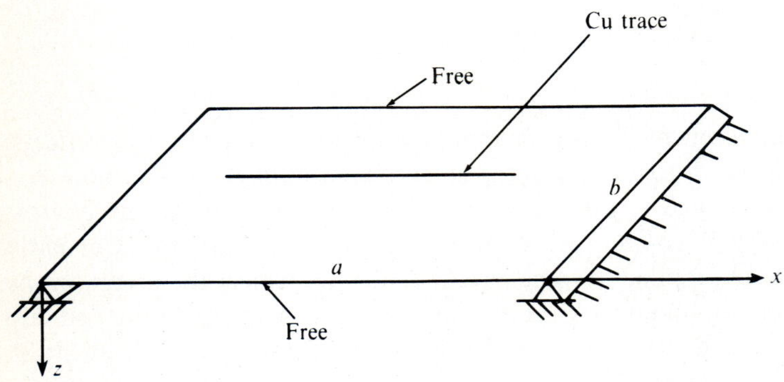

FIGURE 10.23
Simply supported circuit board.

The bending moment produces flexural stresses written as

$$\sigma_b = 6KM/h^2 \tag{10.70}$$

The symbol K is the stress concentration that is due to the geometric discontinuities on the board. Usually the plated through holes produce the largest value of $K = 3$. If we combine Eqs. (10.70) and (a) and let $K = 3$, we obtain the maximum stress as

$$\sigma_{max} = \left(18Dw_0/h^2\right)(\pi/a)^2 \tag{10.71}$$

This result is correct, but it is not in a form that can be interpreted easily since w_0 is not known. Note that from Eqs. (10.62) and (10.63) we can write

$$w_0 = (2\pi)^{1/2} g a_{in}^{*}/\omega_n^{3/2} \tag{10.72}$$

Next, we use Eq. (10.53) and Table 10.2 to show

$$\omega_n = (\pi/a)^2(D/\rho)^{1/2} \tag{b}$$

Substituting Eqs. (10.72) and (b) into Eq. (10.71) yields

$$\sigma_{max} = 18\sqrt{2}\, D^{1/4} g a_{in}^{*} a \rho^{3/4}/(\pi^{1/2}h^2) \tag{c}$$

Recall that $\rho = \gamma h/g$ and $D = 4Eh^3/45$ with $\nu = \frac{1}{4}$ and substitute these equalities in Eq. (c) to obtain a result that can be interpreted in terms of basic dimensions and properties:

$$\sigma_{max} = 7.84 E^{1/4} g^{1/4} a_{in}^{*} a \gamma^{3/4}/h^{1/2} \tag{10.73}$$

Examination of Eq. (10.73) shows that the stresses increase linearly with the input acceleration a^* and the dimension a. The maximum stress decreases with $\sqrt{h}$ and increases with E and density γ raised to powers of $\frac{1}{4}$ and $\frac{3}{4}$, respectively.

To examine the importance of these bending stresses on the performance of the circuit boards, consider the following characterization of a board in vibration:

$$E = 3 \times 10^6 \text{ psi}$$

$$g = 386 \text{ in./s}^2$$

$$a_{in}^{*} = 2\ G$$

$$a = 12 \text{ in.} \quad \text{and} \quad h = 0.060 \text{ in.}$$

$$\gamma = 0.065 \text{ lb/in.}^3$$

Substitution of these values leads to $\sigma_{max} = 6400$ psi. Since the tensile strength of the common circuit board material, epoxy-reinforced with fiberglass, is about 80 ksi, the safety factor $\mathbb{S}$ due to static deformation is 12.5. The *S-N* diagram for the glass fiber-reinforced epoxy shown in Fig. 10.24 indicates that the endurance strength $S_e = 16$ ksi. The safety factor for an infinite life with this 2 G exposure is

$$\mathbb{S} = S_e/\sigma_{max} = 16/6.4 = 2.5 \tag{d}$$

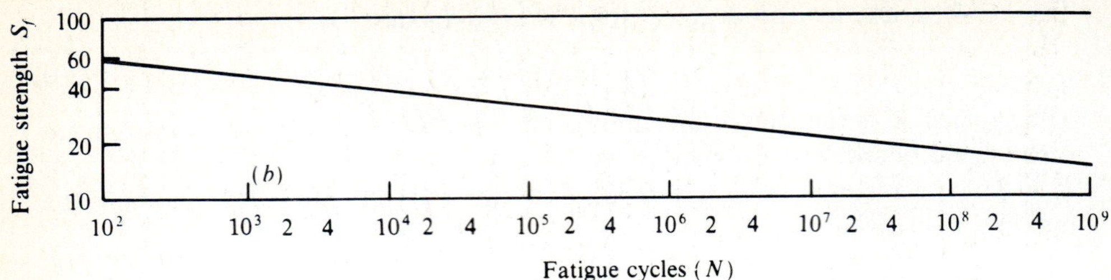

FIGURE 10.24
S-N diagram for a typical fiber glass–epoxy circuit board material.

This example represents a severe vibratory input on a large and poorly supported circuit board. The resulting stresses were relatively low and the safety factors were more than adequate. We may generalize and conclude the epoxy–glass circuit boards can be designed to be fatigue resistant. Indeed, structural failure of the boards is rarely observed. The most common form of damage that does occur in many cases is with the card connectors. The board displacements and rotations cause wear of the pin–contact combination and under more severe vibration, pin breakage will occur.

This generalization should not be extended to circuit substrates fabricated from ceramics. The strength of ceramics, in either tensile or fatigue, is difficult to specify as a deterministic quantity, and failure due to vibration-induced deformation would be difficult to predict. We are fortunate that ceramics used in chip carriers, hybrids and circuit boards are usually small in size. This small size together with their extremely high modulus gives components with very high ω_n, which resist deformation.

10.6.3 Stresses in Copper Signal Traces

In the previous discussion we showed that the stresses induced in the PCB even under severe vibration conditions were relatively low in magnitude. However, this fact does not imply that the stresses in the copper traces that are bonded to the circuit board are also small. To explore this question, let us consider the same circuit board as shown in Fig. 10.23 with the x coordinate along the longitudinal axis of the board. The strain in the x direction on the board is

$$\varepsilon_x = (1/E)(\sigma_x - \nu\sigma_y) \tag{10.74}$$

Since the circuit board is modeled as a plate where $\varepsilon_y = 0$, we may write

$$\sigma_y = \nu\sigma_x \tag{10.75}$$

Next, combine Eqs. (10.74) and (10.75) to give

$$\varepsilon_x = \left[(1 - \nu^2)/E\right]\sigma_x \tag{10.76}$$

The copper trace is bonded to the top surface of the circuit board as illustrated in Fig. 10.23. We assume that the strain in the circuit board is transmitted to the copper trace without change in magnitude and

$$(\varepsilon)_{cu} = (\varepsilon)_{ge} \tag{10.77}$$

where the subscripts cu and ge refer to copper and glass-epoxy. If we take into account the uniaxial geometry of the copper trace, Hooke's law indicates

$$\sigma_{cu} = E_{cu}(\varepsilon_x)_{cu} \tag{10.78}$$

Substituting Eqs. (10.76) and (10.77) into Eq. (10.78) gives

$$\sigma_{cu} = (E_{cu}/E_{ge})(1 - \nu_{ge}^2)(\sigma_x)_{ge} \tag{10.79}$$

This expression relates the stresses developed in the copper trace to those occurring in the circuit board. Let us return to the example used in Section 10.6.2 and note that $E_{cu} = 16 \times 10^6$ psi, $E_{ge} = 3 \times 10^6$ psi, $\nu_{ge} = \frac{1}{4}$ and $(\sigma_x)_{ge} = 6400$ psi. Substituting these values in Eq. (10.79) gives $\sigma_{cu} = 5 \times 6400 = 32{,}000$ psi. Since the elastic modulus of copper is much higher than that of the glass-reinforced epoxy, we develop much larger stresses (a factor of about 5) in the copper trace. This stress is significant relative to the tensile strength of the copper and, indeed, it is larger than the endurance strength S_e. Reference to the *S-N* diagram for cold-drawn copper presented in Fig. 10.18 shows that the anticipated life for the copper trace is about 10^4 cycles. If this circuit board were installed in a product exposed to 2 *G* at the resonance frequency for more than a few minutes, fatigue failures of the circuit traces could be anticipated.

10.7 LEAD WIRE FAILURES ON VIBRATING PRINTED CIRCUIT BOARDS

As the printed circuit board vibrates, the components mounted on the board are subjected to stress from two different effects. First, the mass of the component is subjected to an acceleration that produces a force *P* normal to the surface of the board as shown in Fig. 10.25*a*. The body of the component is kept in equilibrium with reactive forces developed in the lead wires. Second, the surface of the board flexes, which tends to bend the leads back and forth at their junction with the board. We will consider these two effects individually in determining component deflections and lead wire stresses.

Components with lead wires such as resistors, capacitors, DIPs or flat packs are modeled as frames. The body of the component is so rigid relative to the lead wire that we can assume that the deflection of the component is due entirely to the deformation of the lead wires. This assumption permits us to represent the component with a lead wire frame as shown in Fig. 10.25*b*. Since the leads are soldered to the boards and reinforced with a solder fillet, the leads are considered to be built-in when defining the end conditions for the frame. The built-in end condition requires that the leads remain locally perpendicular to the board and

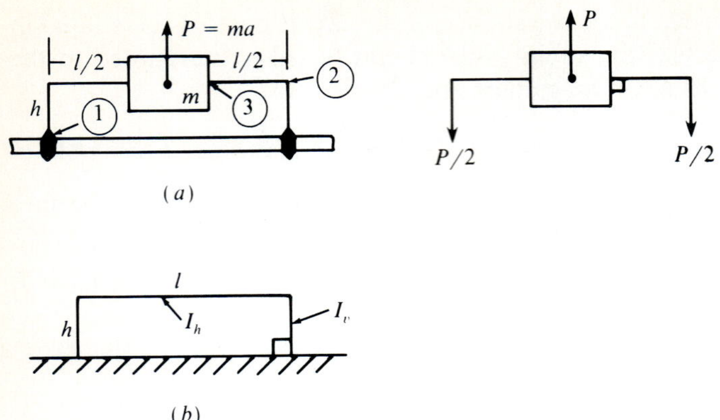

FIGURE 10.25
Modeling a leaded component with a wire frame. (*a*) Load P due to component mass. (*b*) The wire frame representation.

that they rotate with the board as it deflects during exposure to the vibratory input.

The effect of the lead wire deformation and stress is to produce fatigue failure of the leads, cracking of the solder joint or rupture of the seal between the lead and the component body. The three critical locations are shown in Fig. 10.25*a*. We will determine the stresses at these locations due to the two types of loading by using classical methods of frame analysis. The problem is tedious because the frame is statically indeterminate and it is complex because we have loads produced by both component mass and circuit board deflection. Let us divide the combined problem, as shown in Fig. 10.26*a*, into two simpler problems illustrated in Fig. 10.26*b*. Note, that the relative rotation of the leads θ is given by

$$\theta = (\partial w/\partial x)_B - (\partial w/\partial x)_A \tag{10.80}$$

The worst case situation for lead rotation is when the component straddles the center line of the circuit board and the slopes at points A and B are equal in magnitude and opposite in sign. In this case, $\theta = 2(\partial w/\partial x)_B$.

The force P is given by

$$P = \mathbb{F} a_{\text{in}}^{*} W \tag{10.81}$$

The force transmissibility coefficient $\mathbb{F}$ can be approximated by

$$\mathbb{F} = 1.5\sqrt{f_n} \tag{10.82}$$

where f_n is the natural frequency of the component that is determined from Eq. (10.42). Substituting Eq. (10.82) into Eq. (10.81) gives

$$P = 1.5\sqrt{f_n}\, a_{\text{in}}^{*} W \tag{10.83}$$

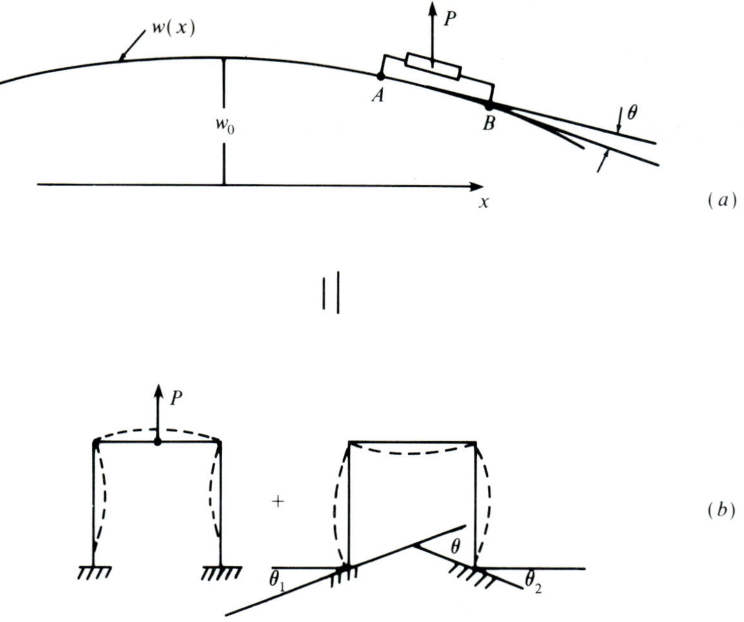

FIGURE 10.26
Circuit board component interaction. (*a*) Circuit board deflection relative to the component. (*b*) Wire frame modeling with the load P and the rotation θ.

Consider first the wire frame with the central load P and use statically indeterminate methods to find the unknown reactions. The free-body diagrams presented in Fig. 10.27 define these unknown reactions as Q and M_1 at the built-in end. Castigliano's theorem, which will be described in the next section, can be employed to determine Q and M_1. The other moments and deflections follows from elementary statics and strength of materials. The results for the

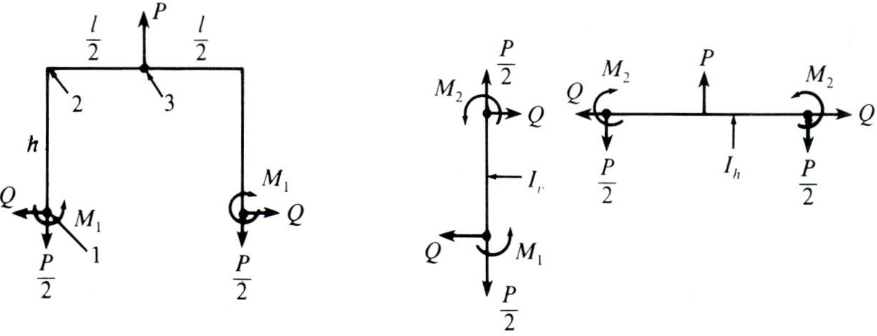

FIGURE 10.27
Free-body diagrams of a wire frame with the load P.

frame subjected to the load P described in Fig. 10.27 are

$$
\begin{aligned}
Q^P &= (3P\ell)/[8h(C+2)] \\
M_1^P &= (P\ell/8)/(C+2) \\
M_2^P &= (P\ell/4)/(C+2) \\
M_3^P &= (P\ell/2)(C+1)/(C+2) \\
\delta_3^P &= [P\ell^3/(96EI_h)][(C+1)/(C+2)] \\
C &= (h/\ell)(I_h/I_v)
\end{aligned}
\tag{10.84}
$$

Consider the second part of the problem, namely, the moments induced in the wire frame due to the rotation θ_1 imposed on the ends of the frame at the attachment points. The free-body diagrams defining the unknown moments and forces are presented in Fig. 10.28. The use of Castigliano's theorem gives the moment M_1 as

$$
\begin{aligned}
M_1^\theta &= [2\theta_1(3+2C)EI_v]/[h(2+C)] \\
Q^\theta &= (M_1/h)(1-C) + (4EI_h\theta_1)/(h\ell) \\
M_2^\theta &= (4EI_h\theta_1/\ell) - CM_1 \\
M_3^\theta &= M_2^\theta \\
\delta_3^\theta &= (\theta_1\ell/4)(1/2C)
\end{aligned}
\tag{10.85}
$$

Superposition of Eqs. (10.84) and (10.85) gives the combined moments, forces and deflections associated with P and θ_1 as

$$
\begin{aligned}
M_1 &= M_1^P + M_1^\theta \quad \text{etc.} \\
Q &= Q^P + Q^\theta \\
\delta_3 &= \delta_3^P + \delta_3^\theta
\end{aligned}
\tag{10.86}
$$

An exercise is given to illustrate the application of these equations. The stresses are determined from the moments by using the flexure formula $\sigma = Mc/I$.

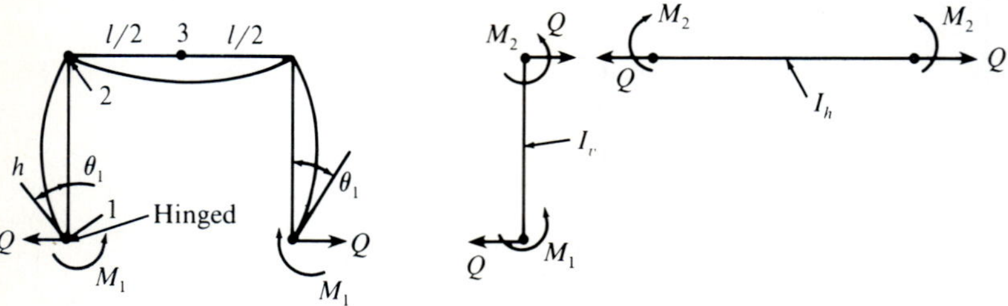

FIGURE 10.28
Free-body diagrams of a wire frame with moments M_1 applied to give rotations of θ_1 at the support points.

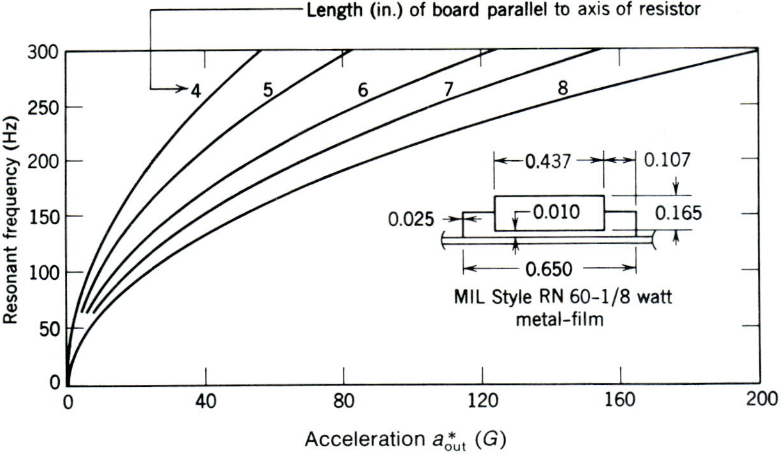

FIGURE 10.29
Maximum acceleration a^*_{out} that a circuit board can tolerate with infinite fatigue life. (*After D. S. Steinberg [13].*)

These stresses are then compared to the fatigue strength S_f of the lead wire material for the number of cycles accumulated in fatigue. If $S_f \geq \sigma$, the lead wire will not fail in fatigue.

Steinberg has prepared several fragility curves for axial lead components using the aforementioned approach. These fragility curves show the allowable accelerations a^*_{out} that can be imposed on a circuit board without inducing a fatigue failure of the leads. A typical fragility curve, presented in Fig. 10.29, shows that a^*_{out} depends on the natural frequency f_n and the length of the circuit board. The longer circuit boards permit higher G levels because the slopes dw/dx are smaller for a specified center displacement w_0. The smaller slopes reduce θ_1 and, in turn, M_1 and σ are reduced to extend the fatigue life.

10.7.1 Stresses in Solder Joints

The stresses in the solder joint are determined from the loading M_1 and $P/2$, which are imposed on the joint as defined by the geometry presented in Fig. 10.30*a*. The solder joint illustrated in this figure is for a double-sided circuit board. The solder has formed a fillet above the board, which has a critical location about one wire diameter d_w above the surface of the board where failure usually occurs. Because the lead wire and solder have different moduli of elasticity, the solution for the stresses must account for the composite nature of the joint materials. The approach taken here is to assume that the moment M_1 is totally constrained by the stresses in the lead wire. This approach is justified because the modulus of the solder is small relative to the Kovar or nickel wires

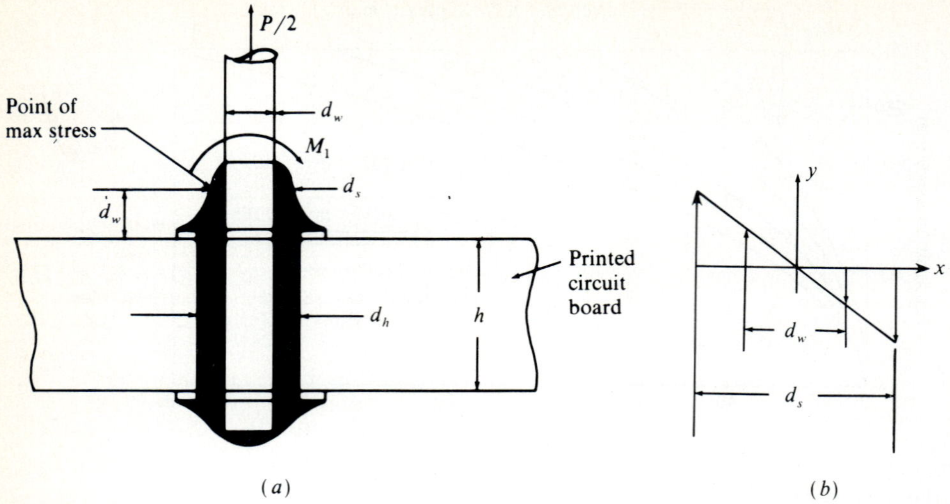

FIGURE 10.30
(*a*) A solder joint on a double-sided printed circuit board. (*b*) Strain distribution at the critical location.

commonly employed. This assumption lets us write

$$\sigma_w = M_1 c / I_w = 32 M_1 / \pi d_w^3 \tag{10.87}$$

The strain in the lead wire is given by

$$\varepsilon_w = \sigma / E_w = 32 M_1 / (E_w \pi d_w^3) \tag{10.88}$$

In bending, plane sections remain plane, which leads to the observation that the strain across the composite joint is linearly distributed as shown in Fig. 10.30*b*. From similar triangles we can write the relation for the maximum strain in the solder as

$$\varepsilon_s = (d_s / d_w) \varepsilon_w \tag{10.89}$$

and the stress as

$$\sigma_s = E_s \varepsilon_s = (d_s / d_w)(E_s / E_w)(32 M_1) / (\pi d_w^3) \tag{10.90}$$

The failure initiation site on a solder joint is usually at a position where $d_s = 3d_w/2$, which leads to a simplification of Eq. (10.90):

$$\sigma_s = (E_s / E_w)(48 M_1) / (\pi d_w^3) \tag{10.91}$$

In addition to the bending stresses, shear stresses τ_s due to the axial force are developed: We determine the average shear stress by

$$\tau_s = F / A_s \tag{10.92}$$

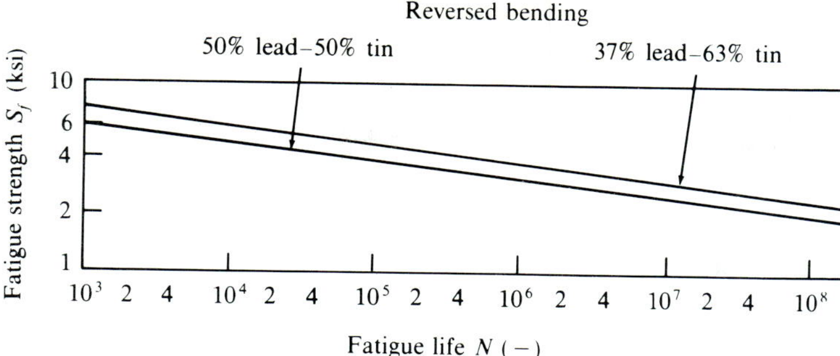

FIGURE 10.31
S-N diagram for common solders.

Note that $F = P/2$ and the shear area A_s is given by

$$A_s = \pi d_h(h + 2d_w) \tag{10.93}$$

where d_h is the diameter of the plated through hole and h is the board thickness. Combining Eqs. (10.92) and (10.93) gives

$$\tau_s = P/[2\pi h d_h(h + 2d_w)] \tag{10.94}$$

The bending stresses and shear stresses combine to give maximum principal and shear stress given by

$$\sigma_{\max} = \sigma_s/2 + \left[(\sigma_s/2)^2 + \tau_s^2\right]^{1/2} \tag{10.95}$$

$$\tau_{\max} = \left[(\sigma_s/2)^2 + \tau_s^2\right]^{1/2} \tag{10.96}$$

The maximum stress is an alternating stress due to reverse bending that occurs in vibratory exposure of the printed circuit board. The *S-N* diagrams for some common solders are presented in Fig. 10.31. The safety factor for the solder joint is given by

$$\mathbb{S} = S_f/\sigma_{\max} \tag{10.97}$$

Since the solder joint may contain voids, inclusions and geometric imperfections, a safety factor of at least 3 is advisable to account for the variability encountered in dealing with a very large number of solder joints exposed to a vibratory environment.

10.8 THE THEOREM OF CASTIGLIANO

Many components and/or devices are attached to boards or chassis using thin metal leads that may be formed into many different shapes. Examples are the gull wing and *J* leads on chip carriers. The elastic stresses and the deflection in these leads may be determined for any configuration by using Castigliano's theorem.

This theorem states that a deflection δ is given by

$$\delta = \partial U/\partial P \tag{10.98a}$$

and the rotation θ is

$$\theta = \partial U/\partial M_1 \tag{10.98b}$$

where U is the total strain energy in the lead wire. Note that the deflection δ is at the load point and in the direction of the load P. Similarly the rotation θ is at the point of application of the moment M_1 and with the same sense as the moment.

Strain energy that is stored in the lead wires may be due to tensile, shear and bending loads. However, for long thin members like lead wires, the contribution of the tension and shear loads to the strain energy is negligible when compared to that of the bending moments. For this reason we will consider the strain energy due only to the bending moment as

$$U = \int_0^L [M^2/(2EI)]\, dx \tag{10.99}$$

If we consider E and I constant along the length L of the lead wire and substitute Eq. (10.99) into Eqs. (10.98) we obtain

$$\delta = (1/EI)\int_0^L M(\partial M/\partial P)\, dx \tag{10.100}$$

$$\theta = (1/EI)\int_0^L M(\partial M/\partial M_1)\, dx \tag{10.101}$$

To show the application of this method consider the bent cantilever beam presented in Fig. 10.32. The beam is made from wire with a stiffness EI. The loads P and Q are applied at point A. Let us determine the vertical deflection δ_A at point A due to the bending. It is clear from Eq. (10.98a) that

$$\delta_A = \partial U/\partial Q \tag{a}$$

If we neglect the strain energy due to axial and shear loading of the wire bent,

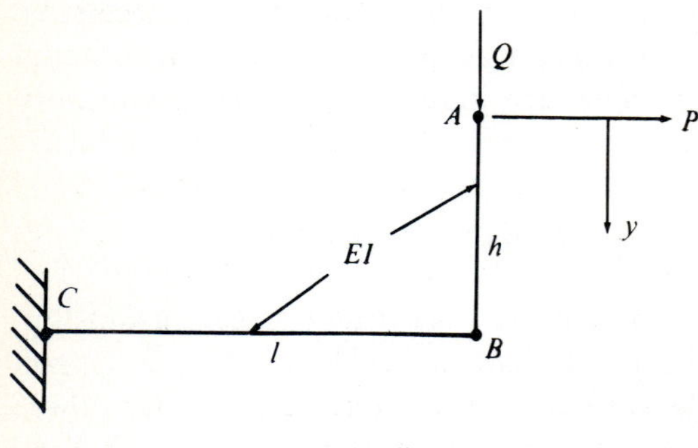

FIGURE 10.32
A bent cantilever beam.

Eq. (a) leads to

$$\delta_A = (1/EI)\left[\int_0^h M_{A-B}(\partial M_{A-B}/\partial Q)\, dy + \int_0^\ell M_{B-C}(\partial M_{B-C}/\partial Q)\, dx \right. \quad \text{(b)}$$

We need two terms in Eq. (b) because the equation for the moment is discontinuous. We can write the equations for the moments M_{A-B} and M_{B-C} as

$$M_{A-B} = Py \qquad \text{and} \quad \partial M_{A-B}/\partial Q = 0 \quad \text{(c)}$$

$$M_{B-C} = Ph + Qx \quad \text{and} \quad \partial M_{B-C}/\partial Q = x \quad \text{(d)}$$

Substituting Eqs. (c) and (d) into Eq. (b) gives

$$\delta_A = (1/EI)\int_0^\ell (Phx + Qx)\, dx \quad \text{(e)}$$

Note that the first term in Eq. (b) vanished because $\partial M_{A-B}/\partial Q = 0$. Integration of Eq. (e) gives

$$\delta_A = [1/(6EI)]\left[3Ph\ell^2 + 2Q\ell^3\right] \quad (10.102)$$

The application of Castigliano's theorem has provided a simple method to determine the deflection in beams and frames with any shape. Because the attachments for most electronic components are formed by bending wire or ribbon into any configuration that facilitates assembly and soldering, the importance of this method of analysis is evident.

Castigliano's theorem can also be employed to determine the unknown forces or moments that occur in structural members that are statically indeterminate. For example, the frame shown in Fig. 10.27 is statically indeterminate. The unknown moment M_1 and force Q at the support can be determined by using the boundary conditions that the horizontal deflection δ and the rotation θ at the support are zero. Substitution of these boundary conditions into Eqs. (10.100) and (10.101) gives two additional relations that can be used with the equations of equilibrium to solve for all of the unknowns M_1, M_2 and Q.

10.9 FASTENERS

Threaded fasteners, bolts and screws, are used in large numbers in the assembly of electronic equipment. When the unit must be disassembled for repair or service, screws are often used; however, if the fastening is to be permanent, rivets are usually preferred. We will review here the standards followed in specifying a fastener and define the thread forms in common usage. We should note that screws are so common that they can be treated like a commodity. As a commodity they do not have any distinguishing characteristics and can be supplied by any firm that can produce screws to the specification standards. The advantage of being able to specify screws as commodities is in reducing fastener cost to a minimum.

There are two standards in common usage, namely, the Unified and the SI thread. The basic form of the thread for the Unified system, presented in Fig.

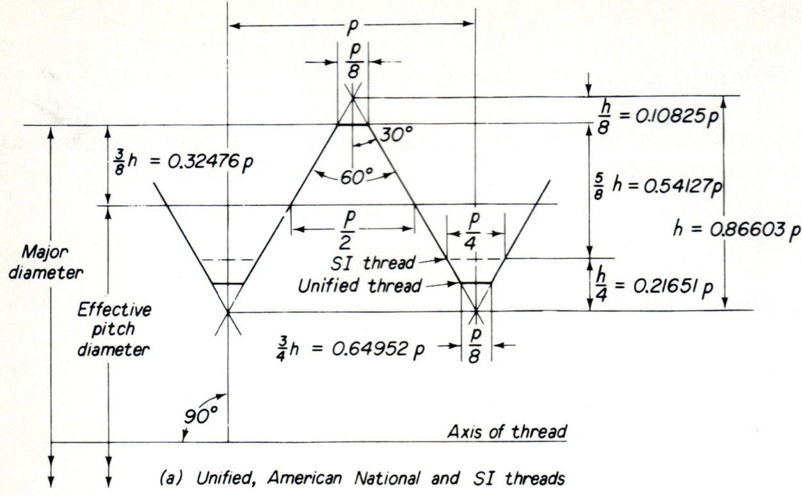

FIGURE 10.33
Unified American National and SI thread forms.

TABLE 10.3
Dimensions of Unified screw threads

Size	d Major diameter (in.)	Coarse-thread series: Threads per inch	Coarse-thread series: Pitch diameter (in.)	Coarse-thread series: A_t Stress area (in.^{2})	Fine-thread series: Threads per inch	Fine-thread series: Pitch diameter (in.)	Fine-thread series: A_t Stress area (in.^{2})
0	0.0600				80	0.0519	0.0018
1	0.0730	64	0.0629	0.0026	72	0.0640	0.0028
2	0.0860	56	0.0744	0.0037	64	0.0759	0.0039
3	0.0990	48	0.0855	0.0049	56	0.0874	0.0052
4	0.1120	40	0.0958	0.0060	48	0.0985	0.0066
5	0.1250	40	0.1088	0.0080	44	0.1102	0.0083
6	0.1380	32	0.1177	0.0091	40	0.1218	0.0102
8	0.1640	32	0.1437	0.0140	36	0.1460	0.0147
10	0.1900	24	0.1629	0.0175	32	0.1697	0.0200
12	0.2160	24	0.1889	0.0242	28	0.1928	0.0258
1/4	0.2500	20	0.2175	0.0318	28	0.2268	0.0364
5/16	0.3125	18	0.2764	0.0524	24	0.2854	0.0580
3/8	0.3750	16	0.3344	0.0775	24	0.3479	0.0878
7/16	0.4375	14	0.3911	0.1063	20	0.4050	0.1187
1/2	0.5000	13	0.4500	0.1419	20	0.4675	0.1599
9/16	0.5625	12	0.5084	0.182	18	0.5264	0.203
5/8	0.6250	11	0.5660	0.226	18	0.5889	0.256
3/4	0.7500	10	0.6850	0.334	16	0.7094	0.373
7/8	0.8750	9	0.8028	0.462	14	0.8286	0.509
1	1.0000	8	0.9188	0.606	12	0.9459	0.663

10.33, has a 60° included angle between adjacent threads. The crest and the root are usually rounded although they are shown as flats in the drawing of the thread forms. The SI threads are similar in that they have the 60° include angle and the same size flats for the crest and the roots. However, the pitch, the distance between adjacent threads, is specified in millimeters. Since the pitch for the Unified and SI threads differs, the screws with these two different thread forms are not interchangeable.

Listings of the standard dimensions for the most commonly used sizes of the Unified and SI threads are shown in Tables 10.3 and 10.4, respectively. We note in both tables that coarse and fine threads are available. The coarse threads are always used in the softer metals like aluminum or in plastics. The fine thread is used only in steel. The stress area A_t is determined by using the average of the pitch and minor diameters to compute the effective cross-sectional area at the threads. The bolt area A_b, not shown in the tables, is determined using the major diameter.

TABLE 10.4
Dimensions of SI threads (all dimensions in millimeters)

	Coarse-thread series			Fine-thread series		
Nominal major diameter, d	Pitch p	Tensile-stress area, A_t	Minor-diameter area, A_r	Pitch p	Tensile-stress area, A_t	Minor-diameter area, A_r
1.6	0.35	1.27	1.07			
2	0.04	2.07	1.79			
2.5	0.45	3.39	2.98			
3	0.5	5.03	4.47			
3.5	0.6	6.78	6.00			
4	0.7	8.78	7.75			
5	0.8	14.2	12.7			
6	1	20.1	17.9			
8	1.25	36.6	32.8	1	39.2	36.0
10	1.5	58.0	52.3	1.25	61.2	56.3
12	1.75	84.3	76.3	1.25	92.1	86.0
14	2	115	104	1.5	125	116
16	2	157	144	1.5	167	157
20	2.5	245	225	1.5	272	259
24	3	353	324	2	384	365
30	3.5	561	519	2	621	596
36	4	817	759	2	915	884
42	4.5	1120	1050	2	1260	1230
48	5	1470	1380	2	1670	1630
56	5.5	2030	1910	2	2300	2250
64	6	2680	2520	2	3030	2980
72	6	3460	3280	2	3860	3800
80	6	4340	4140	1.5	4850	4800
90	6	5590	5360	2	6100	6020
100	6	6990	6740	2	7560	7470

The standards allow one to specify a threaded fastener with efficiency. For example, 8-32 UNC is sufficient to identify a Unified coarse thread, number 8 screw with 32 threads/in. Of course it is also necessary to indicate the length of the screw, the length of the threaded portion and the type of head. With the SI fasteners the threads are designated by diameter and pitch, as M3 × 0.5. This designation specifies a metric screw 3 mm in diameter with threads on a 0.5 mm pitch.

10.9.1 Strength of Fasteners

Threaded fasteners are produced by automatic machines that cold form the head and roll the threads. The screw materials are usually low to medium carbon steels and the elevated strengths are achieved by heat treating. The strength is designated by the SAE grade number as indicated in Table 10.5, with numbers ranging from 1 for the softer lower strength screws to 8 for the harder and higher strength products. Some of the grades have line markings as indicated in Table 10.5. These lines, embossed on the hex heads, permit the assessment of the strength of the material either before or after installation of the bolt.

The proof strength S_p of the bolt material is expressed in terms of the proof load P_p as

$$S_p = P_p/A_t \tag{10.103}$$

The proof load is the maximum force that can be applied to a screw before it

TABLE 10.5
Strength of and grade marking for bolts and screws

SAE grade no.	Proof strength S_p (ksi)	Yield strength S_y (ksi)	Tensile strength S_u (ksi)	Elongation (%)	Hardness Rockwell	Nominal diameter	Grade marking
1	33	36	60	18	B70/B100	$\frac{1}{4}$–$1\frac{1}{2}$	None
2	55	57	74	18	B80/B100	$\frac{1}{4}$–$\frac{3}{4}$	None
4	65	100	115	10	C22/C32	$\frac{1}{4}$–$1\frac{1}{2}$	None
5	85	92	120	14	C25/C34	$\frac{1}{4}$–1	
5.1	85	—	120	14	C25/C40	#6–$\frac{5}{8}$	
5.2	85	92	120	14	C26/C36	$\frac{1}{4}$–1	
7	105	115	133	12	C28/C34	$\frac{1}{4}$–$1\frac{1}{2}$	
8	120	130	150	12	C33/C39	$\frac{1}{4}$–$1\frac{1}{2}$	
8.1	120	130	150	10	C32/C38	$\frac{1}{4}$–$1\frac{1}{2}$	None
8.2	120	130	150	10	C35/C42	$\frac{1}{4}$–1	

undergoes a permanent deformation. In most instances the proof strength is between 80 and 90% of the yield strength.

10.9.2 Bolt Preload

In assembly, the bolts are torqued to tighten the fastener and to draw together the mating surfaces that make up the joint. In effect, the fastener clamps the joint together. The magnitude of clamping pressure depends on the preload applied through the fastener. The amount of preload applied is controlled by the torque used in tightening the assembly. The relationship between the preload F_i and the torque T is given by

$$T = 0.20F_i d \tag{10.104}$$

where d is the major diameter of the bolt.

The amount of preload that is to be applied to the fastener depends on the type of loading imposed on the joint. For static loading a high preload is recommended with

$$0.6P_p \leq F_i \leq 0.9P_p \tag{10.105}$$

With fatigue loading the preload is usually lower and is dependent on the proportioning of the externally applied load between the bolt and the joint. We will cover load proportioning in the next section.

10.10 FASTENED JOINTS LOADED IN TENSION

The joint formed when two or more mating surfaces are fastened together can be loaded in either tension or shear. The method of analysis of the fastener differs significantly with the type of loading. Let us consider the joint shown in Fig. 10.34, with a tensile load P applied so as to separate the mating surfaces. As the load is applied both the bolt and the joint respond by deflecting an amount δ. The surface of the joint remains in contact and $\delta_b = \delta_j$. This equality leads directly to

$$P_b/k_b = P_j/k_j \tag{10.106}$$

where P_b is the proportion of the load P that is carried by the bolt and P_j is the proportion carried by the joint that is obtained by reducing the interface pressure. The k's are the spring rates given by

$$k_b = A_b E_b/L_b \quad \text{and} \quad k_j = A_j E_j/L_j \tag{10.107}$$

The area of the joint depends on the thickness of the edges being clamped and may be approximated by

$$A_j = (\pi/4)\left(d_j^2 - d_b^2\right) \tag{10.108}$$

where

$$d_j = \left(3d_b + L_j\right)/2 \tag{10.109}$$

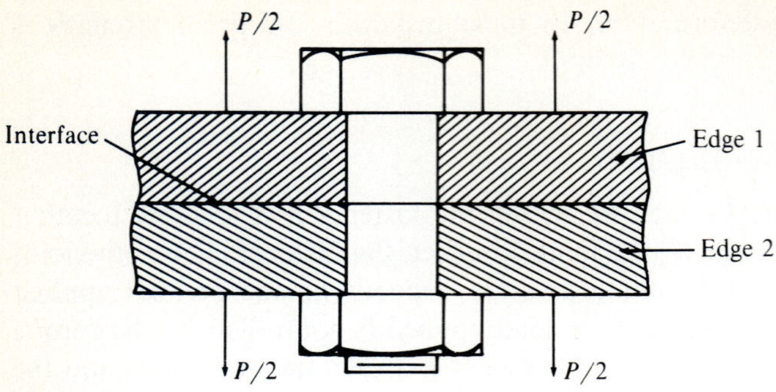

FIGURE 10.34
A bolted joint subjected to a tension load P that tends to separate the interface.

Noting that $P = P_b + P_j$ and using Eq. (10.106) gives

$$P_b = [k_b/(k_b + k_j)]P = CP \tag{10.110a}$$

$$P_j = [k_j/(k_b + k_j)]P = (1 - C)P \tag{10.110b}$$

Examination of Eqs. (10.110) shows that the applied load is partitioned in accordance with the relative stiffness of the two components. The stiffer component carries the larger share of the load. Next, we superimpose the preload to obtain

$$F_b = P_b + F_i \quad \text{and} \quad F_j = P_j - F_i \tag{10.111}$$

The force F_j in the joint must always remain negative to insure a clamping pressure that keeps the mating surfaces in contact. If P_j becomes greater than the bolt preload F_i, the joint will open and the entire load P must be carried by the bolt. If this situation occurs, the fastening system almost always fails. The bolt force F_b produces stresses that can vary with the load P and produce cyclic loading of the bolt. We will consider the effect of these cyclic stresses on the fatigue behavior of the bolt in the next section.

10.10.1 Stress and Fatigue Analysis

The stresses produced in the bolt by the externally applied load P may be determined from

$$\sigma_b = F_b/A_t \tag{10.112}$$

If we consider that P varies with time as shown in Fig. 10.35*a*, then the forces F_b and the stresses σ_b will vary with time in the manner illustrated in Fig. 10.35*b*. We are concerned with two different effects of the fluctuating stresses. First, the maximum value of σ_b must be less than the proof stress of the screw to avoid static failure by permanently deforming the screw. The safety factor for static

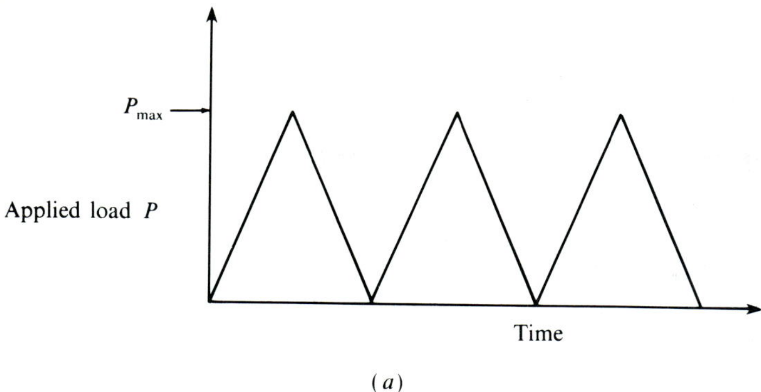

(*a*)

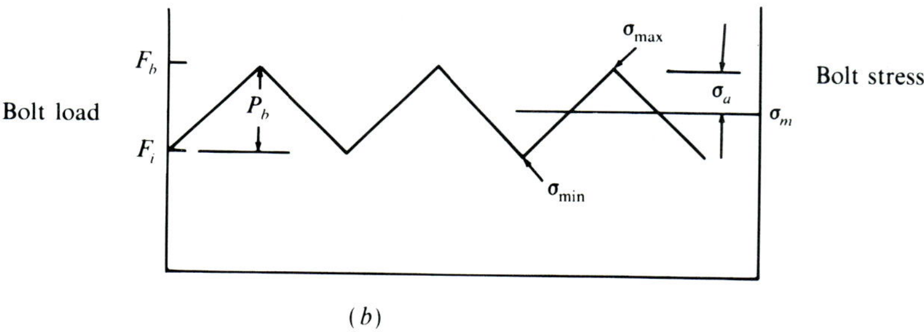

(*b*)

FIGURE 10.35
(*a*) Load applied to the joint. (*b*) Load and stress acting on the bolt.

failure by a single cycle of overload is

$$\mathbb{S} = S_p/(\sigma_b)_{\max} \tag{10.113}$$

Second, the stress σ_b must be limited to avoid fatigue failure. The fatigue analysis involves determining the mean and alternating stresses defined as

$$\sigma_m = (\sigma_{\max} + \sigma_{\min})/2 \quad \text{and} \quad \sigma_a = (\sigma_{\max} - \sigma_{\min})/2 \tag{10.114}$$

where the subscripts m and a refer to mean and alternating components of stress as defined in Fig. 10.35*b*. Combining Eqs. (10.111), (10.110a) and (10.112) with Eq. (10.114) gives

$$\sigma_m = (CP + 2F_i)/2A_t \quad \text{and} \quad \sigma_a = CP/2A_t \tag{10.115}$$

The modified Goodman theory accounts for the effect of both the mean and the alternating stresses acting simultaneously on the fatigue strength. The equation for the modified Goodman line is

$$\sigma_m/S_u + \sigma_a/S_e = 1/\mathbb{S} \tag{10.116}$$

If we use the approximate relation between the endurance strength S_e and the tensile strength S_u that $S_u = 2S_e$ and substitute Eq. (10.115) into Eq. (10.116), we obtain

$$\mathbb{S} = 2A_tS_u/(3CP + 2F_i) \tag{10.117}$$

Inspection of this relation shows that the safety factor $\mathbb{S}$ for fatigue is increased by selecting larger screws (higher A_t) made from higher strength materials. The effect of increasing the preload F_i is to decrease the safety factor. This result does not imply that very low preload should be employed. Remember that F_i must be greater than the maximum value of P_j to keep the joint from separating.

The safety factor employed should be relatively high to account for the effects of the stress concentration at the root of the threads. Because this stress concentration is about 3 for rolled threads, safety factors of 4–5 are commonly employed in designing joint assemblies subjected to fluctuating tensile loads.

10.11 FASTENED JOINTS LOADED IN SHEAR

Design of joints that place the screws in shear rather than tension are preferred for two reasons. First, the stress concentration associated with the sharp radii at the roots of the threads are avoided. Second, the detrimental effects due to the loss of preload are eliminated. Since bolted assemblies tend to loosen when subject to vibration, avoiding the effects of loss of preload is important. When the joint is loaded in shear it is necessary to keep the screws in place with preload, but the preload is not required for load partitioning.

Consider the lap joint fastened with four screws, illustrated in Fig. 10.36. Since the applied load P does not pass through the bolt center, shear stresses are produced by both the direct shear force V and the moment M that exist at the joint. Equilibrium relations give

$$V = P \quad \text{and} \quad M = PL \tag{10.118}$$

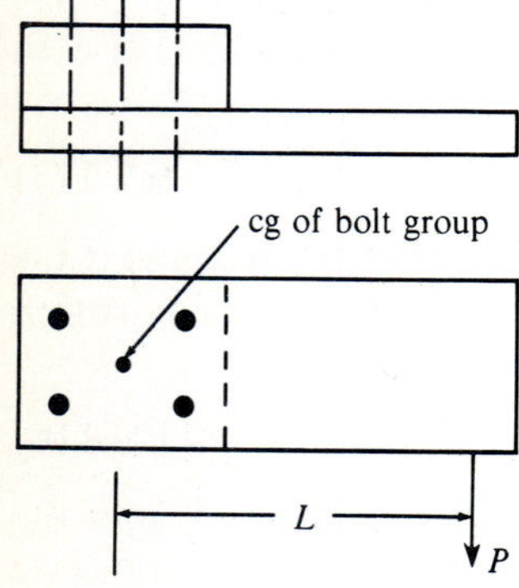

FIGURE 10.36
A fastened joint loaded in shear.

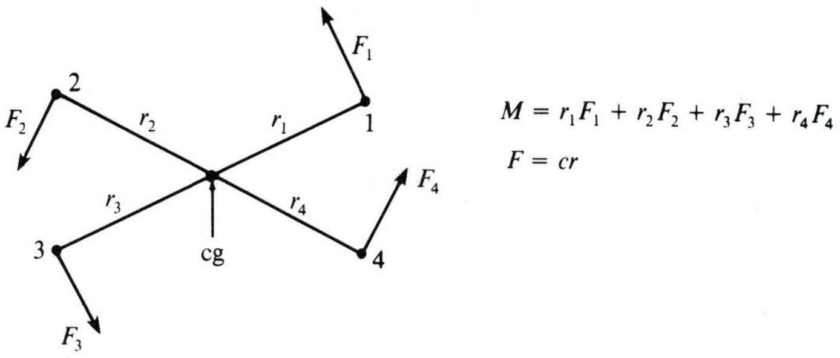

FIGURE 10.37
Shear forces $F_1 \cdots F_4$ developed due to M.

The shear stress τ^* due to the direct shear load V is

$$\tau^* = V/(kA_b) \tag{10.119}$$

where k is the number of bolts in the bolt group. In writing Eq. (10.119), we have assumed that the shear load is uniformly distributed over the k bolts.

The moment M is resisted by developing the forces F_1, F_2, F_k as shown in Fig. 10.37. The moment force relation is

$$M = F_1r_1 + F_2r_2 + \cdots + F_kr_k \tag{10.120}$$

Next, it is assumed that the force F is proportional to the radial position of the individual bolt from the center of gravity (cg) of the bolt group. This assumption leads to

$$F = cr \tag{10.121}$$

Combining Eqs. (10.120) and (10.121) gives the force F_k as

$$F_k = Mr_k/(r_1^2 + r_2^2 + \cdots + r_k^2) \tag{10.122}$$

The stress τ^{**} due to the moment M is then

$$\tau_k^{**} = F_k/A_{bk} \tag{10.123}$$

Note that F_k and τ_k are maximums for the bolt furthest removed from the cg of the bolt group. It is this bolt that determines if failure in shear will occur. We combine the contributions of the two shear stresses by using vector addition to obtain

$$\tau_{max} = \left[(\tau_y^*)^2 + (\tau_x^{**} + \tau_y^{**})^2\right]^{1/2} \tag{10.124}$$

The safety factor in shear for the joint is

$$\mathbb{S} = \tau_{max}/S_{ys} \tag{10.125}$$

where the yield strength in shear $S_{ys} = S_y/2$ according to the Tresca theory of yielding.

The preceding derivation discounted the effect of friction forces in providing resistance to either V or M. This is a conservative approach, which is recommended. Friction forces will decrease the maximum shear stress, but in exposure to vibrations the bolts tend to loosen and the benefits of the friction resistance are lost.

REFERENCES

1. "Navy Manufacturing Screening Program," NAVMAT P-9492, Navy Dept., May 1979.
2. Tuskin, W.: "Recipe for Reliability: Shake and Bake," *IEEE Spectrum*, pp. 37–42, Dec. 1986.
3. "Definition of the Shipping Environment," ASTM Standard D4169.
4. Bresk, F. C. and K. Irving: "Application of Product Fragility Information in Product Design," Lansmont Corporation, Pacific Grove, Calif.
5. "Environmental Test Methods and Engineering Guidelines," MIL Standard 810D, July 19, 1983.
6. MIL Standard-167-1. (Ships), Mechanical Vibrations of Shipboard Equipment (Type I-Environmental and Type II-Internally Excited), May 1, 1974.
7. "Test Requirements for Space Vehicles," MIL Standard-1540B (USAF), October 10, 1982.
8. MIL Standard-901D. Shock Tests, HI (High Impact): Shipboard Machinery, Equipment and Systems, Requirements for, March 17, 1989.
9. "Barry Controls Application Selection Guide," Bulletin C7-13, Barry Wright Corp., 1983.
10. Manson, S. S.: "Fatigue, A Complex Subject—Some Simple Approximations," *Expl. Mech.*, vol. 5, no. 7, p. 193, 1965.
11. Leissa, A. W.: "Vibration of Plates," NASA SP-160, pp. 41–160, 1969.
12. Janich, R.: "Die Naherungsweise Berechnung der Eigenfrequenzen von rechteckigen Platten bei verschiedenen Randbedingungen," *Die Bautechnik*, vol. 3, pp. 93–99, 1962.
13. Steinberg, D. S.: *Vibration Analysis for Electronic Equipment*, John Wiley, New York, p. 281, 1973.
14. Steinberg, D. S.: "Circuit Components vs. *G* Forces," *Machine Design*, Oct. 14, 1971.
15. Juvinall, R. C.: *Engineering Considerations of Stress, Strain and Strength*, McGraw-Hill, New York, 1967.
16. Sines, G. and J. L. Waisman: *Metal Fatigue*, McGraw-Hill, New York, 1959.

EXERCISES

10.1. Write an engineering brief describing a manufacturing screening test. Indicate the advantages and disadvantages of the test. Outline the test procedures.

10.2. You are to implement a temperature cycling test as a part of a manufacturing screen. The product being screened has nearly 1000 devices and/or parts. Specify the number of cycles and indicate the number of failures anticipated on each cycle. If it takes 2 hr to cycle the temperature and another 1.5 h to identify, locate and replace the defective part, estimate the average time from the start of the test until the product is deemed to be defect-free.

10.3. Will a manufacturing screening test eliminate all of the latent defects.

10.4. You are shipping 1000 units of a product each week by truck to a distribution point. One of the circuit boards in the product has a natural frequency of 30 Hz. Determine the acceleration that may be encountered in transit.

10.5. Describe an application of an electronic system that you are familiar with that is exposed to severe shock or vibrations on a daily basis. Examine that product and indicate any special packaging features that have been used to mitigate the effects of this environment.

10.6. A transformer is mounted to an enclosure using a bracket that can be modeled as a cantilever beam. The static deflection of the beam due to the weight of the transformer is 0.02 in. Determine the natural frequency f_n for this assembly.

10.7. Beginning with Eq. (10.7), verify Eq. (10.9).

10.8. Beginning with Eq. (10.7), verify Eq. (10.10).

10.9. Beginning with Eq. (10.7), verify Eq. (10.11).

10.10. Describe why the critically damped and overdamped conditions are not a major concern in protecting electronic equipment from vibratory environments.

10.11. Beginning with Eq. (10.14), verify Eq. (10.15).

10.12. In vibration with damping $\omega_{\text{nd}} \leq \omega_d$. Prepare a graph showing $\omega_{\text{nd}}/\omega_n$ for $0 < d < 1$.

10.13. Beginning with Eq. (10.18), verify Eqs. (10.19a) and (10.19b).

10.14. Prepare a graph showing the transmission coefficient $\mathbb{A}$ as a function of the damping ratio d if $r = 1$.

10.15. Prepare a graph showing $\mathbb{A}/\mathbb{F}$ as a function of r if d is equal to 0.01, 0.02, 0.05, 0.10, 0.2 and 0.5. Comment on the significance of these curves.

10.16. Why do we need two transmission coefficients to describe resonance effects on the response of the systems.

10.17. Beginning with Eq. (10.34), verify Eq. (10.38) and then proceed to verify Eqs. (10.39) and (10.40).

10.18. Using the data presented in Fig. 10.14*b* for the 372-1 isolator, determine the spring rate. If four of these isolators are used to support a 70 lb assembly, determine the natural frequency of the isolated system. Find the transmissibility coefficient if the forced disturbance is at a frequency of 170 Hz.

10.19. Prepare an engineering brief describing the results of the analysis of the isolator defined in Exercise 10.18.

10.20. The assembly of Exercise 10.18 is housed in a case 8 in. high, 20 in. wide and 18 in. deep, and is to be permanently mounted on a shelf that is located in a trailer. Prepare a sketch showing the arrangement of the isolators and the mounting of the isolators to the shelf and the case.

10.21. The 10 W wire wound resistor defined in Fig. 10.17*a* is supported by two posts that are staked into a PCB. The distance from the body to each post is 0.4 in. Find the natural frequency of the resistor after soldering the leads to the posts.

10.22. Determine the transmissibility coefficient $\mathbb{F}$ for the resistor of Exercise 10.21 if the damping ratio $d = 0.02$ and the driving frequency is 33 Hz. Next, find the bending stress induced during vibration.

10.23. Determine the coefficients a and b in Eq. (10.47) for nickel and cold drawn copper.

10.24. Determine the safety factor $\mathbb{S}$ for cold drawn copper lead wires subjected to a reverse bending stress of 20 ksi. We anticipate that the lead wires will be exposed to 10^5 cycles of loading.

10.25. A qualification test specifies sweeping from 10–2000 Hz and back a total of 10 times. The system you have designed has a circuit board with $f_n = 180$ Hz.

Determine the number of cycles at high load level that the board must withstand if the sweep rate is 1 octave/min.

10.26. Using Eq. (10.52) prepare a graph that shows the strain range $\Delta\varepsilon$ for cold drawn copper wire as a function of N for $S_u = 80$ ksi and $RA = 0.3$. Plot a second curve for a softer copper wire where $S_u = 40$ ksi and $RA = 0.5$. Comment on the results.

10.27. Customers are returning product that malfunctions because of broken wires. An inspection of these units reveals that the failures are due to nicks that occur on about 1 wire in 200. The wires that fail are always solid conductors. Prepare an engineering brief describing two or more approaches for resolving this problem.

10.28. Prepare a sketch showing four different boundary constraint combinations that are commonly used to model circuit boards in vibration. Why do you believe that these support conditions are so popular. Identify these support conditions in Table 10.2.

10.29. Would you model an edge connector as a simple support or as a clamped condition. Explain why? A wedge lock retainer provides what type of support.

10.30. Determine the flexural rigidity D of a circuit board 0.60 in. thick that is fabricated from epoxy–glass with $E = 3.5 \times 10^6$ psi.

10.31. Prepare a graph showing the flexural rigidity D as a function of thickness h that ranges from 0.01–0.125 in. Let $E = 0.5$, 1, 2 and 5×10^6 psi.

10.32. If the circuit board of Exercise 10.30 is clamped along its 5 in. edge, simply supported along the opposite edge and free along its two 12 in. edges, determine its natural frequency.

10.33. Examine Table 10.2 and identify the case that gives the highest ω_n if $a/b = 3$. Also, identify the case that gives the lowest natural frequency.

10.34. Determine the natural frequency for the circuit board illustrated in case no. 16. Let $a/b = 4$, $a = 18$ in., $E = 3 \times 10^6$ psi, $h = 0.125$ in. and $\gamma = 0.065$ lb/in.^{3}.

10.35. Write an equation comparable to Eq. (10.56) that describes the deflection for a circuit board with two opposing edges free, one edge clamped and the other simply supported.

10.36. Write the equation describing the deflection of the plate corresponding to case no. 10 in Table 10.2. Reduce this equation to show the deflection along each center line.

10.37. For the boundary conditions given in Exercise 10.32 find the acceleration $\ddot{w}_{max}$.

10.38. For the circuit board defined in Exercises 10.30 and 10.32, determine the maximum displacement if $\mathbb{A} = f_n^{1/2}$. Repeat the determination, but use another method for establishing $\mathbb{A}$ by letting $d = 0.04$.

10.39. Use Eq. (10.63) in Eq. (10.62) to eliminate $\mathbb{A}$. Then describe the dependency of $w_{\omega n}$ on ω_n at the resonance condition.

10.40. Examine Fig. 10.21 and explain why the accelerometers are mounted adjacent to the supports. Describe the signal you would expect from the accelerometers.

10.41. A vendor describes tests of a new conformal coating that has increased the damping ratio d from 0.04 to 0.07. Prepare an engineering brief for management describing the advantages and disadvantages of this coating.

10.42. A heat frame consisting of a sheet of aluminum is bonded to the circuit board of Exercise 10.30. Determine the change in the flexural rigidity of the board. If the boundary conditions are the same as those described in Exercise 10.31, find the new value of f_n. If the edges of the heat frame are clamped so that the free edges are treated as fixed, find the value of f_n.

10.43. Prepare an engineering brief describing the effects of heat frames on the vibration of PCBs.

10.44. Verify Eq. (10.64).

10.45. Examine the effect of the aspect ratio (a/b) of a circuit board having an area A. Consider the case where the board is simply supported on all four edges. Prepare a graph showing the normalized natural frequency as a function of (a/b).

10.46. The customer is the Navy, with a shipboard application where the maximum driving frequency is 50 Hz. Prepare an engineering brief that describes a design plan for the circuit boards to withstand a 2 hr vibration dwell at 2 G. Introduce the concept of fragility level in your plan.

10.47. Determine the increase in the flexural rigidity of the circuit board shown in Fig. 10.22, if $h = 0.060$ in., and the ribs are aluminum strips 0.125×0.5 in. in size. The width $b = 6$ in. and E for the circuit board is 4×10^6 psi. Determine the enhancement in ω_n due to the ribs.

10.48. Prepare an engineering brief that explains the damping enhancement provided by viscous coatings on vibrating plates. The brief should refer to a library search where the effects on coatings on vibrating metal panels have been characterized.

10.49. Write the equation for the displacement $w(x)$ of the circuit boards described in case nos. 6, 12, 15 and 18 of Table 10.2.

10.50. Begin with Eq. (10.67) and verify Eq. (10.68).

10.51. Derive an equation for the maximum stress in a circuit board [similar to Eq. (10.73)] for case 6 in Table 10.2.

10.52. Consider a circuit board supported as in case no. 6 and subject to a 5 G input acceleration. The board thickness is set at 0.08 in. and the aspect ratio is set at $a/b = 2$. Determine the size of the board if the maximum stresses are to be limited to 3000 psi. Where on the board will this maximum stress occur?

10.53. Determine the natural frequency of a module consisting of a ceramic card (alumina) $0.04 \times 4 \times 5$ in. in size bonded to a heat frame that is made from copper sheet 0.625 in. thick. The heat frame is tightly clamped to a cold rail with wedge lock retainers. Constraint conditions are closely described by case no. 5, Table 10.2.

10.54. A strain gage mounted on the glass-reinforced portion of a circuit board indicates a peak strain of 0.0032 when the system is subjected to a dwell at resonance frequency. The resonance frequency is 90 Hz. Estimate the time of dwell prior to circuit malfunctions due to fatigue failure of the copper traces on the circuit board.

10.55. Equation (10.42) gives the frequency f_n of an axial leaded component soldered to two rigid posts. Derive a similar relation for an axial leaded component with bent leads soldered to a circuit board.

10.56. Determine the force P for a 5 W resistor supported as shown in Fig. 10.17*b* with $L/2 = 0.15$ in. for an input acceleration ranging from 1–10 G.

10.57. Verify Eqs. (10.84).

10.58. Verify Eqs. (10.85).

10.59. The axial leaded component defined in Fig. 10.29 is mounted straddling the center line of a circuit board that is being exposed to a 5 G input acceleration. The displacement of the circuit board is given by $w = w_0 \sin(\pi x/a)$ where $a = 6$ in. If the degree of damping of the board is 0.05, determine the maximum stress in the leads. How do your results compare to those due to Steinberg in Fig. 10.29?

10.60. Use the results of Fig. 10.29 for a 6 in. long circuit board with a resonance frequency of 150 Hz. Find the deflection w at the center of the circuit board that cannot be exceeded if infinite life of the component leads is to be insured.

10.61. Prepare a graph showing the stresses developed in a soldered joint as a function of M_1. Consider wire diameters of 20, 25 and 30 mil in the solution.

10.62. A solder joint is exposed to shear stresses of 800 psi and bending stresses of 1200 psi. Determine the maximum principal stress and the maximum shear stress. If the solder joint has an endurance strength of $S_e = 2000$ psi, determine the safety factor based on infinite life.

10.63. For the beam shown in Fig. 10.32, determine the horizontal displacement of point A.

10.64. For the beam shown in Fig. 10.32, determine the rotation of end at point A. Also, find the vertical displacement at point B.

10.65. For the beam in Fig. 10.32, determine the spring rate in both the horizontal and vertical directions.

10.66. For the semicircular beam shown in Fig. E10.66, find the deflection of point A in the direction of the load P. Also, find the horizontal deflection of point A.

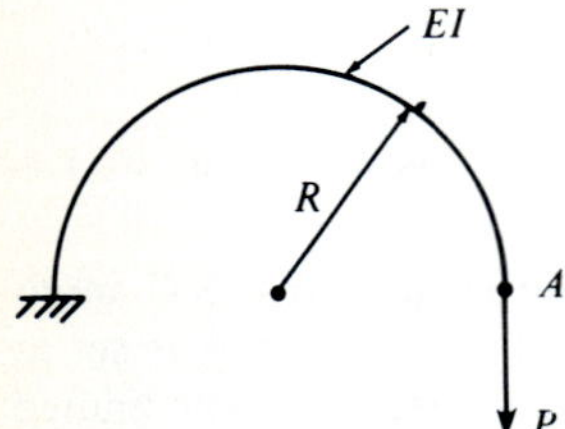

FIGURE E10.66

10.67. Verify Eq. (10.102).

10.68. Explain why rivets are usually preferred as the fastener when it is not necessary to provide for the opportunity for disassembly.

10.69. Take out one of the screws holding the cover of your PC and describe the screw in terms of the terminology presented in Section 10.9.

10.70. Prepare an engineering brief giving arguments for using a grade 5 bolt material rather than a grade 1. Can you support this position even if the loads do not require a high strength fastener.

10.71. Sketch a curve showing the cost of a fastener as a function of the size. Include in this estimate the labor associated with tapping a hole for the screw and the assembly operation. Would you ever use a 1-72 UNF?

10.72. Specify the torque to be applied to a 1/4-20 UNC grade 5 bolt if it is used to clamp a joint subjected to a static load.

10.73. Determine the spring rates k_b and k_j for a joint designed with a 10-32 screw that is clamping to two steel flanges, each 0.125 in. thick.

10.74. If the joint of Exercise 10.73 is subjected to a load that varies from 0–500 lb in a vibratory application, find the maximum, minimum, mean and alternating stresses acting on the screw. If a preload of $F_i = P_p/2$ is applied to the screw, determine the safety factor for static loading and for fatigue loading.

10.75. Verify Eq. (10.117).

10.76. Derive a relation like Eq. (10.117), but introduce the effect of the stress concentration, due to the root of the thread, acting to elevate the alternating stress. Also, assume that the stress concentration does not affect the mean stress.

10.77. Prepare an engineering brief that justifies elimination of the friction force in deriving Eq. (10.122). Include in this brief the effect of 0.0001 in. of vibration wear under the head of the bolt and of the nut. Consider that the total joint thickness is 0.4 in.

10.78. Determine the maximum load that the bolts can support in the cantilever beam with the lap joint described in Fig. E10.78. Describe two methods for increasing this load. Which method would cost more.

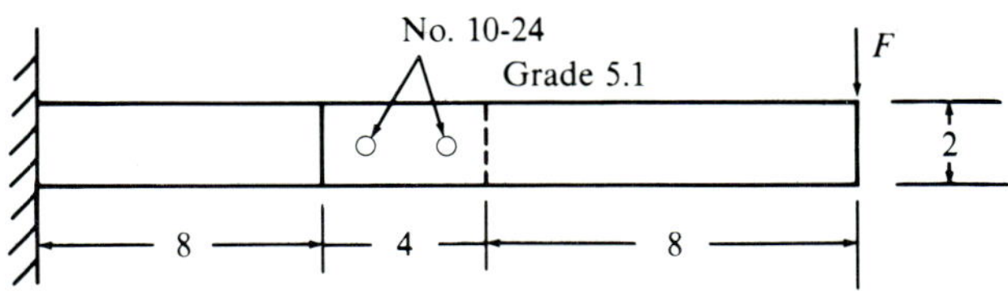

FIGURE E10.78

INDEX